Customizing AutoCAD 2002

Sham Tickoo

Professor
Department of Manufacturing Engineering Technologies
Purdue University Calumet
Hammond, Indiana
U.S.A.

Contributing Authors

Gregory Neff

Associate Professor
Department of Manufacturing Engineering Technologies
Purdue University Calumet

CADCIM Technologies

(www.cadcim.com)
USA

autodesk Press

Customizing AutoCAD 2002®
Sham Tickoo

Business Unit Director:
Alar Elken

Executive Editor:
Sandy Clark

Acquisitions Editor:
James DeVoe

Development Editor:
John Fisher

Editorial Assistant:
Jasmine Hartman

Executive Marketing Manager:
Maura Theriault

Channel Manager:
Mary Johnson

Marketing Coordinator:
Karen Smith

Executive Production Manager:
Mary Ellen Black

Production Manager:
Larry Main

Production Editor:
Tom Stover

Art/Design Coordinator:
Mary Beth Vought

COPYRIGHT © 2002 Thomson Learning.

Printed in Canada
1 2 3 4 5 XXX 05 04 03 02 01

For more information contact Autodesk Press,
3 Columbia Circle, PO Box 15015,
Albany, NY 12212-5015.

Or find us on the World Wide Web at
www.autodeskpress.com

All rights reserved. No part of this work covered by the copyright hereon may be reproduced or used in any form or by any means—graphic, electronic, or mechanical, including photocopying, recording, taping, Web distribution or information storage and retrieval systems—without written permission of the publisher.

For permission to use material from this text or product, contact us by
Tel   (800) 730-2214
Fax   (800) 730-2215
www.thomsonrights.com

Library of Congress Cataloging-in-Publication Data
Tickoo, Sham.
Customizing AutoCAD 2002 / Sham Tickoo.
   p. cm.
Includes Index.
ISBN 0-7668-3852-8
1. Computer graphics.
2. AutoCAD. I. Title.

T385 .T527 2001
620'.0042'0285536—dc21
         2001037126

NOTICE TO THE READER

Publisher does not warrant or guarantee any of the products described herein or perform any independent analysis in connection with any of the product information contained herein. Publisher does not assume, and expressly disclaims, any obligation to obtain and include information other than that provided to it by the manufacturer.

The reader is expressly warned to consider and adopt all safety precautions that might be indicated by the activities herein and to avoid all potential hazards. By following the instructions contained herein, the reader willingly assumes all risks in connection with such instructions.

The Publisher makes no representation or warranties of any kind, including but not limited to, the warranties of fitness for particular purpose or merchantability, nor are any such representations implied with respect to the material set forth herein, and the publisher takes no responsibility with respect to such material. The publisher shall not be liable for any special, consequential, or exemplary damages resulting, in whole or part, from the readers' use of, or reliance upon, this material.

Trademarks: Autodesk, the Autodesk logo, and AutoCAD are registered trademarks of Autodesk, Inc., in the USA and other countries. Thomson Learning is a trademark used under license. Online Companion is a trademark and Autodesk Press is an imprint of Thomson Learning. Thomson Learning uses "Autodesk Press" with permission from Autodesk, Inc., for certain purposes. All other trademarks, and/or product names are used solely for identification and belong to their respective holders.

Table of Contents

Preface — xi
Dedication — xv

Chapter 1: Template Drawings

Creating Template Drawings	1-1
Standard Template Drawings	1-2
Loading a Template Drawing	1-7
Customizing Drawings with Layers and Dimensioning Specifications	1-7
Customizing Drawings with Layout	1-12
Customizing Drawings with Viewports	1-15
Customizing Drawings According to Plot Size and Drawing Scale	1-17

Chapter 2: Script Files and Slide Shows

What are Script Files?	2-1
SCRIPT Command	2-4
RSCRIPT Command	2-9
DELAY Command	2-9
RESUME Command	2-10
Command Line Switches	2-11
Invoking a Script File when Loading AutoCAD	2-11
What Is a Slide Show?	2-20
What are Slides?	2-20
MSLIDE Command	2-20
VSLIDE Command	2-21
Preloading Slides	2-23
Slide Libraries	2-25
Slide Shows with Rendered Images	2-29

Chapter 3: Creating Linetypes and Hatch Patterns

Standard Linetypes	3-1
Linetype Definition	3-2
Elements of Linetype Specification	3-3
Creating Linetypes	3-3
Alignment Specification	3-9
Ltscale Command	3-9
LTSCALE Factor for Plotting	3-12
Alternate Linetypes	3-12
Modifying Linetypes	3-13
Current Linetype Scaling (CELTSCALE)	3-16
Complex Linetypes	3-17
Hatch Pattern Definition	3-25
How Hatch Works	3-27
Simple Hatch Pattern	3-28
Effect of Angle and Scale Factor on Hatch	3-29
Hatch Pattern with Dashes and Dots	3-30
Hatch with Multiple Descriptors	3-33
Saving Hatch Patterns in a Separate File	3-37
Custom Hatch Pattern File	3-38

Chapter 4: Customizing the ACAD.PGP File

What is the ACAD.PGP File?	4-1
Sections of the ACAD.PGP File	4-7
REINIT Command	4-10

Chapter 5: Pull-down, Shortcut, and Partial Menus and Customizing Toolbars

AutoCAD Menu	5-1
Standard Menus	5-3
Writing a Menu	5-3
Loading Menus	5-10
Restrictions	5-12
Cascading Submenus in Menus	5-12
Shortcut and Context Menus	5-18
Submenus	5-22
Loading Menus	5-23
Partial Menus	5-27
Accelerator Keys	5-32
Toolbars	5-34
Menu-Specific Help	5-38
Customizing the Toolbars	5-39

Chapter 6: Image Tile Menus

Image Tile Menus	6-1
Submenus	6-2
Writing an Image Tile Menu	6-3
Slides for Image Tile Menus	6-4
Loading Menus	6-9
Restrictions	6-10
Image Tile Menu Item Labels	6-10

Chapter 7: Button and Auxiliary Menus

Button Menus	7-1
Writing Button Menus	7-2
Special Handling of Button Menus	7-5
Submenus	7-8
Loading Menus	7-9
Auxiliary Menus	7-12

Chapter 8: Tablet Menus

Standard Tablet Menu	8-2
Features of a Tablet Menu	8-2
Customizing a Tablet Menu	8-3
Writing a Tablet Menu	8-4
Tablet Configuration	8-6
Loading Menus	8-7
Tablet Menus with Different Block Sizes	8-9
Assigning Commands to a Tablet	8-12
Automatic Menu Swapping	8-13

Chapter 9: Screen Menus

Screen Menu	9-1
Loading Menus	9-6
Submenus	9-7
Multiple Submenus	9-15
Long Menu Definitions	9-30
Menu Command Repetition	9-32
Automatic Menu Swapping	9-33
Menuecho System Variable	9-33
Menus for Foreign Languages	9-34
Use of Control Characters in Menu Items	9-34
Special Characters	9-35
Command Definition Without Enter and Space	9-37
Menu Items with Single Object Selection Mode	9-38
Use of AutoLISP in Menus	9-39
DIESEL Expressions in Menus	9-40

Customizing AutoCAD 2002

Chapter 10: Customizing the Standard AutoCAD Menu

The Standard AutoCAD Menu	10-1
Submenus	10-18
Customizing Tablet Area-1	10-21
Submenus	10-26
Customizing Tablet Area-2	10-30
Customizing Tablet Area-3	10-30
Customizing Tablet Area-4	10-33
Customizing Buttons and Auxiliary Menus	10-34
Customizing Pull-down and Shortcut Menus	10-40
Cascading Submenus in Menus	10-40
Shortcut Menus	10-44
Submenus	10-44
Customizing IMAGE TILE Menus	10-45
Submenus	10-46
Image Tile Menu Item Labels	10-46
Customizing the Screen Menu	10-49
Submenus	10-49

Chapter 11: Shapes and Text Fonts

Shape Files	11-1
Shape Description	11-1
Vector Length and Direction Encoding	11-2
Special Codes	11-5
Standard Codes	11-6
Code 000: End of Shape Definition	11-6
Code 001: Activate Draw Mode	11-7
Code 002: Deactivate Draw Mode	11-7
Code 003: Divide Vector Lengths	11-9
Code 004: Multiply Vector Lengths	11-9
Codes 005 and 006: Location Save/Restore	11-9
Code 007: Subshape	11-11
Code 008: X-Y Displacement	11-11
Code 009: Multiple X-Y Displacements	11-11
Code 00A or 10: Octant Arc	11-11
Code 00B or 11: Fractional Arc	11-12
Code 00C or 12: Arc Definition by Displacement and Bulge	11-14
Code 00D or 13: Multiple Bulge-Specified Arc	11-14
Code 00E or 14: Flag Vertical Text	11-15
Text Font Files	11-19

Chapter 12: AutoLISP

About AutoLISP	12-1
Mathematical Operations	12-2
Incremented, Decremented, and Absolute Numbers	12-4
Trigonometric Functions	12-4
Relational Statements	12-6
defun, setq, getpoint, and Command Functions	12-8
Loading an AutoLISP Program	12-13
getcorner, getdist, and setvar Functions	12-16
list Function	12-20
car, cdr, and cadr Functions	12-20
graphscr, textscr, princ, and terpri Functions	12-22
getangle and getorient Functions	12-26
getint, getreal, getstring, and getvar Functions	12-28
polar and sqrt Functions	12-30
itoa, rtos, strcase, and prompt Functions	12-33
Flowcharts	12-38
Conditional Functions	12-38

Chapter 13: Visual LISP

Visual LISP	13-1
Overview of Visual LISP	13-2
Starting Visual LISP	13-2
Using Visual LISP Text Editor	13-3
Visual LISP Console	13-6
Using Visual LISP Text Editor	13-9
Visual LISP Formatter	13-12
Debugging the Program	13-14
General Recommendtions for writing the LISP File	13-17
Tracing Variables	13-28
Visual LISP Error Codes and Messages	13-29

Chapter 14: Visual LISP: Editing the Drawing Database

Editing the Drawing Database	14-1
ssget	14-1
ssget "X"	14-4
Group Codes for ssget "X"	14-4
sslength	14-5
ssname	14-6
ssadd	14-6
ssdel	14-7
entget	14-7
assoc	14-8
cons	14-8

Customizing AutoCAD 2002

subst	14-8
entmod	14-9
How the Database is Retrieved and Edited	14-10
Some More Functions to Retrieve Entity Data	14-12

Chapter 15: Programmable Dialog Boxes Using Dialog Control Language

Dialog Control Language	15-1
Dialog Box	15-2
Dialog Box Components	15-2
Button and Text Tiles	15-4
Tile Attributes	15-5
Predefined Attributes	15-6
key, label, and is_default Attributes	15-6
fixed_width and alignment Attributes	15-8
Loading a DCL File	15-9
Displaying a New Dialog Box	15-10
Use of Standard Button Subassemblies	15-12
AutoLISP Functions	15-13
Managing Dialog Boxes with AutoLISP	15-14
Row and Boxed Row Tiles	15-16
Column, Boxed Column, and Toggle Tiles	15-17
Mnemonic Attribute	15-17
AutoLISP Functions	15-20
Predefined Radio Button, Radio Column, Boxed Radio Column, and Radio Row Tiles	15-25
Edit Box Tile	15-31
width and edit_width Attributes	15-31
Slider and Image Tiles	15-34
AutoLISP Functions	15-39

Chapter 16: DIESEL: A String Expression Language

DIESEL	16-1
Status Line	16-1
MODEMACRO System Variable	16-3
Customizing the Status Line	16-3
Macro Expressions Using DIESEL	16-4
Using AutoLISP with MODEMACRO	16-6
Diesel Expressions in Menus	16-8
MACROTRACE System Variable	16-9
Diesel String Functions	16-10

Chapter 17: Visual Basic

About Visual Basic	17-1
Objects	17-2
Add Method	17-2
AddCircle	17-3
AddLine	17-3
AddArc	17-4
AddText	17-4
Finding Help on Methods and Properties	17-4
Loading and Saving VBA Projects	17-6
GetPoint, GetDistance, and GetAngle Methods	17-10
GetPoint Method	17-10
GetDist Method	17-10
GetAngle Method	17-10
PolarPoint and AngleFromXAxis Methods	17-14
PolarPoint Method	17-14
AngleFromXAxis	17-14
Additional VBA Examples	17-19

Chapter 18: Accessing External Databases

Understanding Databases	18-1
AutoCAD Database Connectivity	18-3
Database Configuration	18-3
dbConnect Manager	18-4
Viewing and Editing Table Data from AutoCAD	18-6
Creating Links with Graphical Objects	18-8
Creting Labels	18-12
AutoCAD SQL Environment (ASE)	18-16
AutoCAD Query Editor	18-17
Forming Selection Sets Using Link Select	18-22
Conversion of ASE Links to AutoCAD 2002 Format	18-24

Chapter 19: Geometry Calculator

Geometry Calculator	19-1
Real, Integer, and Vector Expressions	19-1
Numeric Functions	19-3
Using Snap Modes	19-3
Obtaining the Radius of an Object	19-4
Locating a Point on a Line	19-5
Obtaining an Angle	19-6
Locating the Intersection Point	19-8
Applications of the Geometry Calculator	19-9
Using AutoLISP Variables	19-12
Filtering X, Y, and Z Coordinates	19-13

Converting Units 19-14
Additional Functions 19-14

Index

Appendices
The followings appendices are available on the author's Web site, **www.cadcim.com** or **www.calumet.purdue.edu/public/mets/tickoo/index.htm**

Appendix A: System Requirements and AutoCAD Installation
Appendix B: AutoCAD Commands
Appendix C: AutoCAD System Variables

Preface

AutoCAD, developed by Autodesk Inc., is the most popular PC-CAD system available in the market. Nearly 2.1 million people in 80 countries around the world are using AutoCAD to generate various kinds of drawings. In 2000 the market share of AutoCAD grew to about 78 percent, making it the worldwide standard for generating drawings. Also, AutoCAD's open architecture has allowed third-party developers to write application software that has significantly added to its popularity. For example, the author of this book has developed a software package "SMLayout" for sheet metal products that generates flat layout of various geometrical shapes such as transitions, intersections, cones, elbows, and tank heads. Several companies in Canada and the United States are using this software package with AutoCAD to design and manufacture various products. AutoCAD has also provided facilities that allow users to customize AutoCAD to make it more efficient and therefore increase their productivity.

The purpose of this book is to unravel the customizing power of AutoCAD and explain it in a way that is easy to understand. Every customizing technique is thoroughly explained with examples and illustrations that make it easy to comprehend the customizing concepts of AutoCAD. When you are done reading this book, you will be able to generate a Template drawing, write script files, edit existing menus, write your own menus, write shape and text files, create new linetypes and hatch patterns, define new commands, write programs in the AutoLISP programming language, edit the existing drawing database, create your own dialog boxes using DCL, customize the status line using DIESEL, and edit the Program Parameter file (ACAD.PGP). In the process, you will discover some new applications of AutoCAD that are unique and might have a significant effect on your drawings. You will also get a better idea of why AutoCAD has become such a popular software package and an international standard in PC-CAD.

To use this book, you do not need to be an AutoCAD expert or a programmer. If you know the basic AutoCAD commands, you will have no problem in understanding the material presented in this book. The book contains a detailed description of various customizing techniques that you can use to customize your system. Every chapter has several examples that illustrate some possible applications of these customizing techniques. At the end of each chapter are some exercises that provide a challenge to the user to solve the problems on his/her own. In a class situation, these exercises can be assigned to students to test their understanding of the material explained in the chapter. The chapters on AutoLISP programming assume that the user has no programming background and therefore all commands have been thoroughly explained in a way that makes programming easy to understand and interesting to learn. All chapters, except Slide Shows and Editing the Drawing Database, are independent and can be read in any order and used without reading the rest of the book. The user needs only to read the chapters on Script Files before the chapter on Slide Shows and the chapter on AutoLISP before the chapter on Editing the Drawing Database. However, in order to get a good understanding of customizing techniques, it is recommended to start from Chapter 1 and then progress through the chapters. AutoCAD Release 14 features are indicated by asterisk (*) at the end of the feature. The following is a summary of each chapter.

Chapter 1: Template Drawings
This chapter explains how to create a Template drawing and how to standardize the information that is common to all drawings. It also describes how to create a Template drawing with paper space and predefined viewports.

Chapter 2: Script Files and Slide Shows
This chapter introduces the user to script files and how to utilize them to group AutoCAD commands in a predetermined sequence to perform a given operation. This chapter also explains how to use script files to create a slide show that can be used for product presentation.

Chapter 3: Creating Linetypes and Hatch Patterns
This chapter explains how to create a new linetype and how to edit the linetype file, ACAD.LIN. This chapter also describes the techniques of creating a new hatch pattern and the effect of hatch scale and hatch angle on hatch.

Chapter 4: Customizing the ACAD.PGP File
This chapter explains the use of AutoCAD's Program Parameter file (ACAD.PGP) to define aliases for the operating system commands and some of the AutoCAD commands.

Chapter 5: Pull-down, Shortcut and Partial Menus and Customizing Toolbars
This chapter explains how to write pull-down, shortcut, and partial menus. The chapter also explains how to customize the toolbars. Several examples have been given with thorough explanation.

Chapter 6: Image Tile Menus
This chapter explains the Image tile menus and how to write an Image tile menu. It also discusses submenus and how to make slides for the Image tile menu.

Chapter 7: Buttons and Auxiliary Menus
This chapter deals with buttons and auxiliary menus and how to assign AutoCAD commands to different buttons of a multi-button pointing device.

Chapter 8: Tablet Menus
This chapter explains how to write a tablet menu, and how to load other menus from the tablet menu. Advantages of the tablet menu, design of the tablet menu, and how AutoCAD assigns commands to different blocks of the tablet menu are also discussed.

Chapter 9: Screen Menus
This chapter describes the procedure to write a screen menu with multiple submenus and how to load Image tile or pull-down menus from the screen menu.

Chapter 10: Customizing the Standard AutoCAD Menu
This chapter describes how to edit and change various menu sections of the standard AutoCAD menu, ACAD.MNU. It also contains information about submenus and how to load different submenus.

Chapter 11: Shapes and Text Fonts
This chapter explains what shapes are and how to create shape and text fonts. It also contains a detailed description of special codes and their application to creating shapes and text fonts.

Chapter 12: AutoLISP
This chapter explains different AutoLISP functions and how to use these functions to write a program. It also introduces the user to basic programming techniques and use of relational and conditional statements in a program.

Chapter 13: Visual LISP
This chapter explains how to start Visual LISP in AutoCAD, use Visual LISP text editor, run Visual LISP programs, use Visual LISP console and formatter, and how to debug Visual LISP programs.

Chapter 14: Visual LISP: Editing the Drawing Database
This chapter describes those AutoLISP functions that allow a user to edit the drawing database.

Chapter 15: Programmable Dialog Boxes Using Dialog Control Language
This chapter is an introduction to Dialog Control Language and its applications in customizing the existing dialog boxes and writing new dialog boxes. It also explains the use of AutoLISP in controlling the dialog boxes.

Chapter 16: DIESEL: A String Expression Language
This chapter describes the DIESEL string expression language and its application in customizing the status line by altering the value of AutoCAD system variable MODEMACRO.

Chapter 17: Visual Basic
This chapter describes how to install the AutoCAD preview VBA, load and run sample VBA projects, utilize the visual basic editor, use AutoCAD objects and object properties, and apply and use AutoCAD methods.

Chapter 18: Accessing External Database
This chapter explains how to create a database from the information associated with the AutoCAD objects like blocks. The information extracted can then be arranged in rows and columns.

Chapter 19: Geometry Calculator
The **Geometry Calculator** is an ADS application that can be used as an online calculator. This chapter explains how the calculator can be used to evaluate vector, real, and integer expressions. It can also access the existing geometry by using the first three characters of the standard AutoCAD object snap functions like MID, CEN, END, etc.

DEDICATION

To teachers, who make it possible to disseminate knowledge
to enlighten the young and curious minds
of our future generations

To students, who are dedicated to learning new technologies
and making the world a better place to live

Thanks

To the faculty and students of the METS department
of Purdue University Calumet for their cooperation

CADSoft Technologies for their valuable input

To Bill Fane for performing the technical edit
and Liz Kingslien for copy edit

Author's web sites
Faculty
Please contact the author at stickoo@calumet.purdue.edu to access the Web site for faculty that contains the following:
1. Drawings used in examples and exercises for every chapter.
2. Istructor's Guide containing solution to problems and answers to questions for every chapter.

Students
You can download drawing exercises, tutorials, program listings, and other special topics by accessing the Web site **www.cadcim.com** or www.calumet.purdue.edu/public/mets/tickoo/index.htm.

Chapter 1

Template Drawings

Learning Objectives

After completing this chapter, you will be able to:
- *Create template drawings.*
- *Load template drawings using dialog boxes and the command line.*
- *Do an initial drawing setup.*
- *Customize drawings with layers and dimensioning specifications.*
- *Customize drawings with layouts, viewports and paper space.*

CREATING TEMPLATE DRAWINGS

One way to customize AutoCAD is to create template drawings that contain initial drawing setup information and, if desired, visible objects and text. When the user starts a new drawing, the settings associated with the template drawing are automatically loaded. If you start a new drawing from scratch, AutoCAD loads default setup values. For example, the default limits are (0.0,0.0), (12.0,9.0) and the default layer is 0 with white color and continuous linetype. Generally, these default parameters need to be reset before generating a drawing on the computer using AutoCAD. A considerable amount of time is required to set up the layers, colors, linetypes, limits, snaps, units, text height, dimensioning variables, and other parameters. Sometimes, border lines and a title block may also be needed.

In production drawings, most of the drawing setup values remain the same. For example, the company title block, border, layers, linetypes, dimension variables, text height, LTSCALE, and other drawing setup values do not change. You will save considerable time if you save these values and reload them when starting a new drawing. You can do this by making template drawings, which can contain the initial drawing setup information, set according to company specifications. They can also contain a border, title block, tolerance table, block definitions, floating viewports in the paper space, and perhaps some notes and instructions that are common to all drawings.

THE STANDARD TEMPLATE DRAWINGS

The AutoCAD software package comes with standard template drawings like acad.dwt, acadiso.dwt, ansi a.dwt, din a.dwt, iso a4.dwt, jisa3.dwt. The ansi, din, and iso template drawings are based on the drawing standards developed by ANSI (American National Standards Institute), DIN (German), and ISO (International Organization for Standardization). When you start a new drawing and you are using traditional startup dialog, AutoCAD displays the **Create New Drawing** dialog box on the screen. To load the template drawing, select the **Use a Template** button and AutoCAD will display the list of standard template drawings. From this list you can select any template drawing according to your requirements. If you want to start a drawing with default settings, select the **Start from Scratch** button in the **Create New Drawing** dialog box. The following are some of the system variables, with the default values that are assigned to a new drawing:

System Variable Name	Default Value
CHAMFERA	0.5000
CHAMFERB	0.5000
COLOR	Bylayer
DIMALT	Off
DIMALTD	2
DIMALTF	25.4
DIMPOST	None
DIMASO	On
DIMASZ	0.18
FILLETRAD	0.5000
GRID	0.5000
GRIDMODE	0
ISOPLANE	Left
LIMMIN	0.0000,0.0000
LIMMAX	12.0000,9.0000
LTSCALE	1.0
MIRRTEXT	1 (Text mirrored like other objects)
TILEMODE	1 (On)

Example 1

Create a template drawing using **Advanced Setup** of the **Create Drawings tab** with the following specifications. The name of the template drawing is **PROTO1**.

Units	Engineering with precision 0'-0.00"
Angle	Decimal degrees with precision 0.
Angle Direction	Counterclockwise
Area	144'x96'

Step 1

Choose **New** from the **File** menu to display the **AutoCAD 2002 Today** window as shown in Figure 1-1. In this dialog box choose the **Create Drawings** tab and then choose the **Wizards** option. Now, select the **Advanced Setup** option.

Template Drawings 1-3

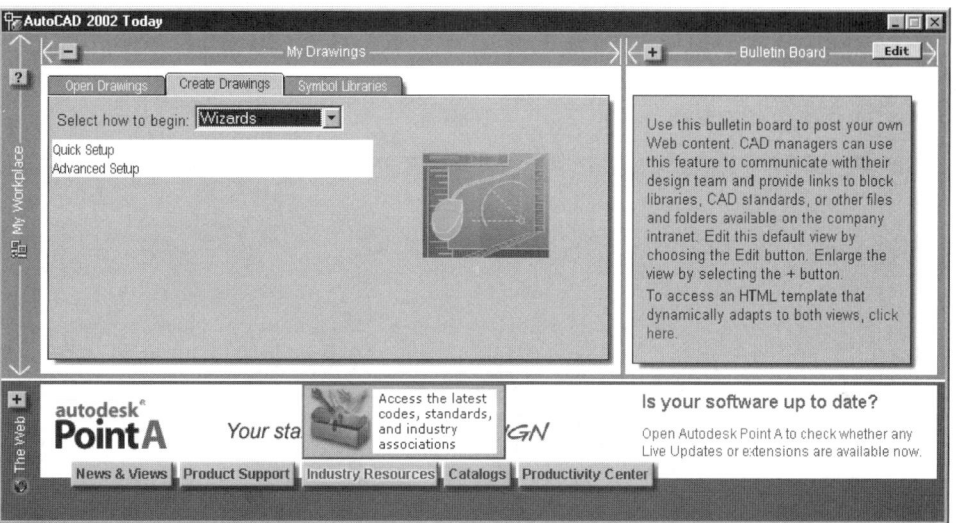

Figure 1-1 AutoCAD 2002 Today window

Step 2
When you select the **Advanced Setup** option, the **Units** page of the **Advanced Setup** dialog box is displayed as shown in Figure 1-2. Select the **Engineering** radio button. Select **0'-0.00"** precision from the **Precision** drop-down list and then choose the **Next** button.

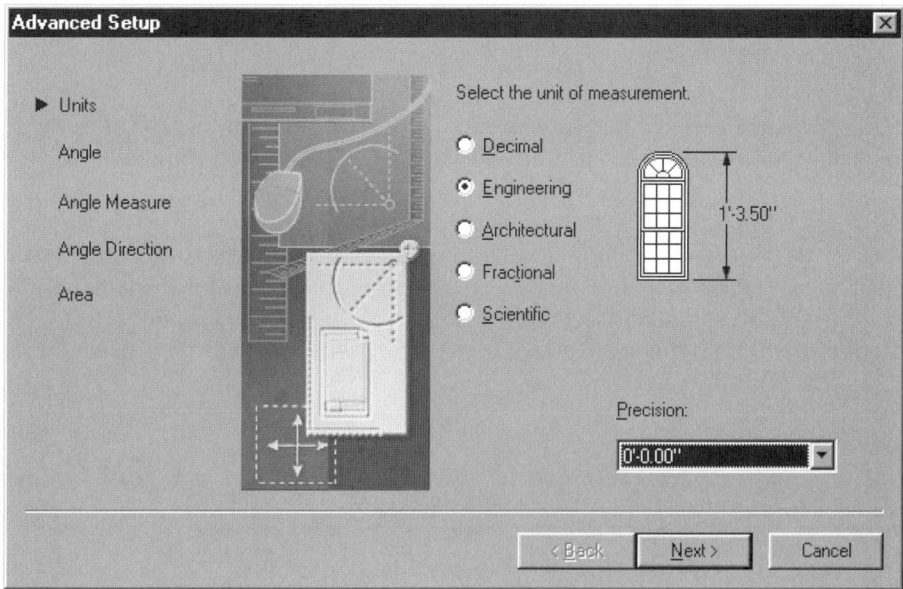

Figure 1-2 Advanced Setup dialog box to define Units

Step 3

Now, the **Angle** page of the **Advanced Setup** dialog box is displayed as shown in Figure 1-3. In this dialog box, select the **Decimal Degrees** radio button and select **0** from the **Precision** drop-down list and then choose the **Next** button.

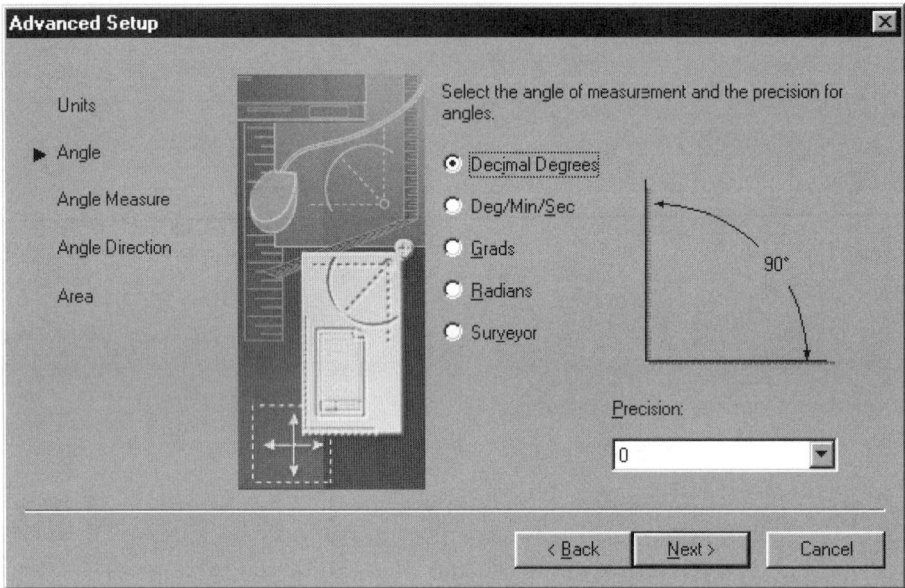

Figure 1-3 Advanced Setup dialog box to define Angle

Step 4

The **Angle Measure** page of the **Advanced Setup** dialog box is displayed. Select the **East** radio button. Choose the **Next** button to display the **Angle Direction** page.

Step 5

Select the **Counterclockwise** radio button and then choose the **Next** button. The **Area** page is displayed. Specify the area as 144' and 96' by entering the value of width and length as **144'** and **96'** in the **Width** and **Length** edit boxes and then choose the **Finish** button. Use the **All** option of the **ZOOM** command to display the new limits on the screen. Save the template drawing as **PROTO1.DWT**.

Note

*If you want to customize only units and area, you can use the **Quick Setup** in the **Create New Drawing** dialog box.*

Example 2

Create a template drawing with the following specifications. The name of the template drawing is PROTO2.

  Limits   18.0,12.0

Template Drawings

Snap	0.25
Grid	0.50
Text height	0.125
Units	2 digits to the right of decimal point
	Decimal degrees
	2 digits to the right of decimal point
	0 angle along positive X axis (east)
	Angle positive if measured counterclockwise

Step 1
Start AutoCAD and select the **Start from Scratch** button in the **Create Drawings** tab of **AutoCAD 2002 Today** window. You can also invoke the **AutoCAD 2002 Today** window (Figure 1-4) by selecting **New** in the **File** menu or entering **NEW** at the AutoCAD Command prompt.

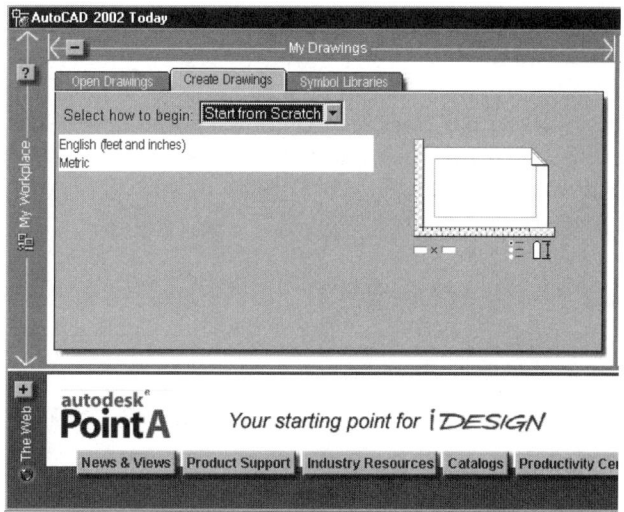

Figure 1-4 AutoCAD 2002 Today window

Once you are in the **Drawing Window**, use the following AutoCAD commands to set up the values.

Step 2
Setting limits, snap, grid, and text size
The **LIMITS** command can be invoked by choosing **Drawing Limits** from the **Format** menu or by entering **LIMITS** at the Command prompt.

    Command: **LIMITS**
    Specify lower left corner or [ON/OFF] <0.00,0.00>: **0,0**
    Specify upper right corner <12.0,9.0>: **18.0,12.0**

After you set the limits, use the **ZOOM** command with the **All** option to display the new limits on the screen.

To set the **SNAP** and **GRID**, right click the **Snap** or **Grid** button in the status bar to display the shortcut menu. Choose the **Settings** in the shortcut menu to display the **Drafting Settings** dialog box. You can also choose the **Object Snap Settings** button from the **Object Snap** toolbar to display the **Drafting Settings** dialog box. Choose the **SNAP** and **GRID** tab. Enter **0.25** and **0.25** in the **Snap X spacing** and **Snap Y spacing** edit boxes respectively. Enter **0.5** and **0.5** in the **Grid X spacing** and **Grid Y spacing** edit boxes respectively. Then choose **OK**. You can also use **SNAP** and **GRID** commands to set these values.

Size of the text can be changed by entering **TEXTSIZE** at the Command prompt.

    Command: **TEXTSIZE**
    Enter new value for TEXTSIZE <0.2000>: **0.125**

Step 3
Setting units

You can use the **Drawing Units** dialog box (Figure 1-5) to set the units. To invoke the **Drawing Units** dialog box, choose **Units** in the **Format** menu or enter **UNITS** at the Command prompt. In the **Angle** area choose **Decimal Degrees** from the **Type** drop-down list and choose **0.00** from the **Precision** drop-down list. Also select the **Clockwise** radio button from the **Angle** area. Select the **Direction** button to display the **Direction Control** dialog box (Figure 1-6) and select the **East** radio button.

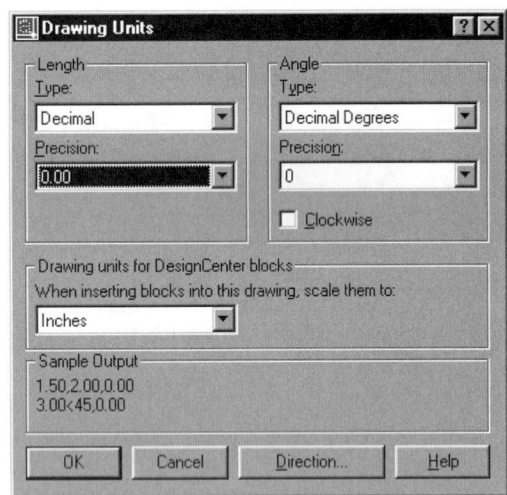

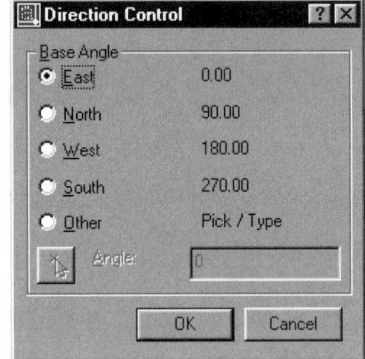

Figure 1-5 Drawing Units dialog box   *Figure 1-6 Direction Control dialog box*

Step 4

Now, save the drawing as **PROTO1.DWT** using AutoCAD's **SAVEAS** command. You must select template (*DWT) from the list box in the dialog box. This drawing is now saved as PROTO1.DWT on the default drive. You can also save this drawing on a floppy diskette in drives A or B.

LOADING A TEMPLATE DRAWING

You can use the template drawing any time you want to start a new drawing. To use the preset values of the template drawing, start AutoCAD or select the **NEW** button from the Standard toolbar. AutoCAD displays the **AutoCAD 2002 Today** window. You can also start a new drawing by selecting the **New** option from the **File** menu. To load the template drawing, select the **Template** option in the **Create Drawing** tab of **AutoCAD 2002 Today** window, Figure 1-7. AutoCAD will display the list of template drawings. From this list, select **PROTO1** template drawing. AutoCAD will start a new drawing that will have the same setup as that of template drawing, **PROTO1**.

You can have several template drawings, each with a different setup. For example, **PROTOB** for a 18" by 12" drawing, **PROTOC** for a 24" by 18" drawing. Each template drawing can be created according to user-defined specifications. You can then load any of these template drawings as discussed previously.

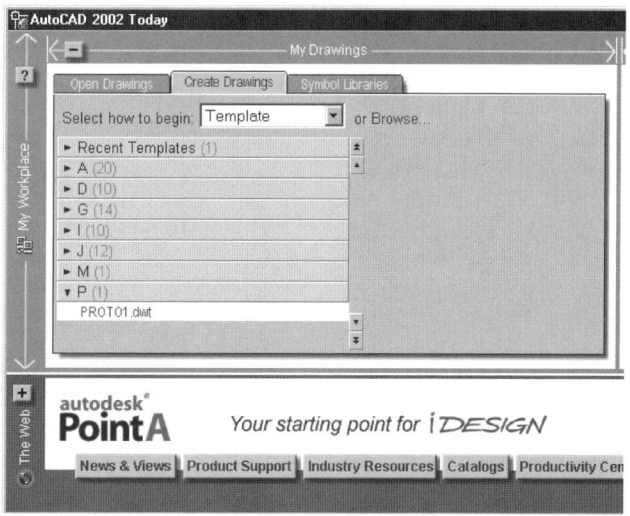

Figure 1-7 AutoCAD 2002 Today window with the list of template drawings

Note
*You can also create a template drawing by entering **NEW** at the Command prompt and keeping the **FILEDIA** system variable **0**.*

CUSTOMIZING DRAWINGS WITH LAYERS AND DIMENSIONING SPECIFICATIONS

Most production drawings need multiple layers for different groups of objects. In addition to layers, it is a good practice to assign different colors to different layers to control the line width at the time of plotting. You can generate a template drawing that contains the desired number of layers with linetypes and colors according to your company specifications. You can then use this template drawing to make a new drawing. The next example illustrates the procedure used for customizing a drawing with layers, linetypes, and colors.

Example 3

Create a template drawing (**PROTO3**) that has a border and the company's title block, as shown in Figure 1-8. In addition to this, you want the following initial drawing setup:

Limits	48.0,36.0
Text height	0.25
PLINE width	0.02
Ltscale	4.0

DIMENSIONS
Overall dimension scale factor 4.0
Dimension text above the extension line
Dimension text aligned with dimension line

LAYERS

Layer Names	Line Type	Color
0	Continuous	White
OBJ	Continuous	Red
CEN	Center	Yellow
HID	Hidden	Blue
DIM	Continuous	Green
BOR	Continuous	Magenta

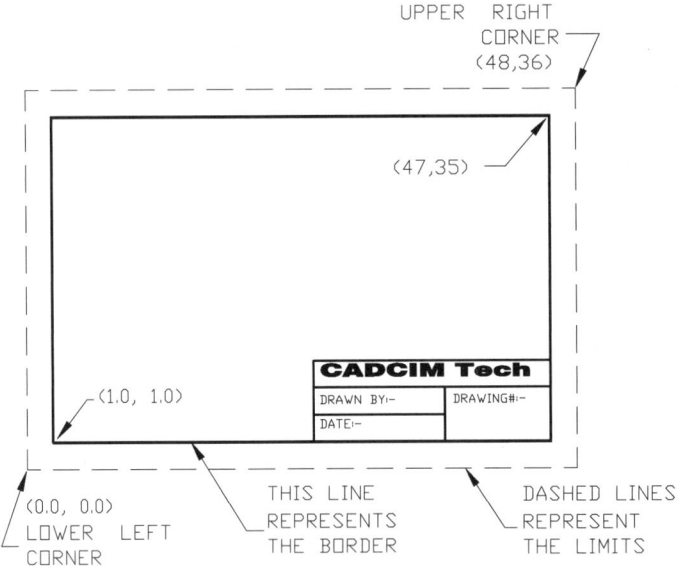

Figure 1-8 Template drawing for Example 3

Template Drawings

Step 1
Setting limits, text size, polyline width, polyline and linetype scaling

Start a new drawing with default parameters. You can do this by selecting the **Start from Scratch** option in the **Create Drawings** tab of the **AutoCAD Today** window. Once you are in the drawing editor, use the AutoCAD commands to set up the values as given for this example. Also, draw a border and a title block as shown in Figure 1-8. In this figure the hidden lines indicate the drawing limits. The border lines are 1.0 units inside the drawing limits. For the border lines, use a polyline of width 0.02 units. Use the following procedure to produce the prototype drawing for Example 3:

The **LIMITS** command can be invoked by choosing **Drawing Limits** from the **Format** menu or by entering **LIMITS** at the Command prompt.

    Command: **LIMITS**
    Specify lower left corner or [ON/OFF] <0.00,0.00>: **0,0**
    Specify upper right corner <12.0,9.0>: **48.0,36.0**

Size of the text can be changed by entering **TEXTSIZE** at the Command prompt.

    Command: **TEXTSIZE**
    Enter new value for TEXTSIZE <0.2000>: **0.25**

Polyline width can be changed by entering **PLINEWID** at the Command prompt.

    Command: **PLINEWID**
    Enter new value for PLINEWID <0.0000>: **0.02**

To draw the border, use the **PLINE** command. You can invoke the **PLINE** command by choosing the **Polyline** button from the **Draw** toolbar or by entering **PLINE** at the Command prompt.

    Command: **PLINE**
    Specify start point: **1.0,1.0**
    Current line-width is **0.02**
    Specify next point or [Arc/Close/Halfwidth/Length/Undo/Width]:**47,1**
    Specify next point or [Arc/Close/Halfwidth/Length/Undo/Width]:**47,35**
    Specify next point or [Arc/Close/Halfwidth/Length/Undo/Width]:**1,35**
    Specify next point or [Arc/Close/Halfwidth/Length/Undo/Width]:**C**

Linetype scale can be changed by entering **LTSCALE** at the Command prompt.

    Command: **LTSCALE**
    Enter new linetype scale factor<Current>: **4.0**

Step 2
Setting dimensioning parameters

You can use the **Dimension Style Manager** dialog box (Figure 1-9) to set the dimension variables. Choose the **Dimension Style** button from the **Dimension** toolbar or choose **Style** from the

Dimension menu to invoke the **Dimension Style Manager** dialog box. You can also invoke this dialog box by entering **DIMSTYLE** at the Command prompt. Choose the **New** button from the **Dimension Style Manager** dialog box. The **Create New Dimension Style** dialog box is displayed as shown in the Figure 1-10. Specify new style name as **MYDIM1** in the **New Style Name** edit box and then choose the **Continue** button. The **New Dimension Style** dialog box will be displayed. In the **New Dimension Style** dialog box choose the **Modify** button to display the **New Dimension Style:MYDIM1** dialog box (Figure 1-11).

Overall dimension scale factor
To specify dimension scale factor, choose the **Fit** tab of the **Modify Dimension Style** dialog box. Set the value in the **Use overall scale of** as **4** in the **Scale for Dimension Features** area (Figure 1-11).

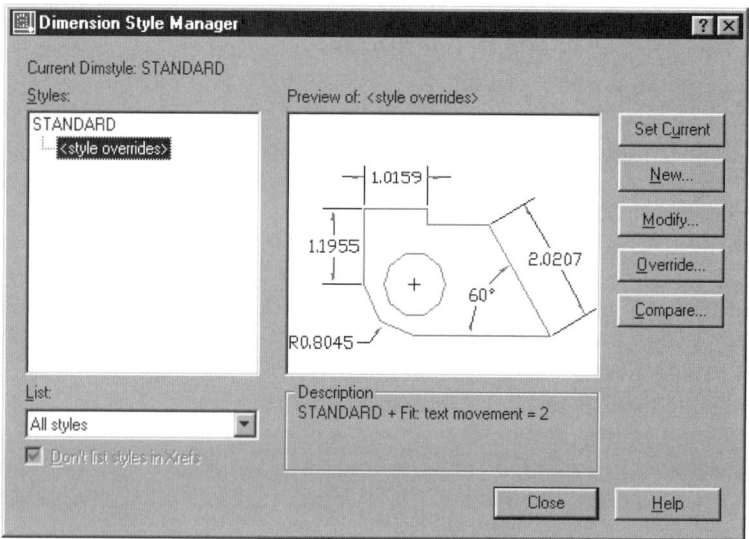

Figure 1-9 **Dimension Style Manager** *dialog box*

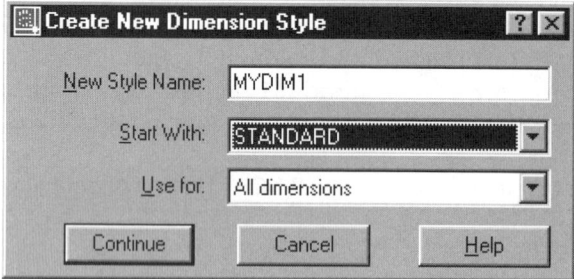

Figure 1-10 **Create New Dimension Syle** *dialog box*

Dimension text over the dimension line
Choose the **Text** tab of the **Modify Dimension Style** dialog box. Select the **Above** option from the **Vertical** drop-down list in the **Text Placement** area.

Template Drawings

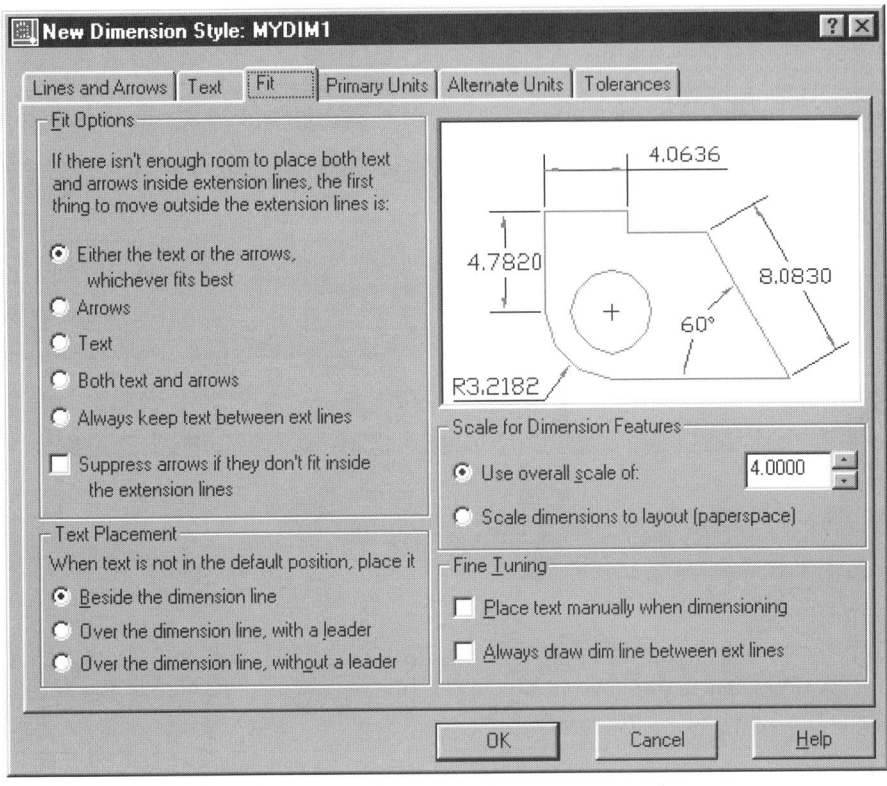

Figure 1-11 New Dimension Style dialog box

Dimension text aligned with dimension line
In the **Text Alignment area of Text tab** choose the **Aligned with the dimension line** radio button and then choose **OK**. In the **Dimension Style Manager** dialog box, choose the **Close** button to exit.

Step 3
Setting layers
Choose the **Layers** button from the **Object Properties** toolbar or choose **Layer** from the **Format** menu to invoke **Layer Properties Manager** dialog box as shown in Figure 1-12. You can also invoke the **Layer Properties Manager** dialog box by entering **LAYER** at the Command prompt. Choose the **New** button in the **Layer Properties Manager** dialog box and rename **Layer1** as **OBJ**. Select the color swatch of the **OBJ** layer to display the **Select Color** dialog box. Choose the **Red** color and choose **OK,** so the red color will be assigned to the OBJ layer. Again choose the **New** button in the **Layer Properties Manager** dialog box and rename the **Layer1** as **CEN**. Choose the linetype swatch to display the **Select Linetype** dialog box. If the different linetypes are not already loaded, choose the **Load** button to display the **Load or Reload Linetypes** dialog box. Choose the **CENTER** linetype from the **Available Linetypes** area and choose OK. Again the **Select Linetype** dialog box will reappear. Choose the **CENTER** linetype from the **Loaded Linetypes** area and choose **OK**. Choose the color swatch to display the **Select Color** dialog box. Choose the **Yellow** color and choose **OK,** so the color yellow and linetype center will be assigned

to the layer Cen. Similarly different linetypes and different colors can be set for different layers mentioned in the example.

You can also use the **-LAYER** command to set the layers and linetypes from the Command prompt.

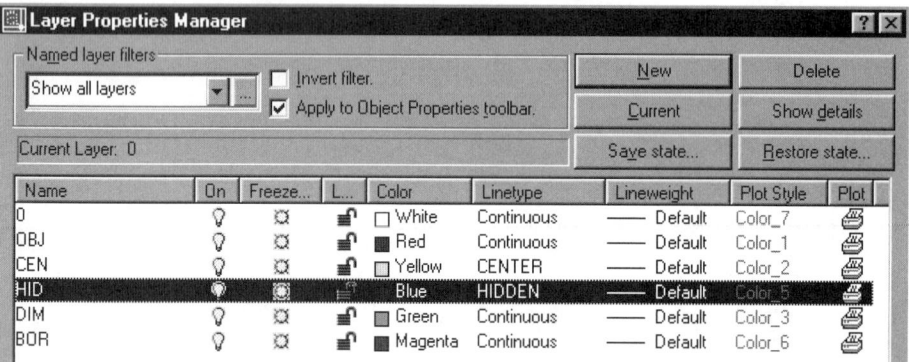

Figure 1-12 Layer Properties Manager dialog box

Step 4
Next, add the title block and the text as shown in Figure 1-8. After completing the drawing, save it as PROTO3.DWT. You have created a template drawing (PROTO3) that contains all of the information given in Example 3.

CUSTOMIZING A DRAWING WITH LAYOUT

The Layout (paper space) provides a convenient way to plot multiple views of a 3D drawing or multiple views of a regular 2D drawing. It takes quite some time to set up the viewports in model space with different vpoints and scale factors. You can create prototype drawings that contain predefined viewport settings, with vpoint and other desired information. Now if you create a new drawing, or insert a drawing, the views are automatically generated. The following example illustrates the procedure for generating a prototype drawing with paper space and model space viewports.

Example 4

Create a template drawing as shown in Figure 1-13 with four views in Layout3 (Paper space) that display front, top, side, and 3D views of the object. The plot size is 10.5 by 8 inches. The plot scale is 0.5 or 1/2" = 1". The paper space viewports should have the following vpoint setting:

Viewports	Vpoint	View
Top right	1,-1,1	3D view
Top left	0,0,1	Top view
Lower right	1,0,0	Right side view
Lower left	0,-1,0	Front view

Step 1
Start AutoCAD and create a new drawing. Use the following commands to set up various parameters.

Template Drawings 1-13

The first step is to create a layout. To create a layout you can use the **LAYOUT** command. You can also right-click on **Model** or any **Layout** tab to display the shortcut menu. From the shortcut menu select **New layout**.

> Command: **LAYOUT**
> Enter layout option [Copy/Delete/New/Template/Rename/SAveas/Set/?] <set>: **N**
> Enter new Layout name <Layout3>: **Layout3**

Step 2
The next step is to select the new layout (Layout 3) tab. When you select this tab, AutoCAD displays the **Page Setup-Layout3** dialog box. Select the **Plot Device** tab and then select the printer or plotter that you want to use. In this example HP LaserJet4000 is used.

Step 3
Next, select the **Layout Settings** tab and select the paper size that is supported by the selected plotting device. In this example the paper size is 8.5x11. Select the **OK** button to accept the settings and exit the dialog box. AutoCAD displays the new layout (Layout3) on the screen with default viewport. Use the **ERASE** command to erase this viewport.

Step 4
The next step is to set up a layer (VIEW) for viewports and assign it a color (green). To invoke the **Layer Properties Manager** dialog box, choose the **Layers** button from the **Object Properties** toolbar or choose **Layer** from the **Format** menu. You can also invoke the **Layer Properties Manager** dialog box by entering **LAYER** at the Command prompt. Now the **Layer Properties Manager** dialog box is displayed. Choose the **New** button and name the **Layer1** as **VIEW**. Select the color swatch of the **VIEW** layer to display the **Select Color** dialog box. Select the color **Green** and Choose the **OK** button. This color will be assigned to **View** layer. Also, make the **VIEW** layer current and then select the **OK** button to exit.

Layers can also be set at the Command prompt by entering **-LAYER** at the Command prompt.

Step 5
To create four viewports, use the **MVIEW** command, Figure 1-13. In order to invoke the **MVIEW** command choose **Viewports > 4Viewport** from the **View** menu or you can directly enter **MVIEW** command at the Command prompt. Then switch to model space to zoom the display to half the size.

> Command: **MVIEW**
> Specify corner of viewport or
> [ON/OFF/Fit/Hideplot/Lock/Object/Polygonal/Restore/2/3/4] <Fit>:**4**
> Specify first corner or [Fit] <Fit>: **0.25,0.25**
> Specify opposite corner: **10.25,7.75**

Choose **Paper** button in the status bar to activate the model space or enter **MSPACE** at the Command prompt.

> Command: **MSPACE** (or **MS**)

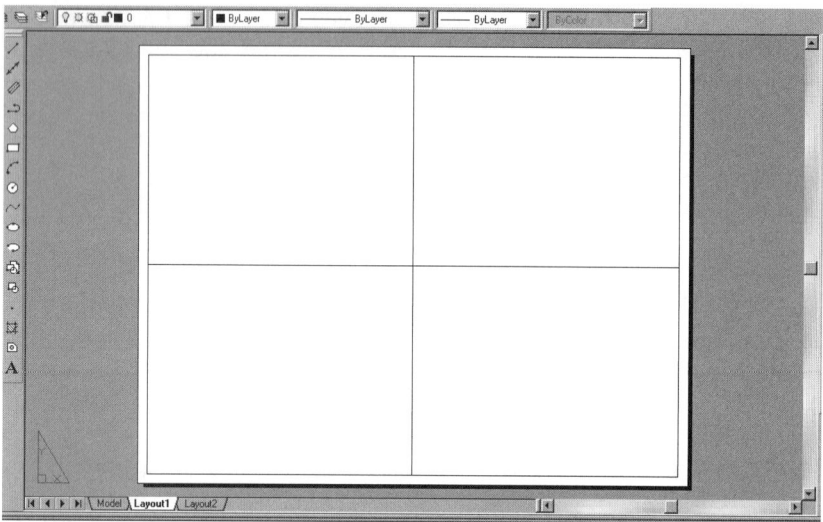

Figure 1-13 *Paper space with four viewports*

Make the first viewport active by selecting a point in the viewport and then use the **ZOOM** command to specify the paper space scale factor to 0.5. The **ZOOM** command can be invoked by choosing **Zoom > Scale** from the **View** menu or by entering **ZOOM** at the Command prompt.

Command: **ZOOM**
Specify corner of window, enter a scale factor (nX or nXP), or
[All/Center/Dynamic/Extents/Previous/Scale/Window] <real time>: **0.5XP**

Now, make the next viewport active and specify the scale factor. Do the same for the remaining viewports.

Step 6
The next step is to change the vpoints of different paper space viewports by using the **VPOINT** command. The **VPOINT** command can be invoked by choosing **3D Views > VPOINT** from the **View** menu or by entering **VPOINT** at the Command prompt. The vpoint values for different viewports are shown in Example 5. To set the view point for the lower-left viewport the Command prompt sequence is as follows:

Command: **VPOINT**
Current view direction: VIEWDIR=0.0000,0.0000,1.0000
Specify a view point or [Rotate] <display compass and tripod>: **0,-1,0**

Similarly use the **VPOINT** command to set the vpoint of other viewports.

Step 7
Use the **Model** button in the status bar to change to paper space and then set a new layer PBORDER with yellow color. Make the PBORDER layer current, draw a border, and if needed a

Template Drawings

title block using the **PLINE** command. You can also change to paper space by entering **PSPACE** at the Command prompt.

The **PLINE** command can be invoked by choosing the **Polyline** button from the **Draw** toolbar or by choosing **Polyline** from the **Draw** menu. The **PLINE** Command can also be invoked by entering **PLINE** at the Command prompt.

Command: **PLINE**
Specify start point: **0,0**
Current line-width is 0.0000
Specify next point or [Arc/Close/Halfwidth/Length/Undo/Width]: **0,8.0**
Specify next point or [Arc/Close/Halfwidth/Length/Undo/Width]: **10.5,8.0**
Specify next point or [Arc/Close/Halfwidth/Length/Undo/Width]: **10.5,0**
Specify next point or [Arc/Close/Halfwidth/Length/Undo/Width]: **C**

Step 8
The last step is to select the **Model** tab (or change the **TILEMODE** to 1) and save the prototype drawing. To test the layout that you just created, make the 3D drawing as shown in Figure 1-17 or make any 3D object. Now, if you switch to **Layout 3** tab, you will find four different views of the object (Figure 1-14). If the object views do not appear in the viewports, use the **PAN** commands to position the views in the viewports. You can freeze the VIEW layer so that the viewports do not appear on the drawing. Now you can plot this drawing from The Layout3 with a plot scale factor of 1:1 and the size of the plot will be exactly as specified.

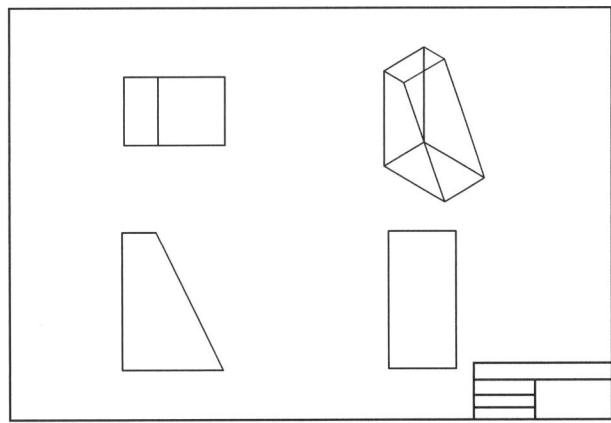

Figure 1-14 Four views of the 3D object in paper space

CUSTOMIZING DRAWINGS WITH VIEWPORTS

In certain applications you might need multiple model space viewport configurations to display different views of an object. This involves setting up the desired viewports and then changing the viewpoint for different viewports. You can create a prototype drawing that contains a required number of viewports and the viewpoint information. Now, if you insert a 3D object in one of the viewports of the prototype drawing, you will automatically get different views of the object without setting viewports or viewpoints. The following example illustrates the procedure for creating a prototype drawing with a standard number (four) of viewports and viewpoints.

Example 5

Create a prototype drawing with four viewports, as shown in Figure 1-15. The viewports should have the following viewpoints (vpoints):

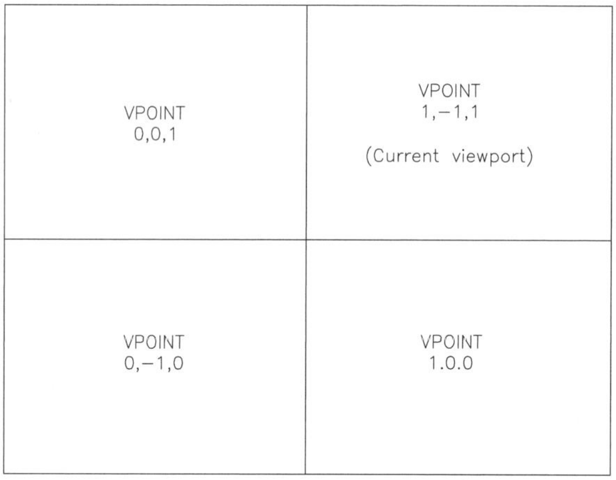

```
+---------------------+---------------------+
|                     |                     |
|      VPOINT         |      VPOINT         |
|      0,0,1          |      1,-1,1         |
|                     |                     |
|                     |  (Current viewport) |
|                     |                     |
+---------------------+---------------------+
|                     |                     |
|                     |                     |
|      VPOINT         |      VPOINT         |
|      0,-1,0         |      1,0,0          |
|                     |                     |
|                     |                     |
+---------------------+---------------------+
```

Figure 1-15 Viewports with different viewpoints

Viewports	Vpoint	View
Top right	1,-1,1	3D view
Top left	0,0,1	Top view
Lower right	1,0,0	Right side view
Lower left	0,-1,0	Front view

Step 1
Start AutoCAD and create a new drawing from scratch. Use the following commands to set the viewports and vpoints.

Step 2
Setting viewports
Viewports and corresponding viewpoints can be set with the **VPORTS** command. You can also choose the **Display Viewports Dialog** button from the **Viewports** toolbar or choose **Viewports > New Viewports** from the **View** menu to display the **Viewports** dialog box as shown in Figure 1-16. Choose **Four:Equal** viewports from the **Standard Viewports** area. In the **Preview** area four equal viewports are displayed. Select **3D** from the **Setup** drop-down list. The four viewports with the different viewpoints will be displayed in the **Preview** area as Top, Front, Right and SE Isometric respectively. **Top** represents the viewpoints as (0,0,1), **Front** represents the viewpoints as (0,-1,0), **Right** represents the viewpoints as (1,0,0) and **SE Isometric** represents the view-

Template Drawings

points as (1,-1,1) respectively. Choose the **OK** button. Save the drawing as **PROTO5.DWT**.

Viewports and viewpoints can also be set by entering **-VPORTS** and **VPOINT** at the Command prompt respectively.

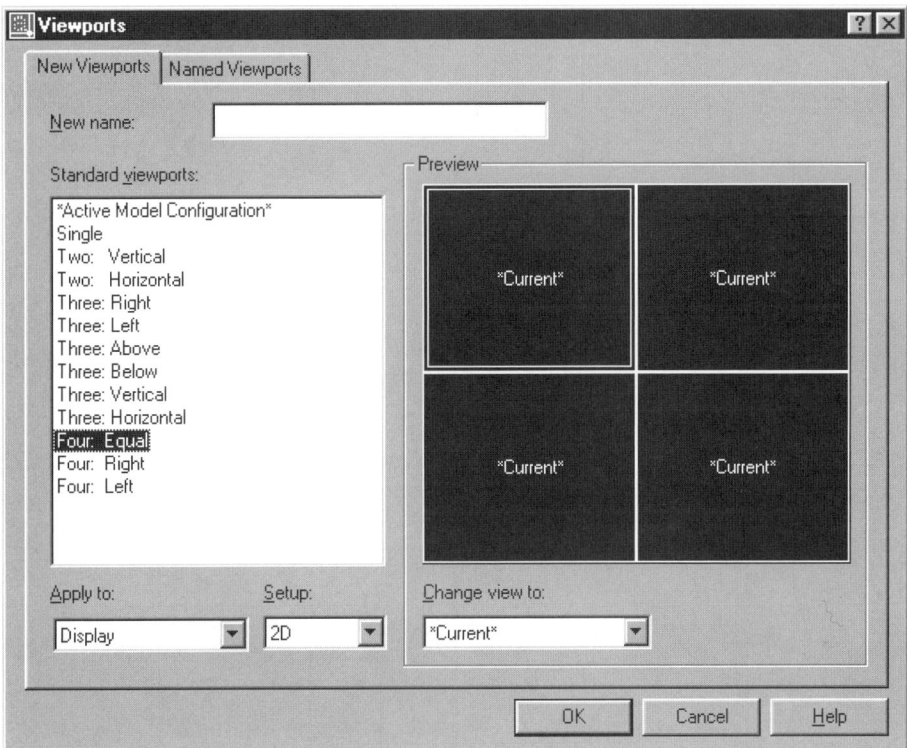

Figure 1-16 Viewports dialog box

Step 3
Now, start a new drawing and draw the 3D tapered block as shown in Figure 1-17.

Step 4
Again, start a new drawing, TEST, using the prototype drawing PROTO5. Make the top right viewport current and insert or create a drawing shown in Figure 1-17. Four different views will be automatically displayed on the screen as shown in Figure 1-18.

CUSTOMIZING DRAWINGS ACCORDING TO PLOT SIZE AND DRAWING SCALE

For controlling the plot area, it is recommended to use layouts. You can make the drawing of any size and then use the layout to specify the sheet size and then draw the border and title block. However you can also plot a drawing in the model space and set up the system variables so that the plotted drawing is to your specifications. You can generate a template drawing

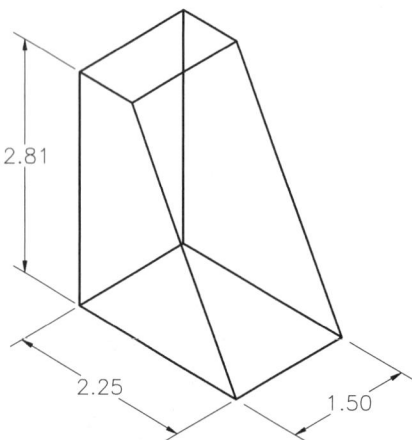

Figure 1-17 3D tapered block

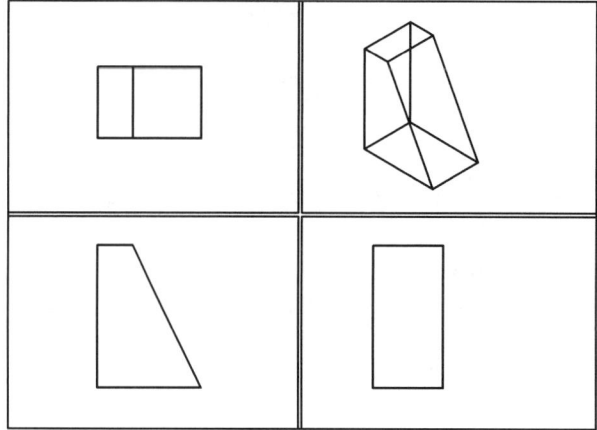

Figure 1-18 Different views of 3D tapered block

according to plot size and scale. For example, if the scale is 1/16" = 1' and the drawing is to be plotted on a 36" by 24" area, you can calculate drawing parameters like limits, **DIMSCALE**, and **LTSCALE** and save them in a template drawing. This will save considerable time in the initial drawing setup and provide uniformity in the drawings. The next example explains the procedure involved in customizing a drawing according to a certain plot size and scale. (Note, you can also use the paper space to specify the paper size and scale.)

Example 6

Create a template drawing (**PROTO6**) with the following specifications:

    Plotted sheet size   36" by 24" (Figure 1-19)

Template Drawings

Scale	1/8" = 1.0'
Snap	3'
Grid	6'
Text height	1/4" on plotted drawing
Linetype scale	Calculate
Dimscale factor	Calculate
Units	Architectural
	Precision, 16-denominator of smallest fraction
	Angle in degrees/minutes/seconds
	Precision, 0d00'
	Direction control, base angle, east
	Angle positive if measured counterclockwise
Border	Border should be 1" inside the edges of the plotted drawing sheet, using PLINE 1/32" wide when plotted (Figure 1-19)

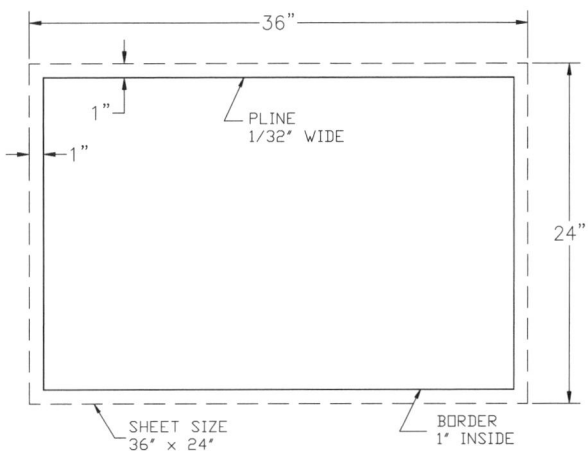

Figure 1-19 Border of template drawing

Step 1
Calculating limits, text height, linetype scale, dimension scale and polyline width

In this example, you need to calculate some values before you set the parameters. For example, the limits of the drawing depend on the plotted size of the drawing and the scale of the drawing. Similarly, **LTSCALE** and **DIMSCALE** depend on the plot scale of the drawing. The following calculations explain the procedure for finding the values of limits, ltscale, dimscale, and text height.

Limits

Given:
Sheet size   36" x 24"
Scale   1/8" = 1'
         or 1" = 8'
Calculate:
XLimit

YLimit
Since sheet size is 36" x 24" and scale is 1/8"=1'
Therefore, XLimit = 36 x 8' = 288'
YLimit = 24 x 8' = 192'

Text height

Given:
Text height when plotted = 1/4"
Scale   1/8" = 1'
Calculate:
Text height
Since scale is 1/8" = 1'
or 1/8" = 12"
or   1" = 96"
Therefore, scale factor = 96
Text height = 1/4" x 96
= 24" = 2'

Linetype scale and dimension scale

Known:
Since scale is 1/8" = 1'
or 1/8" = 12"
or   1" = 96"

Calculate:
Ltscale and Dimscale
Since scale factor = 96
Therefore, LTSCALE = Scale factor = 96
Similarly, DIMSCALE = 96
(All dimension variables, like DIMTXT and DIMASZ, will be multiplied by 96.)

Polyline Width

Given:
Scale is 1/8" = 1'
Calculate:
PLINE width
Since scale is 1/8" = 1'
or   1" = 8'
or   1" = 96"

Therefore,
PLINE width = 1/32 x 96
= 3"

After calculating the parameters, use the following AutoCAD commands to set up the drawing, then save the drawing as **PROTO6.DWT.**

Template Drawings

Step 2
Setting units
Start a new drawing and choose **Units** from the **Format** menu or enter **UNITS** at the Command prompt to display the **Drawing Units** dialog box. Choose **Architectural** from the **Type** drop-down list in the **Length** area. Choose **0'-01/16"** from the **Precision** drop-down list. Make sure the **Clockwise** radio button in the **Angle** area is not checked. Select **Deg/Min/Sec** from the **Type** drop-down list and select **0d00** from the **Precision** drop-down list in the **Angle** area. Now choose the **Direction** button to display the **Directional Control** dialog box. Choose the **East** radio button if it is not selected in the **Base Angle** area and then choose **OK**.

Step 3
Setting limits, snap and grid, textsize, linetype scale, dimension scale, dimension style and pline

To set the **LIMITS**, select **Drawing Limits** from the **Format** menu or enter **LIMITS** at the Command prompt.

Command: **LIMITS**
Specify lower left corner or [ON/OFF] <0'-0",0'-0">:**0,0**
Specify upper right corner <1'-0",0'-9">: **288',192'**

To set the **SNAP** and **GRID**, right-click on the **Snap** or **Grid** button in the status bar to invoke the shortcut menu. In the shortcut menu choose the **Settings** to display the **Drafting Settings** dialog box. You can also choose the **Object Snap Settings** button from the **Object Snap** toolbar to display the **Drafting Settings** dialog box. In the dialog box choose the **Snap and Grid** tab. Enter **3'** and **3'** in the **Snap X spacing** and **Snap Y spacing** edit boxes respectively. Enter **6'** and **6'** in the **Grid X spacing** and **Grid Y spacing** edit boxes respectively. Then choose **OK**.

You can also set these values by entering **SNAP** and **GRID** at the Command prompt.

The size of the text can be changed by entering **TEXTSIZE** at the Command prompt.

Command: **TEXTSIZE**
Enter new value for TEXTSIZE <current>: **2'**

To set the **LTSCALE**, choose the **Linetype** from the **Format** menu or enter **LINETYPE** at the Command prompt to invoke the **Linetype Manager** dialog box. Choose the **Show details** button. Specify the **Global scale factor** as **96** in the **Global scale factor** edit box.

You can also change the scale of the linetype by entering **LTSCALE** at the Command prompt.

To set the **DIMSTYLE**, choose the **Dimension Style** button from the **Dimension** toolbar or choose **Style** from the **Dimension** menu to invoke the **Dimension Style Manager** dialog box. Choose the **New** button from the **Dimension Style Manager** dialog box to invoke the **Create New Dimension** Style dialog box. Specify the new style name as **MYDIM2** in the **New Style Name** edit box and then choose the **Continue** button. The **New Dimension Style** dialog box will be displayed. Choose the **Modify** button to display the **New Dimension Style: MYDIM2** dialog box. Now choose the **Fit** tab. Set the value in the **Use overall scale of** as **96** in the **Scale for Dimension Features** area. Now choose the **OK** button to again display the **Dimension Style Manager** dialog box. Choose **Close** to exit the dialog box.

You can invoke **PLINE** command by choosing the **Polyline** button from the **Draw** toolbar or enter **PLINE** at the Command prompt.

Command: **PLINE**
Specify start point: **8',8'**
Current line-width is **0.0000**
Specify next point or [Arc/Close/Halfwidth/Length/Undo/Width]:**W**
Specify starting width<0.00>: **3**
Specify ending width<0'-3">: **ENTER**
Specify next point or [Arc/Close/Halfwidth/Length/Undo/Width]: **280',8'**
Specify next point or [Arc/Close/Halfwidth/Length/Undo/Width]: **280',184'**
Specify next point or [Arc/Close/Halfwidth/Length/Undo/Width]: **8',184'**
Specify next point or [Arc/Close/Halfwidth/Length/Undo/Width]: **C**

Now save the drawing as **PROTO6.DWT**.

Self-Evaluation Test

Answer the following questions and then compare your answers to the correct answers given at the end of this chapter.

1. The template drawings are stored in _____.

2. To use a template file, select the _____ option in the **Create Drawing** tab of _____ dialog box.

3. To start a drawing with default setup, select the _____ option in the **Create Drawing** tab of _____ dialog box.

4. If plot size is 36" x 24", and the scale is 1/2" = 1', then XLimit = _____ and YLimit = _____.

5. You can use AutoCAD's _____ command to set up a viewport in paper space.

Review Questions

Answer the following questions:

1. The default value of **DIMSCALE** is _____.

2. The default value for **DIMTXT** is _____.

Template Drawings　　　　　　　　　　　　　　　　　　　　　　　　　　　　　　1-23

3. The default value for **SNAP** is _____.

4. Architectural units can be selected by using AutoCAD's _____ or _____ commands.

5. Name three standard template drawings that come with AutoCAD software _____ , _____ , and _____.

6. If the plot size is 24" x 18", and the scale is 1 = 20, the XLimit = _____ and YLimit = _____.

7. If the plot size is 200 x 150 and limits are (0.00,0.00) and (600.00,450.00), the **LTSCALE** factor = _____.

8. _____ provides a convenient way to plot multiple views of a 3D drawing or multiple views of a regular 2D drawing.

9. You can use AutoCAD's _____ command to change to paper space.

10. You can use AutoCAD's _____ command to change to model space.

11. The values that can be assigned to **TILEMODE** are _____ and _____.

12. In the model space, if you want to reduce the display size by half, the scale factor you enter in the **ZOOM**-Scale command is _____.

Exercises

Exercise 1   *General*

Create a template drawing (**PROTOE1**) with the following specifications:

Units	Architectural with precision 0'-0 1/16
Angle	Decimal Degrees with precision 0.
Base angle	East.
Angle direction	Counterclockwise.
Limits	48' x 36'

Exercise 2   *General*

Create a template drawing (**PROTOE2**) with the following specifications:

Limits	36.0,24.0
Snap	0.5
Grid	1.0
Text height	0.25
Units	Decimal

Precision 0.00
Decimal degrees
Precision 0
Base angle, East
Angle positive if measured counterclockwise

Exercise 3 — *General*

Create a template drawing **PROTOE3** with the following specifications:

Limits	48.0,36.0
Text height	0.25
PLINE width	0.03
Ltscale	4.0
Dimscale	4.0
Plot size	10.5 x 8

LAYERS

Layer Names	Line Type	Color
0	Continuous	White
OBJECT	Continuous	Green
CENTER	Center	Magenta
HIDDEN	Hidden	Blue
DIM	Continuous	Red
BORDER	Continuous	Cyan

Exercise 4 — *General*

You want to set up a prototype drawing with the following specifications (the name of the drawing is **PROTOE4**).

Limits	36.0,24,0
Border	35.0,23.0
Grid	1.0
Snap	0.5
Text height	0.15
Units	Decimal (up to 2 places)
Ltscale	1
Current layer	Object

LAYERS

Layer Name	Linetype	Color
0	Continuous	White
Object	Continuous	Red
Hidden	Hidden	Yellow
Center	Center	Green
Dim	Continuous	Blue

Template Drawings

Border	Continuous	Magenta
Notes	Continuous	White

This prototype drawing should have a border line and title block as shown in Figure 1-20.

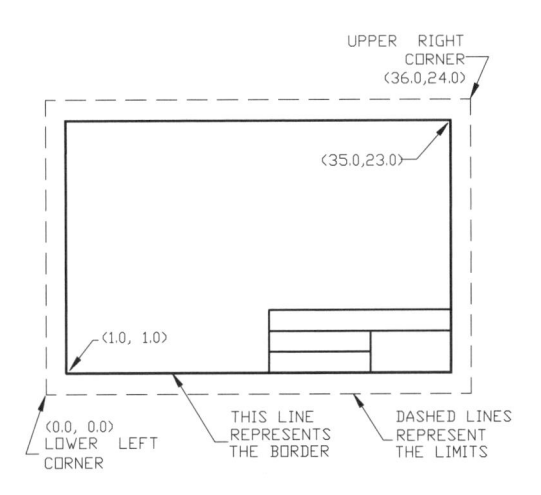

Figure 1-20 Prototype drawing

Exercise 5 *General*

Create a template drawing (**PROTOE5**) with the following specifications:

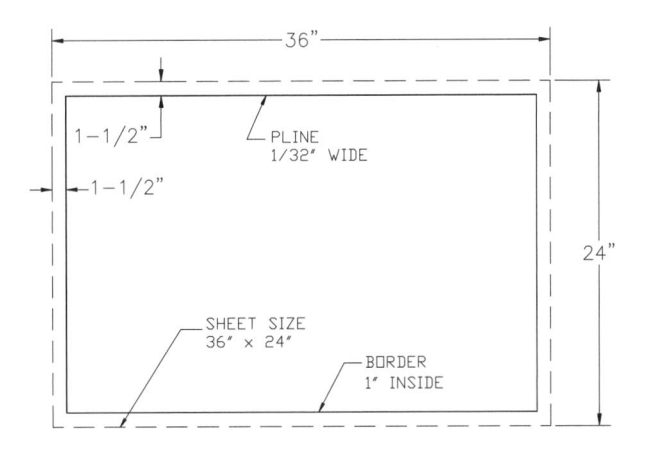

Figure 1-16 Drawing for Exercise 5

Plotted sheet size   36" x 24"   (Figure 1-21)
Scale   1/2" = 1.0'

Text height	1/4" on plotted drawing
Ltscale	24
Dimscale	24
Units	Architectural
	32-denominator of smallest fraction to display
	Angle in degrees/minutes/seconds
	Precision 0d00"00"
	Angle positive if measured counterclockwise
Border	Border is 1-1/2" inside the edges of the plotted drawing sheet, using PLINE 1/32" wide when plotted.

Exercise 7 — *General*

Create a prototype drawing with the following specifications (the name of the drawing is **PROTOE7**)

Plotted sheet size	24" x 18" (Figure 1-21)
Scale	1/2"=1.0'
Border	The border is 1" inside the edges of the plotted drawing sheet, using PLINE 0.05" wide when plotted (Figure 1-21)

Dimension text over the dimension line
Dimensions aligned with the dimension line
Calculate overall dimension scale factor
Enable the display of alternate units
Dimensions to be associative.

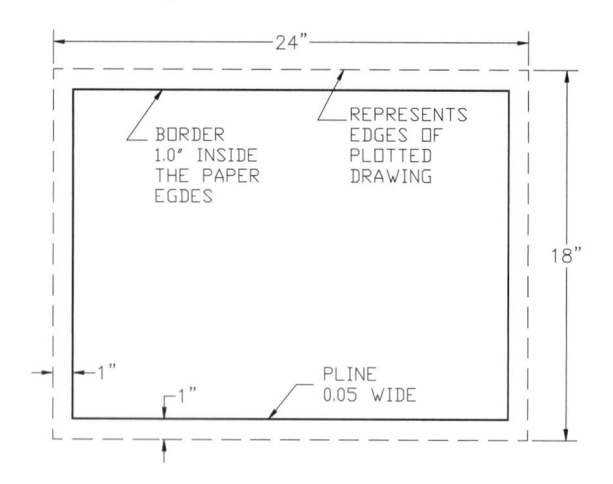

Figure 1-21 Prototype drawing

Answers to Self-Evaluation Test

1 - .dwt, 2 - Use A Template, 3 - Start From Scratch, 4 - 72"x48", 5 - Mview.

Chapter 2

Script Files and Slide Shows

Learning Objectives

After completing this chapter, you will be able to:
- *Write script files and use the **SCRIPT** command to run script files.*
- *Use the **RSCRIPT** and **DELAY** commands in script files.*
- *Invoke script files when loading AutoCAD.*
- *Create a slide show.*
- *Preload slides when running a slide show.*

WHAT ARE SCRIPT FILES?

AutoCAD has provided a facility called **script files** that allows you to combine different AutoCAD commands and execute them in a predetermined sequence. The commands can be written as a text file using any text editor like Notepad or AutoCAD's **EDIT** command (if the **ACAD.PGP** file is present and **EDIT** is defined in the file). These files, generally known as script files, have extension **.SCR** (example: **PLOT1.SCR**). A script file is executed with the AutoCAD **SCRIPT** command.

Script files can be used to generate a slide show, do the initial drawing setup, or plot a drawing to a predefined specification. They can also be used to automate certain command sequences that are used frequently in generating, editing, or viewing a drawing. Scripts cannot access dialog boxes or menus. When commands that open the file and plot dialog boxes are issued from a script file, AutoCAD runs the command line version of the command instead of opening the dialog box.

Example 1

Write a script file that will perform the following initial setup for a drawing (file name **SCRIPT1.SCR**). It is assumed that the drawing will be plotted on 12x9 size paper (Scale factor for plotting = 4).

Ortho	On		Zoom	All
Grid	2.0		Text height	0.125
Snap	0.5		Ltscale	4.0
Limits	0,0	48.0,36.0	Dimscale	4.0

Step 1: Understanding commands and prompt entries

Before writing a script file, you need to know the AutoCAD commands and the entries required in response to the command prompts. To find out the sequence of the prompt entries, you can type the command at the keyboard and then respond to different prompts. The following is a list of AutoCAD commands and prompt entries for Example 1.

Command: **ORTHO**
Enter mode [ON/OFF] <OFF>: **ON**

Command: **GRID**
Specify grid spacing(X) or [ON/OFF/Snap/Aspect] <1.0>: **2.0**

Command: **SNAP**
Specify snap spacing or [ON/OFF/Aspect/Rotate/Style/Type] <1.0>: **0.5**

Command: **LIMITS**
Reset Model space limits:
Specify lower left corner or [ON/OFF] <0.0,0.0>: **0,0**
Specify upper right corner <12.0,9.0>: **48.0,36.0**

Command: **ZOOM**
Specify corner of window, enter a scale factor (nX or nXP), or
[All/Center/Dynamic/Extents/Previous/Scale/Window] <real time>: **A**

Command: **TEXTSIZE**
Enter new value for TEXTSIZE <0.02>: **0.125**

Command: **LTSCALE**
Enter new linetype scale factor <1.0000>: **4.0**

Command: **DIMSCALE**
Enter new value for DIMSCALE <1.0000>: **4.0**

Step 2: Writing the script file

Once you know the AutoCAD commands and the required prompt entries, you can write the script file using any text editor like **Notepad**.

Script Files and Slide Shows 2-3

You can also use AutoCAD's **EDIT** command. As you enter the **EDIT** command, AutoCAD prompts you to enter the file to edit. Press enter in response to the prompt to display the **MS-DOS Editor**. Write the script file in the **MS-DOS Editor**. The following file is a listing of the script file for Example 1:

```
ORTHO
ON
GRID
2.0
SNAP
0.5
LIMITS
0,0
48.0,36.0
ZOOM
ALL
TEXTSIZE
0.125
LTSCALE
4.0
DIMSCALE 4.0
```

Notice that the commands and the prompt entries in this file are in the same sequence as mentioned before. You can also combine several statements in one line, as shown in the following list:

```
;This is my first script file, SCRIPT1.SCR
ORTHO ON
GRID 2.0
SNAP 0.5
LIMITS 0,0 48.0,36.0 ZOOM ALL
TEXTSIZE 0.125
LTSCALE 4.0
DIMSCALE 4.0
```

Save the script file as **SCRIPT1.SCR** on A or C drive and exit the text editor. Notice the space between the commands and the prompt entries. For example, between **ORTHO** command and **ON** there is a space. Similarly, there is a space between **GRID** and **2.0**.

> **Note**
>
> *In the script file, a space is used to terminate a command or a prompt entry. Therefore, spaces are very important in these files. Make sure there are no extra spaces, unless they are required to press ENTER more than once.*
>
> *After you change the limits, it is a good practice to use the **ZOOM** command with the **All** option to display the new limits on the screen.*

AutoCAD ignores and does not process any lines that begin with a semicolon (;). This allows you to put comments in the file.

SCRIPT COMMAND

The AutoCAD **SCRIPT** command allows you to run a script file while you are in the drawing editor. Choose the **Run Script** button from the **Tools** toolbar to invoke the **Select Script File** dialog box as shown in the Figure 2-1. You can also invoke the **Select Script File** dialog box by entering **SCRIPT** at the Command prompt. You can enter the name of the script file or you can accept the default file name. The default script file name is the same as the drawing name. If you want to enter a new file name, type the name of the script file **without** the file extension (**.SCR**). (The file extension is assumed and need not be included with the file name.)

Step 3: Running the script file

To run the script file of Example 1, invoke the **SCRIPT** command, select the file **SCRIPT1**, and then choose the **Open** button in the **Select Script File** dialog box (Figure 2-1) . You will see the changes taking place on the screen as the script file commands are executed.

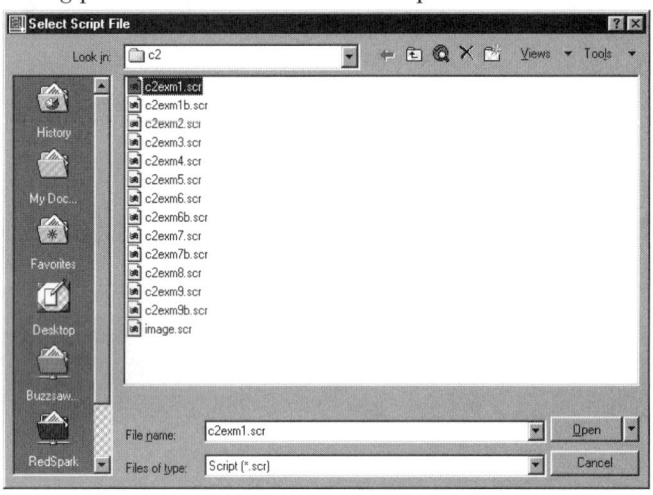

Figure 2-1 Select Script File dialog box

You can also enter the name of the script file at the Command prompt by setting **FILEDIA**=0. The format of the **SCRIPT** command is:

    Command: **FILEDIA**
    Enter new value for FILEDIA <1>: **0**
    Command: **SCRIPT**
    Enter script file name <current>: *Script file name.*

Example 2

Write a script file that will set up the following layers with the given colors and linetypes (file name **SCRIPT2.SCR**).

Script Files and Slide Shows   2-5

Layer Names	Color	Linetype	Line Weight
Object	Red	Continuous	default
Center	Yellow	Center	default
Hidden	Blue	Hidden	default
Dimension	Green	Continuous	default
Border	Magenta	Continuous	default
Hatch	Cyan	Continuous	0.05

Step 1: Understanding commands and prompt entries
As mentioned earlier, you need to know the AutoCAD commands and the required prompt entries before writing a script file. For Example 2, you need the following commands to create the layers with the given colors and linetypes:

Command: **-LAYER**
Enter an option
[?/Make/Set/New/ON/OFF/Color/Ltype/LWeight/Plot/Freeze/Thaw/LOck/Unlock]: **N**
Enter name list for new layer(s): **OBJECT,CENTER,HIDDEN,DIM,BORDER, HATCH**

Enter an option
[?/Make/Set/New/ON/OFF/Color/Ltype/LWeight/Plot/Freeze/Thaw/LOck/Unlock]: **L**
Enter loaded linetype name or [?] <Continuous>: **CENTER**
Enter name list of layer(s) for linetype "CENTER" <0>: **CENTER**

Enter an option
[?/Make/Set/New/ON/OFF/Color/Ltype/LWeight/Plot/Freeze/Thaw/LOck/Unlock]: **L**
Enter loaded linetype name or [?] <Continuous>: **HIDDEN**
Enter name list of layer(s) for linetype "HIDDEN" <0>: **HIDDEN**

Enter an option
[?/Make/Set/New/ON/OFF/Color/Ltype/LWeight/Plot/Freeze/Thaw/LOck/Unlock]: **C**
Enter color name or number (1-255): **RED**
Enter name list of layer(s) for color 1 (red) <0>:**OBJECT**

Enter an option
[?/Make/Set/New/ON/OFF/Color/Ltype/LWeight/Plot/Freeze/Thaw/LOck/Unlock]: **C**
Enter color name or number (1-255): **YELLOW**
Enter name list of layer(s) for color 2 (yellow) <0>: **CENTER**

Enter an option
[?/Make/Set/New/ON/OFF/Color/Ltype/LWeight/Plot/Freeze/Thaw/LOck/Unlock]: **C**
Enter color name or number (1-255): **BLUE**
Enter name list of layer(s) for color 5 (blue)<0>: **HIDDEN**
Enter an option
[?/Make/Set/New/ON/OFF/Color/Ltype/LWeight/Plot/Freeze/Thaw/LOck/Unlock]: **C**
Enter color name or number (1-255): **GREEN**
Enter name list of layer(s) for color 3 (green)<0>: **DIM**

Enter an option
[?/Make/Set/New/ON/OFF/Color/Ltype/LWeight/Plot/Freeze/Thaw/LOck/Unlock]: **C**
Enter color name or number (1-255): **MAGENTA**
Enter name list of layer(s) for color 6 (magenta)<0>: **BORDER**

Enter an option
[?/Make/Set/New/ON/OFF/Color/Ltype/LWeight/Plot/Freeze/Thaw/LOck/Unlock]: **C**
Enter color name or number (1-255): **CYAN**
Enter name list of layer(s) for color 4 (cyan)<0>: **HATCH**

Enter an option
[?/Make/Set/New/ON/OFF/Color/Ltype/LWeight/Plot/Freeze/Thaw/LOck/Unlock]:**LW**
Enter lineweight (0.0mm - 2.11mm):0.05
Enter name list of layers(s) for lineweight 0.05mm <0>:**HATCH**
[?/Make/Set/New/ON/OFF/Color/Ltype/LWeight/Plot/Freeze/Thaw/LOck/Unlock]:
(RETURN)

Step 2: Writing the script file

The following file is a listing of the script file that creates different layers and assigns the given colors and linetypes to these layers:

```
;This script file will create new layers and
;assign different colors and linetypes to layers
LAYER
NEW
OBJECT,CENTER,HIDDEN,DIM,BORDER,HATCH
L
CENTER
CENTER
L
HIDDEN
HIDDEN
C
RED
OBJECT
C
YELLOW
CENTER
C
BLUE
HIDDEN
C
GREEN
DIM
C
MAGENTA
BORDER
```

Script Files and Slide Shows 2-7

    C
    CYAN
    HATCH
                 (This is a blank line to terminate the **LAYER** command. End of script file.)

Save the script file as **SCRIPT2.SCR**.

Step 3: Running the script file
To run the script file of Example 2, choose the **Run Script** button from the **Tools** menu or enter **SCRIPT** at the Command prompt to invoke the **Select Script File** dialog box. Choose the **SCRIPT2.SCR** and then choose **Open**. You can also enter the **SCRIPT** command and the name of the script file at the Command prompt by setting **FILEDIA**=0.

Example 3

Write a script file that will rotate the circle and the line, as shown in Figure 2-2, around the lower endpoint of the line through 45-degree increments. The script file should be able to produce a continuous rotation of the given objects with a delay of two seconds after every 45-degree rotation (file name **SCRIPT3.SCR**).

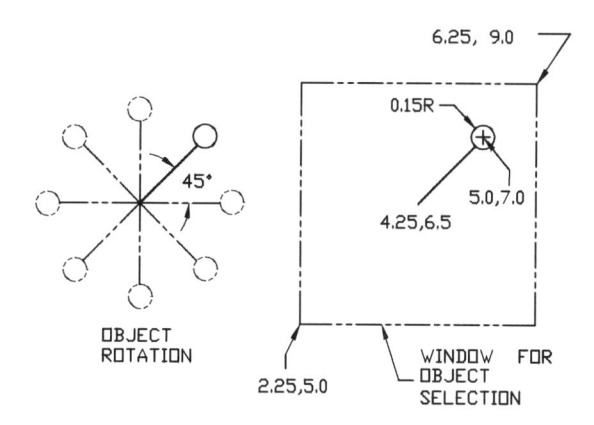

Figure 2-2 Line and circle rotated through 45-degree increments

Step 1: Understanding commands and prompt entries
Before writing the script file, enter the required commands and the prompt entries at the keyboard. Write down the exact sequence of the entries in which they have been entered to perform the given operations. The following is a listing of the AutoCAD command sequence needed to rotate the circle and the line around the lower endpoint of the line:

    Command: **ROTATE**
    Current positive angle in UCS: ANGDIR=counterclockwise ANGBASE=0
    Select objects: **W**   *(Window option to select object)*
    Specify first corner: **2.25, 5.0**
    Specify opposite corner: **6.25, 9.0**
    Select objects: [Enter]

Specify base point: **4.25,6.5**
Specify rotation angle or [Reference]: **45**

Step 2: Writing the script file

Once the AutoCAD commands, command options, and their sequences are known, you can write a script file. As mentioned earlier, you can use any text editor to write a script file. The following file is a listing of the script file that will create the required rotation of the circle and line of Example 3. The line numbers and *(Blank line for Return)* are not a part of the file. They are shown here for reference only.

```
ROTATE                                    1
W                                         2
2.25,5.0                                  3
6.25,9.0                                  4
         (Blank line for Return.)         5
4.25,6.5                                  6
45                                        7
```

Line 1
ROTATE
In this line, **ROTATE** is an AutoCAD command that rotates the objects.

Line 2
W
In this line, W is the Window option for selecting the objects that need to be edited.

Line 3
2.25,5.0
In this line, 2.25 defines the X coordinate and 5.0 defines the Y coordinate of the lower left corner of the object selection window.

Line 4
6.25,9.0
In this line, 6.25 defines the X coordinate and 9.0 defines the Y coordinate of the upper right corner of the object selection window.

Line 5
Line 5 is a blank line that terminates the object selection process.

Line 6
4.25,6.5
In this line, 4.25 defines the X coordinate and 6.5 defines the Y coordinate of the base point for rotation.

Line 7
45
In this line, 45 is the incremental angle for rotation.

Script Files and Slide Shows
2-9

> **Note**
> *One of the limitations of the script files is that all the information has to be contained within the file. These files do not let you enter information. For instance, in Example 3, if you want to use the Window option to select the objects, the Window option (W) and the two points that define this window must be contained within the script file. The same is true for the base point and all other information that goes in a script file. There is no way that a script file can prompt you to enter a particular piece of information and then resume the script file, unless you embed AutoLISP commands to prompt for user input.*

Step 3: Running the script file

Choose the **Run Script** button from the **Tools** menu or enter **SCRIPT** at the Command prompt to invoke the **Select Script File** dialog box. Choose the **SCRIPT3.SCR** and then choose **Open**. You can see the changes taking place on the screen. You can also open a script file by entering **SCRIPT** at the Command and setting **FILEDIA**=0.

RSCRIPT COMMAND

The AutoCAD **RSCRIPT** command allows the user to execute the script file indefinitely until canceled. It is a very desirable feature when the user wants to run the same file continuously. For example, in the case of a slide show for a product demonstration, the **RSCRIPT** command can be used to run the script file again and again until it is terminated by pressing the ESC (Escape) key from the keyboard. Similarly, in Example 3, the rotation command needs to be repeated indefinitely to create a continuous rotation of the objects. This can be accomplished by adding **RSCRIPT** at the end of the file, as in the following file:

```
ROTATE
W
2.25,5.0
6.25,9.0
        (Blank line for Return.)
4.25,6.5
45
RSCRIPT
```

The **RSCRIPT** command on line 8 will repeat the commands from line 1 to line 7, and thus set the script file in an indefinite loop. The script file can be stopped by pressing the ESC or the BACKSPACE key.

> **Note**
> *You cannot provide conditional statements in a script file to terminate the file when a particular condition is satisfied unless you use the AutoLISP functions in the script file.*

DELAY COMMAND

In the script files, some of the operations happen very quickly and make it difficult to see the operations taking place on the screen. It might be necessary to intentionally introduce a pause between certain operations in a script file. For example, in a slide show for a product demonstration, there must be a time delay between different slides so that the audience has enough time to see them. This is accomplished by using the AutoCAD **DELAY** command, which introduces a delay before the next command is executed. The general format of the **DELAY** command is:

Command: DELAY Time
        Where **Command** ------ AutoCAD command prompt
                   DELAY ---------- **DELAY** command
                   Time ------------- Time in milliseconds

The **DELAY** command is to be followed by the delay time in milliseconds. For example, a delay of 2,000 milliseconds means that AutoCAD will pause for approximately two seconds before executing the next command. It is approximately two seconds because computer processing speeds vary. The maximum time delay you can enter is 32,767 milliseconds (about 33 seconds). In Example 3, a two-second delay can be introduced by inserting a **DELAY** command line between line 7 and line 8, as in the following file listing:

ROTATE
W
2.25,5.0
6.25,9.0
        (Blank line for Return.)
4.25,6.5
45
DELAY 2000
RSCRIPT

The first seven lines of this file rotate the objects through a 45-degree angle. Before the **RSCRIPT** command on line 8 is executed, there is a delay of 2,000 milliseconds (about two seconds). The **RSCRIPT** command will repeat the script file that rotates the objects through another 45-degree angle. Thus, a slide show is created with a time delay of two seconds after every 45-degree increment.

RESUME COMMAND

If you cancel a script file and then want to continue it, you can do so by using the AutoCAD **RESUME** command.

    Command: **RESUME**

The **RESUME** command can also be used if the script file has encountered an error that causes it to be suspended. The **RESUME** command will skip the command that caused the error and continue with the rest of the script file. If the error occurred when the command was in progress, use a leading apostrophe with the **RESUME** command (**'RESUME**) to invoke the **RESUME** command in transparent mode.

    Command: **'RESUME**

COMMAND LINE SWITCHES

The command line switches can be used as arguments to the acad.exe file that launches AutoCAD. You can also use the **Options** dialog box to set the environment or by adding a set of environment variables in the autoexec.bat file. The command line switches and environment variables override the values set in the **Options** dialog box for the current session only. These switches do not alter the system registry. The following is the list of the command line switches:

Switch	Function
/c	Controls where AutoCAD stores and searches for the hardware configuration file. The default file is acad 2002.cfg.
/s	Specifies which directories to search for support files if they are not in the current directory
/b	Designates a script to run after AutoCAD starts
/t	Specifies a template to use when creating a new drawing
/nologo	Starts AutoCAD without first displaying the logo screen
/v	Designates a particular view of the drawing to be displayed upon start-up of AutoCAD
/r	Reconfigures AutoCAD with the default device configuration settings
/p	Specifies the profile to use on start-up

INVOKING A SCRIPT FILE WHEN LOADING AUTOCAD

The script files can also be run when loading AutoCAD, without getting into the drawing editor. The format of the command for running a script file when loading AutoCAD is:

"Drive**Program Files\ACAD2002\acad.exe**" [existing-drawing] [/t template] [/v view] /b Script-file

In the following example, AutoCAD will open the existing drawing (Mydwg1) and then run the script file (Setup) through the Run dialog box as shown in Figure 2-3.

Example
"C:\Program Files\ACAD2002\acad.exe" Mydwg1 /b Setup

Where **ACAD2002** ----- AutoCAD2002 subdirectory containing AutoCAD system files
acad.exe --------- ACAD command to start AutoCAD
MyDwg1 -------- Existing drawing file name
Setup ------------- Name of the script file

In the following example, AutoCAD will start a new drawing with the default name (Drawing), using the template file temp1, and then run the script file (Setup).

Example
"C:\Program Files\ACAD2002\acad.exe" /t temp1 /b Setup

Where **temp1** ------------ Existing template file name
Setup ------------- Name of the script file

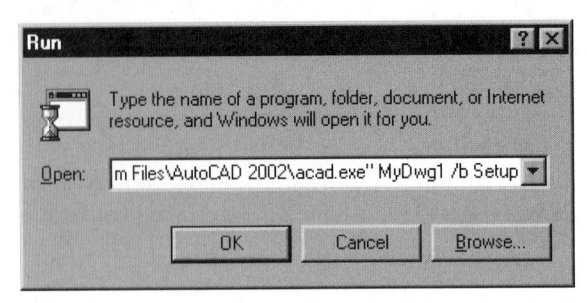

*Figure 2-3 Invoking script file when loading AutoCAD using the **Run** dialog box*

or

"C:\ProgramFiles\ACAD2002\acad.exe"/t temp1 "C:\MyFolder"/b Setup

> Where C**\Program Files\ACAD2002\acad.exe**   Path name for acad.exe
> **C:\MyFolder** --- Path name for the Setup script file

In the following example, AutoCAD will start a new drawing with the default name (Drawing), and then run the script file (Setup).

Example
"C:\Program Files\ACAD 2002\acad.exe" /b Setup
> Where **Setup**------------- Name of the script file

Here, it is assumed that the AutoCAD system files are loaded in the AutoCAD 2002 directory.

> **Note**
> *For invoking a script file when loading AutoCAD, the drawing file or the template file specified in the command must exist in the search path. You cannot start a new drawing with a given name. You can also use any template drawing file that is found in the template directory to run a script file through the **Run** dialog box.*
>
> *You should avoid abbreviations to prevent any confusion. For example, a C can be used as a close option when you are drawing lines. It can also be used as a command alias for drawing a circle. If you use both of these in a script file, it might be confusing.*

Example 4

Write a script file that can be invoked when loading AutoCAD and create a drawing with the following setup (filename SCRIPT4.SCR):

Grid	3.0
Snap	0.5
Limits	0,0
	36.0,24.0
Zoom	All
Text height	0.25

Script Files and Slide Shows   2-13

Ltscale	3.0	
Dimscale	3.0	

Layers

Name	Color	Linetype
Obj	Red	Continuous
Cen	Yellow	Center
Hid	Blue	Hidden
Dim	Green	Continuous

Step 1: Writing the script file

First, write a script file and save the file under the name **SCRIPT4.SCR**. The following file is a listing of this script file that does the initial setup for a drawing:

```
GRID 3.0
SNAP 0.5
LIMITS 0,0 36.0,24.0 ZOOM ALL
TEXTSIZE 0.25
LTSCALE 3
DIMSCALE 3.0
LAYER NEW
OBJ,CEN,HID,DIM
L CENTER CEN
L HIDDEN HID
C RED OBJ
C YELLOW CEN
C BLUE HID
C GREEN DIM
```
(Blank line for ENTER.)

Step 2: Loading the script file through the run dialog box

After you have written and saved the file, quit the drawing editor. To run the script file, SCRIPT4, select Start, Run, and then enter the following command line:

"C:\Program Files\ACAD 2002\acad.exe" /t EX4 /b SCRIPT4

Where **acad.exe** --------- ACAD to load AutoCAD
       EX4 -------------- Drawing filename
       SCRIPT4 ------- Name of the script file

Here it is assumed that the template file (EX4) and the script file (SCRIPT4) is on C drive. When you enter this line, AutoCAD is loaded and the file EX4.DWT is opened. The script file, SCRIPT4, is then automatically loaded and the commands defined in the file are executed.

In the following example, AutoCAD will start a new drawing with the default name (Drawing), and then run the script file (SCRIPT4) (Figure 2-4).

Example
C:\AutoCAD 2002\ACAD /b SCRIPT4
        Where **SCRIPT4** ------- Name of the script file

Here, it is assumed that the AutoCAD system files are loaded in the AutoCAD 2002 directory.

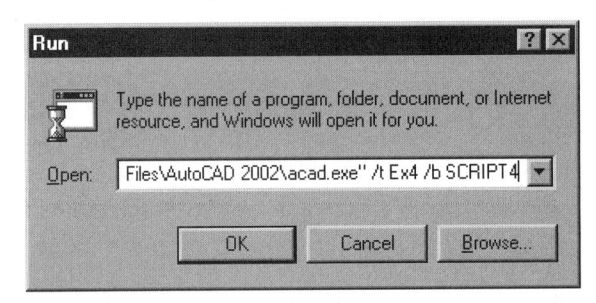

*Figure 2-4 Invoking script file when loading AutoCAD using the **Run** dialog box*

Example 5

Write a script file that will plot a 36" by 24" to maximum plot size, using your system printer/plotter. Use the Window option to select the drawing to be plotted.

Step 1: Understanding commands and prompt entries
Before writing a script file to plot a drawing, find out the plotter specifications that must be entered in the script file to obtain the desired output. To determine the prompt entries and their sequences to set up the plotter specifications, enter the AutoCAD **-PLOT** command at the keyboard. Note the entries you make and their sequence (the entries for your printer or plotter will probably be different). The following is a listing of the plotter specifications with the new entries:

    Command: **-PLOT**
    Detailed plot configuration? [Yes/No] <No>: **Yes**
    Enter a layout name or [?] <Model>: Enter
    Enter an output device name or [?] <HP LaserJet 4000 Series PCL 6>: Enter
    Enter paper size or [?] <Letter (8 1/2 x 11 in)>: Enter
    Enter paper units [Inches/Millimeters] <Inches>: **I**
    Enter drawing orientation [Portrait/Landscape] <Landscape>: **L**
    Plot upside down? [Yes/No] <No>: **N**
    Enter plot area [Display/Extents/Limits/View/Window] <Display>: **W**
    Enter lower left corner of window <0.000000,0.000000>: **0,0**
    Enter upper right corner of window <0.000000,0.000000>: **36,24**
    Enter plot scale (Plotted Inches=Drawing Units) or [Fit] <Fit>: **F**
    Enter plot offset (x,y) or [Center] <0.00,0.00>: **0,0**
    Plot with plot styles? [Yes/No] <Yes>: **Yes**
    Enter plot style table name or [?] (enter . for none) <>: **.**
    Plot with lineweights? [Yes/No] <Yes>: **Y**
    Scale lineweights with plot scale? [Yes/No] <No>: **N**

Script Files and Slide Shows

Plot paper space first? [Yes/No] <No>: N
Remove hidden lines? [Yes/No] <No>: N
Write the plot to a file [Yes/No] <N>: N
Save changes to layout [Yes/No]? <N>N
Proceed with plot [Yes/No] <Y>: Y

Step 2: Writing the script file

Now you can write the script file by entering the responses to these prompts in the file. The following file is a listing of the script file that will plot a 36" by 24" drawing on 9" by 6" paper after making the necessary changes in the plot specifications. The comments on the right are not a part of the file.

Plot
y
          (Blank line for ENTER, selects default layout.)
          (Blank line for ENTER, selects default printer.)
          (Blank line for ENTER, selects the default paper size.)
I
L
N
w
0,0
36,24
F
0,0
Y
.   (Enter . for none)
Y
N
N
N
N
N
Y

Saving and running the script file for this example is the same as that which has been described for previous examples. You can use a blank line to accept the default value for a prompt. A blank line in the script file will cause a Return. However, you must not accept the default plot specifications because the file might have been altered by another user or by another script file. Therefore, always enter the actual values in the file so that when you run a script file, it does not take the default values.

Exercise 1   *General*

Write a script file that will plot a 288' by 192' drawing on a 36" x 24" sheet of paper. The drawing scale is 1/8" = 1'. (The filename is SCRIPT9.SCR. In this example assume that AutoCAD is configured for the HPGL plotter and the plotter description is HPGL-Plotter.)

Example 6

Write a script file to animate a clock with continuous rotation of the second hand (longer needle) through 5 degree and the minutes hand (shorter needle) through 2 degree clockwise around the center of the clock, (Figure 2-5). The specifications are given below.

Specification for the rim made of donut.
Color of Donut	Blue
Inside diameter of Donut	8.0
Outside diameter of Donut	8.4
Center point of Donut	5,5

Specification for the digit mark made of polyline.
Color of the digit mark	Green
Start point of Pline	5,8.5
Initial width of Pline	0.5
Final width of Pline	0.5
Height of Pline	0.5

Specification for second hand (long needle) made of polyline.
Color of the second hand	Red
Start point of Pline	5,5
Initial width of Pline	0.5
Final width of Pline	0.0
Length of Pline	3.5
Rotation of the second hand	5 degree clockwise

Specification for minute hand (shorter needle) made of polyline.
Color of the minute hand	Cyan
Start point of Pline	5,5
Initial width of Pline	0.35
Final width of Pline	0.0
Length of Pline	3.0
Rotation of the minute hand	2 degree clockwise

Step 1: Understanding the commands and prompt entries for creation of the clock

For this example you can create two script files and the files need to be linked. The first script file will demonstrate the creation of the clock on the screen. The next script file will demonstrate the rotation of the needles of the clock.

First write a script file that deals with the creation of the clock as follows and save the file under the name **CLOCK.SCR**.

    Command: **COLOR**
    Enter default object color<BYLAYER>: **Blue**
    Command: **DONUT**
    Specify inside diameter of donut<0.5>: **8.0**

Script Files and Slide Shows 2-17

Figure2-5 Drawing for Example 6

Specify outside diameter of donut<0.5>: **8.4**
Specify center of donut or <exit>: **5,5**
Specify center of donut or <exit>: Enter
Command: **COLOR**
Enter default object color<Blue>: **Green**
Command: **PLINE**
Specify start point: **5,8.5**
Specify next point or [Arc/Close/Halfwidth/Length/Undo/Width]: **Width**
Specify starting width<0.00>: **0.25**
Specify starting width<0.25>: **0.25**
Specify next point or [Arc/Close/Halfwidth/Length/Undo/Width]: **@0.25<270**
Specify next point or [Arc/Close/Halfwidth/Length/Undo/Width]: Enter
Command: **ARRAY**
Select objects: **Last**
Select objects: Enter
Enter the type of array[Rectangular/Polar]<R>: **Polar**
Specify center point of array: **5,5**
Enter the number of items in the array: **12**
Specify the angle to fill(+= ccw, -=cw)<360>: **360**
Rotate arrayed objects ? [Yes/No]<Y>: **Y**
Command: **COLOR**
Enter default object color<Green>: **RED**
Command: **PLINE**
Specify start point: **5,5**
Specify next point or [Arc/Close/Halfwidth/Length/Undo/Width]: **Width**
Specify starting width<0.5>: **0.5**
Specify ending width<0.5>: **0**
Specify next point or [Arc/Close/Halfwidth/Length/Undo/Width]: **@3.5<0**
Specify next point or [Arc/Close/Halfwidth/Length/Undo/Width]: Enter
Command: **COLOR**

Enter default object color<Red>: **Cyan**
Command: **PLINE**
Specify start point: **5,5**
Specify next point or [Arc/Close/Halfwidth/Length/Undo/Width]: **Width**
Specify starting width<0.5>: **0.35**
Specify ending width<0.35>: **0**
Specify next point or [Arc/Close/Halfwidth/Length/Undo/Width]: **@3<90**
Specify next point or [Arc/Close/Halfwidth/Length/Undo/Width]: Enter
Command: **SCRIPT**
ROTATE.SCR

Now you can write the script file by entering the responses to these prompts in the file **CLOCK.SCR**. Listing of the script file is as follows.

Color
Blue
Donut
8.0
8.4
5,5
                Blank line for ENTER
Color
Green
Pline
5,8.5
W
0.25
0.25
@0.25<270
                Blank line for ENTER
Array
L
                Blank line for ENTER
P
5,5
12
360
Y
Color
Red
Pline
5,5
W
0.5
0
@3.5<0
                Blank line for ENTER

Script Files and Slide Shows 2-19

    Color
    Cyan
    Pline
    5,5
    w
    0.35
    0
    @3<90
                              Blank line for ENTER
    Script
    ROTATE.scr   (Name of the script file that will cause rotation)

Step 2: Understanding the commands and its sequences for rotation of the needles

The last line in the above script file is ROTATE.SCR. This is the name of the script file that will rotate the clock hands. Before writing the script file, enter ROTATE command and respond to command prompts that will cause the desired rotation. The following is a listing of the AutoCAD command sequence needed to rotate the objects.

    Command:**ROTATE**
    Select objects:**L**
    Select objects: [Enter]
    Specify base point:5,5
    Specify rotation angle or [Reference]:-2
    Command:**ROTATE**
    Select objects:**C**
    Specify first corner:3,3
    Specify other corner:7,7
    Select objects:**Remove**
    Remove objects:**L**
    Remove objects: [Enter]
    Specify base point:5,5
    Specify rotation angle or [Reference]:-5

Now you can write the script file by entering the responses to these prompts in the file **ROTATE.SCR**. The following is the listing of the script file that will rotate the clock hands.

    Rotate
    L
                                 Blank line for ENTER
    5,5
    -2
    Rotate
    c
    3,3
    7,7
    R
    L

Blank line for ENTER

5,5
-5
Rscript

Save the above script file as **ROTATE.SCR**. Now run the script file **CLOCK.SCR**. Since this file is linked with **ROTATE.SCR**, it will automatically run **ROTATE.SCR** after running **CLOCK.SCR**. Also note that ROTATE.SCR file must be saved to a directory in the AutoCAD support file search path, or the last line of the CLOCK.SCR must include a fully-resolved path to ROTATE.SCR, or AutoCAD would not find it.

WHAT IS A SLIDE SHOW?

AutoCAD provides a facility using script files to combine the slides in a text file and display them in a predetermined sequence. In this way, you can generate a slide show for a slide presentation. You can also introduce a time delay in the display so that the viewer has enough time to view a slide.

A drawing or parts of a drawing can also be displayed by using the AutoCAD display commands. For example, you can use **ZOOM**, **PAN**, or other commands to display the details you want to show. If the drawing is very complicated, it takes quite some time to display the desired information, and it may not be possible to get the desired views in the right sequence. However, with slide shows you can arrange the slides in any order and present them in a definite sequence. In addition to saving time, this will also help to minimize the distraction that might be caused by constantly changing the drawing display. Also, some drawings are confidential in nature and you may not want to display some portions or views of them. You can send a slide show to a client without losing control of the drawings and the information that is contained in them.

WHAT ARE SLIDES?

A **slide** is the snapshot of a screen display; it is like taking a picture of a display with a camera. The slides do not contain any vector information like AutoCAD drawings, which means that the entities do not have any information associated with them. For example, the slides do not retain any information about the layers, colors, linetypes, start point, or endpoint of a line or viewpoint. Therefore, slides cannot be edited like drawings. If you want to make any changes in the slide, you need to edit the drawing and then make a new slide from the edited drawing.

MSLIDE COMMAND

Slides are created by using the AutoCAD **MSLIDE** command at the Command prompt. If **FILEDIA** is set to 1, the **MSLIDE** command displays the **Create Slide File** dialog box (Figure 2-6) on the screen. You can enter the slide file name in this dialog box. If **FILEDIA** is set to 0, the command will prompt you to enter the slide file name.

Command: **MSLIDE**
Enter name of slide file to create <Default>: *Slide file name.*

Example
Command: **MSLIDE**

Script Files and Slide Shows 2-21

Slide File: <Drawing1> **SLIDE1**
  Where **Drawing1** ------- Default slide file name
         SLIDE1 --------- Slide file name

In the preceding example, AutoCAD will save the slide file as **SLIDE1.SLD**.

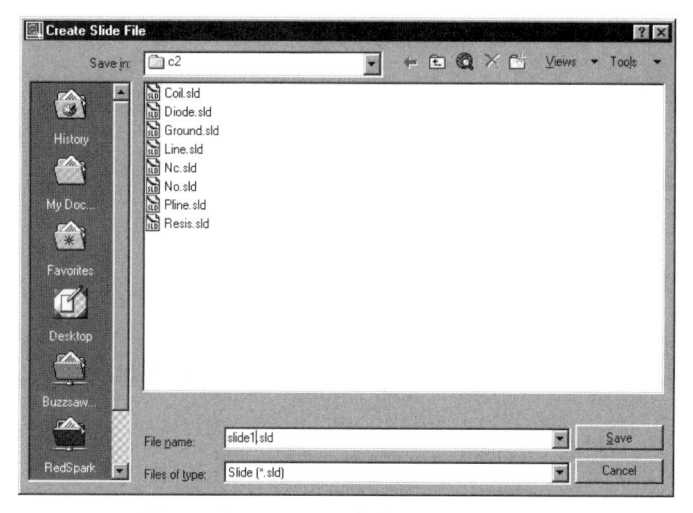

Figure 2-6 Create Slide File dialog box

Note
*In model space, you can use the **MSLIDE** command to make a slide of the existing display in the current viewport.*

If you are in the paper space viewport, you can make a slide of the display in the paper space that includes any floating viewports.

*When the viewports are not active, the **MSLIDE** command will make a slide of the current screen display.*

VSLIDE COMMAND

To view a slide, use the **VSLIDE** command at the Command prompt. The **Select Slide File** dialog box is displayed as shown in the Figure 2-7. Choose the file you want to view and then choose **OK**. The corresponding slide will be displayed on the screen. If the FILEDIA is 0, the slides that you want to view can be directly entered at the Command prompt.

  Command: **VSLIDE**
  Enter name of slide file to create<Default>: *Name*.

  Example
  Command: **VSLIDE**
  Slide file <Drawing1>: SLIDE1

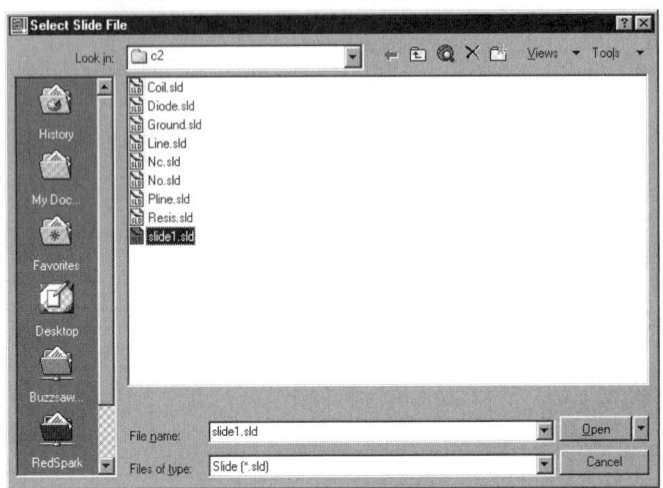

Where **Drawing1** ------- Default slide file name
 SLIDE1 --------- Name of slide file

Figure 2-7 Select Slide Files dialog box

Note
*After viewing a slide, you can use the AutoCAD **REDRAW** command, role the wheel on a wheel mouse, or pan with a wheel mouse to remove the slide display and return to the existing drawing on the screen.*

*Any command that is automatically followed by a redraw will also display the existing drawing. For example, AutoCAD **GRID**, **ZOOM ALL**, and **REGEN** commands will automatically return to the existing drawing on the screen.*

You can view the slides on high-resolution or low-resolution monitors. Depending on the resolution of the monitor, AutoCAD automatically adjusts the image. However, if you are using a high-resolution monitor, it is better to make the slides on the same monitor to take full advantage of that monitor.

Example 7

Write a script file that will create a slide show of the following slide files, with a time delay of 15 seconds after every slide (Figure 2-8).

SLIDE1, SLIDE2, SLIDE3, SLIDE4

Step 1: Creating the slides

The first step in a slide show is to create the slides using the **MSLIDE** command. The **MSLIDE** command will invoke the **Create Slide File** dialog box. Enter the name of the slide as **SLIDE1** and choose the **Save** button to exit the dialog box. Similarly other slides can be created and correspondingly saved. Figure 2-8 shows the drawings that have been saved as slide files **SLIDE1, SLIDE2, SLIDE3,** and **SLIDE4**. The slides must be saved to a directory in AutoCAD's search path or the script would not find them.

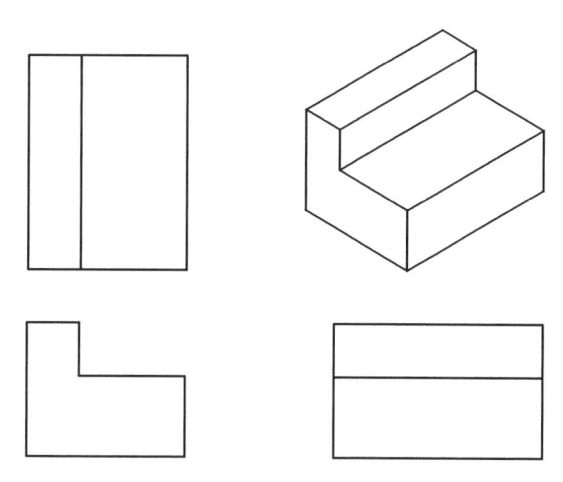

Figure 2-8 Slides for slide show

Step 2: Writing the script file
The second step is to find out the sequence in which you want these slides to be displayed, with the necessary time delay, if any, between slides. Then you can use any text editor or the AutoCAD **EDIT** command (provided the **ACAD.PGP** file is present and **EDIT** is defined in the file) to write the script file with the extension **.SCR**.

The following file is a listing of the script file that will create a slide show of the slides in Figure 2-8. The name of the script file is **SLDSHOW1**.

    VSLIDE SLIDE1
    DELAY 15000
    VSLIDE SLIDE2
    DELAY 15000
    VSLIDE SLIDE3
    DELAY 15000
    VSLIDE SLIDE4
    DELAY 15000

Step 3: Running the script file
To run this slide show, choose the **Run Script** from the **Tools** menu or enter **SCRIPT** at the Command comma Command prompt to invoke the **Select Script File** dialog box. Choose **SLDSHOW1** and choose **Open**. You can see the changes taking place on the screen.

PRELOADING SLIDES
In the script file of Example 7, VSLIDE SLIDE1 in line 1 loads the slide file, **SLIDE1**, and displays it on screen. After a pause of 15,000 milliseconds, it starts loading the second slide file, **SLIDE2**. Depending on the computer and the disk access time, you will notice that it takes some time to load the second slide file; the same is true for the other slides. To avoid the delay in loading the slide files, AutoCAD has provided a facility to preload a slide while

viewing the previous slide. This is accomplished by placing an asterisk (*) in front of the slide file name.

VSLIDE SLIDE1	*(View slide, SLIDE1.)*
VSLIDE *SLIDE2	*(Preload slide, SLIDE2.)*
DELAY 15000	*(Delay of 15 seconds.)*
VSLIDE	*(Display slide, SLIDE2.)*
VSLIDE *SLIDE3	*(Preload slide, SLIDE3.)*
DELAY 15000	*(Delay of 15 seconds.)*
VSLIDE	*(Display slide, SLIDE3.)*
VSLIDE *SLIDE4	
DELAY 15000	
VSLIDE	
DELAY 15000	
RSCRIPT	*(Restart the script file.)*

Example 8

Write a script file to generate a continuous slide show of the following slide files, with a time delay of two seconds between slides SLD1, SLD2, SLD3

The slide files are located in different subdirectories, as shown in Figure 2-9.

```
                    C:
                    |
              PROGRAM FILES
                    |
                ACAD 2002
                    |
       ┌────────────┼────────────┐
   SUBDIR1(SLD1)  SUBDIR2(SLD2)  SUBDIR3(SLD3)
```

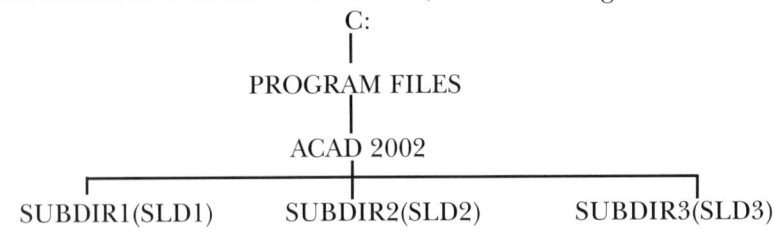

Figure 2-9 *Subdirectories of the C drive*

Where **C:** -------------------- Root directory.
PROGRAM FILES ------ *Root directory.*
ACAD 2002 ---------------- *Subdirectory where the AutoCAD files are loaded.*
SUBDIR1 ----------------- *Drawing subdirectory.*
SUBDIR2 ----------------- *Drawing subdirectory.*
SUBDIR3 ----------------- *Drawing subdirectory.*
SLD1 -------------------- *Slide file in SUBDIR1 subdirectory.*
SLD2 -------------------- *Slide file in SUBDIR2 subdirectory.*
SLD3 -------------------- *Slide file in SUBDIR3 subdirectory.*

The following file is the listing of the script files that will generate a slide show for the slides in Example 8:

VSLIDE "C:/Program Files/ACAD 2002/SUBDIR1/SLD1.SLD"
DELAY 2000
VSLIDE "C:/Program Files/ACAD 2002/SUBDIR2/SLD2.SLD"

Script Files and Slide Shows

    DELAY 2000
    VSLIDE "C:/Program Files/ACAD 2002/SUBDIR3/SLD3.SLD"
    DELAY 2000
    RSCRIPT

Line 1
VSLIDE "C:/Program Files/ACAD 2002/SUBDIR1/SLD1.SLD"
In this line, the AutoCAD command **VSLIDE** loads the slide file **SLD1**. The path name is mentioned along with the command **VSLIDE**. If the path name directory contains spaces then the path name must be enclosed in quotes.

Line 2
DELAY 2000
This line uses the AutoCAD **DELAY** command to create a pause of approximately two seconds before the next slide is loaded.

Line 3
VSLIDE "C:/Program Files/ACAD 2002/SUBDIR2/SLD2.SLD"
In this line, the AutoCAD command **VSLIDE** loads the slide file **SLD2**, located in the subdirectory **SUBDIR2**. If the slide file is located in a different subdirectory, you need to define the path with the slide file.

Line 5
VSLIDE "C:/Program Files/ACAD 2002/SUBDIR3/SLD3.SLD"
In this line, the **VSLIDE** command loads the slide file SLD3, located in the subdirectory SUBDIR3.

Line 7
RSCRIPT
In this line, the **RSCRIPT** command executes the script file again and displays the slides on the screen. This process continues indefinitely until the script file is canceled by pressing the ESC key or the BACKSPACE key.

SLIDE LIBRARIES

AutoCAD provides a utility, SLIDELIB, which constructs a library of the slide files. The format of the SLIDELIB utility command is:

    SLIDELIB (Library filename) <(Slide list filename)

    Example
    SLIDELIB SLDLIB <SLDLIST
             Where   **SLIDELIB** ------ AutoCAD's SLIDELIB utility
                         SLDLIB --------- Slide library filename
                         SLDLIST ------- List of slide filenames

The SLIDELIB utility is supplied with the AutoCAD software package. You can find this utility (SLIDELIB.EXE) in the support subdirectory. The slide file list is a list of the slide filenames that you want in a slide show. It is a text file that can be written by using any text editor or AutoCAD's **EDIT** command (provided ACAD.PGP file is present and **EDIT** is defined in the file). The slide files in the slide file list should not contain any file extension (.SLD). However, if you want to add a file extension it should be .SLD.

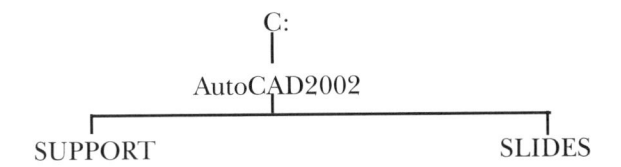

```
            C:
            |
        AutoCAD2002
        |           |
     SUPPORT      SLIDES
```

The slide file list can be also created by using the following command, if you have DOS version 5.0 or above. You can take help of make directory (md) or change directory (cd) while making or changing directories.

C:\AutoCAD2002\SLIDES>**DIR *.SLD/B>SLDLIST**

In this example assume that the name of the slide file list is **SLDLIST** and all slide files are in the SLIDES subdirectory. To use this command to create a slide file list, all slide files must be in the same directory.

When you use the SLIDELIB utility, it reads the slide filenames from the file that is specified in the slide list and the file is then written to the file specified by the library. In the Example 9, the SLIDELIB utility reads the slide filenames from the file SLDLIST and writes them to the library file SLDLIB:

C:\>**SLIDELIB SLDLIB <SLDLIST**

Note
*You **cannot** edit a slide library file. If you want to change anything, you have to create a new list of the slide files and then use the SLIDELIB utility to create a new slide library.*

*If you edit a slide while the slide is displayed on the screen, the slide is not edited. Instead, the current drawing that is behind the slide will be edited. Therefore, do not use any editing commands while you are viewing a slide. Use the **VSLIDE** and **DELAY** commands only when viewing a slide.*

The path name is not saved in the slide library; therefore, if you have more than one slide with the same name, even though they are in different subdirectories, only one slide will be saved in the slide library.

Example 9

Use AutoCAD's SLIDELIB utility to generate a continuous slide show of the following slide files with a time delay of 2.5 seconds between the slides. (The filenames are: SLDLIST for slide list file, SLDSHOW1 for slide library, SHOW1 for script file.)

front, top, rside, 3dview, isoview

Script Files and Slide Shows

The slide files are located in different subdirectories as shown in Figure 2-10.

```
                        C
                        |
                    Dwg-Files
                        |
        _____|_____
        |               |               |
      Proj-A          Proj-B        Slide-Files
      front           3dview          sldlist
      top             isoview         slideli.exe
      rside
```

Where C ------------------ (C drive.)
       Dwg-Files ------- (Subdirectory where drawing files are located)
       Proj-A ------------ (Drawing subdirectory.)
       Proj-B ------------ (Drawing subdirectory.)
       Slide-Files ------ (Directory where Slidelib.exe and sldlist are copied)

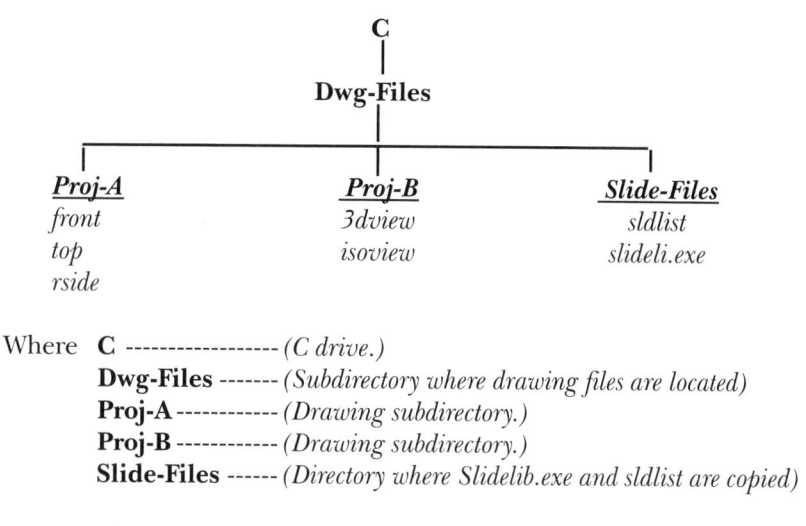

Figure 2-10 Drawing subdirectories of C drive

Step 1

The first step is to create a list of the slide file names with the drive and the directory information. Assume that you are in the **Slide-Files** subdirectory. You can use a text editor or AutoCAD's EDIT function to create a list of the slide files that you want to include in the slide show. These files do not need a file extension. However, if you choose to give them a file extension, it should be .SLD. The following file is a listing of the file SLDLIST for Example 9:

    c:\Dwg-Files\Proj-A\front
    c:\Dwg-Files\Proj-A\top
    c:\Dwg-Files\Proj-A\rside
    c:\Dwg-Files\Proj-B\3dview
    c:\Dwg-Files\Proj-B\isoview

Step 2

The second step is to use AutoCAD's SLIDELIB utility program to create the slide library. The name of the slide library is assumed to be **sldshow1** for this example. Before creating the slide library, copy the slide list file (SLDLIST) and the SLIDELIB utility from the support directory to the Slide-Files directory. This way all the needed files are in one directory. Enter **SHELL** command at AutoCAD Command prompt and then press the ENTER key at the OS Command prompt. The AutoCAD Shell Active dialog box will be displayed on the screen, Figure 2-11. You can also use Windows DOS box by choosing Programs > MS-DOS Prompt.

    Command: **SHELL**
    OS Command: Press ENTER

Now enter the following to run the SLIDELIB utility to create the slide library. Here it is assumed that Slide-Files directory is current.

C:\Dwg-Files\Slide-Files>SLIDELIB sldshow1 <sldlist

Where **SLIDELIB** ------ AutoCAD's SLIDELIB utility
sldshow1 -------- Slide library
sldlist ----------- Slide file list

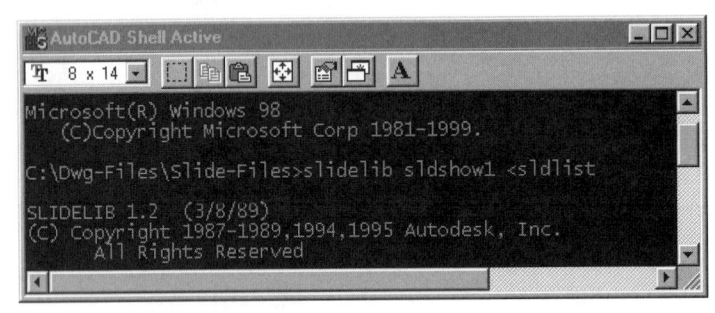

Figure 2-11 *AutoCAD Shell Active dialog box*

Step 3
Now you can write a script file for the slide show that will use the slides in the slide library. The name of the script file for this example is assumed to be SHOW1.

> VSLIDE sldshow1(front)
> DELAY 2500
> VSLIDE sldshow1(top)
> DELAY 2500
> VSLIDE sldshow1(rside)
> DELAY 2500
> VSLIDE sldshow1(3dview)
> DELAY 2500
> VSLIDE sldshow1(isoview)
> DELAY 2500
> RSCRIPT

Step 4
Invoke the **Select Script File** (Figure 2-12) dialog box by choosing the **Run Script** button from the **Tools** menu or enter the **SCRIPT** command at the Command prompt. You can also enter the **SCRIPT** command from the Command prompt after setting the system variable FILEDIA to 0.

> Command: **SCRIPT**
> Enter script file name<default>: **SHOW1**

SLIDE SHOWS WITH RENDERED IMAGES
Slide shows can also contain rendered images. In fact, use of rendered images is highly recommended for a more dynamically pleasing, entertaining, and interesting show. AutoCAD provides the following AutoLISP function for using rendered images during a slide show presentation.

Script Files and Slide Shows

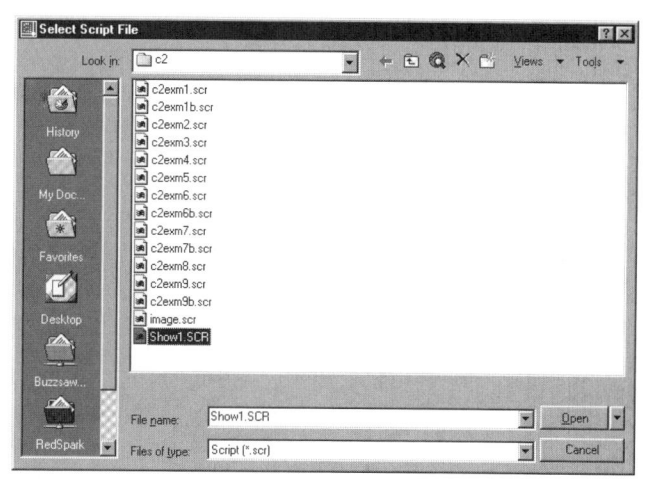

*Figure 2-12 Selecting script file from **Select Script File** dialog box*

(C:REPLAY FILENAME TYPE [<XOFF> <YOFF> <XSIZE> <YSIZE>])
   Where:

        C:REPLAY ---------------- AutoLISP function calling AutoCAD's **Replay** command.

        FILENAME --------------- Name of rendered file, without the extension. The filename is to be enclosed in quotes. Include directory path and one forward slash (/) for directory structures.

        TYPE -------------------- The rendered file type (e.g., Bmp, Tif, Tga). The file type should also be enclosed in quotes.

        XOFF -------------------- The rendered file's "X" coordinate offset from 0,0.

        YOFF -------------------- The rendered file's "Y" coordinate offset from 0,0.

        XSIZE -------------------- The size of the rendered file to be displayed in the "X" (horizontal) direction measured in pixels from the offset position. This information can generally be obtained when creating a rendered file with AutoCAD's **SAVEIMG** command.

        YSIZE -------------------- The size of the rendered file to be displayed in the "Y" (vertical) direction measured in pixels from the offset position. This information can generally be obtained when creating a rendered file with AutoCAD's **SAVEIMG** command.

AutoCAD ships with a file called BIGLAKE.TGA as an example of a background texture map, Figure 2-13. To display this rendered image enter the following at the Command prompt. Note: The "biglake.tga" file must be in AutoCAD's support directory's search path.

(C:REPLAY "C:/PROGRAM FILES/AutoCAD 2002//TEXTURES/BIGLAKE" "TGA" 150 50 944 564)
  Where:

C:REPLAY:	----------------	Function for AutoCAD to perform - issue the *Replay* command.
"C:/PRO../BIGLAKE"	--	Name of the rendered image file, including drive and directory.
"TGA"	--------------------	The rendered image file type.
150	--------------------	The "X" offset for the image displayed.
50	--------------------	The "Y" offset for the image displayed.
944	--------------------	The "X" size (horizontal) for the image displayed measured (in pixels) from the offset position.
564	--------------------	The "Y" size (vertical) for the image displayed measured (in pixels) from the offset position.

Combining Slides and Rendered Images

The following is an example of a script file (replay.scr) that combines the use of AutoCAD slides and rendered images (*.bmp files) into a seamless slide show presentation. The "F_" files are the wire FRAME images, while the "R_" files are the RENDERED images. It is assumed that the BMP image files are on A drive.

```
VSLIDE A:\SLIDES\F_ROLL
DELAY 3000
(C:REPLAY "A:/RENDERS/R_ROLL" "BMP" 0 0 944 564)
DELAY 3000
VSLIDE A:\SLIDES\F_BKCASE
DELAY 3000
(C:REPLAY "A:/RENDERS/R_BKCASE" "BMP" 0 0 944 564)
DELAY 3000
VSLIDE A:\SLIDES\F_MOUSE
DELAY 3000
(C:REPLAY "A:/RENDERS/R_MOUSE" "BMP" 0 0 944 564)
DELAY 3000
VSLIDE A:\SLIDES\F_TABLE2
DELAY 3000
(C:REPLAY "A:/RENDERS/R_TABLE2" "BMP" 0 0 944 564)
DELAY 3000
RSCRIPT
```

Script Files and Slide Shows

Figure 2-13 Displaying a rendered image using a script file

Self-Evaluation Test

Answer the following questions and then compare your answers to the correct answers given at the end of this chapter.

SCRIPT FILES

1. AutoCAD has provided a facility of _____ that allows you to combine different AutoCAD commands and execute them in a predetermined sequence.

2. Before writing a script file, you need to know the AutoCAD _____ and the _____ required in response to the command prompts.

3. The AutoCAD _____ command is used to run a script file.

4. In a script file, the _____ is used to terminate a command or a prompt entry.

5. The **DELAY** command is to be followed by _____ in milliseconds.

SLIDE SHOWS

6. Slides do not contain any _____ information, which means that the entities do not have any information associated with them.

7. Slides _____ edited like a drawing.

8. Slides can be created using the AutoCAD _____ command.

9. To view a slide, use the AutoCAD _____ command.

10. AutoCAD provides a utility that constructs a library of the slide files. This is done with AutoCAD's utility program called _____.

Review Questions

Answer the following questions.

SCRIPT FILES

1. The _____ files can be used to generate a slide show, do the initial drawing setup, or plot a drawing to a predefined specification.

2. In a script file, you can _____ several statements in one line.

3. When you run a script file, the default script file name is the same as the _____ name.

4. When you run a script file, type the name of the script file without the file _____.

5. One of the limitations of script files is that all the information has to be contained _____ the file.

6. The AutoCAD _____ command allows you to re-execute a script file indefinitely until the command is canceled.

7. You cannot provide a _____ statement in a script file to terminate the file when a particular condition is satisfied.

8. The AutoCAD _____ command introduces a delay before the next command is executed.

9. If the script file was canceled and you want to continue the script file, you can do so by using the AutoCAD _____ command.

SLIDE SHOWS

10. AutoCAD provides a facility through _____ files to combine the slides in a text file and display them in a predetermined sequence.

Script Files and Slide Shows 2-33

11. A _____ can also be introduced in the script file so that the viewer has enough time to view a slide.

12. Slides are the _____ of a screen display.

13. In model space, you can use the **MSLIDE** command to make a slide of the _____ display in the _____ viewport.

14. If you are in paper space, you can make a slide of the display in paper space that _____ any floating viewports.

15. If you want to make any change in the slide, you need to _____ the drawing, then make a new slide from the edited drawing.

16. If the slide is in the slide library and you want to view it, the slide library name has to be _____ with the slide filename.

17. You cannot _____ a slide library file. If you want to change anything, you have to create a new list of the slide files and then use the _____ utility to create a new slide library.

18. The path name _____ be saved in the slide library. Therefore, if you have more than one slide with the same name, although with different subdirectories, only one slide will be saved in the slide library.

Exercises

SCRIPT FILES

Exercise 2 *General*

Write a script file that will do the following initial setup for a drawing:

Grid	2.0
Snap	0.5
Limits	0,0
	18.0,12.0
Zoom	All
Text height	0.25
Ltscale	2.0

Overall dimension scale factor is 2
Aligned dimension text with the dimension line
Dimension text above the dimension line
Size of the center mark is 0.75

Exercise 3
General

Write a script file that will set up the following layers with the given colors and linetypes (filename SCRIPTE3.SCR):

Contour	Red	Continuous
SPipes	Yellow	Center
WPipes	Blue	Hidden
Power	Green	Continuous
Manholes	Magenta	Continuous
Trees	Cyan	Continuous

Exercise 4
General

Write a script file that will do the following initial setup for a new drawing:

Limits	0,0 24,18
Grid	1.0
Snap	0.25
Ortho	On
Snap	On
Zoom	All
Pline width	0.02
PLine	0,0 24,0 24,18 0,18 0,0
Ltscale	1.5
Units	Decimal units
	Precision 0.00
	Decimal degrees
	Precision 0
	Base angle East (0.00)
	Angle measured counterclockwise
Layers	

Name	**Color**	**Linetype**
Obj	Red	Continuous
Cen	Yellow	Center
Hid	Blue	Hidden
Dim	Green	Continuous

Exercise 5
General

Write a script file that will plot a given drawing according to the following specifications. (Use the plotter for which your system is configured and adjust the values accordingly.)

Plot, using the Window option
Window size (0,0 24,18)
Do not write the plot to file
Size in inch units

Script Files and Slide Shows

Plot origin (0.0,0.0)
Maximum plot size (8.5,11 or the smallest size available on your printer/plotter)
90 degree plot rotation
No removal of hidden lines
Plotting scale (Fit)

Exercise 6 — *General*

Write a script file that will continuously rotate a line in 10-degree increments around its midpoint (Figure 2-14). The time delay between increments is one second.

Exercise 7 — *General*

Write a script file that will continuously rotate the following arrangements (Figure 2-15). One set of two circles and one line should rotate clockwise while the other set of two circles and one line should rotate counterclockwise. Assume the rotation to be 5-degrees around the intersection of the lines for both sets of arrangements. Specifications are given below.

Start point of the horizontal line	2,4
End point of the horizontal line	8,4
Center point of circle at the start point of horizontal line	2,4
Diameter of the circle	1.0
Center point of circle at the end point of horizontal line	8,4
Diameter of circle	1.0
Start point of the vertical line	5,1
End point of the vertical line	5,7
Center point of circle at the start point of the vertical line	5,1
Diameter of the circle	1.0
Center point of circle at the end point of the vertical line	5,7
Diameter of the circle	1.0

(Select one set of two circles and one line and create one group. Similarly select another set of two circles and one line and create another group. Rotate one group clockwise and another group counterclockwise.)

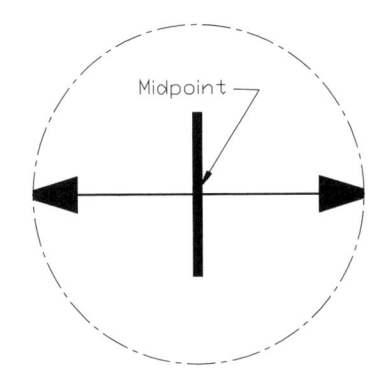

Figure 2-14 Drawing for Exercise 6

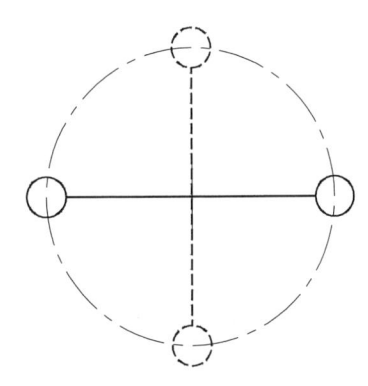

Figure 2-15 Drawing for Exexrcise 7

SLIDE SHOWS

Exercise 8 — *General*

Make the slides shown in Figure 2-16 and write a script file for a continuous slide show. Provide a time delay of 5 seconds after every slide. (You do not have to use the slides shown in Figure 2-16; you can use any slides of your choice.)

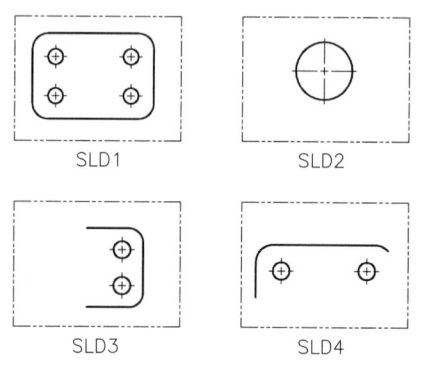

Figure 2-16 *Slides for slide show*

Exercise 9 — *General*

From Exercise 8, list the slides in a file SLDLIST2 and create a slide library file SLDLIB2. Then write a script file SHOW2 using the slide library with a time delay of five seconds after every slide.

Exercise 10 — *General*

Write a script file to generate a continuous slide show of the following slide files with a time delay of three seconds between the slides.

    SLDX1, SLDX2, SLDX3
    SLDX1, SLDX2, SLDX3
    SLDZ1, SLDZ2, SLDZ3

```
                    C:
                    |
                   ACAD
        _____|_____
       |            |            |
    SUBDIR1     SUBDIR2       SUBDIR3

    (SLDX1)     (SLDY1)       (SLDZ1)
    (SLDX2)     (SLDY2)       (SLDZ2)
    (SLDX3)     (SLDY3)       (SLDZ3)
```

Figure 2-17 *Subdirectories of C drive*

Script Files and Slide Shows 2-37

Assume that the slide files are located in different subdirectories as shown in Figure 2-16 and ACAD is the current subdirectory.

Where	
C:	*(C drive.)*
ACAD	*(Subdirectory where the AutoCAD files are loaded.)*
SUBDIR1	*(Drawing subdirectory.)*
SUBDIR1	*(Drawing subdirectory.)*
SUBDIR1	*(Drawing subdirectory.)*

Answers to Self-Evaluation Test

1- **SCRIPT** files, 2- Commands, options, 3- **SCRIPT**, 4- blank space, 5- time, 6- vector, 7- cannot be, 8- **MSLIDE**, 9- **VSLIDE**, 10- S**LIDELIB**

Chapter 3

Creating Linetypes and Hatch Patterns

Learning Objectives

After completing this chapter, you will be able to:
Create Linetypes:
- *Write linetype definitions.*
- *Create different linetypes.*
- *Create linetype files.*
- *Determine **LTSCALE** for plotting the drawing to given specifications.*
- *Define alternate linetypes and modify existing linetypes.*
- *Create string and shape complex linetypes.*

Create Hatch Patterns:
- *Understand hatch pattern definition.*
- *Create new hatch patterns.*
- *Determine the effect of angle and scale factor on hatch.*
- *Create hatch patterns with multiple descriptors.*
- *Save hatch patterns in a separate file.*
- *Define custom hatch pattern file.*

STANDARD LINETYPES

The AutoCAD software package comes with a library of standard linetypes that has 38 different standard linetypes and 7 complex linetypes, including ISO linetypes. These linetypes are saved in the **ACAD.LIN** file. You can modify existing linetypes or create new ones.

LINETYPE DEFINITION

All linetype definitions consist of two parts: **header line and pattern line**.

Header Line

The **header line** consists of an asterisk (*) followed by the name of the linetype and the linetype description. The name and the linetype description should be separated by a comma. If there is no description, the comma that separates the linetype name and the description is not required.

The format of the header line is:

* Linetype Name, Description

Example
*HIDDENS,__ __ __ __ __ __

Where * ------------------ Asterisk sign
HIDDENS ------ Linetype name
, -------------------- Comma
__ __ __ __ ------ Linetype description

All linetype definitions require a linetype name. When you want to load a linetype or assign a linetype to an object, AutoCAD recognizes the linetype by the name you have assigned to the linetype definition. The names of the linetype definition should be selected to help the user recognize the linetype by its name. For example, the linetype name LINEFCX does not give the user any idea about the type of line. However, a linetype name like DASHDOT gives a better idea about the type of line that a user can expect.

The linetype description is a textual representation of the line. This representation can be generated by using dashes, dots, and spaces at the keyboard. The graphic is used by AutoCAD when you want to display the linetypes on the screen by using the AutoCAD **LINETYPE** command with the ? option or by using the dialog box. The linetype description cannot exceed 47 characters.

Pattern Line

The **pattern line** contains the definition of the line pattern. The definition of the line pattern consists of the alignment field specification and the linetype specification. The alignment field specification and the linetype specification are separated by a comma.

The format of the pattern line is:

Alignment Field Specification, Linetype Specification

Example
A,.75,-.25,.75

Creating Linetypes and Hatch Patterns 3-3

```
Where  A ------------------ Alignment field specification
       , -------------------- Comma
       .75,-.25,.75 ----- Linetype specification
```

The letter used for alignment field specification is A. This is the only alignment field supported by AutoCAD; therefore, the pattern line will always start with the letter A. The linetype specification defines the configuration of the dash-dot pattern to generate a line. The maximum number for dash length specification in the linetype is 12, provided the linetype pattern definition fits on one 80-character line.

ELEMENTS OF LINETYPE SPECIFICATION

All linetypes are created by combining the basic elements in a desired configuration. There are three basic elements that can be used to define a linetype specification.

```
Dash         (Pen down)
Dot          (Pen down, 0 length)
Space        (Pen up)
```

Example

_____ . _____ . _____ . _____

```
Where  . -------------------- Dot (pen down with 0 length)
       Blank space -------------- Space (pen up)
       _____ ------------------ Dash (pen down with specified length)
```

The dashes are generated by defining a positive number. For example, .5 will generate a dash 0.5 units long. Similarly, spaces are generated by defining a negative number. For example, -.2 will generate a space 0.2 units long. The dot is generated by defining a 0 length.

Example
A,.5,-.2,0,-.2,.5
```
       Where  0 ------------------ Dot (zero length)
              -.2 ----------------- Length of space (pen up)
              .5 ------------------ Length of dash (pen down)
```

CREATING LINETYPES

Before creating a linetype, you need to decide the type of line you want to generate. Draw the line on a piece of paper and measure the length of each element that constitutes the line. You need to define only one segment of the line, because the pattern is repeated when you draw a line. Linetypes can be created or modified by any one of the following methods:

Using a text editor like Notepad
Adding a new linetype in the ACAD.LIN file
Using the AutoCAD LINETYPE command

The following example, Example 3, explains how to create a new linetype by using a text editor, adding a linetype to the **ACAD.LIN** file, and using the AutoCAD **LINETYPE** command.

Example 1

Create linetype DASH3DOT (Figure 3-1) with the following specifications:

Length of the first dash 0.5
Blank space 0.125
Dot
Blank space 0.125
Dot
Blank space 0.125
Dot
Blank space 0.125

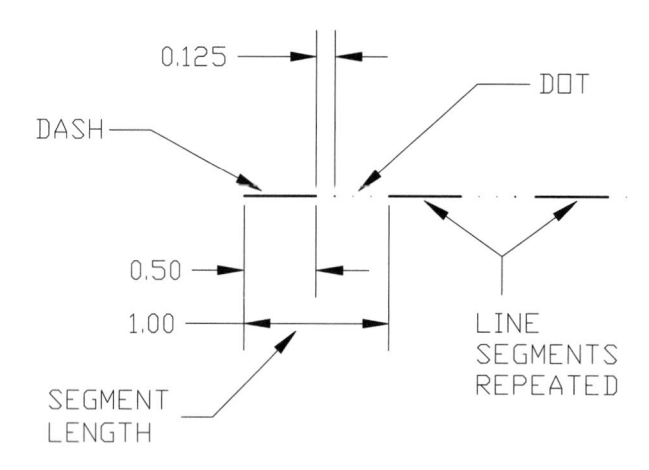

Figure 3-1 Linetype specifications of DASH3DOT

Using a Text Editor

Step 1: Writing definition of linetype

You can start a new linetype file and then add the line definitions to this file. Use any text editor like Notepad to start a new file (NEWLT.LIN) and then add the following two lines to the file to define the DASH3DOT linetype. The name and the description must be separated by a comma (,). The description is optional. If you decide not to give one, omit the comma after the linetype name DASH3DOT.

    *****DASH3DOT,___ . . . ___ . . . ___
    A,.5,-.125,0,-.125,0,-.125,0,-.125

Save it as NEWLT.LIN in AutoCAD's Support directory.

Creating Linetypes and Hatch Patterns 3-5

Step 2: Loading the linetype

To load this linetype, choose **Linetype** from the **Format** menu to display the **Linetype Manager** dialog box as shown in Figure 3-2. Choose the **Load** button in the **Linetype Manager** dialog box to display the **Load or Reload Linetypes** dialog box. Choose the **File** button in the **Load or Reload Linetypes** dialog box to display the **Select Linetype File** dialog box. Choose the **NEWLT.LIN** file in the **Select Lintype File** dialog box and then choose **Open**. Again the **Load or Reload Linetypes** dialog box is displayed. Choose the **DASH3DOT** linetype in the **Available Linetypes** area and then choose **OK**. Again, the **Linetype Manager** dialog box is displayed. Choose the **DASH3DOT** linetype and then choose the **Current** button to make the selected linetype current. Then choose **OK**.

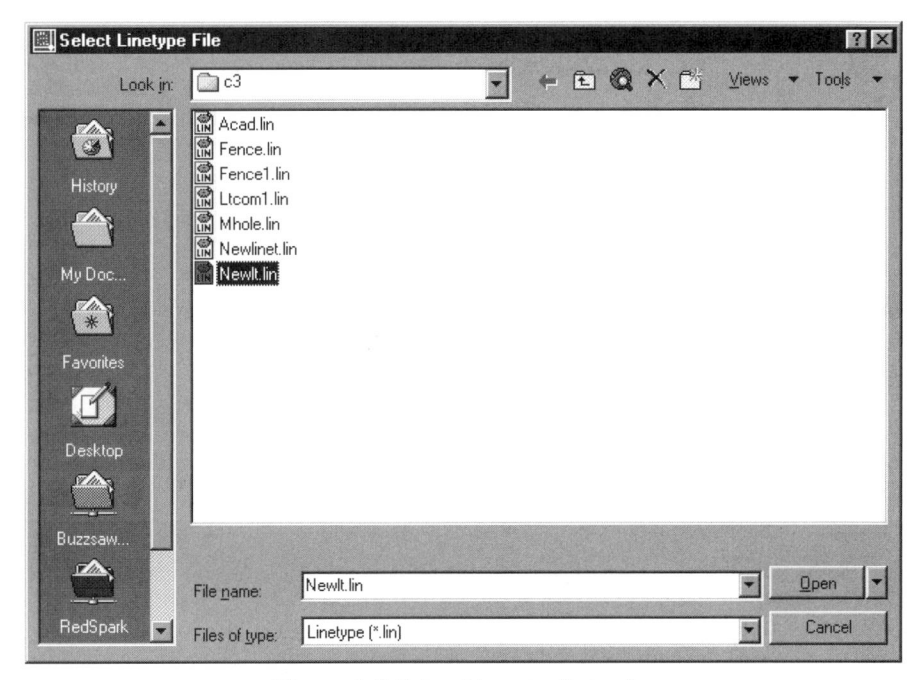

Figure 3-2 Select Linetype dialog box

Adding a New Linetype in the ACAD.LIN File

Step 1: Adding a new linetype in the ACAD.LIN

You can also use a text editor (like Notepad) to create a new linetype. Using the text editor, load the file and insert the lines that define the new linetype. The following file is a partial listing of the **ACAD.LIN** file after adding a new linetype to the file:

```
*BORDER,__ __ . __ __ . __ __ . __ __ . __ __ .
A,.5,-.25,.5,-.25,0,-.25
*BORDER2,_._._._._._._._._._._._._._.
A,.25,-.125,.25,-.125,0,-.125
*BORDERX2,____ ____ . ____ ____ . ____ ____ .
A,1.0,-.5,1.0,-.5,0,-.5
```

```
*CENTERX2,____ __ ____ __ ____ __ __
A,2.5,-.5,.5,-.5
*DASHDOT,__ . __ . __ . __ . __ . __ . __ .
A,.5,-.25,0,-.25
*DOTX2,. . . . . . . . . . . . . .
A,.25,-.125
*HIDDEN2,_ _ _ _ _ _ _ _ _ _ _ _ _ _ _ _ _ _ _ _
A,.125,-.0625
*HIDDENX2,___ ___ ___ ___ ___ ___ ___ ___ ___ ___
A,.5,-.25
*PHANTOM,____ __ __ ____ __ __ ____ __ __
A,1.25,-.25,.25,-.25,.25,-.25
*PHANTOMX2,_____ ___ ___ _____
```

```
*GAS_LINE,Gas line ----GAS----GAS----GAS----GAS----GAS----GAS--
A,.5,-.2,["GAS",STANDARD,S=.1,R=0.0,X=-0.1,Y=-.05],-.25
*ZIGZAG,Zig zag /\/\/\/\/\/\/\/\/\/\/\/\/\/\/\/\
A,.0001,-.2,[ZIG,ltypeshp.shx,x=-.2,s=.2],-.4,[ZIG,ltypeshp.shx,r=180,x=.2,s=.2],-.2
```
*DASH3DOT,___ . . . ___ . . . ___
A,.5,-.125,0,-.125,0,-.125,0,-.125

The last two lines of this file define the new linetype, DASH3DOT. The first line contains the name DASH3DOT and the description of the line (___ . . .___). The second line contains the alignment and the pattern definition.

Step 2: Loading the linetype
Save the file and then load the linetype using the AutoCAD **LINETYPE** command. The procedure of loading the linetype is the same as described earlier in this example. The lines and polylines that this linetype will generate are shown in Figure 3-3.

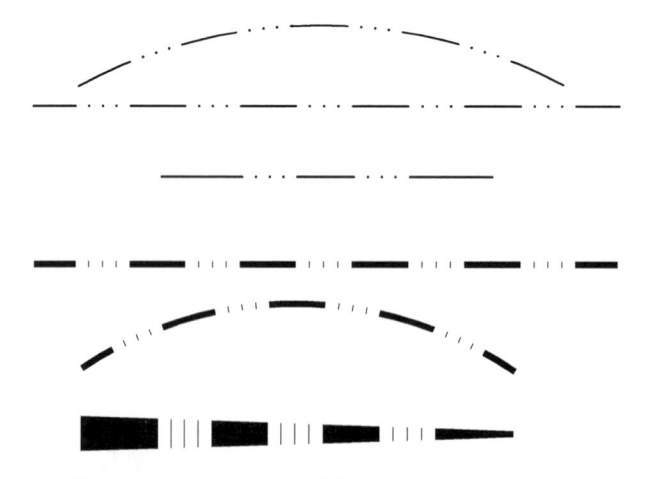

Figure 3-3 *Lines created by linetype DASH3DOT*

Creating Linetypes and Hatch Patterns 3-7

> **Note**
> *If you change the LTSCALE factor, all lines that are drawn in the drawing are affected by the new ratio.*

Using the AutoCAD Linetype Command

Step 1: Creating a linetype

To create a linetype using the AutoCAD **LINETYPE** command, first make sure that you are in the drawing editor. Then enter the **-LINETYPE** command and select the Create option to create a linetype.

    Command: **-LINETYPE**
    Enter an option [?/Create/Load/Set]: **C**

Enter the name of the linetype and the name of the library file in which you want to store the definition of the new linetype.

    Enter name of linetype to create: **DASH3DOT**

If **FILEDIA**=1, the **Create or Append Linetype File** dialog box (Figure 3-4) will appear on the screen. If **FILEDIA**=0, AutoCAD will prompt you to enter the name of the file.

    Enter linetype file name for new linetype definition <default>: **Acad**

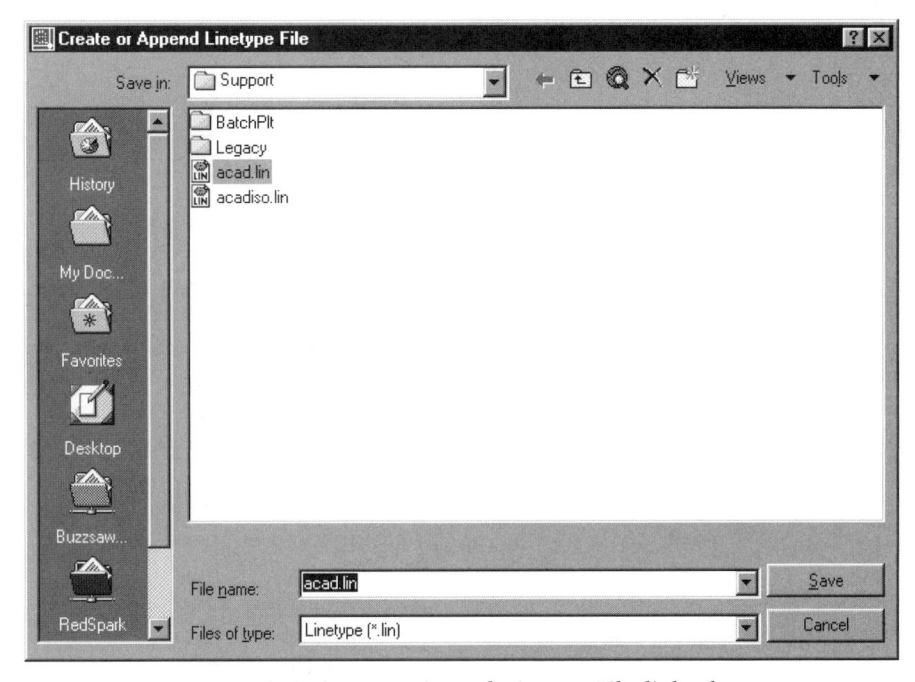

Figure 3-4 Create or Append Linetype File dialog box

If the linetype already exists, the following message will be displayed on the screen:
Wait, checking if linetype already defined...
"Linetype" already exists in this file. Current definition is:
alignment, dash-1, dash-2, _____.
Overwrite?<N>

If you want to redefine the existing line style, enter Y; otherwise, type N or press RETURN to choose the default value of N. You can then repeat the process with a different name of the linetype. After entering the name of the linetype and the library file name, AutoCAD will prompt you to enter the descriptive text and the pattern of the line.

Descriptive text: **DASH3DOT,____ . . . ____ . . . ____
Enter linetype pattern (on next line):
A,.5,-.125,0,-.125,0,-.125,0,-.125

Descriptive Text

**DASH3DOT,____ . . . ____ . . . ____

For the descriptive text, you have to type an asterisk (*) followed by the name of the linetype. For Example 1, the name of the linetype is DASH3DOT. The name *DASH3DOT can be followed by the description of the linetype; the length of this description cannot exceed 47 characters. In this example, the description is dashes and dots ____ . . . ____. It could be any text or alphanumeric string. The description is displayed on the screen when you list the linetypes.

Pattern

A,.5,-.125,0,-.125,0,-.125,0,-.125

The line pattern should start with an alignment definition. Currently, AutoCAD supports only one type of alignment—A. Therefore, it is automatically displayed on the screen when you select the **LINETYPE** command with the Create option. After entering **A** for pattern alignment, you must define the pen position. A positive number (.5 or 0.5) indicates a "pen-down" position, and a negative number (-.25 or -0.25) indicates a "pen-up" position. The length of the dash or the space is designated by the magnitude of the number. For example, 0.5 will draw a dash 0.5 units long, and -0.25 will leave a blank space of 0.25 units. A dash length of 0 will draw a dot (.). Here are the pattern definition elements for Example 1:

.5	pen down	0.5 units long dash
-.125	pen up	.125 units blank space
0	pen down	dot
-.125	pen up	.125 units blank space
0	pen down	dot
-.125	pen up	.125 units blank space
0	pen down	dot
-.125	pen up	.125 units blank space

Creating Linetypes and Hatch Patterns 3-9

After you enter the pattern definition, the linetype (DASH3DOT) is automatically saved in the **ACAD.LIN** file

Step 2: Loading the linetype
You can use the LINETYPE command to load the linetype or choose Linetype in the Format pulldown menu. The linetype (DASH3DOT) can also be loaded using the AutoCAD **-LINETYPE** command and selecting the **Load** option.

ALIGNMENT SPECIFICATION

The alignment specifies the pattern alignment at the start and the end of the line, circle, or arc. In other words, the line would always start and end with the dash (___). The alignment definition "A" requires that the first element be a dash or dot (pen down), followed by a negative (pen up) segment. The minimum number of dash segments for alignment A is two. If there is not enough space for the line, AutoCAD will draw a continuous line.

For example, in the linetype DASH3DOT of Example 1, the length of each line segment is 1.0 (.5 + .125 + .125 + .125 + .125 = 1.0). If the length of the line drawn is less than 1.00, the line will be drawn as a continuous line (Figure 3-5). If the length of the line is 1.00 or greater, the line will be drawn according to DASH3DOT linetype. AutoCAD automatically adjusts the length of the dashes and the line will always start and end with a dash. The length of the starting and ending dashes will be at least half the length of the dash as specified in the file. If the length of the dash as specified in the file is 0.5, the length of the starting and ending dashes will be at least 0.25. To fit a line that starts and ends with a dash, the length of these dashes can also increase as shown in Figure 3-5.

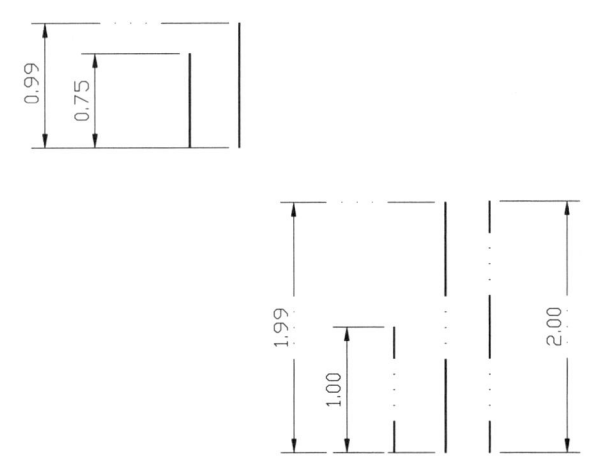

Figure 3-5 Alignment of linetype DASH3DOT

LTSCALE COMMAND

As we mentioned previously, the length of each line segment in the DASH3DOT linetype is 1.0 (.5 + .125 + .125 + .125 + .125 = 1.0). If you draw a line that is less than 1.0 units long,

AutoCAD will draw a single dash that looks like a continuous line (Figure 3-6). This problem can be rectified by changing the linetype scale factor variable **LTSCALE** to a smaller value. This can be accomplished by using AutoCAD's **LTSCALE** command:

Command: **LTSCALE**
Enter new linetype scale factor <default>: *New value.*

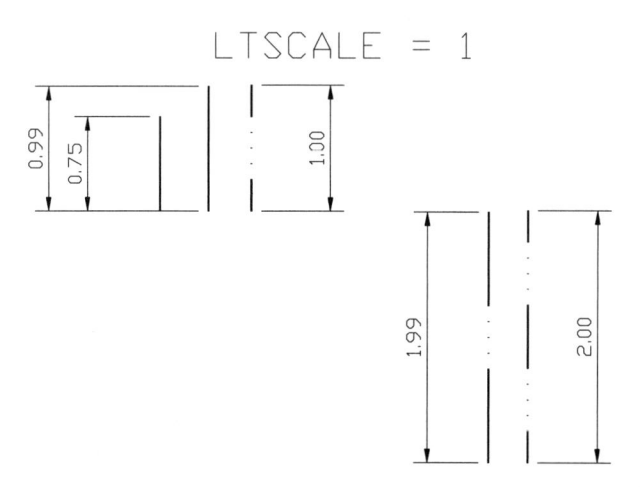

LTSCALE = 1

Figure 3-6 Alignment when Ltscale = 1

The default value of the **LTSCALE** variable is 1.0. If the LTSCALE is changed to 0.75, the length of each segment is reduced by 0.75 (1.0 x 0.75 = 0.75). Then, if you draw a line 0.75 units or longer, it will be drawn according to the definition of DASH3DOT (___ . . . ___) (Figures 3-7 and 3-8).

LTSCALE = 0.99

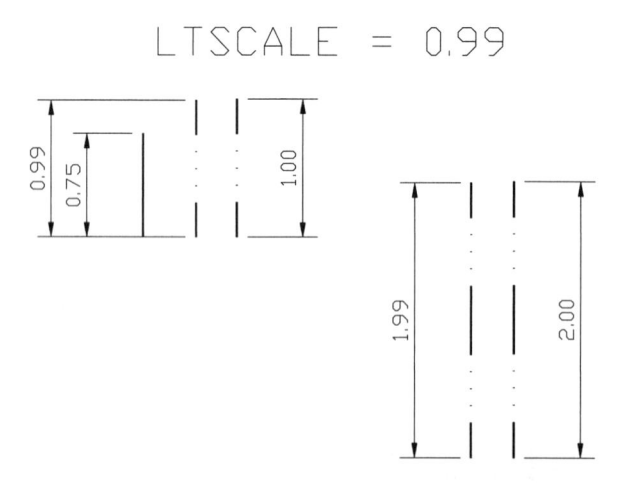

Figure 3-7 Alignment when Ltscale = 0.99

Creating Linetypes and Hatch Patterns

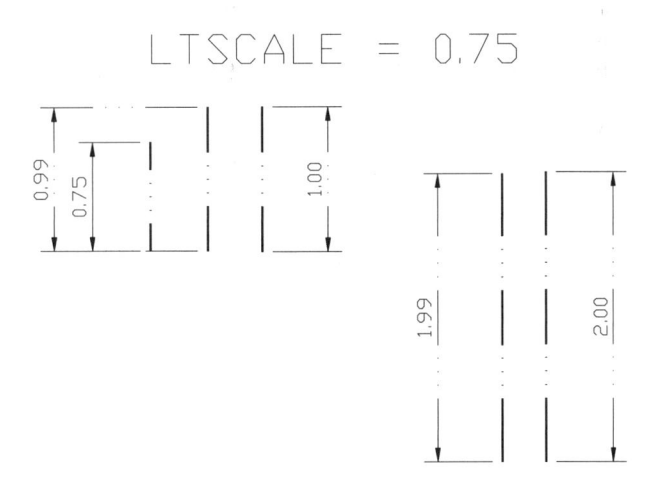

Figure 3-8 Alignment when Ltscale = 0.75

The appearance of the lines is also affected by the limits of the drawing. Most of the AutoCAD linetypes work fine for a drawing that has the limits 12,9. Figure 3-9 shows a line of linetype DASH3DOT that is four units long and the limits of the drawing are 12,9. If you increase the limits to 48,36 the lines will appear as continuous lines. If you want the line to appear the same as before **on the screen**, the LTSCALE should be changed. Since the limits of the drawing have increased four times, the LTSCALE should also increase by the same amount. If you change the scale factor to four, the line segments will also increase by a factor of four. As shown in Figure 3-9, the length of the starting and the ending dash has increased to one unit.

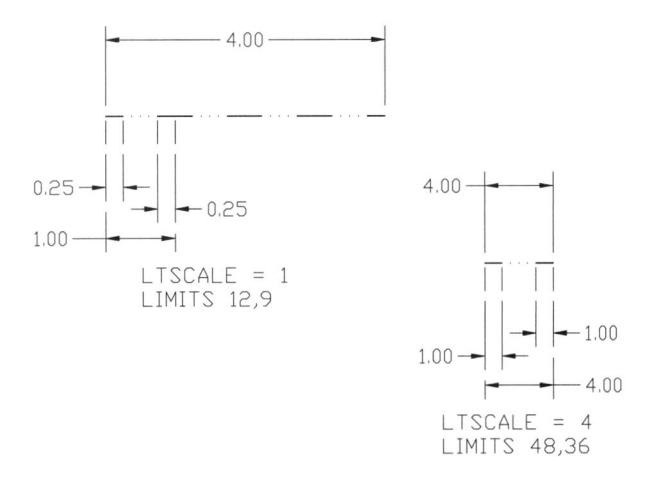

Figure 3-9 Linetype DASH3DOT before and after changing the LTSCALE

In general, the approximate LTSCALE factor for **screen display** can be obtained by dividing the X-limit of the drawing by the default X-limit (12.00). However, **it is recommended that the linetype scale must be set according to plot scale** discussed in the next section.

LTSCALE factor for SCREEN DISPLAY = X-limits of the drawing/12.00
Example
    Drawing limits are 48,36
    LTSCALE factor for screen display = 48/12 = 4

    Drawing sheet size is 36,24 and scale is 1/4" = 1'
    LTSCALE factor for screen display = 12 x 4 x (36 / 12) = 144

LTSCALE FACTOR FOR PLOTTING

The LTSCALE factor for plotting depends on the size of the sheet you are using to plot the drawing. For example, if the limits are 48 by 36, the drawing scale is 1:1, and you want to plot the drawing on a 48" by 36" size sheet, the LTSCALE factor is 1. If you check the specification of a hidden line in the ACAD.LIN file, the length of each dash is 0.25. Therefore, when you plot a drawing with 1:1 scale, the length of each dash in a hidden line is 0.25.

However, if the drawing scale is 1/8" = 1' and you want to plot the drawing on a 48" by 36" paper, the LTSCALE factor must be 96 (8 x 12 = 96). The length of each dash in the hidden line will increase by a factor of 96 because the LTSCALE factor is 96. Therefore, the length of each dash will be 24 units (0.25 x 96 = 24). At the time of plotting, the scale factor for plotting must be 1:96 to plot the 384' by 288' drawing on a 48" by 36" size paper. Each dash of the hidden line that was 24" long on the drawing will be 0.25 (24/96 = 0.25) inch long when plotted. Similarly, if the desired text size on the paper is 1/8", the text height in the drawing must be 12" (1/8 x 96 = 12").

Ltscale Factor for PLOTTING = Drawing Scale

Sometimes your plotter may not be able to plot a 48" by 36" drawing or you might like to decrease the size of the plot so that the drawing fits within a specified area. To get the correct dash lengths for hidden, center, or other lines, you must adjust the LTSCALE factor. For example, if you want to plot the previously mentioned drawing in a 45" by 34" area, the correction factor is:

    Correction factor   = 48/45
                                   = 1.0666
    New LTSCALE factor   = LTSCALE factor x Correction factor
                                   = 96 x 1.0666
                                   = 102.4

New Ltscale Factor for PLOTTING = Drawing Scale x Correction Factor

Note
*If you change the **LTSCALE** factor, all lines are affected by the new ratio.*

ALTERNATE LINETYPES

One of the problems with the LTSCALE factor is that it affects all the lines in the drawing. As shown in Figure 3-10(a), the length of each segment in all DASH3DOT type lines is

Creating Linetypes and Hatch Patterns   3-13

approximately equal, no matter how long the lines. You might want to have a small segment length if the lines are small and a longer segment length if the lines are long. You can accomplish this by using CELTSCALE (discussed later in this chapter) or by defining an alternate linetype with a different segment length. For example, you can define a linetype DASH3DOT and DASH3DOTX with different line pattern specifications.

```
*DASH3DOT,____ . . . ____ . . . ____ . . . ____
A,0.5,-.125,0,-.125,0,-.125,0,-.125
*DASH3DOTX,_____  . . .  _____
A,1.0,-.25,0,-.25,0,-.25,0,-.25
```

In DASH3DOT linetype the segment length is one unit, whereas in DASH3DOTX linetype the segment length is two units. You can have several alternate linetypes to produce the lines with different segment lengths. Figure 3-10(b) shows the lines generated by DASH3DOT and DASH3DOTX.

Note
Although you might have used different linetypes with different segment lengths, the lines will be affected equally when you change the LTSCALE factor. For example, if the LTSCALE factor is 0.5, the segment length of DASH3DOT line will be 0.5 and the segment length of DASH3DTX will be 1.0 units.

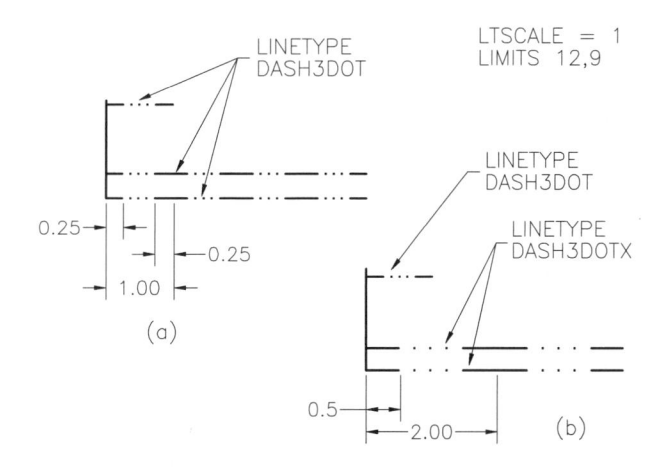

Figure 3-10 Linetypes generated by DASH3DOT and DASH3DOTX

MODIFYING LINETYPES

You can also modify the linetypes that are defined in the ACAD.LIN file. You must save a copy of the original ACAD.LIN file before making any changes to it. You need a text editor, such as Notepad, to modify the linetype. You can also use the EDIT function of DOS, or AutoCAD's **EDIT** command (provided the ACAD.PGP file is present and EDIT is defined in the file). For example, if you want to change the dash length of the border linetype from 0.5 to 0.75, load the file, then edit the pattern line of the border linetype. The following file is a partial listing of the ACAD.LIN file after changing the border and centerx2 linetypes.

;; AutoCAD Linetype Definition file, Version 2.0
;; Copyright 1991, 1992, 1993, 1994, 1996 by Autodesk, Inc.
;;
***BORDER,__ __ . __ __ . __ __ . __ __ . __ __ . __ __ . __ __ .**
A,.75,-.25,.75,-.25,0,-.25
*BORDER2,Border (.5x) __.__.__.__.__.__.__.__.__.__.
A,.25,-.125,.25,-.125,0,-.125
*BORDERX2,Border (2x) ____ ____ . ____ ____ . ____
A,1.0,-.5,1.0,-.5,0,-.5

*CENTER,Center ____ _ ____ _ ____ _ ____ _ ____
A,1.25,-.25,.25,-.25
*CENTER2,Center (.5x) ___ _ ___ _ ___ _ ___ _ ___ _ ___
A,.75,-.125,.125,-.125
***CENTERX2,Center (2x) _____ __ _____ __ ____**
A,3.5,-.5,.5,-.5

*DASHDOT,Dash dot __ . __ . __ . __ . __ . __ . __
A,.5,-.25,0,-.25
*DASHDOT2,Dash dot (.5x) _._._._._._._._._._._._.
A,.25,-.125,0,-.125
*DASHDOTX2,Dash dot (2x) ____ . ____ . ____ . ____
A,1.0,-.5,0,-.5

*DASHED,Dashed __ __ __ __ __ __ __ __ __ __ _
A,.5,-.25
*DASHED2,Dashed (.5x) _ _ _ _ _ _ _ _ _ _ _ _ _ _
A,.25,-.125
*DASHEDX2,Dashed (2x) ____ ____ ____ ____ ____ __
A,1.0,-.5

*DIVIDE,Divide ____ . . ____ . . ____ . . ____ . . ____
A,.5,-.25,0,-.25,0,-.25
*DIVIDE2,Divide (.5x) __..__..__..__..__..__..__..__.._
A,.25,-.125,0,-.125,0,-.125
*DIVIDEX2,Divide (2x) _____ . . _____ . . _
A,1.0,-.5,0,-.5,0,-.5

*DOT,Dot . . . .
A,0,-.25
*DOT2,Dot (.5x) ....
A,0,-.125
*DOTX2,Dot (2x) .   .   .   .
A,0,-.5

*HIDDEN,Hidden __ __ __ __ __ __ __ __ __ __ __ __
A,.25,-.125
*HIDDEN2,Hidden (.5x) _ _ _ _ _ _ _ _ _ _ _ _ _ _ _ _

Creating Linetypes and Hatch Patterns 3-15

```
A,.125,-.0625
*HIDDENX2,Hidden (2x) ____  ____  ____  ____  ____  ____  ____
A,.5,-.25

*PHANTOM,Phantom _____  __  __  _____  __  __  _____
A,1.25,-.25,.25,-.25,.25,-.25
*PHANTOM2,Phantom (.5x) ___ _ _ ___ _ _ ___ _ _ ___ _ _
A,.625,-.125,.125,-.125,.125,-.125
*PHANTOMX2,Phantom (2x) _____   ____   ____   _
A,2.5,-.5,.5,-.5,.5,-.5
;;  ISO 128 (ISO/DIS 12011) linetypes
;;  The size of the line segments for each defined ISO line, is
;;  defined for a usage with a pen width of 1 mm. To use them with
;;  the other ISO predefined pen widths, the line has to be scaled
;;  with the appropriate value (e.g. pen width 0,5 mm -> ltscale 0.5).
;;
*ACAD_ISO02W100,ISO dash __ __ __ __ __ __ __ __ __ __ __ __
A,12,-3
*ACAD_ISO04W100,ISO long-dash dot ____ . ____ . ____ . ____ . _
A,24,-3,.5,-3

*ACAD_ISO14W100,ISO dash triple-dot __ . . . __ . . . __ . . . _
A,12,-3,.5,-3,.5,-3,.5,-3
*ACAD_ISO15W100,ISO double-dash triple-dot __ __ . . . __ __ . .
A,12,-3,12,-3,.5,-3,.5,-3,.5,-3
;;  Complex linetypes

;;  Complex linetypes have been added to this file.
;;  These linetypes were defined in LTYPESHP.LIN in
;;  Release 13, and are incorporated in ACAD.LIN in Release 14

;;  These linetype definitions use LTYPESHP.SHX.
;;
*FENCELINE1,Fenceline circle ——0——0——0——0——0——0——
A,.25,-.1,[CIRC1,ltypeshp.shx,x=-.1,s=.1],-.1,1
*FENCELINE2,Fenceline square ——[]——[]——[]——[]——[]——
A,.25,-.1,[BOX,ltypeshp.shx,x=-.1,s=.1],-.1,1
*TRACKS,Tracks -|-|-|-|-|-|-|-|-|-|-|-|-|-|-|-|-|-|-|-|-|-|-|-
A,.15,[TRACK1,ltypeshp.shx,s=.25],.15
*BATTING,Batting SSSSSSSSSSSSSSSSSSSSSSSSSSSSSSSSSSSSSSSSSS
A,.0001,-.1,[BAT,ltypeshp.shx,x=-.1,s=.1],-.2,[BAT,ltypeshp.shx,r=180,x=.1,s=.1],-.1
*HOT_WATER_SUPPLY,Hot water supply —— HW —— HW —— HW ——
A,.5,-.2,["HW",STANDARD,S=.1,R=0.0,X=-0.1,Y=-.05],-.2
*GAS_LINE,Gas line ——GAS——GAS——GAS——GAS——GAS——GAS—
A,.5,-.2,["GAS",STANDARD,S=.1,R=0.0,X=-0.1,Y=-.05],-.25
*ZIGZAG,Zig zag /\/\/\/\/\/\/\/\/\/\/\/\/\/\/\/\
A,.0001,-.2,[ZIG,ltypeshp.shx,x=-.2,s=.2],-.4,[ZIG,ltypeshp.shx,r=180,x=.2,s=.2],-.2
```

Example 2

Create a new file, **NEWLINET.LIN**, and define a linetype, VARDASH, with the following specifications:

    Length of first dash 1.0
    Blank space 0.25
    Length of second dash 0.75
    Blank space 0.25
    Length of third dash 0.5
    Blank space 0.25
    Dot
    Blank space 0.25
    Length of next dash 0.5
    Blank space 0.25
    Length of next dash 0.75

Step 1: Writing definition of linetype

Use a text editor and insert the following lines that define the new linetype **VARDASH**.

```
*VARDASH,——— —— — . —— ————
A,1,-.25,.75,-.25,.5,-.25,0,-.25,.5,-.25,.75,-.25
```

Step 2: Loading the linetype

You can use the **LINETYPE** command to load the linetype or choose Linetype in the **Format** pulldown menu. The type of lines that this linetype will generate are shown in Figure 3-11.

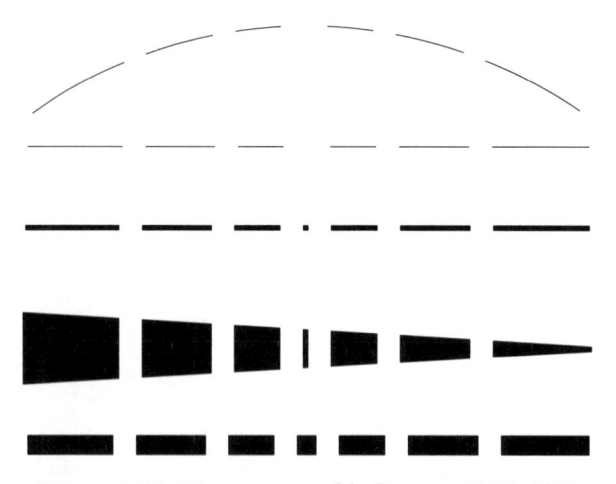

Figure 3-11 *Lines generated by linetype VARDASH*

CURRENT LINETYPE SCALING (CELTSCALE)

Like **LTSCALE**, the **CELTSCALE** system variable controls the linetype scaling. The difference is that **CELTSCALE** determines the current linetype scaling. For example, if you set the **CELTSCALE** to 0.5, all lines drawn after setting the new value for **CELTSCALE** will have the

Creating Linetypes and Hatch Patterns 3-17

linetype scaling factor of 0.5. The value is retained in the **CELTSCALE** system variable. The first line (a) in Figure 3-12 is drawn with the **CELTSCALE** factor of 1 and the second line (b) is drawn with the **CELTSCALE** factor of 0.5. The length of the dashes is reduced by a factor of 0.5 when the **CELTSCALE** is 0.5.

The **LTSCALE** system variable controls the global scale factor. For example, if **LTSCALE** is set to 2, all lines in the drawing will be affected by a factor of 2. The net scale factor is equal to the product of **CELTSCALE** and **LTSCALE**. Figure 3-12(c) shows a line that is drawn with **LTSCALE** of 2 and **CELTSCALE** of 0.25. The net scale factor is = **LTSCALE** x **CELTSCALE** = 2 x 0.25 = 0.5.

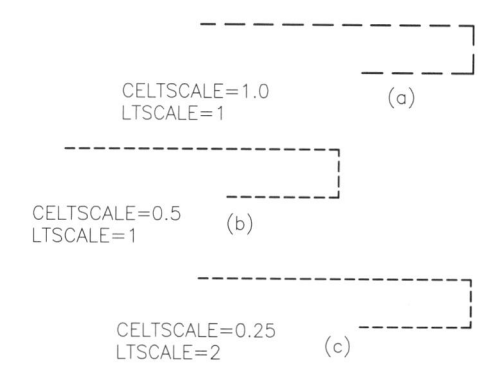

*Figure 3-12 Using **CELTSCALE** to control current linetype scaling*

> **Note**
> *You can change the current linetype scale factor of a line by using the **Properties** dialog box that can be invoked by choosing the **Properties** tool in the **Standard** toolbar. You can also use the **CHANGE** command and then select the ltScale option.*

COMPLEX LINETYPES

AutoCAD has provided a facility to create complex linetypes. The complex linetypes can be classified into two groups: string complex linetype and shape complex linetype. The difference between the two is that the string complex linetype has a text string inserted in the line, whereas the shape complex linetype has a shape inserted in the line. The facility of creating complex linetypes increases the functionality of lines. For example, if you want to draw a line around a building that indicates the fence line, you can do it by defining a complex linetype that will automatically give you the desired line with the text string (Fence). Similarly, you can define a complex linetype that will insert a shape (symbol) at predefined distances along the line.

Creating a String Complex Linetype

When writing the definition of a string complex linetype, the actual text and its attributes must be included in the linetype definition. The format of the string complex linetype is:

   ["String", Text Style, Text Height, Rotation, X-Offset, Y-Offset]

String. It is the actual text that you want to insert along the line. The text string must be enclosed in quotation marks (" ").

Text Style. This is the name of the text style file that you want to use for generating the text string. The text style must be predefined.

Text Height. This is the actual height of the text, if the text height defined in the text style is 0. Otherwise, it acts as a scale factor for the text height specified in the text style. In Figure 3-13, the height of the text is 0.1 units.

Rotation. The rotation can be specified as an absolute or relative angle. In absolute rotation the angle is always measured with respect to the positive X axis, no matter what AutoCAD's direction setting. The absolute angle is represented by letter "a".

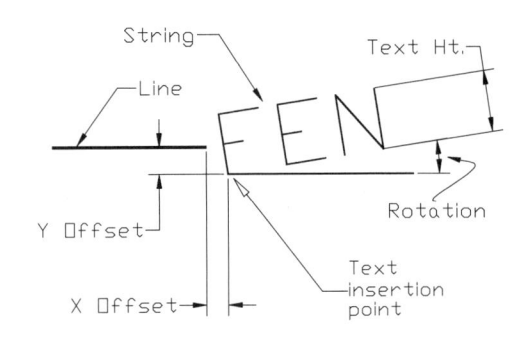

Figure 3-13 The attributes of a string complex linetype

In relative rotation the angle is always measured with respect to orientation of dashes in the linetype. The relative angle is represented by the letter "r". The angle can be specified in radians (r), grads (g), or degrees (d). The default is degrees.

X-Offset. This is the distance of the lower left corner of the text string from the endpoint of the line segment measured along the line. If the line is horizontal, then the X-Offset distance is measured along the X axis. In Figure 3-13, the X-Offset distance is 0.05.

Y-Offset. This is the distance of the lower left corner of the text string from the endpoint of the line segment measured perpendicular to the line. If the line is horizontal, then the Y-Offset distance is measured along the Y axis. In Figure 3-13, the Y-Offset distance is -0.05. The distance is negative because the start point of the text string is 0.05 units below the endpoint of the first line segment.

Example 3

In the following example, you will write the definition of a string complex linetype that consists of the text string "Fence" and line segments. The length of each line segment is 0.75. The height of the text string is 0.1 units, and the space between the end of the text string and the following line segment is 0.05 (Figure 3-14).

Step 1: Determining the line specifications

Before writing the definition of a new linetype, it is important to determine the line specification. One of the ways this can be done is to actually draw the lines and the text the way you want them to appear in the drawing. Once you have drawn the line and the text to your satisfaction, measure the distances needed to define the string complex linetype. In this example, the values are given as follows:

Text string=	Fence
Text style=	Standard
Text height=	0.1
Text rotation=	0
X-Offset=	0.05

Creating Linetypes and Hatch Patterns 3-19

Y-Offset= -0.05
Length of the first line segment= 0.75
Distance between the line segments= 0.575

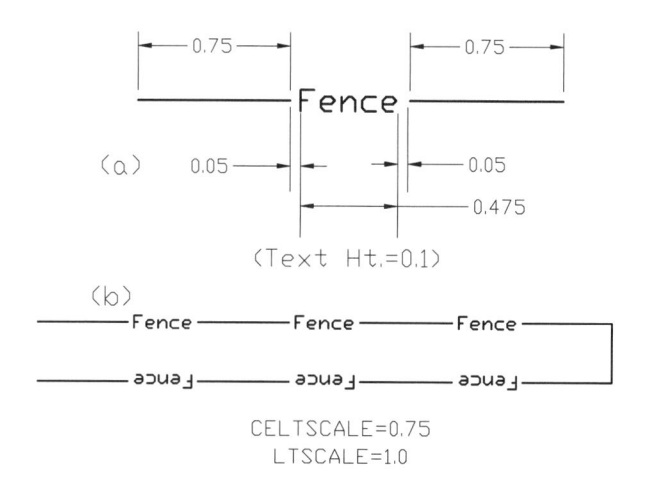

Figure 3-14 The attributes of a string complex linetype and line specifications for Example 3

Step 2: Writing the definition of string complex linetype
Use a text editor to write the definition of the string complex linetype. You can add the definition to the AutoCAD **ACAD.LIN** file or create a separate file. The extension of the file must be **.LIN**. The following file is the listing of the **FENCE.LIN** file for Example 3. The name of the linetype is NEWFence1.

*NEWFence1,New fence boundary line
A,0.75,["Fence",Standard,S=0.1,A=0,X=0.05,Y=-0.05],-0.575
or
A,0.75,-0.05,["Fence",Standard,S=0.1,A=0,X=0,Y=-0.05],-0.525

Step 3: Loading the linetype
You can use the **LINETYPE** command to load the linetype or choose Linetype in the **Format** pulldown menu. Draw a line or any object to check if the line is drawn to the given specifications as shown in the Figure 3-15. Notice that the text is always drawn along the X axis. Also, when you draw a line at an angle, polyline, circle, or spline, the text string does not align with the object (Figure 3-15).

Step 4: Aligning the text with the line
In the NEWFence linetype definition, the specified angle is 0 degrees (Absolute angle A = 0). Therefore, when you use the NEWFence linetype to draw a line, circle, polyline, or spline, the text string (Fence) will be at zero degrees. If you want the text string (Fence) to align with the polyline (Figure 3-16), spline, or circle, specify the angle as relative angle (R = 0) in the

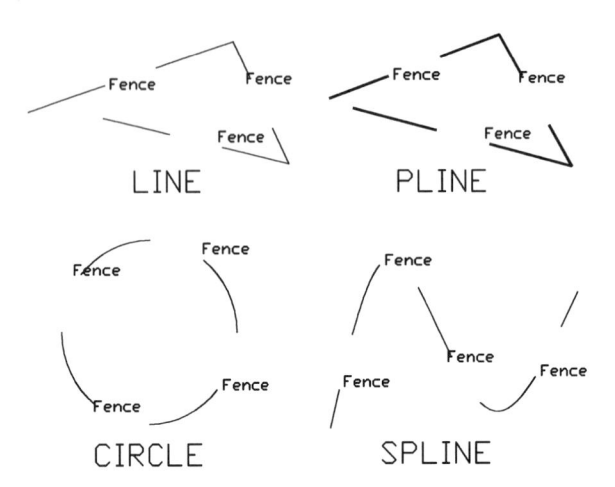

Figure 3-15 *Using string complex linetype with angle A=0*

NEWFence linetype definition. The following is the linetype definition for NEWFence linetype with relative angle R = 0:

    *NEWFence2,New fence boundary line
    A,0.75,["Fence",Standard,S=0.1,R=0,X=0.05,Y=-0.05],-0.575

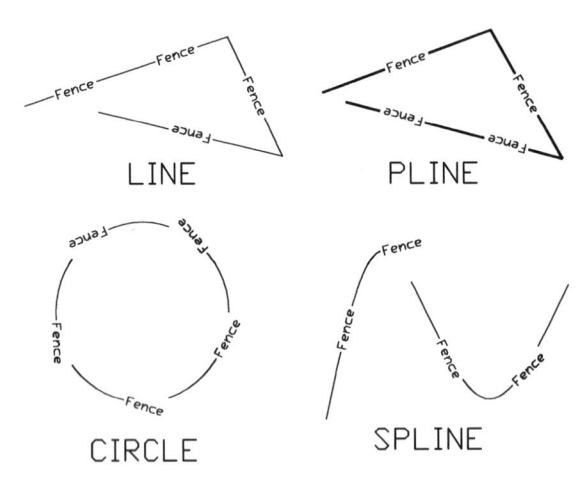

Figure 3-16 *Using a string complex linetype with angle R = 0*

Step 5: Aligning the midpoint of text with the line

In Figure 3-16, you might notice that the text string is not properly aligned with the circumference of the circle. This is because AutoCAD draws the text string in a direction that is tangent to the circle at the text insertion point. To resolve this problem, you must define the middle point of the text string as the insertion point. Also, the line specifications should be

Creating Linetypes and Hatch Patterns 3-21

measured accordingly. Figure 3-17 gives the measurements of the NEWFence linetype with the middle point of the text as the insertion point.

The following is the linetype definition for NEWFence linetype:

*NEWFence3,New fence boundary line
A,0.75,-0.287,["FENCE",Standard,S=0.1,X=-0.237,Y=-0.05],-0.287

Note
If no angle is defined in the line definition, it defaults to angle R = 0. Also, the text does not automatically insert to its midpoint like the regular text with MID justification.

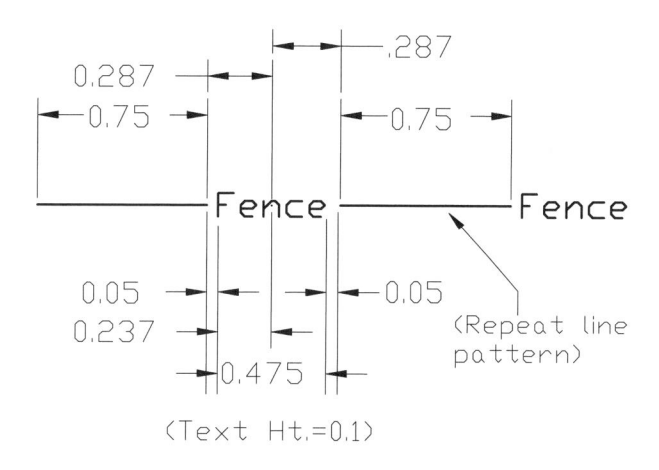

Figure 3-17 Specifications of a string complex linetype with the middle point of the text string as the text insertion point

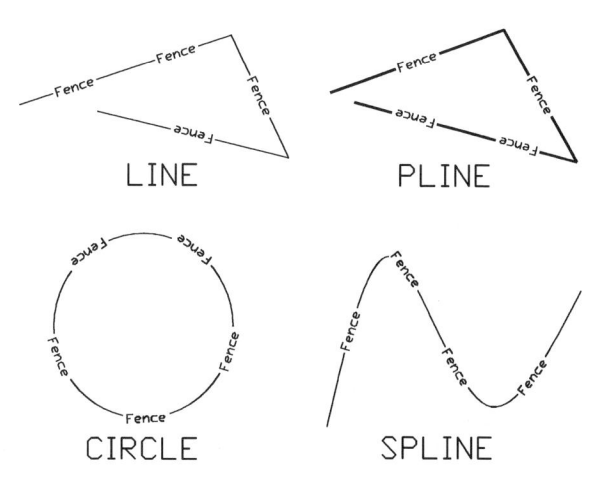

Figure 3-18 Using a string complex linetype with the middle point of the text string as the text insertion point

Creating a Shape Complex Linetype

As with the string complex linetype, when you write the definition of a shape complex linetype, the name of the shape, the name of the shape file, and other shape attributes, like rotation, scale, X-Offset, and Y-Offset, must be included in the linetype definition. The format of the shape complex linetype is:

[Shape Name, Shape File, Scale, Rotation, X-Offset, Y-Offset]

The following is the description of the attributes of Shape Complex Linetype (Figure 3-19).

Shape Name. This is the name of the shape that you want to insert along the line. The shape name must exist; otherwise, no shape will be generated along the line.

Shape File. This is the name of the compiled shape file (.SHX) that contains the definition of the shape being inserted in the line. The name of the subdirectory where the shape file is located must be in the ACAD search path. The shape files (.SHP) must be compiled before using the SHAPE command to load the shape.

Scale. This is the scale factor by which the defined shape size is to be scaled. If the scale is 1, the size of the shape will be the same as defined in the shape definition (.SHP file).

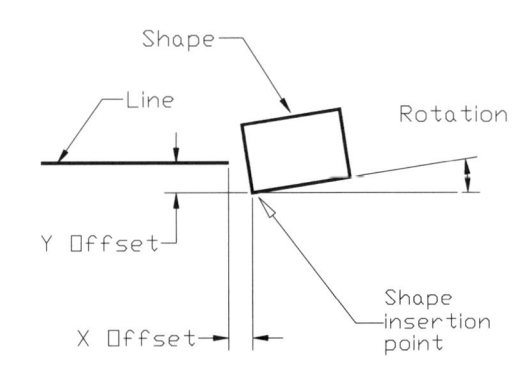

Figure 3-19 *The attributes of a shape complex linetype*

Rotation. The rotation can be specified as an absolute or relative angle. In absolute rotation, the angle is always measured with respect to the positive X axis, no matter what AutoCAD's direction setting. The absolute angle is represented by letter "a." In relative rotation, the angle is always measured with respect to the orientation of dashes in the linetype. The relative angle is represented by the letter "r." The angle can be specified in radians (r), grads (g), or degrees (d). The default is degrees.

X-Offset. This is the distance of the shape insertion point from the endpoint of the line segment measured along the line. If the line is horizontal, then the X-Offset distance is measured along the X axis. In Figure 3-19, the X-Offset distance is 0.2.

Y-Offset. This is the distance of the shape insertion point from the endpoint of the line segment measured perpendicular to the line. If the line is horizontal, then the Y-Offset distance is measured along the Y axis. In Figure 3-19, the Y-Offset distance is 0.

Example 4

In the following example, you will write the definition of a shape complex linetype that consists of the shape (Manhole; the name of the shape is MH) and a line. The scale of the

Creating Linetypes and Hatch Patterns 3-23

shape is 0.1, the length of each line segment is 0.75, and the space between line segments is 0.2.

Step 1: Determining the line specifications
Before writing the definition of a new linetype, it is important to determine the line specifications. One of the ways this can be done is to actually draw the lines and the shape the way you want them to appear in the drawing (Figure 3-20). Once you have drawn the line and the shape to your satisfaction, measure the distances needed to define the shape complex linetype. In this example, the values are as follows:

Shape name   MH
Shape file name MHOLE.SHX   (Name of the compiled shape file.)
Scale   0.1
Rotation   0
X-Offset   0.2
Y-Offset   0
Length of the first line segment = 0.75
Distance between the line segments = 0.2

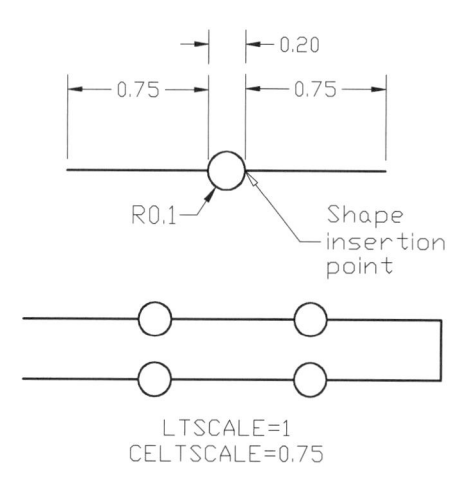

Figure 3-20 The attributes of the shape complex linetype and line specifications for Example 4

Step 2: Writing the definition of the shape
Use a text editor to write the definition of the shape file. The extension of the file must be **.SHP**. The following file is the listing of the **MHOLE.SHP** file for Example 4. The name of the shape is MH. (For details, see Chapter 11, "Shape and Text Fonts.")

```
*215,9,MH
001,10,(1,007),
001,10,(1,071),0
```

Step 3: Compiling the shape
Use the **COMPILE** command to compile the shape file (**.SHP** file). When you use this

command, AutoCAD will prompt you to enter the name of the shape file (Figure 3-21). For this example, the name is **MHOLE.SHP**. The following is the command sequence for compiling the shape file:

Command: **COMPILE**
Enter shape (.SHP) or PostScript font (.PFB) file name: **MHOLE**

You can also compile the shape from the command line by setting **FILEDIA**=0 and then using the **COMPILE** command. The .SHX files must be located in AutoCAD's search path.

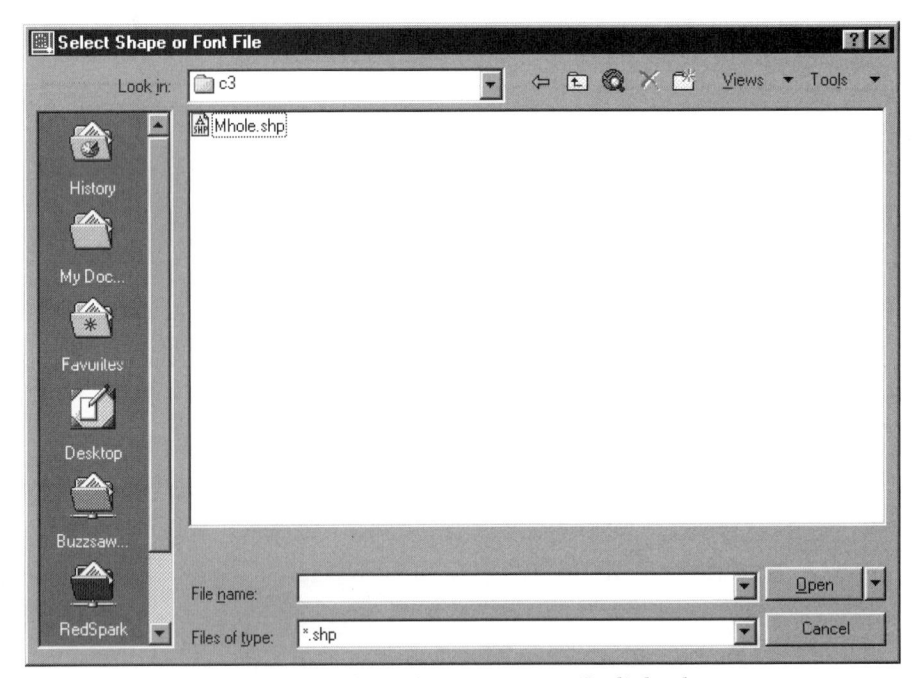

Figure 3-21 Select Shape or Font File dialog box

Step 4: Writing the definition of the shape complex linetype

Use a text editor to write the definition of the shape complex linetype. You can add the definition to the AutoCAD ACAD.LIN file or create a separate file. The extension of the file must be .LIN. The following file is the listing of the **MHOLE.LIN** file for Example 4. The name of the linetype is MHOLE.

```
*MHOLE,Line with Manholes
A,0.75,[MH,MHOLE.SHX,S=0.10,X=0.2,Y=0],-0.2
```

Step 5: Loading the linetype

To test the linetype, load the linetype using the **LINETYPE** command or choose Linetype in the **Format** pulldown menu. Assign the linetype to a layer. Draw a line or any object to check if the line is drawn to the given specifications. The shape is drawn upside down when you draw a line from right to left. Figure 3-22 shows the execution of the linetype **MHOLE.LIN** using line, pline, circle etc.

Creating Linetypes and Hatch Patterns 3-25

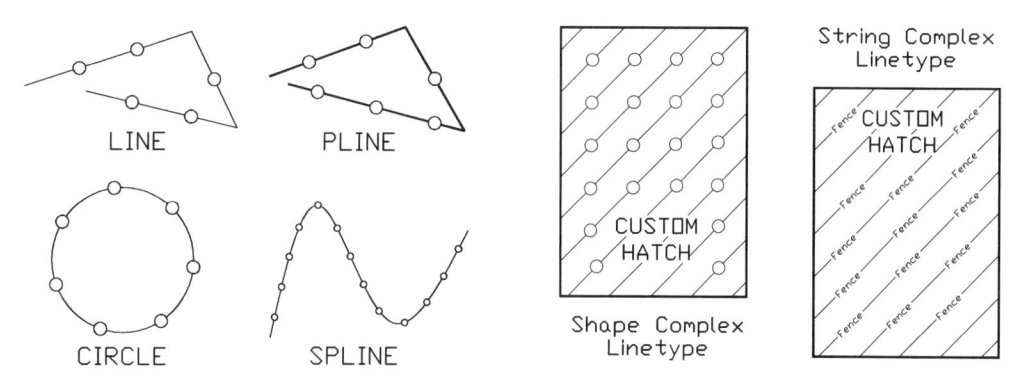

Figure 3-22 Using a shape complex linetype

Figure 3-23 Using shape and string complex linetypes to create custom hatch

HATCH PATTERN DEFINITION

The AutoCAD software comes with a hatch pattern library file, **ACAD.PAT**, that contains 67 hatch patterns. These hatch patterns are sufficient for general drafting work. However, if you need a different hatch pattern, AutoCAD lets you create your own. There is no limit to the number of hatch patterns you can define.

The hatch patterns you define can be added to the hatch pattern library file, **ACAD.PAT**. You can also create a new hatch pattern library file, provided the file contains only one hatch pattern definition, and the name of the hatch is the same as the name of the file. The hatch pattern definition consists of the following two parts: **header line and hatch descriptors**.

Header Line

The **header line** consists of an asterisk (*) followed by the name of the hatch pattern. The hatch name is the name used in the hatch command to hatch an area. After the name, you can give the hatch description, which is separated from the hatch name by a comma (,). The general format of the header line is:

    *HATCH Name [, Hatch Description]
        Where ***** ------------------------------- Asterisk
                 HATCH Name ----------- Name of hatch pattern
                 Hatch Description ------ Description of hatch pattern

The description can be any text that describes the hatch pattern. It can also be omitted, in which case, a comma should not follow the hatch pattern name.

    Example
    *DASH45, Dashed lines at 45 degrees
        Where **DASH45** --------- Hatch name
                 Dashed lines at 45 degrees ------ Hatch description

Hatch Descriptors

The **hatch descriptors** consist of one or more lines that contain the definition of the hatch lines. The general format of the hatch descriptor is:

Angle, X-origin, Y-origin, D1, D2 [,Dash Length.....]

Where **Angle** ------------ Angle of hatch lines
X-origin --------- X coordinate of hatch line
Y-origin --------- Y coordinate of hatch line
D1 ---------------- Displacement of second line (Delta-X)
D2 ---------------- Distance between hatch lines (Delta-Y)
Length ----------- Length of dashes and spaces (Pattern line definition)

Example
45,0,0,0,0.5,0.5,-0.125,0,-0.125

Where **45** ----------------- Angle of hatch line
0 ------------------- X-Origin
0 ------------------- Y-Origin
0 ------------------- Delta-X
0.5 ---------------- Delta-Y
0.5 ---------------- Dash (pen down)
-0.125 ------------ Space (pen up)
0 ------------------- Dot (pen down)
-0.125 ------------ Space (pen up)
0.5,-0.125,0,-0.125 Pattern line definition

Hatch Angle

X-origin and Y-origin. The hatch angle is the angle that the hatch lines make with the positive X axis. The angle is positive if measured counterclockwise (Figure 3-24), and negative if the angle is measured clockwise. When you draw a hatch pattern, the first hatch line starts from the point defined by X-origin and Y-origin. The remaining lines are generated by offsetting the first hatch line by a distance specified by delta-X and delta-Y. In Figure 3-25(a), the first hatch line starts from the point with the coordinates X = 0 and Y = 0. In Figure 3-25(b) the first line of hatch starts from a point with the coordinates X = 0 and Y = 0.25.

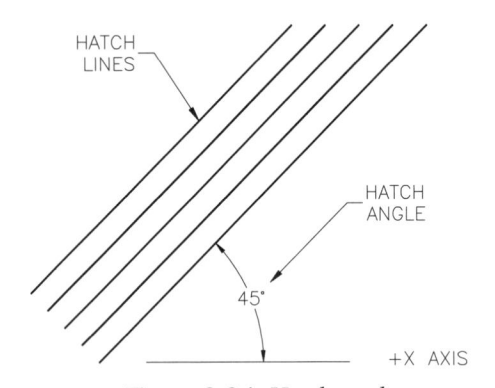

Figure 3-24 Hatch angle

Delta-X and Delta-Y. Delta-X is the displacement of the offset line in the direction in which the hatch lines are generated. For example, if the lines are drawn at a 0-degree angle and delta-X = 0.5, the offset line will be displaced by a distance delta-X (0.5) along the 0-angle direction. Similarly, if the hatch lines are drawn at a 45-degree angle, the offset line will be displaced by a distance delta-X (0.5) along a 45-degree direction (Figure 3-26). Delta-Y is the

Creating Linetypes and Hatch Patterns 3-27

displacement of the offset lines measured perpendicular to the hatch lines. For example, if delta-Y = 1.0, the space between any two hatch lines will be 1.0 (Figure 3-26).

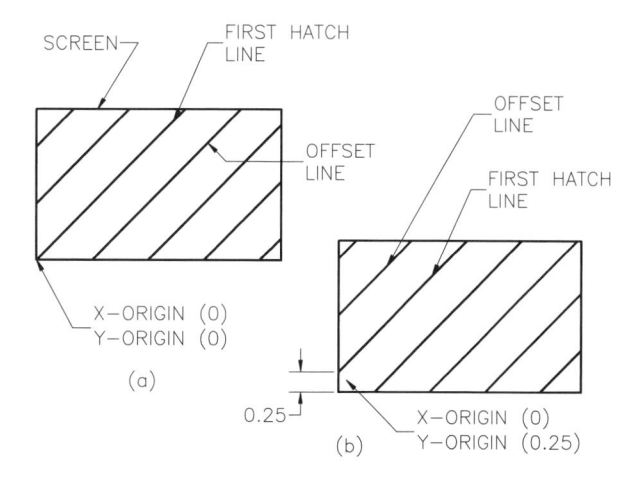

Figure 3-25 X-origin and Y-origin of hatch lines

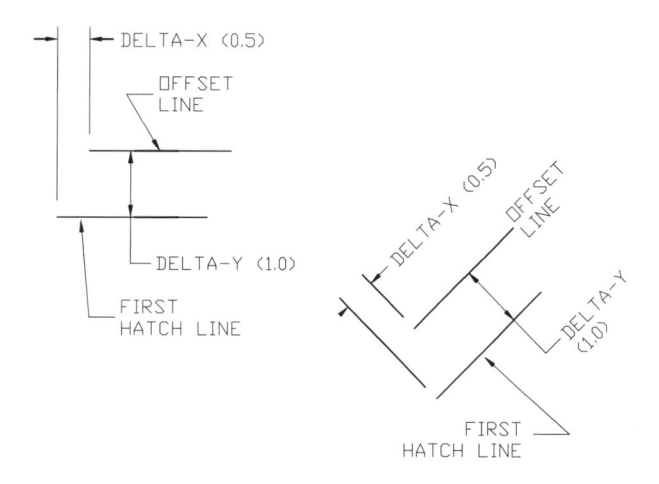

Figure 3-26 Delta-X and delta-Y of hatch lines

HOW HATCH WORKS

When you hatch an area, AutoCAD generates an infinite number of hatch lines of infinite length. The first hatch line always passes through the point specified by the X-origin and Y-origin. The remaining lines are generated by offsetting the first hatch line in both directions. The offset distance is determined by delta-X and delta-Y as shown in Figure 3-26. All selected entities that form the boundary of the hatch area are then checked for intersection with these lines. Any hatch lines found within the defined hatch boundaries are turned on, and the hatch lines outside the hatch boundary are turned off, as shown in Figure 3-27. Since the hatch lines are generated by offsetting, the hatch lines in different areas of the drawing are automatically aligned relative to the drawing's snap origin. Figure 3-27(a) shows the hatch lines as computed by AutoCAD. These lines are not drawn on the screen; they are shown here for illustration only. Figure 3-27(b) shows the hatch lines generated in the circle that was defined as the hatch boundary.

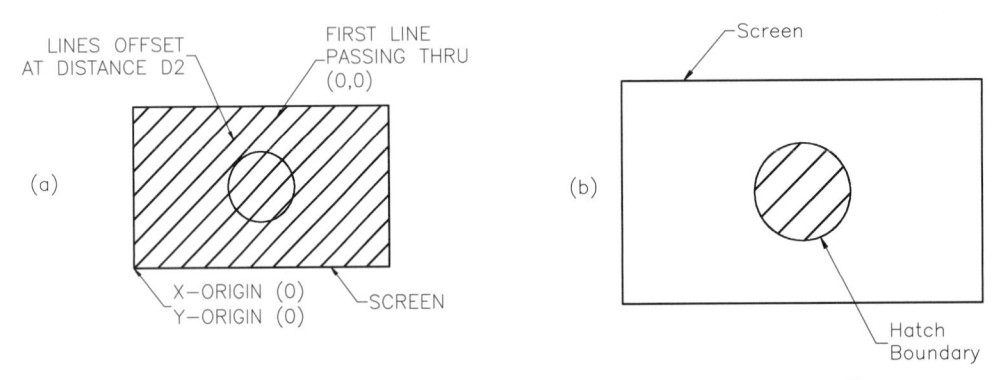

Figure 3-27 Hatch lines outside the hatch boundary are turned off

SIMPLE HATCH PATTERN

It is good practice to develop the hatch pattern specification before writing a hatch pattern definition. For simple hatch patterns it may not be that important, but for more complicated hatch patterns you should know the detailed specifications. Example 5 illustrates the procedure for developing a simple hatch pattern.

Example 5

Write a hatch pattern definition for the hatch pattern shown in Figure 3-28, with the following specifications:

Name of the hatch pattern =	HATCH1
X-Origin =	0
Y-Origin =	0
Distance between hatch lines =	0.5
Displacement of hatch lines =	0
Hatch line pattern =	Continuous

Step 1: Creating the hatch pattern file
This hatch pattern definition can be added to the existing **ACAD.PAT** hatch file. You can use any text editor (like Notepad) to write the file. Load the **ACAD.PAT** file that is located in **AutoCAD2002\SUPPORT** directory and insert the following two lines at the end of the file.

*HATCH1,Hatch Pattern for Example 5
45,0,0,0,.5

Creating Linetypes and Hatch Patterns 3-29

Where 45 ----------------- Hatch angle
       0 ------------------ X-origin
       0 ------------------ Y-origin
       0 ------------------ Displacement of second hatch line
       .5 ----------------- Distance between hatch lines

The first field of hatch descriptors contains the angle of the hatch lines. That angle is 45 degrees with respect to the positive X axis. The second and third fields describe the X and Y coordinates of the first hatch line origin. The first line of the hatch pattern will pass through this point. If the values of the X-origin and Y-origin were 0.5 and 1.0, respectively, then the first line would pass through the point with the X coordinate of 0.5 and the Y coordinate of 1.0, with respect to the drawing origin 0,0. The remaining lines are generated by offsetting the first line, as shown in Figure 3-28.

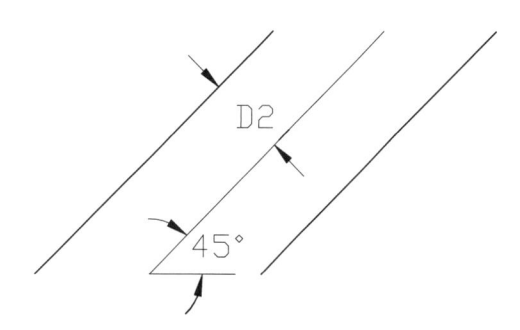

Figure 3-28 Hatch pattern angle and offset distance

Step 2: Loading the hatch pattern

Choose the **Hatch** button from the **Draw** toolbar or choose **Hatch** from the **Draw** menu to display the **Boundary Hatch** dialog box. Make sure Predefined is selected in the **Type:** edit box. Select the hatch pattern name from the drop-down list or choose the **[...]** button adjacent to the **Pattern:** drop-down list to display the **Hatch Pattern Palette** dialog box. Select the hatch pattern file displayed there. Then choose **OK** to display the **Boundary Hatch** dialog box again. Change the **Scale** and **Angle**, if needed, and then hatch an a circle to test the hatch pattern.

The **Boundary Hatch** dialog box can also be invoked by entering **BHATCH** at the Command prompt. Hatching can also be achieved by entering **-HATCH** at the Command prompt.

EFFECT OF ANGLE AND SCALE FACTOR ON HATCH

When you hatch an area, you can alter the angle and displacement of hatch lines you have specified in the hatch pattern definition to get a desired hatch spacing. You can do this by entering an appropriate value for angle and scale factor in the AutoCAD **HATCH** command.

To understand how the angle and the displacement can be changed, hatch an area with the hatch pattern HATCH1 in Example 5. You will notice that the hatch lines have been generated according to the definition of hatch pattern HATCH1. Notice the effect of hatch angle and scale factor on the hatch. Figure 3-29(a) shows a hatch with a 0-degree angle and a scale factor of 1.0. If the angle is 0, the hatch will be generated with the same angle as defined in the hatch pattern definition (45 degrees in Example 5). Similarly, if the scale factor is 1.0, the distance between the hatch lines will be the same as defined in the hatch pattern definition.

Figure 3-29(b) shows a hatch that is generated when the hatch scale factor is 0.5. If you measure the distance between the successive hatch lines, it will be 0.5 x 0.5 = 0.25. Figures 3-29(c) and (d) show the hatch when the angle is 45 degrees and the scale factors are 1.0 and 0.5, respectively.

Scale and Angle can also be set by entering **-HATCH** at the Command prompt.

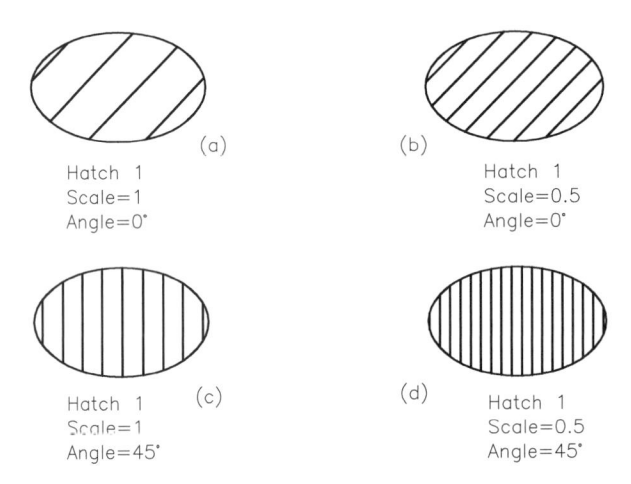

Figure 3-29 Effect of angle and scale factor on hatch

HATCH PATTERN WITH DASHES AND DOTS

The lines you can use in a hatch pattern definition are not restricted to continuous lines. You can define any line pattern to generate a hatch pattern. The lines can be a combination of dashes, dots, and spaces in any configuration. However, the maximum number of dashes you can specify in the line pattern definition of a hatch pattern is six. Example 6 uses a dash-dot line to create a hatch pattern.

Example 6

Write a hatch pattern definition for the hatch pattern shown in Figure 3-30, with the following specifications. Define a new path say C:\Program Files\Hatch1 and save the hatch pattern in that path.

Name of the hatch pattern	HATCH2
Hatch angle =	0
X-origin =	0
Y-origin =	0
Displacement of lines (D1) =	0.25
Distance between lines (D2) =	0.25
Length of each dash =	0.5
Space between dashes and dots =	0.125
Space between dots =	0.125

Creating Linetypes and Hatch Patterns 3-31

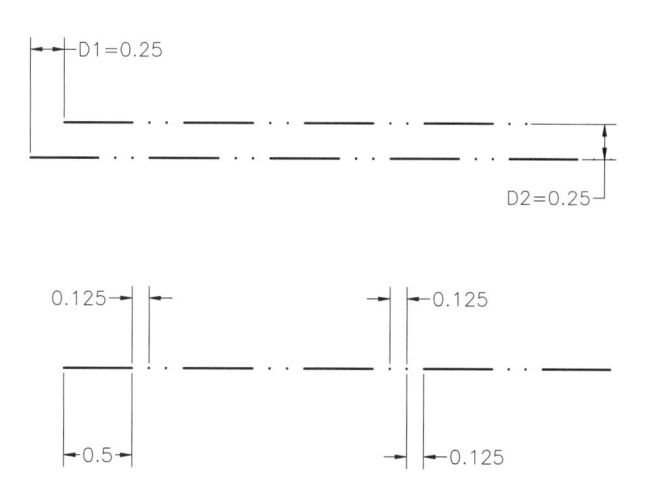

Figure 3-30 Hatch lines made of dashes and dots

Step 1: Writing the definition of a hatch pattern
You can use the AutoCAD **EDIT** command or any text editor (Notepad) to edit the **ACAD.PAT** file. The general format of the header line and the hatch descriptors is:

*HATCH NAME, Hatch Description
Angle, X-Origin, Y-Origin, D1, D2 [,Dash Length.....]

Substitute the values from Example 6 in the corresponding fields of the header line and field descriptor:

*HATCH2,Hatch with dashes and dots
0,0,0,0.25,0.25,0.5,-0.125,0,-0.125,0,-0.125
            Where 0 ------------------- Angle
                   0 ------------------- X-origin
                   0 ------------------- Y-origin
                   0.25 --------------- Delta-X
                   0.25 --------------- Delta-Y
                   0.5 ---------------- Length of dash
                   -0.125 ------------ Space (pen up)
                   0 ------------------- Dot (pen down)
                   -0.125 ------------ Space (pen up)
                   0 ------------------- Dot
                   -0.125 ------------ Space

Specifying a New Path for Hatch Pattern Files
When you enter a hatch pattern name for hatching, AutoCAD looks for that file name in the Support directory or the directory paths specified in AutoCAD. You can specify a new path and directory to store your hatch files.

Create a new folder **Hatch1** in C drive under the Program Files. Save the ACAD.PAT file with hatch pattern **HATCH2** definition in the same subdirectory, Hatch1. Right-click in the drawing area to activate the shortcut menu. Choose **Options** from the shortcut menu to display the **Options** dialog box as shown in Figure 3-31. The **Options** dialog box can also be invoked by choosing **Options** from the **Tools** menu or by directly entering **OPTIONS** at the Command prompt. Choose the **Files** tab in the **Options** dialog box to display the **Search paths, file names and file locations** area. Click on the **plus** sign of the **Support File Search Path** to display the different subdirectories of the **Support File Search Path**. Now choose the **Add** button to display the space to add a new subdirectory. Enter the location of the new subdirectory, C:\Program Files\Hatch1 or click on the **Browse** button to specify the path. Choose the **Apply** button and then choose **OK** to exit the dialog box. Now you have created a subdirectory and you have specified the search path for the hatch files.

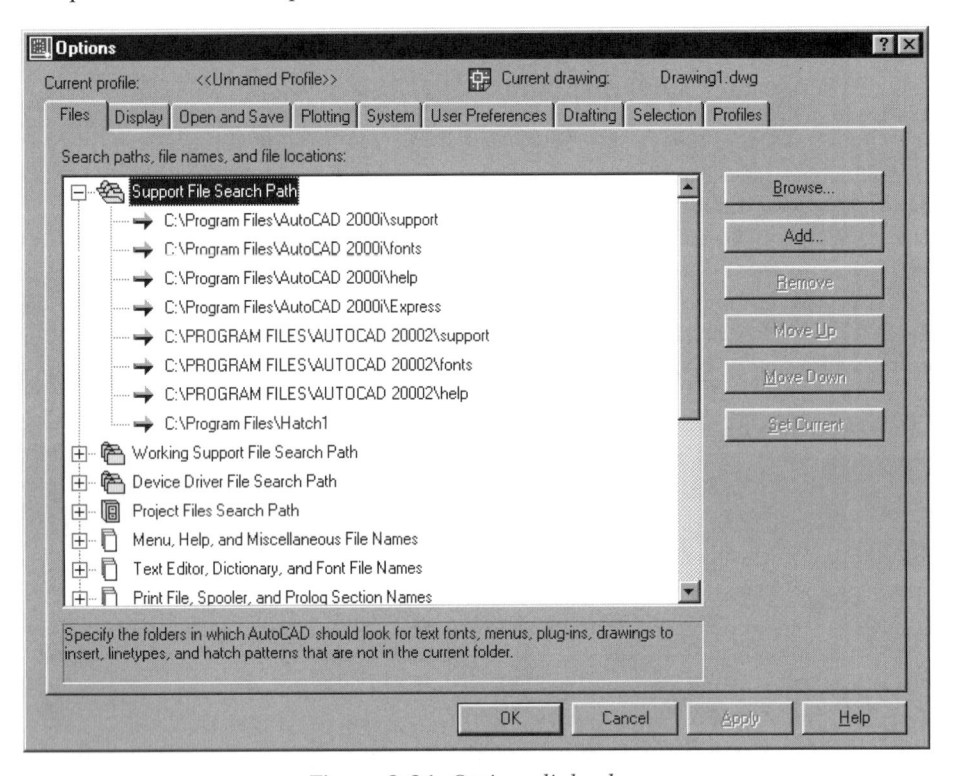

Figure 3-31 Options dialog box

Follow the procedure as described in Example 5 to activate the hatch pattern. The hatch that this hatch definition (Hatch2.pat) will generate is shown in Figure 3-32. Figure 3-32(a) shows the hatch with a 0-degree angle and a scale factor of 1.0. Figure 3-32(b) shows the hatch with a 45-degree angle and a scale factor of 0.5.

Creating Linetypes and Hatch Patterns 3-33

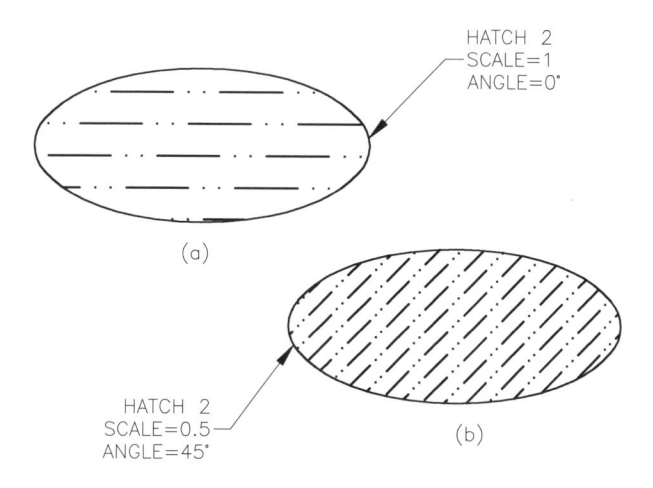

Figure 3-32 Hatch pattern at different angles and scales

HATCH WITH MULTIPLE DESCRIPTORS

Some hatch patterns require multiple lines to generate a shape. For example, if you want to create a hatch pattern of a brick wall, you need a hatch pattern that has four hatch descriptors to generate a rectangular shape. You can have any number of hatch descriptor lines in a hatch pattern definition. It is up to the user to combine them in any conceivable order. However, there are some shapes you cannot generate. A shape that has a nonlinear element, like an arc, cannot be generated by hatch pattern definition. However, you can simulate an arc by defining short line segments because you can use only straight lines to generate a hatch pattern. Example 7 uses three lines to define a triangular hatch pattern.

Example 7

Write a hatch pattern definition for the hatch pattern shown in Figure 3-33, with the following specifications:

Name of the hatch pattern =	HATCH3
Vertical height of the triangle =	0.5
Horizontal length of the triangle =	0.5
Vertical distance between the triangles =	0.5
Horizontal distance between the triangles =	0.5

Each triangle in this hatch pattern consists of the following three elements: a vertical line, a horizontal line, and a line inclined at 45 degrees.

Step 1: Defining specifications for vertical line

For the vertical line, the specifications are (Figure 3-34):

Figure 3-33 Triangle hatch pattern

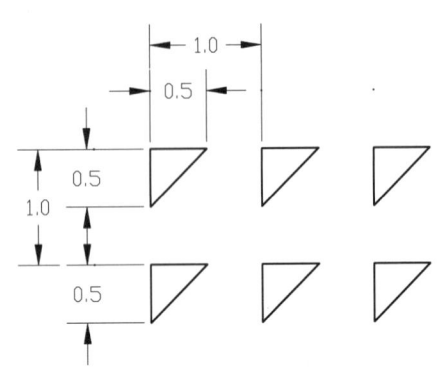

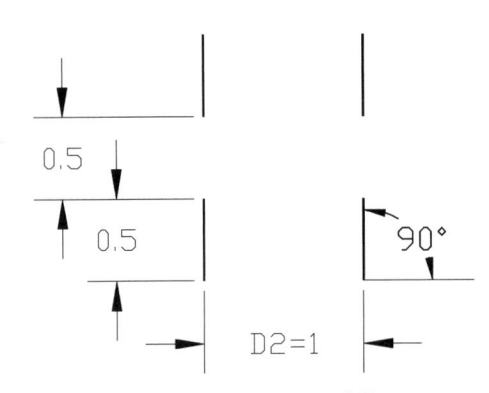

Figure 3-34 Vertical line

Hatch angle =	90 degrees
X-origin =	0
Y-origin =	0
Delta-X (D1) =	0
Delta-Y (D2) =	1.0
Dash length =	0.5
Space =	0.5

Substitute the values from the vertical line specification in various fields of the hatch descriptor to get the following line:

90,0,0,0,1,.5,-.5

```
Where  90 ----------------- Hatch angle
        0 ----------------- X-origin
        0 ----------------- Y-origin
        0 ----------------- Delta-X
        1 ----------------- Delta-Y
       .5 ----------------- Dash (pen down)
      -.5 ----------------- Space (pen up)
```

Step 2: Defining specifications of horizontal line

For the horizontal line (Figure 3-35), the specifications are:

Hatch angle =	0 degrees
X-origin =	0
Y-origin =	0.5
Delta-X (D1) =	0
Delta-Y (D2) =	1.0
Dash length =	0.5
Space =	0.5

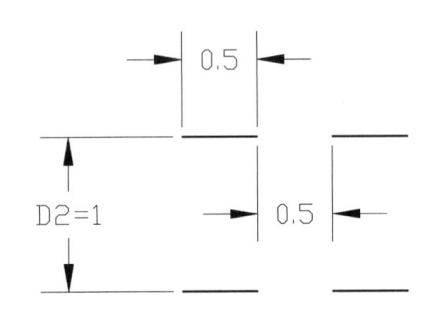

Figure 3-35 Horizontal line

Creating Linetypes and Hatch Patterns 3-35

The only difference between the vertical line and the horizontal line is the angle. For the horizontal line, the angle is 0 degrees, whereas for the vertical line, the angle is 90 degrees. Substitute the values from the vertical line specification to obtain the following line:

0,0,0.5,0,1,.5,-.5

Where **0** ------------------ Hatch angle
             0 ------------------ X-origin
             0.5 ---------------- Y-origin
             0 ------------------ Delta-X
             1 ------------------ Delta-Y
             .5 ----------------- Dash (pen down)
             -.5 ---------------- Space (pen up)

Step 3: Defining specifications of the inclined line

This line is at an angle; therefore, you need to calculate the distances delta-X (D1) and delta-Y (D2), the length of the dashed line, and the length of the blank space. Figure 3-36 shows the calculations to find these values.

Hatch angle =   45 degrees
X-Origin =   0
Y-Origin =   0
Delta-X (D1) =   0.7071
Delta-Y (D2) =   0.7071
Dash length =   0.7071
Space =   0.7071

After substituting the values in the general format of the hatch descriptor, you will obtain the following line:

45,0,0,.7071,.7071,.7071,-.7071

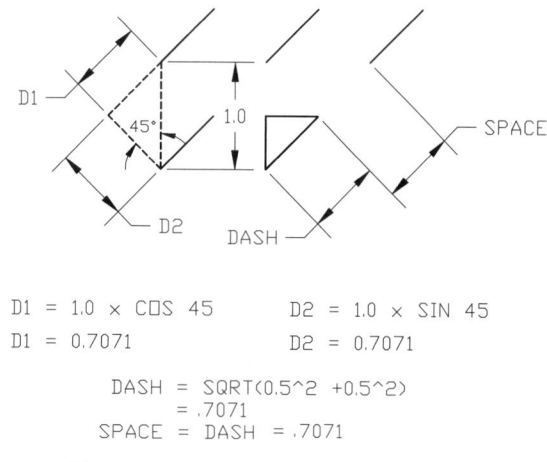

Figure 3-36 Line inclined at 45 degrees

Where 45 ----------------- Hatch angle
       0 ------------------ X-origin
       0 ------------------ Y-origin
       .7071 ------------- Delta-X
       .7071 ------------- Delta-Y
       .7071 ------------- Dash (pen down)
       -.7071 ------------ Space (pen up)

Step 4: Loading the hatch pattern

Now you can combine these three lines and insert them at the end of the **ACAD.PAT** file or you can enter the values in a separate hatch file and save it. You can also use the AutoCAD **EDIT** command to edit the file and insert the lines.

The following file is a partial listing of the **ACAD.PAT** file, after adding the hatch pattern definitions from Examples 5, 6, and 7.

```
*SOLID, Solid fill
45, 0,0, 0,.125
*angle,Angle steel
0, 0,0, 0,.275, .2,-.075
90, 0,0, 0,.275, .2,-.075
*ansi31,ANSI Iron, Brick, Stone masonry
45, 0,0, 0,.125
*ansi32,ANSI Steel
45, 0,0, 0,.375
45, .176776695,0, 0,.375
*ansi33,ANSI Bronze, Brass, Copper
45, 0,0, 0,.25
45, .176776695,0, 0,.25, .125,-.0625
*ansi34,ANSI Plastic, Rubber
45, 0,0, 0,.75
45, .176776695,0, 0,.75
45, .353553391,0, 0,.75
45, .530330086,0, 0,.75
*ansi35,ANSI Fire brick, Refractory material
45, 0,0, 0,.25
45, .176776695,0, 0,.25, .3125,-.0625,0,-.0625
*ansi36,ANSI Marble, Slate, Glass
45, 0,0, .21875,.125, .3125,-.0625,0,-.0625



*swamp,Swampy area
0, 0,0, .5,.866025403, .125,-.875
90, .0625,0, .866025403,.5, .0625,-1.669550806
90, .078125,0, .866025403,.5, .05,-1.682050806
```

Creating Linetypes and Hatch Patterns 3-37

90, .046875,0, .866025403,.5, .05,-1.682050806

0, 0,0, .125,.125, .125,-.125
90, .125,0, .125,.125, .125,-.125
*HATCH1,Hatch at 45 Degree Angle
45,0,0,0,.5
*HATCH2,Hatch with Dashes & Dots:
0,0,0,.25,.25,0.5,-.125,0,-.125,0,-.125
*HATCH3,Triangle Hatch:
90,0,0,0,1,.5,-.5
0,0,0.5,0,1,.5,-.5
45,0,0,.7071,.7071,.7071,-.7071

Load the hatch pattern as described in Example 5 (Hatch3.pat) and test the hatch. Figure 3-37 shows the hatch pattern that will be generated by this hatch pattern (HATCH3). In Figure 3-37(a) the hatch pattern is at a 0-degree angle and the scale factor is 0.5. In Figure 3-37(b) the hatch pattern is at a -45-degree angle and the scale factor is 0.5.

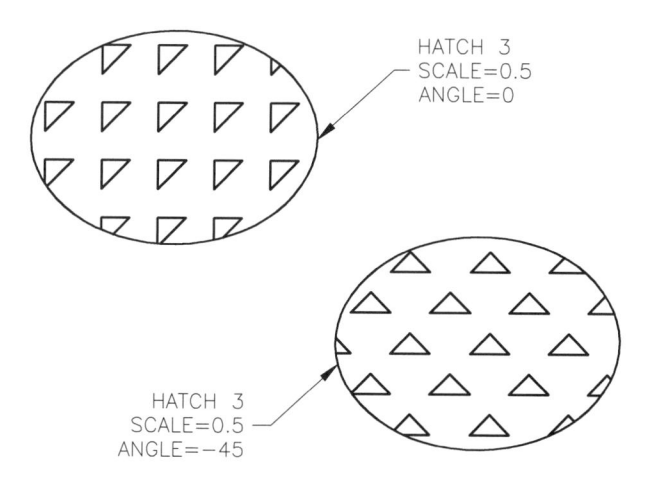

Figure 3-37 Hatch generated by HATCH3 pattern

SAVING HATCH PATTERNS IN A SEPARATE FILE

When you load a certain hatch pattern, AutoCAD looks for that definition in the ACAD.PAT file; therefore, the hatch pattern definitions must be in that file. However, you can add the new pattern definition to a different file and then copy that file to ACAD.PAT. Be sure to make a copy of the original ACAD.PAT file so that you can copy that file back when needed. Assume the name of the file that contains your custom hatch pattern definitions is CUSTOMH.PAT.

1. Copy ACAD.PAT file to ACADORG.PAT
2. Copy CUSTOMH.PAT to ACAD.PAT

If you want to use the original hatch pattern file, copy the ACADORG.PAT file to ACAD.PAT

CUSTOM HATCH PATTERN FILE

As mentioned earlier, you can add the new hatch pattern definitions to the **ACAD.PAT** file. There is no limit to the number of hatch pattern definitions you can add to this file. However, if you have only one hatch pattern definition, you can define a separate file. It has the following three requirements:

1. The name of the file has to be the same as the hatch pattern name.
2. The file can contain only one hatch pattern definition.
3. The hatch pattern name—and, therefore, the hatch file name—should be unique.
4. If you want to save the hatch pattern on the A drive, then the drive letter (A:) should precede the hatch name. For example, if the hatch name is HATCH3, the header line will be ***A:HATCH3, Triangle Hatch:** and the file name **HATCH3.PAT**.

*HATCH3,Triangle Hatch:
90,0,0,0,1,.5,-.5
0,0,0.5,0,1,.5,-.5
45,0,0,.7071,.7071,.7071,-.7071

Note
*The hatch lines can be edited after exploding the hatch with the AutoCAD **EXPLODE** command. After exploding, each hatch line becomes a separate object.*

It is good practice not to explode a hatch because it increases the size of the drawing database. For example, if a hatch consists of 100 lines, save it as a single object. However, after you explode the hatch, every line becomes a separate object and you have 99 additional objects in the drawing.

Keep the hatch lines in a separate layer to facilitate editing of the hatch lines.

Assign a unique color to hatch lines so that you can control the width of the hatch lines at the time of plotting.

Tip
*1. The file or the subdirectory in which hatch patterns have been saved must have been defined in the **Support File Search Path** in the **File** tab of the **Options** dialog box.*

*2. The hatch patterns that you create automatically get added to AutoCAD's slide library as an integral part of AutoCAD 2002 and the hatch patterns are displayed in the **Preview Area** in the **Hatch Pattern Palette** dialog box under the **Boundary Hatch** dialog box. Hence there is no need to create a slide library.*

Creating Linetypes and Hatch Patterns

Self-Evaluation Test

Answer the following questions and then compare your answers to the correct answers given at the end of this chapter.

CREATING LINETYPES

1. The AutoCAD _____ command can be used to change the linetype scale factor.

2. The linetype description should not be more than _____ characters long.

3. A positive number denotes a pen _____ segment.

4. The segment length _____ generates a dot.

5. A negative number denotes a pen _____ segment.

6. The option_____ of linetype Command is used to generate a new linetype.

7. The description in the case of header line is _____ . (optional/necessary)

8. The standard linetypes are stored in the file_____ .

9. The_____ determines the current linetype scaling.

CREATING HATCH PATTERNS

10. The header line consists of an asterisk, the pattern name, and _____ .

11. The **ACAD.PAT** file contains _____ number of hatch pattern definitions.

12. The standard hatch patterns are stored in the file _____ .

13. The first hatch line passes through a point whose coordinates are specified by _____ and _____ .

Review Questions

CREATING LINETYPES

1. The AutoCAD _____ command can be used to create a new linetype.

2. The AutoCAD _____ command can be used to load a linetype.

3. In AutoCAD, the linetypes are saved in the _____ file.

4. AutoCAD supports only _____ alignment field specification.

5. A line pattern definition always starts with _____.

6. A header line definition always starts with _____.

CREATING HATCH PATTERNS

7. The perpendicular distance between the hatch lines in a hatch pattern definition is specified by _____.

8. The displacement of the second hatch line in a hatch pattern definition is specified by _____.

9. The maximum number of dash lengths that can be specified in the line pattern definition of a hatch pattern is _____.

10. The hatch lines in different areas of the drawing will automatically _____ since the hatch lines are generated by offsetting.

11. The hatch angle as defined in the hatch pattern definition can be changed further when you use the AutoCAD _____ command.

12. When you load a hatch pattern, AutoCAD looks for that hatch pattern in the _____ file.

13. The hatch lines can be edited after _____ the hatch by using the AutoCAD _____ command.

Exercises

CREATING LINETYPES

Exercise 1 — *General*

Using the AutoCAD **LINETYPE** command, create a new linetype "DASH3DASH" with the following specifications:

   Length of the first dash 0.75
   Blank space 0.125
   Dash length 0.25
   Blank space 0.125
   Dash length 0.25
   Blank space 0.125
   Dash length 0.25
   Blank space 0.125

Creating Linetypes and Hatch Patterns 3-41

Exercise 2 — *General*

Use a text editor to create a new file, **NEWLT2.LIN**, and a new linetype, DASH2DASH, with the following specifications:

Length of the first dash 0.5
Blank space 0.1
Dash length 0.2
Blank space 0.1
Dash length 0.2
Blank space 0.1

Exercise 3 — *General*

Using AutoCAD's **LINETYPE** command, create a linetype DASH3DOT with the following specifications:

Length of the first dash 0.75
Blank space 0.25
Dot
Blank space 0.25
Dot
Blank space 0.25
Dot
Blank space 0.25

Exercise 4 — *General*

Using AutoCAD's **EDIT** command, create a new file, NEWLINET.LIN, and define a linetype VARDASHX with the following specifications:

Length of first dash 1.5
Blank space 0.25
Length of second dash 0.75
Blank space 0.25
Dot
Blank space 0.25
Length of third dash 0.75

Exercise 5 — *General*

a. Write the definition of a string complex linetype (hot water line) as shown in Figure 3-38(a). To determine the length of the HW text string, you should first draw the text (HW) using any text command and then measure its length.

b. Write the definition of a string complex linetype (gas line) as shown in Figure 3-38(b). Determine the length of the text string as mentioned in part *a*, above.

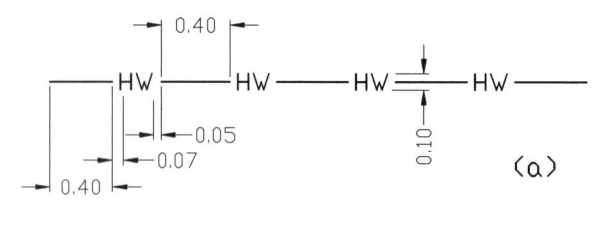

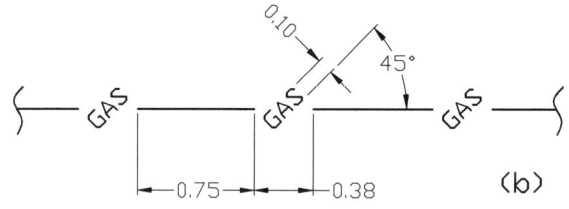

Figure 3-38 Specifications for string a complex linetype

Exercise 6   *General*

Write the shape file for the shape shown in Figure 3-39(a). Compile the shape and use it in defining the shape complex linetype so that you can draw a fence line as shown in Figure 3-39(b). Refer to Chapter 11 for more information about shape files.

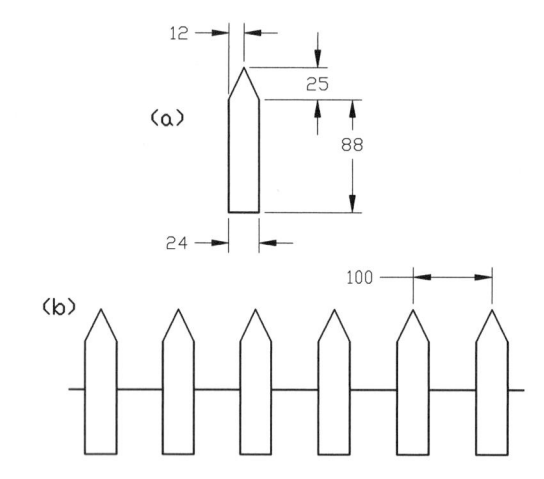

Figure 3-39 Specifications for a shape complex linetype

Creating Hatch Patterns

Exercise 7 — *General*

Determine the hatch pattern specifications and write a hatch pattern definition for the hatch pattern in Figure 3-40.

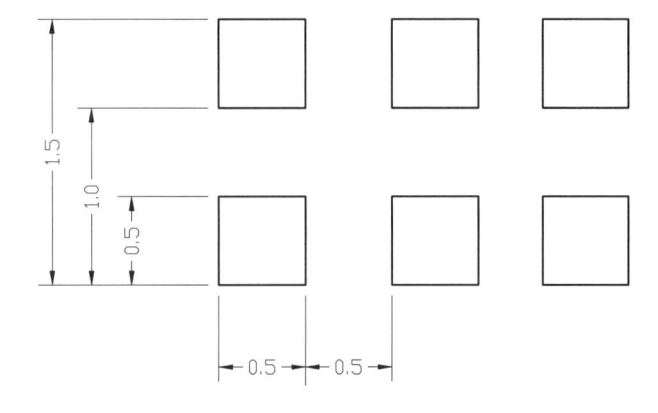

Figure 3-40 Drawing for Exercise 7

Exercise 8 — *General*

Determine the hatch pattern specifications and write a hatch pattern definition for the hatch pattern in Figure 3-41.

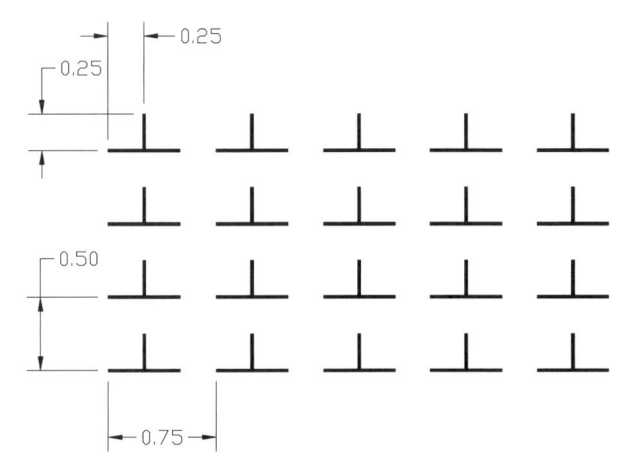

Figure 3-41 Hatch pattern for Exercise 8

Exercise 9   *General*

Determine the hatch specifications and write a hatch pattern definition for the hatch pattern as shown in Figure 3-42. Use this hatch to hatch a circle or rectangle.

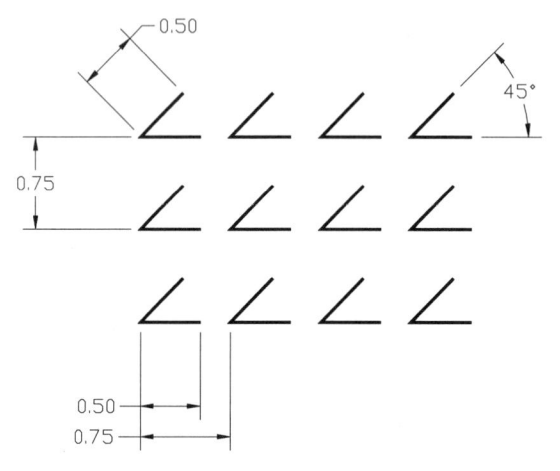

Figure 3-42 Hatch pattern for Exercise 9

Exercise 10   *General*

Determine the hatch specifications and write a hatch pattern definition for the hatch pattern as shown in Figure 3-43. Use this hatch to hatch a circle or rectangle.

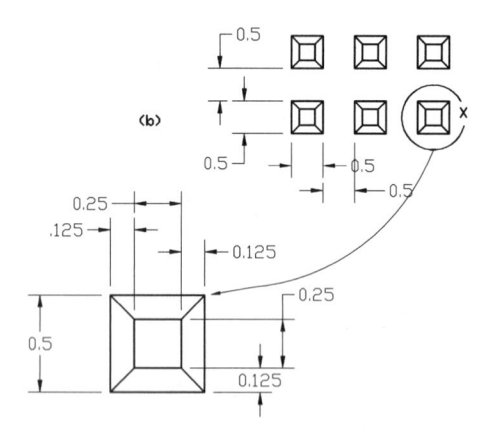

Figure 3-43 Square hatch pattern

Answers to Self-Evaluation Test
1. LTSCALE, **2.** 47, **3.** down, **4.** zero, **5.** up, **6. Create**, **7.** optional, **8. ACAD.LIN**, **9. CELTSCALE**, **10.** Pattern Description, **11.** 67, **12. ACAD.PAT**, **13.** X-origin, Y-origin.

Chapter 4

Customizing the ACAD.PGP File

Learning Objectives
After completing this chapter, you will be able to:
* *Customize the ACAD.PGP file.*
* *Edit different sections of the ACAD.PGP file.*
* *Abbreviate commands by defining command aliases.*
* *Use the REINIT command to reinitialize the PGP file.*

WHAT IS THE ACAD.PGP FILE?

AutoCAD software comes with the program parameters file **ACAD.PGP**, which defines aliases for the operating system commands and some of the AutoCAD commands. When you install AutoCAD, this file is automatically copied on the **AutoCAD 2002\SUPPORT** subdirectory of the hard drive. The **ACAD.PGP** file lets you access the operating system commands from the drawing editor. For example, if you want to delete a file, all you need to do is enter DEL at the Command prompt **(Command: DEL)**, and then AutoCAD will prompt you to enter the name of the file you want to delete.

The file also contains command aliases of some frequently used AutoCAD commands. For example, the command alias for the **LINE** command is L. If you enter L at the Command prompt **(Command: L)**, AutoCAD will treat it as the **LINE** command. The **ACAD.PGP** file also contain comment lines that give you some information about different sections of the file.

The following file is a partial listing of the standard **ACAD.PGP** file. Some of the lines have been deleted to make the file shorter.

; AutoCAD Program Parameters File For AutoCAD 2000
; External Command and Command Alias Definitions
; Copyright (C) 1997-1999 by Autodesk, Inc.

; Each time you open a new or existing drawing, AutoCAD searches
; the support path and reads the first acad.pgp file that it finds.
; -- External Commands --
; While AutoCAD is running, you can invoke other programs or utilities
; such Windows system commands, utilities, and applications.
; You define external commands by specifying a command name to be used
; from the AutoCAD command prompt and an executable command string
; that is passed to the operating system.
; -- Command Aliases --
; You can abbreviate frequently used AutoCAD commands by defining
; aliases for them in the command alias section of acad.pgp.
; You can create a command alias for any AutoCAD command,
; device driver command, or external command.
; Recommendation: back up this file before editing it.
; External command format:
; <Command name>,[<DOS request>],<Bit flag>,[*]<Prompt>,
; The bits of the bit flag have the following meanings:
; Bit 1: if set, don't wait for the application to finish
; Bit 2: if set, run the application minimized
; Bit 4: if set, run the application "hidden"
; Bit 8: if set, put the argument string in quotes
; Fill the "bit flag" field with the sum of the desired bits.
; Bits 2 and 4 are mutually exclusive; if both are specified, only
; the 2 bit is used. The most useful values are likely to be 0
; (start the application and wait for it to finish), 1 (start the
; application and don't wait), 3 (minimize and don't wait), and 5
; (hide and don't wait). Values of 2 and 4 should normally be avoided,
; as they make AutoCAD unavailable until the application has completed.
; Bit 8 allows commands like DEL to work properly with filenames that
; have spaces such as "long filename.dwg". Note that this will interfere
; with passing space delimited lists of file names to these same commands.
; If you prefer multiplefile support to using long file names, turn off
; the "8" bit in those commands.
; Examples of external commands for command windows

```
CATALOG,   DIR /W,        8,File specification: ,
DEL,       DEL,           8,File to delete: ,
DIR,       DIR,           8,File specification: ,
EDIT,      START EDIT,    9,File to edit: ,
SH,        ,              1,*OS Command: ,
SHELL,     ,              1,*OS Command: ,
START,     START,         1,*Application to start: ,
TYPE,      TYPE,          8,File to list: ,
```

External commands area

Customizing the ACAD.PGP File 4-3

```
; Examples of external commands for Windows
; See also the (STARTAPP) AutoLISP function for an alternative method.

EXPLORER, START EXPLORER, 1,,
NOTEPAD,  START NOTEPAD,  1,*File to edit: ,
PBRUSH,   START PBRUSH,   1,,

; Command alias format:
;   <Alias>,*<Full command name>

; The following are guidelines for creating new command aliases.
; 1. An alias should reduce a command by at least two characters.
;    Commands with a control key equivalent, status bar button,
;    or function key do not require a command alias.
;    Examples: Control N, O, P, and S for New, Open, Print, Save.
; 2. Try the first character of the command, then try the first two,
;    then the first three.
; 3. Once an alias is defined, add suffixes for related aliases:
;    Examples: R for Redraw, RA for Redrawall, L for Line, LT for
;    Linetype.
; 4. Use a hyphen to differentiate between command line and dialog
;    box commands.
;    Example: B for Block, -B for -Block.
;
; Exceptions to the rules include AA for Area, T for Mtext, X for Explode.

; -- Sample aliases for AutoCAD commands --
; These examples include most frequently used commands.

3A,     *3DARRAY
3DO,    *3DORBIT
3F,     *3DFACE
3P,     *3DPOLY
A,      *ARC
ADC,    *ADCENTER
AA,     *AREA
AL,     *ALIGN
AP,     *APPLOAD
AR,     *ARRAY
ATT,    *ATTDEF
-ATT,   *-ATTDEF
ATE,    *ATTEDIT
-ATE,   *-ATTEDIT
ATTE,   *-ATTEDIT
B,      *BLOCK
-B,     *-BLOCK
BH,     *BHATCH
```

Command alias section

```
BO,      *BOUNDARY
-BO,     *-BOUNDARY
BR,      *BREAK
C,       *CIRCLE
CH,      *PROPERTIES
-CH,     *CHANGE
CHA,     *CHAMFER
COL,     *COLOR
COLOUR,  *COLOR
CO,      *COPY
D,       *DIMSTYLE
DAL,     *DIMALIGNED
DAN,     *DIMANGULAR
DBA,     *DIMBASELINE
DBC,     *DBCONNECT
DCE,     *DIMCENTER
DCO,     *DIMCONTINUE
DDI,     *DIMDIAMETER
DED,     *DIMEDIT
DI,      *DIST
DIV,     *DIVIDE



L,       *LINE
LA,      *LAYER
-LA,     *-LAYER
LE,      *QLEADER
LEN,     *LENGTHEN
LI,      *LIST
LINEWEIGHT, *LWEIGHT
LO,      *-LAYOUT
LS,      *LIST
LT,      *LINETYPE
-LT,     *-LINETYPE
LTYPE,   *LINETYPE
-LTYPE,  *-LINETYPE
LTS,     *LTSCALE
LW,      *LWEIGHT
M,       *MOVE
MA,      *MATCHPROP
ME,      *MEASURE
MI,      *MIRROR
ML,      *MLINE
MO,      *PROPERTIES
MS,      *MSPACE
MT,      *MTEXT
```

Customizing the ACAD.PGP File

```
MV,      *MVIEW
O,       *OFFSET
OP,      *OPTIONS
ORBIT,   *3DORBIT
OS,      *OSNAP
-OS,     *-OSNAP
P,       *PAN
-P,      *-PAN
PA,      *PASTESPEC
PARTIALOPEN, *-PARTIALOPEN
PE,      *PEDIT
PL,      *PLINE
PO,      *POINT
POL,     *POLYGON
PR,      *OPTIONS
PRCLOSE, *PROPERTIESCLOSE
PROPS,   *PROPERTIES
PRE,     *PREVIEW
PRINT,   *PLOT
PS,      *PSPACE
PU,      *PURGE
R,       *REDRAW
RA,      *REDRAWALL
RE,      *REGEN
REA,     *REGENALL
REC,     *RECTANGLE
REG,     *REGION


SP,      *SPELL
SPL,     *SPLINE
SPE,     *SPLINEDIT
ST,      *STYLE
SU,      *SUBTRACT
T,       *MTEXT
-T,      *-MTEXT
TA,      *TABLET
TH,      *THICKNESS
TI,      *TILEMODE
TO,      *TOOLBAR
TOL,     *TOLERANCE
TOR,     *TORUS
TR,      *TRIM
UC,      *DDUCS
UCP,     *DDUCSP
```

; The following are alternative aliases and aliases as supplied
; in AutoCAD Release 13.

AV,   *DSVIEWER
CP,   *COPY
DIMALI,   *DIMALIGNED
DIMANG,   *DIMANGULAR
DIMBASE, *DIMBASELINE
DIMCONT, *DIMCONTINUE
DIMDIA,   *DIMDIAMETER
DIMED,   *DIMEDIT
DIMTED,   *DIMTEDIT
DIMLIN,   *DIMLINEAR
DIMORD,   *DIMORDINATE
DIMRAD,   *DIMRADIUS
DIMSTY,   *DIMSTYLE
DIMOVER, *DIMOVERRIDE
LEAD,   *LEADER
TM,   *TILEMODE

; Aliases for Hyperlink/URL Release 14 compatibility
SAVEURL, *SAVE
OPENURL, *OPEN
INSERTURL, *INSERT

; Aliases for commands discontinued in AutoCAD 2000:
AAD,   *DBCONNECT
AEX,   *DBCONNECT
ALI,   *DBCONNECT
ASQ,   *DBCONNECT
ARO,   *DBCONNECT
ASE,   *DBCONNECT
DDATTDEF, *ATTDEF
DDATTEXT, *ATTEXT
DDCHPROP, *PROPERTIES
DDCOLOR, *COLOR
DDLMODES, *LAYER
DDLTYPE, *LINETYPE
DDMODIFY, *PROPERTIES
DDOSNAP, *OSNAP
DDUCS,   *UCS

SECTIONS OF THE ACAD.PGP FILE

The contents of the AutoCAD program parameters file (**ACAD.PGP**) can be categorized into three sections. These sections merely classify the information that is defined in the **ACAD.PGP** file. They do not have to appear in any definite order in the file, and they have no section headings. For example, the comment lines can be entered anywhere in the file; the same is true with external commands and AutoCAD command aliases. The **ACAD.PGP** file can be divided into these three sections: **comments, external commands**, and **command aliases**.

Comments

The comments of ACAD.PGP file can contain any number of comment lines and can occur anywhere in the file. Every comment line must start with a semicolon (;) (This is a comment line). Any line that is preceded by a semicolon is ignored by AutoCAD. You should use the comment line to give some relevant information about the file that will help other AutoCAD users to understand, edit, or update the file.

External Command

In the external command section you can define any valid external command that is supported by your system. The information must be entered in the following format:

<Command name>, [OS Command name],<Bit flag>, [*]<Command prompt>,

Command Name. This is the name you want to use to activate the external command from the AutoCAD drawing editor. For example, you can use **goword** as a command name to load the Word program (**Command: goword**). The command name must not be an AutoCAD command name or an AutoCAD system variable name. If the name is an AutoCAD command name, the command name in the **PGP** file will be ignored. Also, if the name is an AutoCAD system variable name, the system variable will be ignored. You should use the command names that reflect the expected result of the external commands. (For example, **hello** is not a good command name for a directory file.) The command names can be uppercase or lowercase.

OS Command Name. The OS Command name is the name of a valid system command that is supported by your operating system. For example, in DOS the command to delete files is DEL; therefore, the OS Command name used in the **ACAD.PGP** file must be DEL. The following is a list of the types of commands that can be used in the PGP file:

OS Commands (del, dir, type, copy, rename, edlin, etc.)
Commands for starting a word processor, or text editors (word, shell, etc.)
Name of the user-defined programs and batch files

Bit flag. This field must contain a number, preferably 8 or 1. The following are the bit flag values and their meaning:

Bit flag set to	Meaning
1	Do not wait for the application to finish
2	Runs the application minimized
4	Runs the application hidden
8	Puts the argument string in quotes

Command Prompt. The command prompt field of the command line contains the prompt you want to display on the screen. It is an optional field that must be replaced by a comma if there is no prompt. If the operating system (OS) command that you want to use contains spaces, the prompt must be preceded by an asterisk (*). For example, the DOS command **EDIT NEW.PGP** contains a space between EDIT and NEW; therefore, the prompt used in this command line must be preceded by an asterisk. The command can be terminated by pressing ENTER. If the **OS** command consists of a single word (DIR, DEL, TYPE), the preceding asterisk must be omitted. In this case you can terminate the command by pressing the SPACEBAR or ENTER.

Command Aliases

It is time-consuming to enter AutoCAD commands at the keyboard because it requires typing the complete command name before pressing ENTER. AutoCAD provides a facility that can be used to abbreviate the commands by defining aliases for the AutoCAD commands. This is made possible by the AutoCAD program parameters file (**ACAD.PGP** file). Each command alias line consists of two fields (**L, *LINE**). The first field (**L**) defines the alias of the command; the second field (***LINE**) consists of the AutoCAD command. The AutoCAD command must be preceded by an asterisk for AutoCAD to recognize the command line as a command alias. The two fields must be separated by a comma. The blank lines and the spaces between the two fields are ignored. In addition to AutoCAD commands, you can also use aliases for AutoLISP command names, provided the programs that contain the definition of these commands are loaded.

Example 1

Add the following external commands and AutoCAD command aliases to the AutoCAD program parameters file (**ACAD.PGP**).

External Commands

Abbreviation	Command Description
GOWORD	This command loads the word processor (Winword) program from the C:\Program Files\Winword.
RN	This command executes the rename command of DOS.
COP	This command executes the copy command of DOS.

Command Aliases Section

Abbreviation	Command	Abbreviation	Command
EL	Ellipse	T	Trim
CO	Copy	CH	Chamfer
O	Offset	ST	Stretch
S	Scale	MI	Mirror

The **ACAD.PGP** file is an ASCII text file. To edit this file you can use the AutoCAD EDIT command (provided the EDIT command is defined in the **ACAD.PGP** file), or any text editor (Notepad or Wordpad). The following is a partial listing of the **ACAD.PGP** file after insertion of the lines for the command aliases of Example 5. **The line numbers are not a part of the**

Customizing the ACAD.PGP File

file; they are shown here for reference only. The lines that have been added to the file are highlighted in bold.

DEL,DEL,	8,File to delete: ,	1
DIR,DIR,	8,File specification ,	2
EDIT, START EDIT,	8,File to edit: ,	3
SH,,	1,*OS Command: ,	4
SHELL,,	1,*OS Command: ,	5
START,START,	1,Application to start: ,	6
		7
GOWORD, START WINWORD,1,,		**8**
RN, RENAME,8,File to rename:,		**9**
COP,COPY,8,File to copy:,		**10**
DIMLIN	*DIMLINEAR	11
DIMORD,	*DIMORDINATE	12
DIMRAD,	*DIMRADIUS	13
DIMSTY,	*DIMSTYLE	14
DIMOVER,	*DIMOVERRIDE	15
LEAD,	*LEADER	16
TM,	*TILEMODE	17
EL,	***ELLIPSE**	**18**
CO,	***COPY**	**19**
O,	***OFFSET**	**20**
S,	***SCALE**	**21**
MI,	***MIRROR**	**22**
ST,	***STRETCH**	**23**

Explanation

Lines 8
GOWORD,   START WINWORD,1,,
In line 8, **goword** loads the word processor program (**winword**). The **winword.exe** program is located in the winword directory under Program Files.

Lines 9 and 10
RN, RENAME,8,File to rename:,
COP,COPY,8,File to copy:,
Line 9 defines the alias for the DOS command **RENAME,** and the next line defines the alias for the DOS command **COPY**. The 8 is a bit flag, and the command prompt **File to rename and File to copy** is automatically displayed to let you know the format and the type of information that is expected.

Lines 18 and 19
EL, *ELLIPSE
CO, *COPY
Line 18 defines the alias (**EL**) for the AutoCAD command **ELLIPSE**, and the next line defines the alias (**CO**) for the **COPY** command. The AutoCAD commands must be preceded by an asterisk. You can put any number of spaces between the alias abbreviation and the AutoCAD command.

> **Note**
> *If a command alias definition duplicates an existing one then the winner is the one that is lower down in the file. For example, in the standard file if you add S, *SCALE to the end of the file then your definition works and the one higher up the file is ignored.*

REINIT COMMAND

When you make any changes in the **ACAD.PGP** file, there are two ways to reinitialize the **ACAD.PGP** file. One is to quit AutoCAD and then reenter it. When you start AutoCAD, the **ACAD.PGP** file is automatically loaded. You can also reinitialize the **ACAD.PGP** file by using the AutoCAD **REINIT** command. The **REINIT** command lets you reinitialize the I/O ports, digitizer, and AutoCAD program parameters file, **ACAD.PGP**. When you enter the **REINIT** command, AutoCAD will display a dialog box (Figure 4-1). To reinitialize the **ACAD.PGP** file, select the corresponding toggle box, and then choose **OK**. AutoCAD will reinitialize the program parameters file **(ACAD.PGP)**, and then you can use the command aliases defined in the file.

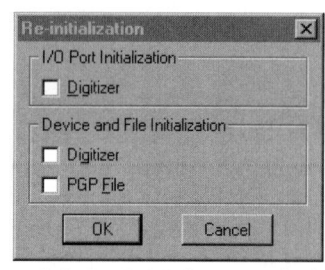

Figure 4-1 Re-initialization dialog box

> **Tip**
> *After you have made changes in the **ACAD.PGP** file, and used it, you should copy and restore the original ACAD.PGP file. This lets other users use the original, unedited file.*

Self-Evaluation Test

Answer the following questions and then compare your answers to the correct answer given at the end of this chapter.

1. One way of reinitializing the **ACAD.PGP** file is to _____ the AutoCAD and then _____ it.

2. The Command used to reinitialize the **ACAD.PGP** file is _____ .

3. In Command alias section, the AutoCAD Command must be preceded by an _____.

4. Every comment line must start with a _____ .

5. The Command alias and the AutoCAD Command must be separated by_____.

Customizing the ACAD.PGP File

Review Questions

Indicate whether the following statements are true or false.

1. The comment section can contain any number of lines. (T/F)

2. AutoCAD ignores any line that is preceded by a semicolon. (T/F)

3. The command alias must not be an AutoCAD command. (T/F)

4. The bit flag field must contain 8. (T/F)

5. In the command alias section, the command alias must be preceded by a semicolon. (T/F)

6. You cannot use aliases for AutoLISP commands. (T/F)

7. The ACAD.PGP file does not come with AutoCAD software. (T/F)

8. The ACAD.PGP file is an ASCII file. (T/F)

Exercises

Exercise 1 *General*

Add the following external commands and AutoCAD command aliases to the AutoCAD program parameters file (**ACAD.PGP**).

External Command Section

Abbreviation	Command Description
MYWORDPAD	This command loads the WORDPAD program that resides in the **Program Files\Accessories** directory.
MYEXCEL	This command loads the EXCEL program that resides in the **Program Fies\Microsoft Office** directory.
CD	This command executes the CHKDSK command of DOS.
FORMAT	This command executes the FORMAT command of DOS.

Command Aliases Section

Abbreviation	Command	Abbreviation	Command
BL	**BLOCK**	LTS	**LTSCALE**
INS	**INSERT**	EXP	**EXPLODE**
DIS	**DISTANCE**	GR	**GRID**
TE	**TIME**		

Answers to the Self-Evaluation Test
1. **QUIT,** Reenter, 2. **REINIT**, 3. Asterisk, 4. Semicolon, 5. Comma.

Chapter 5

Pull-down, Shortcut, and Partial Menus and Customizing Toolbars

Learning Objectives
After completing this chapter, you will be able to:
- *Write pull-down menus.*
- *Load menus.*
- *Write cascading submenus in pull-down menus.*
- *Write cursor menus.*
- *Swap pull-down menus.*
- *Write partial menus.*
- *Define accelerator keys.*
- *Write toolbar definitions.*
- *Write menus to access online help.*
- *Customize toolbars.*

AUTOCAD MENU

The AutoCAD menu provides a powerful tool to customize AutoCAD. The AutoCAD software package comes with a standard menu file named **ACAD.MNU**. When you start AutoCAD, the menu file **ACAD.MNU** is automatically loaded. The AutoCAD menu file contains AutoCAD commands, separated under different headings for easy identification. For example, all draw commands are under **Draw** and all editing commands are under the **Modify** menu. The headings are named and arranged to make it easier for you to locate and access the commands. However, there are some commands that you may never use. Also, some users might like to regroup and rearrange the commands so that it is easier to access those most frequently used. AutoCAD lets the user eliminate rarely used commands from the menu file and define new ones. This is made possible by editing the existing **ACAD.MNU** file or writing a new menu file. There is no limit to the number of files you can write. You can have a separate menu file for each application. For example, you can have separate menu files for mechanical, electrical, and architectural drawings. You can load these menu files any time by using the AutoCAD **MENU** command. The menu files are text files with the extension **.MNU**. These files can be written by using any text editor like Wordpad or Notepad. The menu file can be divided into ten sections, each section identified by a section label. AutoCAD uses the following labels to identify different sections of the AutoCAD menu file:

*****SCREEN**	
*****TABLET(n)**	**n** is from 1 to 4
*****IMAGE**	
*****POP(n)**	**n** is from 1 to 499 (For Shortcut menus n=0 and 500 to 999)
*****BUTTONS(n)**	**n** is from 1 to 4
*****AUX(n)**	**n** is from 1 to 4
*****MENUGROUP**	
*****TOOLBARS**	
*****HELPSTRING**	
*****ACCELERATORS**	

The tablet menu can have up to four different sections. The **POP** menu (pull-down and cursor menu) can have up to 499 sections, and auxiliary and buttons menus can have up to four sections.

Tablet Menus	**Pull-Down and Cursor Menus**
***TABLET1	***POP0
***TABLET2	***POP1
***TABLET3	***POP2
***TABLET4	***POP3
	***POP4
BUTTONS Menus	***POP5
***BUTTONS1	***POP6
***BUTTONS2	***POP7
***BUTTONS3	***POP8
***BUTTONS4	
Auxiliary Menus	
***AUX1	
***AUX2	***POP498
***AUX3	***POP499
***AUX4	

STANDARD MENUS

The menu is a part of the AutoCAD standard menu file, **ACAD.MNU**. The **ACAD.MNU** file is automatically loaded when you start AutoCAD, provided the standard configuration of AutoCAD has not been changed. The menus can be selected by moving the crosshairs to the top of the screen, into the menu bar area. If you move the pointing device sideways, different menu bar titles are highlighted and you can select the desired item by pressing the pick button on your pointing device. Once the item is selected, the corresponding menu is displayed directly under the title (Figure 5-1). The menu can have 499 sections, named as POP1, POP2, POP3, . . ., POP499.

WRITING A MENU

Before you write a menu, you need to design a menu and to arrange the commands the way you want them to appear on the screen. To design a menu, you should select and arrange the commands in a way that provides easy access to the most frequently used

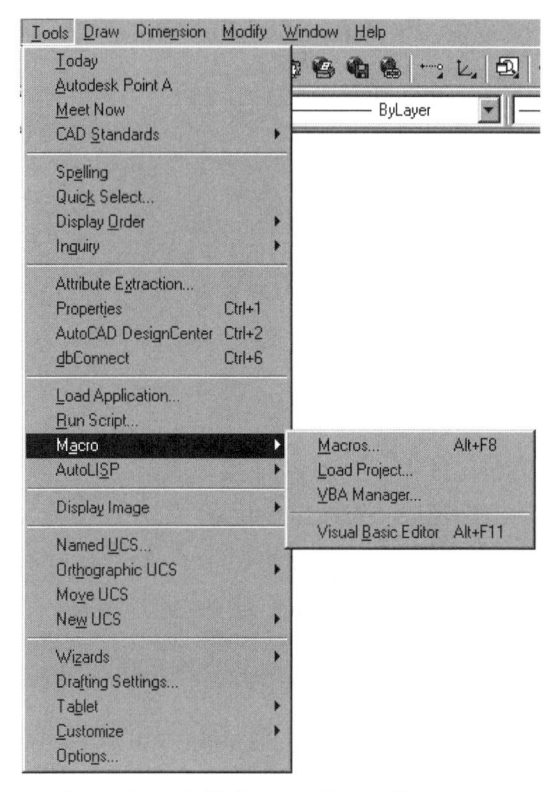

Figure 5-1 Pull-down and cascading menus

commands. A careful design will save a lot of time in the long run. Therefore, consider several possible designs with different command combinations, and then select the one best suited for the job. Suggestions from other CAD operators can prove very valuable.

The second important thing in developing a menu is to know the exact sequence of the commands and the prompts associated with each command. To better determine the prompt entries required in a command, you should enter all the commands and the prompt entries at the keyboard. The following is a description of some of the commands and the prompt entries required for Example 1.

LINE Command

Command: **LINE**

Notice the command and input sequence:

LINE
<RETURN>

CIRCLE (C,R) Command

Command: **CIRCLE**
Specify center point for circle or [3P/2P/Ttr (tan tan radius)]: *Specify center point.*
Specify radius of circle or [Diameter]: *Enter radius.*

Notice the command and input sequence:

CIRCLE
<RETURN>
Center point
<RETURN>
Radius
<RETURN>

CIRCLE (C,D) Command

Command: **CIRCLE**
Specify center point for circle or [3P/2P/Ttr (tan tan radius)]: *Specify center point.*
Specify radius of circle or [Diameter]: **D**
Specify diameter of circle: *Enter diameter.*

Notice the command and input sequence:

CIRCLE
<RETURN>
Center Point
<RETURN>
D
<RETURN>
Diameter
<RETURN>

CIRCLE (2P) Command

Command: **CIRCLE**
Specify center point for circle or [3P/2P/Ttr (tan tan radius)]: **2P**
Specify first end point of circle's diameter: *Specify first point.*
Specify second end point of circle's diameter: *Specify second point.*

Notice the command and input sequence:

CIRCLE
<RETURN>
2P
<RETURN>
Select first point on diameter

Pull-down, Shortcut, and Partial Menus and Customizing Toolbars 5-5

<RETURN>
Select second point on diameter
<RETURN>

ERASE Command

Command: **ERASE**

Notice the command and input sequence:

ERASE
<RETURN>

MOVE Command

Command: **MOVE**

Notice the command and prompt entry sequence:

MOVE
<RETURN>

The difference between the Center-Radius and Center-Diameter options of the **CIRCLE** command is that in the first one the RADIUS is the default, whereas in the second one you need to enter D to use the diameter option. This difference, although minor, is very important when writing a menu file. Similarly, the 2P (two-point) option of the **CIRCLE** command is different from the other two options. Therefore, it is important to know both the correct sequence of the AutoCAD commands and the entries made in response to the prompts associated with those commands.

You can use any text editor (like Notepad) to write the file. You can also use AutoCAD's **EDIT** command to write the menu file. If you use the **EDIT** command, AutoCAD will prompt you to enter the file name you want to edit. The file name can be up to eight characters long, and the file extension must be **.MNU**. If the file name exists, it will be automatically loaded; otherwise a new file will be created. To understand the process of developing a pull-down menu, consider the following example.

> **Note**
> *If AutoCAD's **EDIT** command does not work, check the **ACAD.PGP** file and make sure that this command is defined in the file.*

Example 1

Write a pull-down menu for the following AutoCAD commands:

| LINE | ERASE | REDRAW | SAVE |
| PLINE | MOVE | REGEN | QUIT |

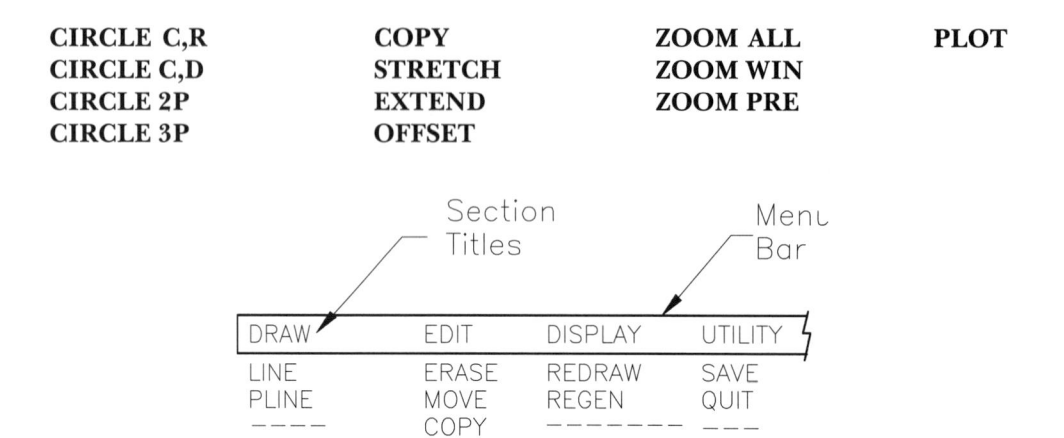

CIRCLE C,R	COPY	ZOOM ALL	PLOT
CIRCLE C,D	STRETCH	ZOOM WIN	
CIRCLE 2P	EXTEND	ZOOM PRE	
CIRCLE 3P	OFFSET		

Figure 5-2 Design of menu

Step 1: Designing the menu

The first step in writing any menu is to design the menu so that the commands are arranged in the desired configuration. Figure 5-2 shows one of the possible designs of this menu. This menu has four different groups of commands; therefore, it will have four sections: POP1, POP2, POP3, and POP4, and each section will have a section label. The following file is a listing of the pull-down menu file for Example 1. **The line numbers are not a part of the file; they are shown here for reference only.**

Step 2: Writing the menu file

```
***POP1                                            1
[DRAW]                                             2
[LINE]*^C^CLINE                                    3
[PLINE]^C^CPLINE                                   4
[--]                                               5
[CIR-C,R]^C^CCIRCLE                                6
[CIR-C,D]^C^CCIRCLE \D                             7
[CIR-2P]^C^CCIRCLE 2P                              8
[CIR-3P]^C^CCIRCLE 3P                              9
***POP2                                           10
[EDIT]                                            11
[ERASE]*^C^CERASE                                 12
[MOVE]^C^CMOVE                                    13
[COPY]^C^CCOPY                                    14
[STRETCH]^C^CSTRETCH;C                            15
[EXTEND]^C^CEXTEND                                16
```

Pull-down, Shortcut, and Partial Menus and Customizing Toolbars 5-7

```
[OFFSET]^C^COFFSET                      17
***POP3                                 18
[DISPLAY]                               19
[REDRAW]'REDRAW                         20
[REGEN]^C^CREGEN                        21
[--]                                    22
[ZOOM-All]^C^C'ZOOM A                   23
[ZOOM-Window]'ZOOM W                    24
[ZOOM-Prev]'ZOOM PREV                   25
[~Exit]^C                               26
***POP4                                 27
[UTILITY]                               28
[SAVE]^C^CSAVE;                         29
[QUIT]^C^CQUIT                          30
[----]                                  31
[PLOT]^C^CPLOT                          32
[EXIT]^C^CEXIT                          33
```

Explanation
Line 1
***POP1
POP1 is the section label for the first pull-down menu. All section labels in the AutoCAD menu begin with three asterisks (***), followed by the section label name, such as POP1.

Line 2
[DRAW]
In this menu item **DRAW** is the menu bar title displayed when the cursor is moved in the menu bar area. The title names should be chosen so you can identify the type of commands you expect in that particular pull-down menu. In this example, all the draw commands are under the title **DRAW** (Figure 5-3), all edit commands are under **EDIT**, and so on for other groups of items. The menu bar title can be of any length depending on the display resolution, display device, and the font size of the text. However, it is recommended to keep them short to accommodate other menu items. If a display device at a certain resolution provides a maximum of 80 characters; to have 16 sections in a single row in the menu bar, the length of each title should not exceed five characters. If the length of the menu bar items exceeds 80 characters, AutoCAD will wrap the items that cannot be accommodated in 80 character space, and display them in the next line. This will result in two menu item lines in the menu bar.

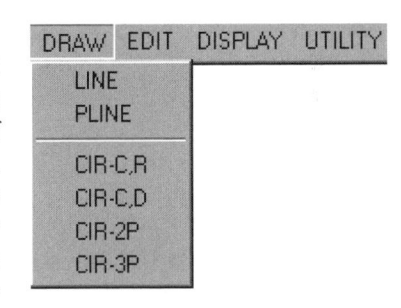

Figure 5-3 Draw pull-doen menu

If the first line in a menu section is blank, the title of that section is not displayed in the menu bar area. Since the menu bar title is not displayed, you **cannot** access that menu. This allows you to turn off the menu section. For example, if you replace **[DRAW]** with a blank line, the **DRAW** section (POP1) of the menu will be disabled; the second section (POP2) will be displayed in its place.

Example
```
***POP1                         Section label
                                Blank line (turns off POP1)
[LINE:]^CLINE                   Menu item
[PLINE:]^CPLINE
[CIRCLE:]^CCIRCLE
```

The menu bar titles are left-justified. If the first title is not displayed, the rest of the menu titles will be shifted to the left. In Example 1, if the **DRAW** title is not displayed in the menu bar area, then the **EDIT**, **DISPLAY**, and **FILE** sections of the menu will move to the left.

Line 3
*^C^CLINE

In this menu item, the command definition starts with an asterisk (*). This feature allows the command to be repeated automatically until it is canceled by pressing ESCAPE, entering CTRL C, or by selecting another menu command. ^C^C cancels the existing command twice; **LINE** is an AutoCAD command that generates lines.

```
*^C^CLINE
        Where  *  ------------------ Repeats the menu item (command)
               ^C^C --------- Cancels existing command twice
               LINE ------------- AutoCAD's LINE command
```

Line 5
[--]

To separate two groups of commands in any section, you can use a menu item that consists of two or more hyphens (--). This line automatically expands to fill the **entire width** of the menu. You can use a blank line in a menu. If any section of a menu (**POP section) has a blank line, it is ignored.

Line 24
[ZOOM-Window]'ZOOM W

In this menu item the single quote (') preceding the **ZOOM** command makes the **ZOOM** Window command transparent. When a command is transparent, the existing command is not canceled. After the **ZOOM** Window command (Figure 5-4), AutoCAD will automatically resume the current operation.

```
[ZOOM-Window]'ZOOM W
        Where  W  ----------------- Window option
               ZOOM ----------- AutoCAD ZOOM command
               '  -------------------- Single quote makes ZOOM transparent
```

Line 26
[~Exit]^C

This menu item is specially meant for canceling the menu. Since this menu item has a tilde (~), the menu item is not available (displayed grayed out), and if you select this item it will not cancel the menu. You can use this feature to disable a menu item or to indicate that the item is not a valid selection. If there is an instruction associated with the item, the instruction

Pull-down, Shortcut, and Partial Menus and Customizing Toolbars 5-9

will not be executed when you select the item. For example, [~OSNAPS]^C^C$S=OSNAPS will not load the OSNAPS submenu on the screen.

Line 29
[SAVE]^C^CSAVE;
In this menu item, the semicolon (;) that follows the **SAVE** command enters RETURN. The semicolon is not required; the command will also work without a semicolon.

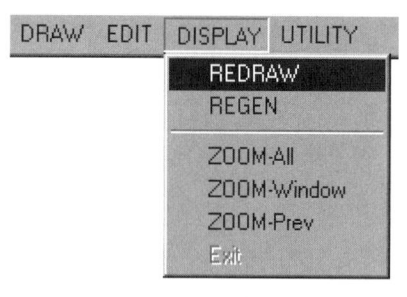

Figure 5-4 Display menu

```
[SAVE]^C^CSAVE;
         Where   SAVE ------------- AutoCAD SAVE command
                 ;    ------------------ Semicolon enters RETURN
```

Line 31
[----]
This menu item has four hyphens. The line will extend to fill the width of the menu. If there is only one hyphen [-], AutoCAD gives a syntax error when you load the menu.

Line 33
[Exit]^C
In this menu item, ^C command definition has been used to cancel the menu. This item provides you with one more option for canceling the menu. This is especially useful for new AutoCAD users who are not familiar with all AutoCAD features. The menu can also be canceled by any of the following actions:

1. Selecting a point.
2. Selecting an item in the screen menu area.
3. Selecting or typing another command.
4. Pressing ESC at the keyboard.
5. Selecting any menu title in the menu bar.

Note
For all menus, the menu items are displayed directly beneath the menu title and are left-justified. If any menu [for example, the rightmost menu (POP16)] does not have enough space to display the entire menu item, the menu will expand to the left to accommodate the entire length of the longest menu item.

You can use // (two forward slashes) for comment lines. AutoCAD ignores the lines that start with //.

The last line in the menu file must be terminated by ENTER or else it will be ignored.

From this example it is clear that every statement in the menu is based on the AutoCAD commands and the information that is needed to complete these commands. This forms the basis for creating a menu file and should be given consideration. Following is a summary of the AutoCAD commands used in Example 1 and their equivalents in the menu file.

AutoCAD Commands	Menu File
Command: **LINE**	[LINE]^C^CLINE

Command: **CIRCLE**
Specify center point for circle or [3P/2P/Ttr (tan tan radius)]:
Specify radius of circle or [Diameter]:   **CIR-C,R]^C^CCIRCLE**

Command: **CIRCLE**
Specify center point for circle or [3P/2P/Ttr (tan tan radius)]:
Specify radius of circle or [Diameter]: D
Specify diameter of circle: [CIR-C,D]^C^CCIRCLE;\D

Command: **CIRCLE**
Specify center point for circle or [3P/2P/Ttr (tan tan radius)]: **2P**
Specify first end point of circle's diameter:
Specify second end point of circle's diameter: [CIR- 2P]^C^CCIRCLE;2P

Command: **ERASE**   [ERASE   ]^C^CERASE

Command: **MOVE**   [MOVE   ]^C^CMOVE

LOADING MENUS

AutoCAD automatically loads the **ACAD.MNC** file when you get into the AutoCAD drawing editor, unless the ACAD.MNU file has been modified or unless a different menu file has been loaded. However, you can also load a different menu file by using the AutoCAD **MENU** command. When you enter the **MENU** command at the Command prompt, AutoCAD displays the **Select Menu File** dialog box (Figure 5-5) on the screen. Select the menu file that you want to load and then choose the **Open** button.

You can also load the menu file from the command line, keeping the **FILEDIA** as **0**.

 Command: **FILEDIA**
 Enter new value for FILEDIA <1>: **0**
 Command: **MENU**

 Enter menu file name or [. (for none)] <ACAD>: **PDM1**
    Where **PDM1** ------------ Name of menu file
        <ACAD> ------- Default menu file

After you enter the **MENU** command, AutoCAD will prompt for the file name. Enter the name of the menu file without the file extension **(.MNU)**, since AutoCAD assumes the extension .MNU. AutoCAD will automatically compile the menu file into **MNC** and **MNR** files. When you load a menu file in windows, AutoCAD creates the following files:

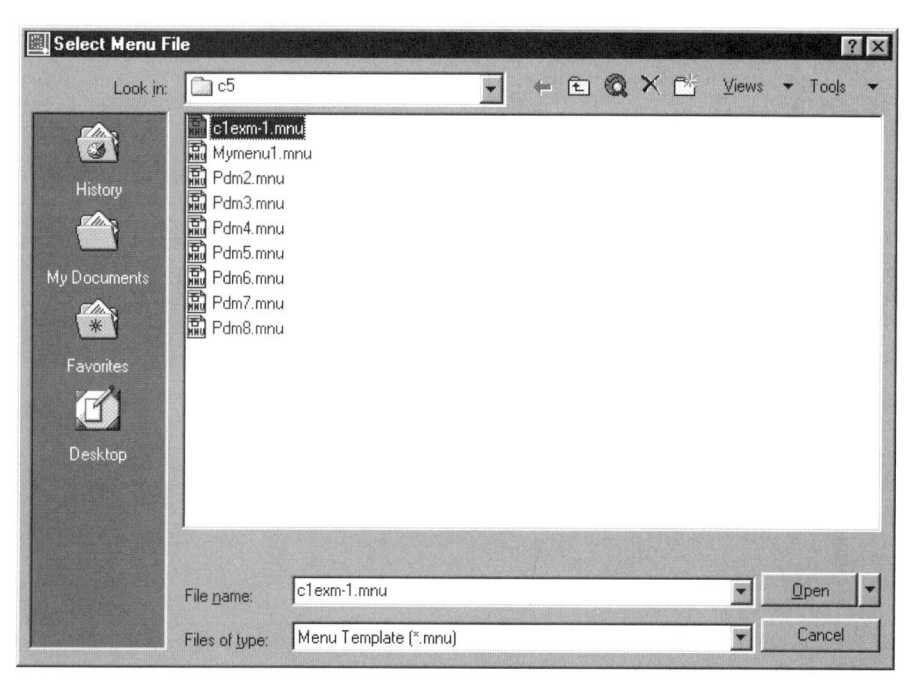

Figure 5-5 Select Menu File dialog box

.mnc and .mnr files — When you load a menu file (.mnu), AutoCAD compiles the menu file and creates **.mnc** and **.mnr** files. The .mnc file is a compiled menu file. The **.mnr** file contains the bitmaps used by the menu.

.mns file — When you load the menu file, AutoCAD also creates an **.mns** file. This is an ASCII file that is the same as the **.mnu** file when you initially load the menu file. Each time you make a change in the contents of the file, AutoCAD changes the **.mns** file.

Note
*After you load the new menu, you cannot use the screen menu, buttons menu, or digitizer because the original menu, **ACAD.MNU**, is not present and the new menu does not contain these menu areas.*

*To activate the original menu again, enter **MENU** at the Command prompt to display the **Select Menu File** dialog box. Select the **ACAD.MNU** file from the **Support** directory and then choose the **Open** button. You can enter the values at the Command prompt if the **FILEDIA** system variable is 1.*

If you need to use input from a keyboard or a pointing device, use the backslash (\). The system will pause for you to enter data.

There should be no space after the backslash (\).

The menu items, menu labels, and command definition can be uppercase, lowercase, or mixed.

You can introduce spaces between the menu items to improve the readability of the menu file. The blank lines are ignored and are not displayed on the screen.

If there are more menu items in the pull-down menu than the number of spaces available, the excess items are not displayed on the screen. For example, if the display device and the screen resolution limits the number of items to 21, items in excess of 21 will not be displayed on the screen and are therefore inaccessible.

If you are using a high-resolution graphics board, you can increase the number of lines that can be displayed on the screen. On some devices this is 80 lines.

RESTRICTIONS

The menus are easy to use and provide a quick access to frequently used AutoCAD commands. However, the menu bar and the menus are disabled during the following commands:

DTEXT Command
After you assign the text height and the rotation angle to a **DTEXT** command, the menu is automatically disabled.

SKETCH Command
The menus are disabled after you set the record increment in the **SKETCH** command.

Exercise 1   *General*

Write a menu for the following AutoCAD commands. The menu design is shown in Figure 5-6.

DRAW	EDIT	DISP/TEXT	TILITY
LINE	FILLET0	TEXT,C	SAVE
PLINE	FILLET	TEXT,L	QUIT
ELLIPSE	CHAMFER	DTEXT,R	END
POLYGON	STRETCH	ZOOM WIN	DIR
DONUT	EXTEND	ZOOM PRE	PLOT
	OFFSET		

CASCADING SUBMENUS IN MENUS

The number of items in a menu or cursor menu can be very large, and sometimes they cannot all be accommodated on one screen. For example, consider a display device that can display a maximum of 21 menu items. If the pull-down menu or the cursor menu has more items than can be displayed, the excess menu items are not displayed on the screen and cannot be accessed. You can overcome this problem by using cascading menus that let you define smaller groups of items within a menu section. When an item is selected, it loads the cascading menu and displays the items, defined in the cascading menu, on the screen.

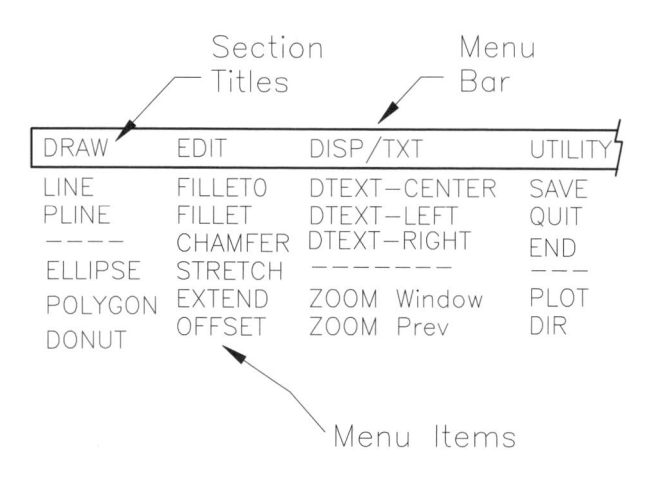

Pull-down, Shortcut, and Partial Menus and Customizing Toolbars   5-13

```
               Section         Menu
              ┌─ Titles       ┌─ Bar
   ┌──────────▼──────────────▼─────────────┐
   │ DRAW      EDIT      DISP/TXT   UTILITY │
   └───────────────────────────────────────┘
     LINE      FILLETO   DTEXT-CENTER  SAVE
     PLINE     FILLET    DTEXT-LEFT    QUIT
     ----      CHAMFER   DTEXT-RIGHT   END
     ELLIPSE   STRETCH   -------       ---
     POLYGON   EXTEND    ZOOM Window   PLOT
     DONUT     OFFSET    ZOOM Prev     DIR
                  ▲
                   ╲
                    ╲ Menu Items
```

Figure 5-6 Menu display for Exercise 1

The cascading feature of AutoCAD allows pull-down and cursor menus to be displayed in a hierarchical order that makes it easier to select submenus. To use the cascading feature in pull-down and cursor menus, AutoCAD has provided some special characters. For example, -> defines a cascaded submenu and <- designates the last item in the menu. The following table lists some of the characters that can be used with the pull-down or cursor menus.

Character	Character Description
--	The item label consisting of two hyphens automatically expands to fill the entire width of the menu. Example: [--]
+	Used to continue the menu item to the next line. This character has to be the last character of the menu item. Example: [Triang:]^C^CLine;1,1;+3,1;2,2;
->	This label character defines a cascaded submenu; it must precede the name of the submenu. Example: [->**Draw**]
<-	This label character designates the last item of the cascaded pull-down or cursor menu. The character must precede the label item. Example: [**<-CIRCLE 3P**]^C^CCIRCLE;3P
<-<-...	This label character designates the last item of the pull-down or cursor menu and also terminates the parent menu. The character must precede the label item. Example: [**<-<-Center Mark**]^C^C_dim;_center

Chapter 5

$(	This label character can be used with the pull-down and cursor menus to evaluate a DIESEL expression. The character must precede the label item. Example: $(if,$(getvar,orthomode),Ortho)
~	This item indicates the label item is not available (displayed grayed out); the character must precede the item. Example: [~Application not available]
!.	When used as a prefix, it displays the item with a check mark.
&	When placed directly before a character, the character is displayed underscored. For example, [W&Block] is displayed as WBlock. It also specifies the character as a menu accelerator key in the pull-down or Shortcut menu.
/c	When placed directly before an item with a character, the character is displayed underscored. For example, [/BWBlock] is displayed as WBlock. It also specifies that the item has a menu accelerator key in the pull-down or Shortcut menu.
\t	The label text to the right of \t is displayed to the right side of the menu.

Consider a display device with a certain resolution and font size that provides space for a maximum of 80 characters. Therefore, if there are 10 menus, the length of each menu title should average eight characters. If the combined length of all menu bar titles exceeds 80 characters, AutoCAD automatically wraps the excess menu items and displays them on the next line in the menu bar. The following is a list of some additional features of the menu.

1. The section labels of the menus are ***POP1 through ***POP16. The menu bar titles are displayed in the menu bar.

2. The menus can be accessed by selecting the menu title from the menu bar at the top of the screen.

3. A maximum of 999 menu items can be defined in the menu. This includes the items that are defined in the submenus. The menu items in excess of 999 are ignored.

4. The number of menu items that can be displayed depends on the display device you are using. If the cursor or the menu contains more items than can be accommodated on the screen, the excess items are truncated. For example, if your system can display only 35 menu items, the menu items in excess of 35 are automatically truncated.

Example 2

Write a pull-down menu for the commands shown in Figure 5-7. The menu must use the AutoCAD cascading feature.

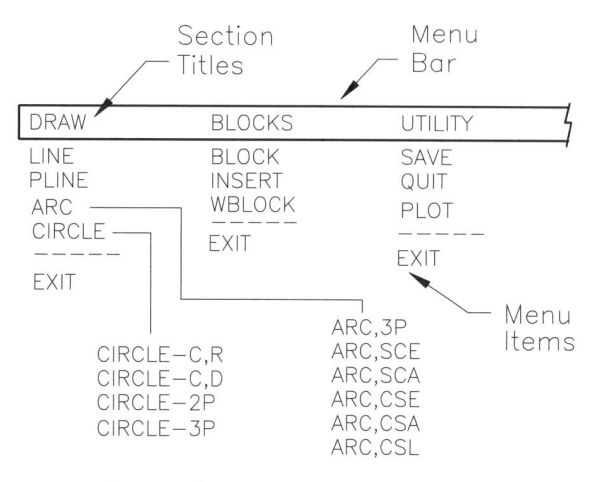

Pull-down, Shortcut, and Partial Menus and Customizing Toolbars   5-15

Figure 5-7 Menu structure for Example 2

Step 1: Writing the menu file
The following file is a listing of the menu for Example 2. **The line numbers are not a part of the menu; they are shown here for reference only.**

***POP1	1
[DRAW]	2
[LINE]^C^CLINE	3
[PLINE]^C^CPLINE	4
[->ARC]	5
[ARC]^C^CARC	6
[ARC,3P]^C^CARC	7
[ARC,SCE]^C^CARC;\C	8
[ARC,SCA]^C^CARC;\C;\A	9
[ARC,CSE]^C^CARC;C	10
[ARC,CSA]^C^CARC;C;\\A	11
[<-ARC,CSL]^C^CARC;C;\\L	12
[->CIRCLE]	13
[CIRCLE C,R]^C^CCIRCLE	14
[CIRCLE C,D]^C^CCIRCLE;\D	15
[CIRCLE 2P]^C^CCIRCLE;2P	16
[<-CIRCLE 3P]^C^CCIRCLE;3P	17
[--]	18
[Exit]^C	19
***POP	20
[BLOCKS]	21
[BLOCK]^C^CBLOCK	22
[INSERT]*^C^CINSERT	23
[WBLOCK]^C^CWBLOCK	24
[--]	25

```
[Exit]^C                                    26
***POP3                                     27
[UTILITY]                                   28
[SAVE]^C^CSAVE                              29
[QUIT]^C^CQUIT                              30
[PLOT]^C^CPLOT                              31
[--]                                        32
[Exit]^C                                    33
```

Explanation
Line 5
[->ARC]
In this menu item, **ARC** is the menu item label that is preceded by the special label characters **->**. These special characters indicate that the menu item has a submenu. The menu items that follow it (Lines 6-12) are the submenu items (Figure 5-8).

Line 12
[<-ARC,CSL]^C^CARC;C;\\L
In this line, the menu item label **ARC,CSL** is preceded by another special label characters, **<-**, which indicates the end of the submenu. The item that contains these characters must be the last menu item of the submenu.

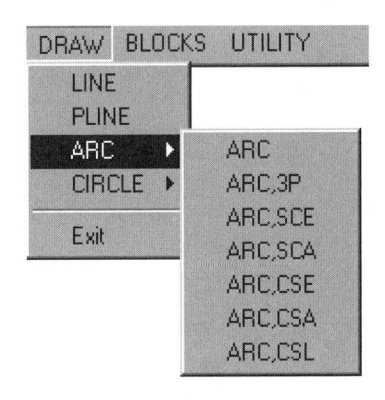

Figure 5-8 Draw menu

Lines 13 and 17
[->CIRCLE]
[<-CIRCLE 3P]^C^CCIRCLE;3P
The special characters -> in front of **CIRCLE** indicate that the menu item has a submenu; the characters <- in front of **CIRCLE 3P** indicate that this item is the last menu item in the submenu. When you select the menu item **CIRCLE** from the menu, it will automatically display the submenu on the side (Figure 5-9).

Step 2: Loading the menu file
Save the menu file with the extension ***.MNU**. Load the menu file using the **MENU** command as described in Example 1.

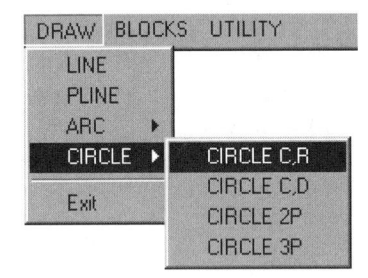

Figure 5-9 Draw menu with CIRCLE cascading menu

Example 3

Write a menu that has the cascading submenus for the commands shown in Figure 5-10.

Writing the menu file
The following file is a listing of the menu for Example 3. **The line numbers are not a part of the menu; they are shown here for reference only.**

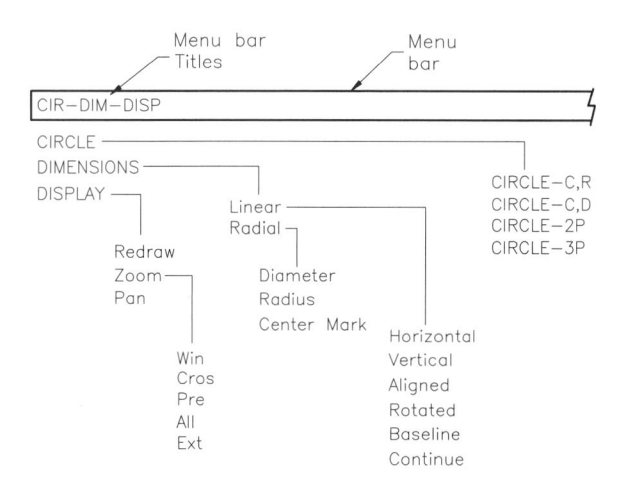

Pull-down, Shortcut, and Partial Menus and Customizing Toolbars 5-17

Figure 5-10 Menu structure for Example 3

```
***POP1                                        1
[DRAW]                                         2
[->CIRCLE]                                     3
  [CIRCLE C,R]^C^C_CIRCLE                      4
  [CIRCLE C,D]^C^C_CIRCLE;\_D                  5
  [CIRCLE 2P]^C^C_CIRCLE;_2P                   6
  [<-CIRCLE 3P]^C^C_CIRCLE;_3P                 7
[->Dimensions]                                 8
  [->Linear]                                   9
  [Horizontal]^C^C_dimlinear                  10
  [Vertical]^C^C_dimlinear                    11
  [Aligned]^C^C_dimaligned                    12
  [Rotated]^C^C_dimrotated                    13
  [Baseline]^C^C_dimbaseline                  14
  [<-Continue]^C^C_dimcontinue                15
  [->Radial]                                  16
  [Diameter]^C^C_dimdiameter                  17
  [Radius]^C^C_dimradius                      18
  [<-<-Center Mark]^C^C_dimcenter             19
[->DISPLAY]                                   20
  [REDRAW]^C^CREDRAW                          21
  [->ZOOM]                                    22
  [...Win]^C^C_ZOOM;_W                        23
  [...Cros]^C^C_ZOOM;_C                       24
  [...Pre]^C^C_ZOOM;_P                        25
  [...All]^C^C_ZOOM;_A                        26
  [<-...Ext]^C^C_ZOOM;_E                      27
[<-PAN]^C^C_Pan                               28
```

Explanation

Lines 8 and 9
[->Dimensions]
[->Linear]
The special label characters **->** in front of the menu item **Dimensions** indicate that it has a submenu, and the characters -> in front of **Linear** indicate that there is another submenu. The second submenu **Linear** is within the first submenu Dimensions (Figure 5-11). The menu items on lines 10 to 15 are defined in the **Linear** submenu, and the menu items Linear and **Radial** are defined in the submenu Dimensions.

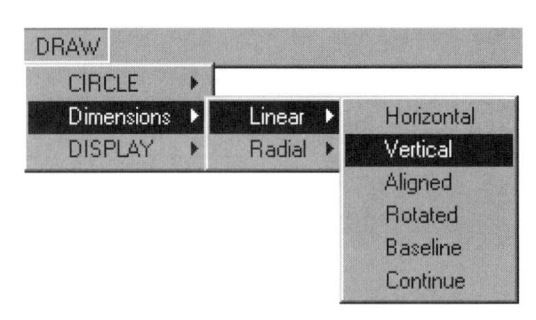

Figure 5-11 Draw menu with cascading menus

Line 16
[->Radial]
This menu item defines another submenu; the menu items on line numbers 17, 18, and 19 are part of this submenu.

Line 19
[<-<-Center Mark]^C^C_dim;_center
In this menu item the special label characters **<-<-** terminate the **Radial** and **Dimensions** (parent submenu) submenus.

Lines 27 and 28
[<-...Ext]^C^C_ZOOM;_E
[<-PAN]^C^C_Pan
The special characters **<-** in front of the menu item **...Ext** terminate the **ZOOM** submenu (Figure 5-12); the special character in front of the menu item **PAN** terminates the **DISPLAY** submenu.

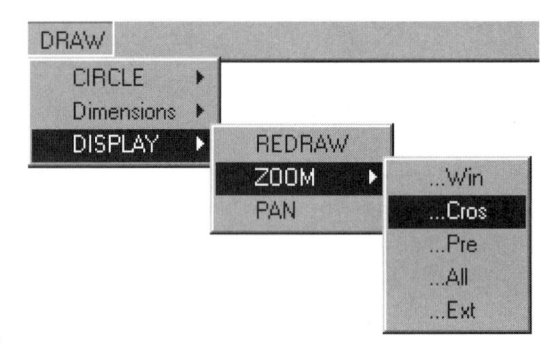

Figure 5-12 Draw pull-down menu with cascading menus

SHORTCUT AND CONTEXT MENUS

The Shortcut menus are similar to the pull-down menus, except the Shortcut menu can contain only 499 menu items compared with 999 items in the pull-down menu. The section label of the Shortcut menu can be ***POP0 and ***POP500 to ***POP 999. The Shortcut menus are displayed near or at the cursor location. Therefore, they can be used to provide convenient and quick access to some of the frequently used commands. The Shortcut menus that are in the upper range are also referred to as Context menus. The following is a list of some of the features of the Shortcut menu.

1. The section label of the Shortcut menu are ***POP0 and ***POP500 to ***POP999. The menu bar title defined under this section label is not displayed in the menu bar.

Pull-down, Shortcut, and Partial Menus and Customizing Toolbars 5-19

2. On most systems, the menu bar title is not displayed at the top of the Shortcut menu. However, for compatibility reasons you should give a dummy menu bar title.

3. The POP0 menu can be accessed through the **$P0=*** menu command. The shortcut menus POP500 through POP999 must be referenced by their alias names. The reserved alias names for AutoCAD use are GRIPS, CMDEFAULT, CMEDIT, and CMCOMMAND. For example, to reference POP500 for grips, use **GRIPS command line under POP500. This command can be issued by a menu item in another menu, such as the button menu, auxiliary menu, or the screen menu. The command can also be issued from an AutoLISP or ADS program.

4. A maximum of 499 menu items can be defined in the Shortcut menu. This includes the items that are defined in the Shortcut submenus. The menu items in excess of 499 are ignored.

5. The number of menu items that can be displayed on the screen depends on the system you are using. If the Shortcut or pull-down menu contains more items than your screen can accommodate, the excess items are truncated. For example, if your system displays 21 menu items, the menu items in excess of 21 are automatically truncated.

6. The system variable **SHORTCUTMENU** controls the availability of Default, Edit, and Command mode shortcut menus. If the value is 0, it restores R14 legacy behavior and disables the Default, Edit, and Command mode shortcut menus. The default value of this variable is 11.

Example 4

Write a Shortcut menu for the following AutoCAD commands using cascading submenus. The menu should be compatible with foreign language versions of AutoCAD. Use the third button of the **BUTTONS** menu to display the cursor menu.

Osnaps	Draw	DISPLAY
Center	Line	REDRAW
Endpoint	PLINE	ZOOM
Intersection	CIR C,R	...Win
Midpoint	CIR 2P	...Cen
Nearest	ARC SCE	...Prev
Perpendicular	ARC CSE	...All
Quadrant		...Ext
Tangent		PAN
None		

Writing the menu file

The following file is a listing of the menu file for Example 4. **The line numbers are not a part of the file; they are for reference only.**

```
***AUX1                                                          1
;                                                                2
$P0=*                                                            3
***POP0                                                          4
[Osnaps]                                                         5
[Center]_Center                                                  6
[End point]_Endp                                                 7
[Intersection]_Int                                               8
[Midpoint]_Mid                                                   9
[Nearest]_Nea                                                   10
[Perpendicular]_Per                                             11
[Quadrant]_Qua                                                  12
[Tangent]_Tan                                                   13
[None]_Non                                                      14
[—]                                                             15
[->Draw]                                                        16
  [Line]^C^C_Line                                               17
  [PLINE]^C^C_Pline                                             18
  [CIR C,R]^C^C_Circle                                          19
  [CIR 2P]^C^C_Circle;_2P                                       20
  [ARC SCE]^C^C_ARC;\C                                          21
  [<-ARC CSE]^C^C_Arc;C                                         22
[—]                                                             23
[->DISPLAY]                                                     24
  [REDRAW]^C^_REDRAW                                            25
  [->ZOOM]                                                      26
    [...Win]^C^C_ZOOM;_W                                        27
    [...Cen]^C^C_ZOOM;_C                                        28
    [...Prev]^C^C_ZOOM;_P                                       29
    [...All]^C^C_ZOOM;_A                                        30
    [<-...Ext]^C^C_ZOOM;_E                                      31
  [<-PAN]^C^C_Pan                                               32
***POP1                                                         33
[SHORTCUTMENU]                                                  34
[SHORTCUTMENU=0]^C^CSHORTCUTMENU;0                              35
[SHORTCUTMENU=1]^C^CSHORTCUTMENU;1                              36
```

Explanation

Line 1
***AUX1
AUX1 is the section label for the first auxiliary menu; *** designates the menu section. The menu items that follow it, until the second section label, are a part of this buttons menu.

Lines 2 and 3
;
$P0=*
The semicolon (;) is assigned to the second button of the pointing device (the first button of

the pointing device is the pick button); the special command **$P0=*** is assigned to the third button of the pointing device.

Lines 4 and 5
***POP0
[Osnaps]
The menu label **POP0** is the menu section label for the Shortcut menu; Osnaps is the menu bar title. The menu bar title is not displayed, but is required. Otherwise, the first item will be interpreted as a title and will be disabled.

Line 6
[Center]_Center
In this menu item, **_Center** is the center object Snap mode. The menu files can be used with foreign language versions of AutoCAD, if AutoCAD commands and the command options are preceded by the underscore (_) character.

After loading the menu, if you press the third button of your pointing device, the Shortcut menu (Figure 5-13) will be displayed at the cursor (screen crosshairs) location. If the cursor is close to the edges of the screen, the Shortcut menu will be displayed at a location that is closest to the cursor position. When you select a submenu, the items contained in the submenu will be displayed, even if the Shortcut menu is touching the edges of the screen display area.

Lines 33 and 34
***POP1
[Draw]
***POP1 defines the first pull-down menu. If no POPn sections are defined or the status line is turned off, the Shortcut menu is automatically disabled.

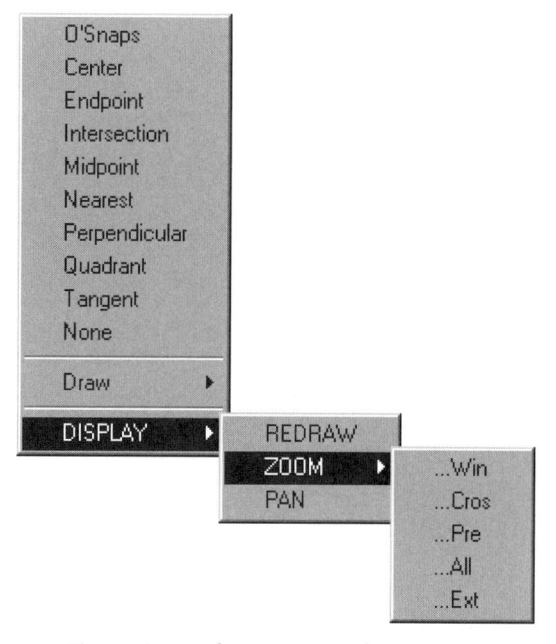

Figure 5-13 Shortcut menu for Example 4

Exercise 2   *General*

Write a menu for the following AutoCAD commands. Use a cascading menu for the **LINE** command options in the menu. (The layout of the menu is shown in Figure 5-14.)

LINE	**ZOOM All**	**TIME**
Continue	**ZOOM Win**	**LIST**
Close	**ZOOM Pre**	**DISTANCE**
Undo	**PAN**	**AREA**

```
    CIRCLE              DBLIST
    ELLIPSE             STATUS
```

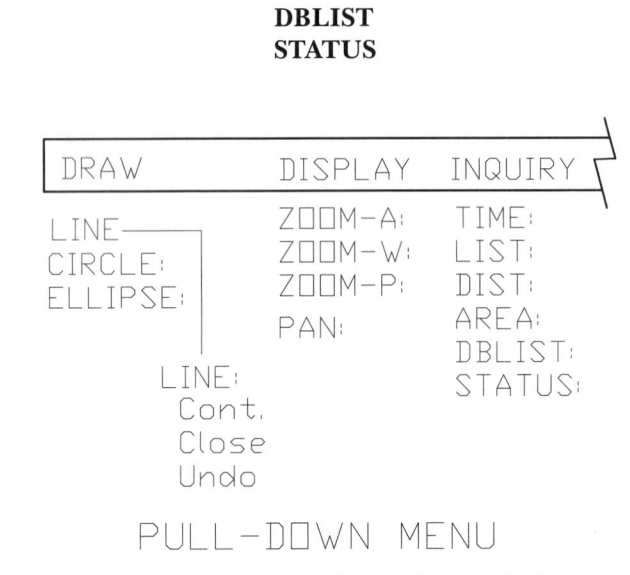

```
        ┌─────────────────────────────────────┐
        │ DRAW       DISPLAY    INQUIRY      │
        └─────────────────────────────────────┘
          LINE─┐     ZOOM-A:    TIME:
          CIRCLE:    ZOOM-W:    LIST:
          ELLIPSE:   ZOOM-P:    DIST:
                │    PAN:       AREA:
                │               DBLIST:
                │               STATUS:
                └─ LINE:
                   Cont.
                   Close
                   Undo

            PULL-DOWN MENU
```

Figure 5-14 Design of menu for Exercise 2

SUBMENUS

The number of items in a pull-down menu or Shortcut menu can be very large and sometimes they cannot all be accommodated on one screen. For example, consider a display device on which the maximum number of items that can be displayed is 21 items, the menu excess items are not displayed on the screen and cannot be accessed. You can overcome this problem by using submenus that let you define smaller groups of items within a menu section. When a submenu is selected, it loads the submenu items and displays them on the screen.

The menus that use AutoCAD's cascading feature are the most efficient and easy to write. The submenus follow a logical pattern that are easy to load and use without causing any confusion. It is strongly recommended to use the cascading menus whenever you need to write the pull-down or the Shortcut menus. However, AutoCAD provides the option to swap the submenus in the menus. These menus can sometimes cause distraction because the original menu is completely replaced by the submenu when swapping the menus.

Submenu Definition

A submenu definition consists of two asterisk signs (**) followed by the name of the submenu. A menu can have any number of submenus and every submenu should have a unique name. The items that follow a submenu, up to the next section label or submenu label, belong to that submenu. Following is the format of a submenu label:

    **Name

        Where ** ----------------- Two asterisk signs (**) designate a submenu
               Name ------------ Name of the submenu

Note
The submenu name can be up to 31 characters long.

The submenu name can consist of letters, digits, and the special characters like: $ (dollar), - (hyphen), and _ (underscore).

The submenu name cannot have any embedded blanks.

The submenu names should be unique in a menu file.

Submenu Reference

The submenu reference is used to reference or load a submenu. It consists of a "$" sign followed by a letter that specifies the menu section. The letter that specifies a menu section is Pn, where n designates the number of the menu section. The menu section is followed by "=" sign and the name of the submenu that the user wants to activate. The submenu name should be without "**". Following is the format of a submenu reference:

$Section=Submenu

Where **$** ------------------- "$" sign
Section ---------- Menu section specifier
= ------------------ "=" sign
Submenu ------- Name of submenu

Example
$P1=P1A

Where **$P1** -------------- P1-Specifies pull-down menu section 1
P1A -------------- Name of submenu

Displaying a Submenu

When you load a submenu in a menu, the submenu items are not automatically displayed on the screen. For example, when you load a submenu P1A that has DRAW-ARC as the first item, the current title of POP1 will be replaced by the DRAW-ARC. But the items that are defined under DRAW-ARC are not displayed on the screen. To force the display of the new items on the screen, AutoCAD uses a special command $Pn=*.

$Pn=*

Where **P** ------------------ P for menu
Pn ---------------- Menu section number (1 to 10)
***** -------------------- Asterisk sign (*)

LOADING MENUS

From the menu, you can load any menu that is defined in the screen or image tile menu sections by using the appropriate load commands. It may not be needed in most of the applications, but if you want to, you can load the menus that are defined in other menu sections.

Loading Screen Menus

You can load any menu that is defined in the screen menu section from the menu by using the following load command:

$S=X $S=LINE

        Where **S** -------------------- S specifies screen menu
                     X ------------------- Submenu name defined in screen menu section
                     LINE ------------- Submenu name defined in screen menu section

The first load command ($S=X) loads the submenu X that has been defined in the screen menu section of the menu file. The X submenu can contain 21 blank lines, so that when it is loaded it clears the screen menu. The second load command ($S=LINE) loads the submenu **LINE** that has also been defined in the screen menu section of the menu file.

Loading an Image Tile Menu

You can also load an image tile menu from the menu by using the following load command:

$I=IMAGE1 $I=*

        Where **$I=Image1** ----- Load the submenu IMAGE1
                     $I=* ------------- Display the dialog box

This menu item consists of two load commands. The first load command $I=IMAGE1 loads the image tile submenu IMAGE1 that has been defined in the image tile menu section of the file. The second load command $I=* displays the new dialog box on the screen.

Example 5

Write a pull-down menu for the following AutoCAD commands. Use submenus for the **ARC** and **CIRCLE** commands.

LINE	BLOCK	QUIT
PLINE	INSERT	SAVE
ARC	WBLOCK	PLOT
ARC 3P		
ARC SCE		
ARC SCA		
ARC CSE		
ARC CSA		
ARC CSL		
CIRCLE		
CIRCLE C,R		
CIRCLE C,D		
CIRCLE 2P		

Step 1: Designing the menu

The layout shown in Figure 5-15 is one of the possible designs for this menu. The **ARC** and **CIRCLE** commands are in separate groups that will be defined as submenus in the menu file.

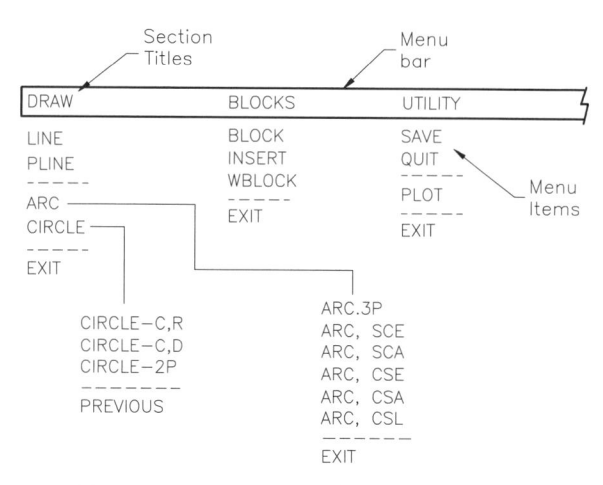

Pull-down, Shortcut, and Partial Menus and Customizing Toolbars 5-25

Figure 5-15 Design of menu for Example 5

Step 2: Writing the menu file
The following file is a listing of the menu for Example 5. The line numbers are not a part of the menu file. They are given here for reference only.

***POP1	1
**P1A	2
[DRAW]	3
[LINE]^C^CLINE	4
[PLINE]^C^CPLINE	5
[--]	6
[ARC]^C^C$P1=P1B $P1=*	7
[CIRCLE]^C^C$P1=P1C $P1=*	8
[--]	9
[Exit]^C	10
	11
**P1B	12
[ARC]	13
[ARC,3P]^C^CARC	14
[ARC,SCE]^C^CARC \C	15
[ARC,SCA]^C^CARC \C \A	16
[ARC,CSE]^C^CARC C	17
[ARC,CSA]^C^CARC C \\A	18
[ARC,CSL]^C^CARC C \\L	19
[--]	20
[Exit]^C	21
	22
**P1C	23
[CIRCLE]	24
[CIRCLE C,R]^C^CCIRCLE	25
[CIRCLE C,D]^C^CCIRCLE \D	26

```
[CIRCLE 2P]^C^CCIRCLE 2P                                               27
[--]                                                                   28
[PREVIOUS]$P1=P1A $P1=*                                                29
                                                                       30
***POP2                                                                31
[BLOCKS]                                                               32
[BLOCK]^C^CBLOCK                                                       33
[INSERT]*^C^CINSERT                                                    34
[WBLOCK]^C^CWBLOCK                                                     35
[--]                                                                   36
[Exit]$P1=P1A $P1=*                                                    37
                                                                       38
***POP3                                                                39
[UTILITY]                                                              40
[SAVE]^C^CSAVE                                                         41
[QUIT]^C^CQUIT                                                         42
[~--]                                                                  43
[PLOT]^C^CPLOT                                                         44
[~--]                                                                  45
[Exit]^C                                                               46
                                                                       47
```

Explanation
Line 2
P1A

**P1A defines the submenu P1A. All the submenus have two asterisk signs (**) followed by the name of the submenu. The submenu can have any valid name. In this example, P1A has been chosen because it is easy to identify the location of the submenu. P indicates that it is a menu, 1 indicates that it is in the first menu (POP1), and A indicates that it is the first submenu in that section.

Line 6
[--]

The two hyphens enclosed in the brackets will automatically expand to fill the entire width of the menu. This menu item cannot be used to define a command. If it does contain a command definition, the command is ignored. For example, if the menu item is [--]^C^CLINE, the command ^C^CLINE will be ignored.

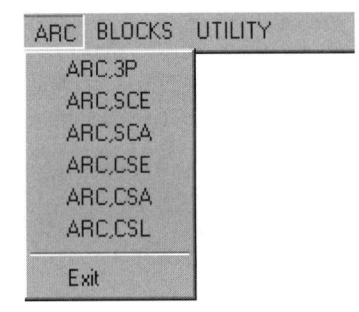

Figure 5-16 Arc pull-down menu replaces Draw menu

Line 7
[ARC]^C^C$P1=P1B $P1=*

In this menu item, $P1=P1B loads the submenu P1B and assigns it to the first menu section (POP1), but the new menu is not displayed on the screen. $P1=* forces the display of the new menu on the screen.

For example, if you select **CIRCLE** from the first menu (POP1), the menu bar title **DRAW** will be replaced by **CIRCLE**, but the new menu is not displayed on the screen. Now, if you select **CIRCLE** from the menu bar, the command defined under the **CIRCLE** submenu will be displayed in the menu. To force the display of the menu that is currently assigned to POP1, you can use AutoCAD's special command **$P1=***. If you select **CIRCLE** from the first menu (POP1), the **CIRCLE** submenu will be loaded and automatically displayed on the screen.

Line 21
[EXIT]^C
When you select this menu item the current menu will be canceled. It will not return to the previous submenu (**DRAW**). If you check the menu bar, it will display **ARC** as the section title, not **DRAW**. Therefore, it is not a good practice to cancel a submenu. It is better to return to the first menu before canceling it or define a command that automatically loads the previous menu and then cancels the menu.

Example
[EXIT]$P1=P1A $P1=* ^C^C

Line 29
[PREVIOUS]$P1=P1A $P1=*
In this menu item $P1=P1A loads the submenu P1A, which happens to be the previous menu in this case. You can also use $P1= to load the previous menu. $P1=* forces the display of the submenu P1A.

Note
When you swap the pull-down menus, the menus will get incremented. For example, if a menu file loads POP1 through POP8 and then a partial menu or a LISP routine inserts a new menu at POP5, the existing pull-down menus will be incremented by one. Thus, $P7=P7A $P= will take the new P7A menu and use it to replace what used to be POP6, but has now been pushed over to POP7.*

Step 3: Loading the menu file
Save the menu file and load it using the **MENU** command as described in earlier examples.

PARTIAL MENUS
AutoCAD has provided a facility that allows users to write their own menus and then load them in the menu bar. For example, in Windows you can write partial menus, toolbars, and definitions for accelerator keys. After you write the menu, AutoCAD lets you load the menu and use it with the standard menu. For example, you could load a partial menu and use it like a menu. You can also unload the menus that you do not want to use. These features make it convenient to use the menus that have been developed by AutoCAD users and developers.

Menu Section Labels
The following is a list of the additional menu section labels.

Section label	Description
***MENUGROUP	Menu file group name
***TOOLBARS	Toolbar definition
***HELPSTRING	Online help
***ACCELERATORS	Accelerator key definitions

Writing Partial Menus
The following example illustrates the procedure for writing a partial menu:

Example 6

In this example you will write a partial menu for Windows. The menu file has two menus, POP1 (MyDraw) and POP2 (MyEdit), as shown in Figure 5-17.

Step 1: Writing the menu file
Use a text editor to write the following menu file. The name of the file is assumed to be **MYMENU1.MNU**. The following is the listing of the menu file for this example:

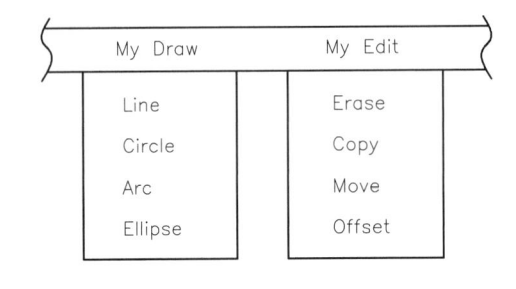

Figure 5-17 Pull-down menus for Example 6

```
***MENUGROU=Menu1                    1
***POP1                              2
[/MMyDraw]                           3
[/LLine]^C^CLine                     4
[/CCircle]^C^CCircle                 5
[/AArc]^C^CArc                       6
[/EEllipse]^C^CEllipse               7
***POP2                              8
[/EMyEdit]                           9
[/EErase]^C^CErase                  10
[/CCopy]^C^CCopy                    11
[/VMove]^C^CMove                    12
[/OOffset]^C^COffset                13
```

Explanation
Line 1
***MENUGROUP=Menu1
MENUGROUP is the section label and the Menu1 is the name tag for the menu group. The MENUGROUP label must precede all menu section definitions. The name of the MENUGROUP (Menu1) can be up to 32 characters long (alphanumeric), excluding spaces and punctuation marks. There is only one MENUGROUP in a menu file. All section labels must be preceded by *** (***MENUGROUP).

Line 2
***POP1
POP1 is the menu section label. The items on line numbers 3 through 7 belong to this section. Similarly, the items on line numbers 9 through 13 belong to the menu section **POP2**.

Line 3
[/MMyDraw]

Pull-down, Shortcut, and Partial Menus and Customizing Toolbars 5-29

/M defines the mnemonic key you can use to activate the menu item. For example, /M will display an underline under the letter M in the text string that follows it. If you enter the letter M, AutoCAD will execute the command defined in that menu item. MyDraw is the menu item label. The text string inside the brackets [ ] , except /M, has no function. It is used for displaying the function name so that the user can recognize the command that will be executed by selecting that item.

Line 4
[/LLine]^C^CLine
In this line, the /L defines the mnemonic key, and the Line that is inside the brackets is the menu item label. ^C^C cancels the command twice, and the Line is the AutoCAD **LINE** command. The part of the menu item statement that is outside the brackets is executed when you select an item from the menu. When you select Line 4, AutoCAD will execute the **LINE** command.

Step 2: Loading the menu file

Save the file as **MYMENU1.MNU**. Choose the **Customize Menu** button from the **Tools** menu or enter the **MENULOAD** command at the Command prompt to invoke the **Menu Customization** dialog box (Figure 5-18). To load the menu file, enter the name of the menu file, **MYMENU1.MNU**, in the **File Name**: edit box. You can also use the **Browse...** option to invoke the **Select Menu File** dialog box. Select the name of the file, and then use the **OK** button to return to the **Menu Customization** dialog box. To load the selected menu file, choose the **LOAD** button. The name of the menu group (**MENU1**) will be displayed in the **Menu Groups** list box.

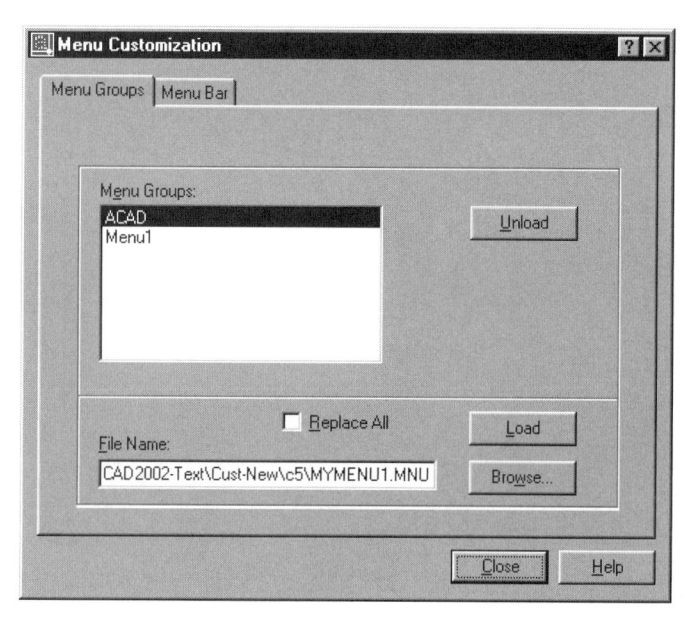

Figure 5-18 **Menu Customization** *dialog box (Menu Groups tab)*

You can also load the menu file from the command line, as follows:

Command: **FILEDIA**
Enter new value for FILEDIA <1>: **0**   *(Disables the file dialog boxes.)*
Command: **MENULOAD**
Enter name of menu file to load: **MYMENU1.MNU**

You can also use the AutoLISP functions to set the **FILEDIA** system variable to 0 and then load the partial menu.

Command: **(SETVAR "FILEDIA" 0)**
Command: **(Command "MENULOAD" "MYMENU1")**

Step 3: Inserting menus in the Menu Bar

In the **Menu Customization** dialog box, select the **Menu Bar** tab to display the menu bar options (Figure 5-19). In the **Menu Groups** list box select Menu1; the menus defined in the menu group (Menu1) will be displayed in the **Menus** list box. In the **Menus** list box select the menu (MyDraw) that you want to insert in the menu bar. In the **Menu Bar** list box select the position where you want to insert the new menu. For example, if you want to insert the new menu in front of Format, select the **Format** menu in the **Menu Bar** list box. Choose the **Insert** button to insert the selected menu (MyDraw) in the menu bar. The menu (MyDraw1) is displayed in the menu bar located at the top of your screen.

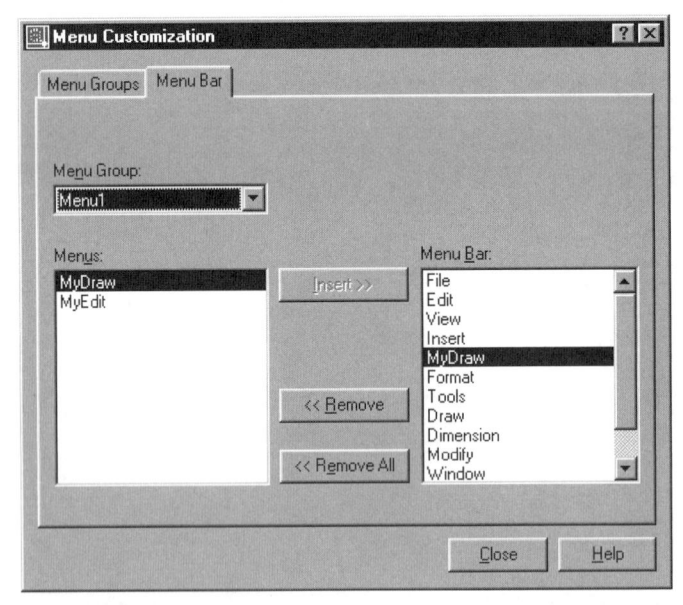

*Figure 5-19 Menu Customization dialog box (**Menu Bar** tab)*

You can also load the menu from the Command line. Once the menu is loaded, use the **MENUCMD** command (AutoLISP function) to display the partial menus.

Pull-down, Shortcut, and Partial Menus and Customizing Toolbars 5-31

Command: **(MENUCMD "P5=+Menu1.POP1")**
Command: **(MENUCMD "P6=+Menu1.POP2")**

After you enter these commands, AutoCAD will display the menu titles in the menu bar, as shown in Figure 5-20. If you select MyDraw, the corresponding menu as defined in the menu file will be displayed on the screen. Similarly, selecting MyEdit will display the corresponding edit menu.

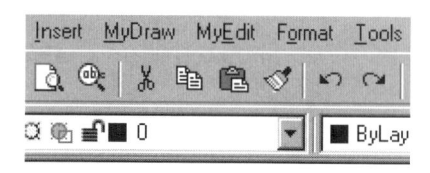

Figure 5-20 MENUCMD command places the menu titles in the menu bar

MENUCMD is an AutoLISP function, and P5 determines where the POP1 menu will be displayed. In this example the POP1 (MyDraw) menu will be displayed as the fifth menu. Menu1 is the MENUGROUP name as defined in the menu file, and POP1 is the menu section label. The MENUGROUP name and the menu section label must be separated by a period (.).

Step 4: Unloading the menu
If you want to unload the menu (MyDraw), select the Menu Bar tab of the **Menu Customization** dialog box. In the **Menu Bar** list box select the menu item that you want to unload. Choose the **Remove** button to remove the selected item.

Step 5: Unloading the menugroup
You can also unload the menu groups (Menu1) using the **Menu Customization** dialog box. Choose the **Customize Menu** button from the **Tools** menu or enter **MENULOAD** or **MENUUNLOAD** at the Command prompt to invoke the **Menu Customization** dialog box. AutoCAD will display the names of the menu files in the **Menu Groups:** list box. Select the **Menu1** menu, and then choose the **Unload** button. AutoCAD will unload the menu group. Choose the **Close** button to exit the dialog box.

You can also unload the menu file from the command line, as follows:

Command: **FILEDIA**
Enter new value for FILEDIA <1>: **0**
Command: **MENUUNLOAD**
Enter the name of a MENUGROUP to unload: **MENU1.MNU**

The **MENUUNLOAD** command unloads the entire menu group. You can also unload an individual menu without unloading the entire menu group by using the following command:

Command: **(MENUCMD "P5=-)**

This command will unload the menu at position five (P5, MyDraw menu). The menu group is still loaded, but the P5 menu is not visible. The menu can also be reinitialized by using the **MENU** command to load the base menu **ACAD.MNU**. This will remove all partial menus and the tag definitions associated with the partial menus.

Accelerator Keys

AutoCAD for Windows also supports user-defined accelerator keys. For example, if you enter **C** at the Command prompt, AutoCAD draws a circle because it is a command alias for a circle as defined in the ACAD.PGP file. You cannot use the **C** key to enter the **COPY** command. To use the **C** key for entering the **COPY** command, you can define the accelerator keys. You can combine the SHIFT key with C in the menu file so that when you hold down the SHIFT key and then press the C key, AutoCAD will execute the **COPY** command. The following example illustrates the use of accelerator keys.

Example 7

In this example you will add the following accelerator keys to the partial menu of Example 6.

CONTROL+"E"	to draw an ellipse (**ELLIPSE** command)
SHIFT+"C"	to copy (**COPY** command)
[CONTROL'Q"]	to quit (**QUIT** command)

Step 1: Writing the accelerators in a file

The following file is the listing of the partial menu file that uses the accelerator keys of Example 7.

```
***MENUGROUP=Menu1
***POP1
**Alias
[/MMyDraw]
[/LLine]^C^CLine
[/CCircle]^C^CCircle
[/AArc]^C^CArc
ID_Ellipse [/EEllipse]^C^CEllipse

***POP2
[/EMyEdit]
[/EErase]^C^CErase
ID_Copy [/CCopy]^C^CCopy
[/OOffset]^C^COffset
[/VMove]^C^CMov

***ACCELERATORS
ID_Ellipse [CONTROL+"E"]
ID_Copy [SHIFT+"C"]
[CONTROL'Q"]^C^CQuit
```

Explanation

This menu file defines three accelerator keys. The **ID_Copy [SHIFT+"C"]** accelerator key consists of two parts. The ID_Copy is the name tag, which must be the same as used earlier in the menu item definition. The SHIF "C" is the label that contains the modifier (SHIFT) and the keyname (C). The keyname or the string, such as "ESCAPE," must be enclosed in quotation marks.

The accelerator keys can be defined in two ways. One way is to give the name tag followed by the label containing the modifier. The modifier is followed by a single character or a special virtual key enclosed in quotation marks [CONTROL+"E"] or ["ESCAPE"]. You can also use the plus sign (+) to concatenate the modifiers [SHIFT + CONTROL + "L"]. The other way of defining an accelerator key is to give the modifier and the key string, followed by a command sequence [CONTROL "Q"]^C^CQuit.

Step 2: Loading the accelerators

Save the file with the extension **.MNU**. Load the file using the **MENU** command. After you load the file, SHIFT+C will enter the **COPY** command and CTRL+E will draw an ellipse. Similarly, Ctrl+Q will cancel the existing command and enter the **QUIT** command. If it does not work, you may have to unload the partial menu from Example 6 because it uses the same menugroup name.

Special Virtual Keys

The following are the special virtual keys. These keys must be enclosed in quotation marks when used in the menu file.

String	Description	String	Description
"F1"	F1 key	"NUMBERPAD0"	0 key
"F2"	F2 key	"NUMBERPAD1"	1 key
"F3"	F3 key	"NUMBERPAD2"	2 key
"F4"	F4 key	"NUMBERPAD3"	3 key
"F5"	F5 key	"NUMBERPAD4"	4 key
"F6"	F6 key	"NUMBERPAD5"	5 key
"F7"	F7 key	"NUMBERPAD6"	6 key
"F8"	F8 key	"NUMBERPAD7"	7 key
"F9"	F9 key	"NUMBERPAD8"	8 key
"F10"	F10 key	"NUMBERPAD9"	9 key
"F11"	F11 key	"UP"	UP-ARROW key
"F12"	F12 key	"DOWN"	DOWN-ARROW key
"HOME"	HOME key	"LEFT"	LEFT-ARROW key
"END"	END key	"RIGHT"	RIGHT-ARROW key
"INSERT"	INS key	"ESCAPE"	ESC key
"DELETE"	DEL key		

Valid Modifiers

The following are the valid modifiers:

String	Description
CONTROL	The CTRL key on the keyboard
SHIFT	The SHIFT key (left or right)
COMMAND	The Apple key on Macintosh keyboards
META	The meta key on UNIX keyboards

Toolbars

The contents of the toolbar and its default layout can be specified in the Toolbar section (***TOOLBARS) of the menu file. Each toolbar must be defined in a separate submenu.

Toolbar Definition. The following is the general format of the toolbar definition:

```
***TOOLBARS
**MYTOOLS1
TAG1 [Toolbar ("tbarname", orient, visible, xval, yval, rows)]
TAG2 [Button ("btnname", id_small, id_large)]macro
TAG3 [Flyout ("flyname", id_small, id_large, icon, alias)]macro
TAG4 [control (element)]
[—]
```

***TOOLBARS** is the section label of the toolbar, and **MYTOOLS1** is the name of the submenu that contains the definition of a toolbar. Each toolbar can have five distinct items that control different elements of the toolbar: TAG1, TAG2, TAG3, TAG4, and separator ([—]). The first line of the toolbar (TAG1) defines the characteristics of the toolbar. In this line, **Toolbar** is the keyword, and it is followed by a series of options enclosed in parentheses. The following describes the available options.

tbarname	This is a text string that names the toolbar. The tbarname text string must consist of alphanumeric characters with no punctuation other than a dash (-) or an underscore (_).
orient	This determines the orientation of the toolbar. The acceptable values are Floating, Top, Bottom, Left, and Right. These values are not case-sensitive.
visible	This determines the visibility of the toolbar. The acceptable values are Show and Hide. These values are not case-sensitive.
xval	This is a numeric value that specifies the X ordinate in pixels. The X ordinate is measured from the left edge of the screen to the left side of the toolbar.
yval	This is a numeric value that specifies the Y ordinate in pixels. The Y ordinate is measured from the top edge of the screen to the top of the toolbar.
rows	This is a numeric value that specifies the number of rows.

The second line of the toolbar (TAG2) defines the button. In this line the **Button** is the key word and it is followed by a series of options enclosed in parentheses. The following is the description of the available options.

Pull-down, Shortcut, and Partial Menus and Customizing Toolbars 5-35

btnname This is a text string that names the button. The text string must consist of alphanumeric characters with no punctuation other than a dash (-) or an underscore (_). This text string is displayed as ToolTip when you place the cursor over the button.

id_small This is a text string that names the ID string of the small image resource (16 by 16 bitmap). The text string must consist of alphanumeric characters with no punctuation other than a dash (-) or an underscore (_). The id_small text string can also specify a user-defined bitmap (Example: ICON_16_CIRCLE). The **bit map images must exist or you just get a smiley face**.

id_big This is a text string that names the ID string of the large image resource (32 by 32 bitmap). The text string must consist of alphanumeric characters with no punctuation other than a dash (-) or an underscore (_). The id_big text string can also specify a user-defined bitmap (Example: ICON_32_CIRCLE).

macro The second line (TAG2), which defines a button, is followed by a command string (macro). For example, the macro can consist of ^C^CLine. It follows the same syntax as that of any standard menu item definition.

The third line of the toolbar (TAG3) defines the flyout control. In this line the **Flyout** is the key word, and it is followed by a series of options enclosed in parentheses. The following describes the available options.

flyname This is a text string that names the flyout. The text string must consist of alphanumeric characters with no punctuation other than a dash (-) or an underscore (_). This text string is displayed as ToolTip when you place the cursor over the flyout button.

id_small This is a text string that names the ID string of the small image resource (16 by 16 bitmap). The text string must consist of alphanumeric characters with no punctuation other than a dash (-) or an underscore (_). The id_small text string can also specify a user-defined bitmap.

id_big This is a text string that names the ID string of the large image resource (32 by 32 bitmap). The text string must consist of alphanumeric characters with no punctuation other than a dash (-) or an underscore (_). The id_big text string can also specify a user-defined bitmap.

icon This is a Boolean key word that determines whether the button displays its own icon or the last icon selected. The acceptable values are **ownicon** and **othericon**. These values are not case-sensitive.

alias The alias specifies the name of the toolbar submenu that is defined with the standard ****aliasname** syntax.

macro	The third line (TAG3), which defines a flyout control, is followed by a command string (macro). For example, the macro can consist of ^C^CCircle. It follows the same syntax as that of any standard menu item definition.

The fourth line of the toolbar (TAG4) defines a special control element. In this line the Control is the key word, and it is followed by the type of control element enclosed in parentheses. The following describes the available control element types.

element	This parameter can have one of the following three values: Layer: This specifies the layer control element. Linetype: This specifies the linetype control element. Color: This specifies the color control element.

The fifth line ([--]) defines a separator.

Example 8

In this example you will write a menu file for a toolbar for the **LINE**, **PLINE**, **CIRCLE**, **ELLIPSE**, and **ARC** commands. The name of the toolbar is MyDraw1 (Figure 5-21).

Step 1: Writing the toolbars

Use any text editor to list the toolbars. The following is the listing of the menu file containing toolbars. In this file listing, ID specifies the name tag.

```
MENUGROUP=M1
***TOOLBARS
**TB_MyDraw1
ID_MyDraw1[_Toolbar("MyDraw1", _Floating, _Hide, 10, 200, 1)]
ID_Line  [_Button("Line", ICON_16_LINE, ICON_32_LINE)]^C^C_line
ID_Pline [_Button("Pline", ICON_16_PLine, ICON_32_PLine)]^C^C_PLine
ID_Circle[_Button("Circle", ICON_16_CirRAD, ICON_32_CirRAD)]^C^C_Circle
ID_ELLIPSE[_Button("Ellipse",ICON_16_EllCEN, ICON_32_EllCEN)]^C^C_ELLIPSE
ID_Arc[_Button("Arc 3Point", ICON_16_Arc3Pt, ICON_32_Arc3Pt)]^C^C_Arc
```

Step 2: Loading the menu file containing toolbars

Save the file with the extension **.MNU**. Use the **MENULOAD** command to load the **MyDraw1** menu group as discussed earlier while loading the partial menus. To display the new toolbar (**MyDraw1**) on the screen, select **Toolbars** from the **View** menu and then select **MyDraw1** in the **Menu Groups** list box. Turn the **MyDraw1** toolbar on in the **Toolbars** list box. MyDraw1 toolbar is displayed on the screen.

You can also load the new toolbar from the command line. After using the **MENULOAD** command to load the MyDraw1 menu group, use the **-TOOLBAR** command to display the MyDraw1 toolbar.

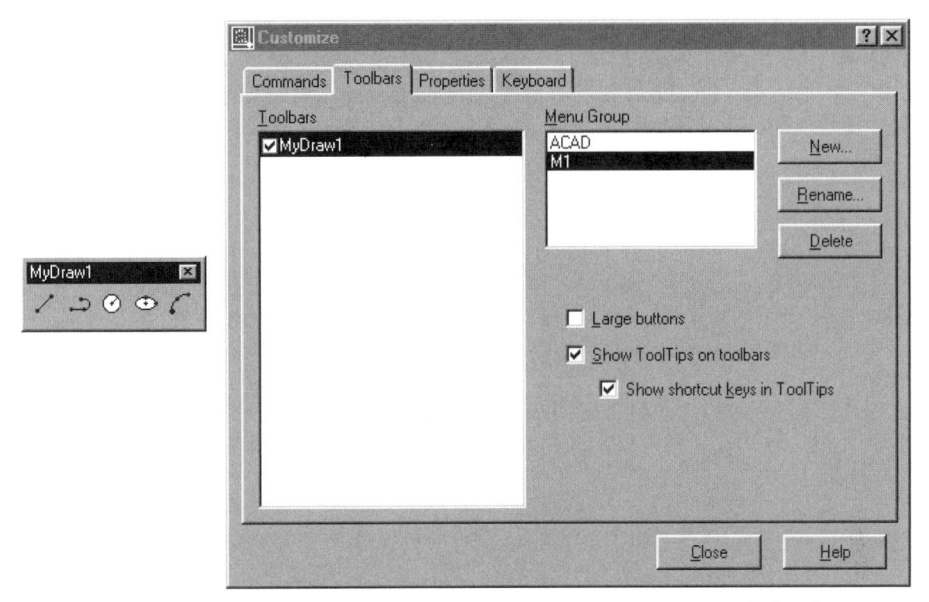

Pull-down, Shortcut, and Partial Menus and Customizing Toolbars   5-37

*Figure 5-21 MyDraw1 toolbar for Example 8 and **Toolbars** dialog box*

Command: **-TOOLBAR**
Enter toolbar name or [ALL]: **MYDRAW1**
Enter an option [Show/Hide/Left/Right/Top/Bottom/Float] <Show>: **S**

Example 9

In this example you will write a menu file for a toolbar with a flyout. The name of the toolbar is MyDraw2 (Figure 5-22), and it contains two buttons, Circle and Arc. When you select the Circle button, it should display a flyout with radius, diameter, 2P, and 3P buttons (Figure 5-23). Similarly, when you select the Arc button, it should display the 3Point, SCE, and SCA buttons.

Step 1: Writing the toolbars menu file
Use any text editor to write the menu file for the toolbars. The following is the listing of the menu file.

```
***Menugroup=M2
***TOOLBARS
**TB_MyDraw2
ID_MyDraw2[_Toolbar("MyDraw2", _Floating, _Show, 10, 100, 1)]
ID_TbCircle[_Flyout("Circle", ICON_16_Circle, ICON_32_Circle, _OtherIcon, M2.TB_Circle)]
ID_TbArc[_Flyout("Arc", ICON_16_Arc, ICON_32_Arc, _OtherIcon, M2.TB_Arc)]
**TB_Circle
ID_TbCircle[_Toolbar("Circle", _Floating, _Hide, 10, 150, 1)]
```

ID_CirRAD[_Button("Circle C,R", ICON_16_CirRAD, ICON_32_CirRAD)]^C^C_Circle
ID_CirDIA[_Button("Circle C,D", ICON_16_CirDIA, ICON_32_CirDIA)]^C^C_Circle;\D
ID_Cir2Pt[_Button("Circle 2Pts", ICON_16_Cir2Pt, ICON_32_Cir2Pt)]^C^C_Circle;2P
ID_Cir3Pt[_Button("Circle 3Pts", ICON_16_Cir3Pt, ICON_32_Cir3Pt)]^C^C_Circle;3P

**TB_Arc
ID_TbArc[_Toolbar("Arc", _Floating, _Hide, 10, 150, 1)]
ID_Arc3PT[_Button("Arc,3Pts", ICON_16_Arc3PT, ICON_32_Arc3PT)]^C^C_Arc
ID_ArcSCE[_Button("Arc,SCE", ICON_16_ArcSCE, ICON_32_ArcSCE)]^C^C_Arc;\C
ID_ArcSCA[_Button("Arc,SCA", ICON_16_ArcSCA, ICON_32_ArcSCA)]^C^C_Arc;\C;\A

Explanation

ID_TbCircle[_Flyout("Circle", ICON_16_Circle, ICON_32_Circle, _OtherIcon, M2.TB_Circle)]

In this line M2 is the MENUGROUP name (***MENUGROUP=M2) and TB_Circle is the name of the toolbar submenu. **M2.TB_Circle** will load the submenu TB_Circle that has been defined in the M2 menugroup. If M2 is missing, AutoCAD will not display the flyout when you select the Circle button.

ID_CirDIA[_Button("Circle C,D", ICON_16_CirDIA, ICON_32_CirDIA)]^C^C_Circle;\D

CirDIA is a user-defined bitmap that displays the Circle-diameter button. If you use any other name, AutoCAD will not display the desired button.

Step 2: Loading the menu file containing the toolbars
To load the toolbar, use the **MENULOAD** command to load the menu file. The **MyDraw1** toolbar will be displayed on the screen.

Figure 5-22 Toolbar for Example 9

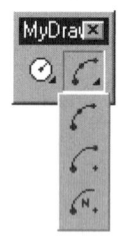

Figure 5-23 Toolbar with flyout

Menu-Specific Help
AutoCAD for Windows allows access to online help. For example, if you want to define a helpstring for the **CIRCLE** and **ARC** commands, the syntax is as follows:

***HELPSTRINGS
ID_Copy *[This command will copy the selected object.]*
ID_Ellipse *[(This command will draw an ellipse.]*

The ***HELPSTRING** is the section label for the helpstring menu section. The lines defined in this section start with a name tag (ID_Copy) and are followed by the label enclosed in square brackets. The Helpstring tag names are limited to 12 characters. When a menu item is highlighted, AutoCAD searches for the name tag for the corresponding entry in the ***HELPSTRINGS section. If there is a match, the text string contained within the label is displayed in the status line.

CUSTOMIZING THE TOOLBARS

AutoCAD has provided several toolbars that should be sufficient for general use. However, sometimes you may need to customize the toolbars so that the commands that you use frequently are grouped in one toolbar. This saves time in selecting commands. It also saves the drawing space because you do not need to have several toolbars on the screen. The following example explains the process involved in creating and editing the toolbars.

Example 10

In this example you will create a new toolbar (MyToolbar1) that has Line, Polyline, Circle (Center, Radius option), Arc (Center, Start, End option), Spline, and Paragraph Text (**MTEXT**) commands. You will also change the image and tooltip of the Line button and perform other operations like deleting toolbars and buttons and copying buttons between toolbars.

Step1
Select the toolbars in the **View** menu. The **Toolbars** dialog box will appear on the screen. You can also invoke this dialog box by entering **TBCONFIG** at AutoCAD's Command prompt.

Step 2
Choose **New...** button in the **Toolbars** dialog box; AutoCAD will display the **New Toolbar** dialog box at the top of the existing (Toolbars) dialog box.

Step 3
Enter the name of the toolbar in the **Toolbar Name** edit box (Example MyToolbar1) and then choose the **OK** button to exit the dialog box. The name of the new toolbar (MyToolbar1) is displayed in the **Customize** dialog box (Figure 5-24).

Step 4
Select the new toolbar so that it is highlighted, if it is not already selected, by selecting the box that is located just to the left of the toolbar name (MyToolbar1). The new toolbar (MyToolbar1) appears on the screen.

Step 5
Choose the **Commands** tab in the **Customize** dialog box (Figure 5-25). The **Categories** and **Commands** lists are displayed in the dialog box.

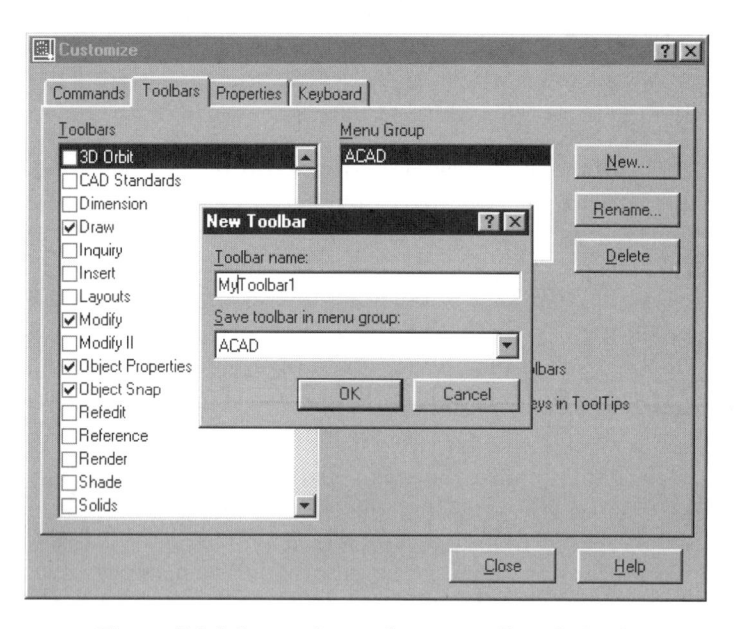

Figure 5-24 Customize and New Toolbar dialog box

Step 6

In the **Categories** list box select the Draw item. The draw commands are displayed in the Commands area of the **Customize** dialog box.

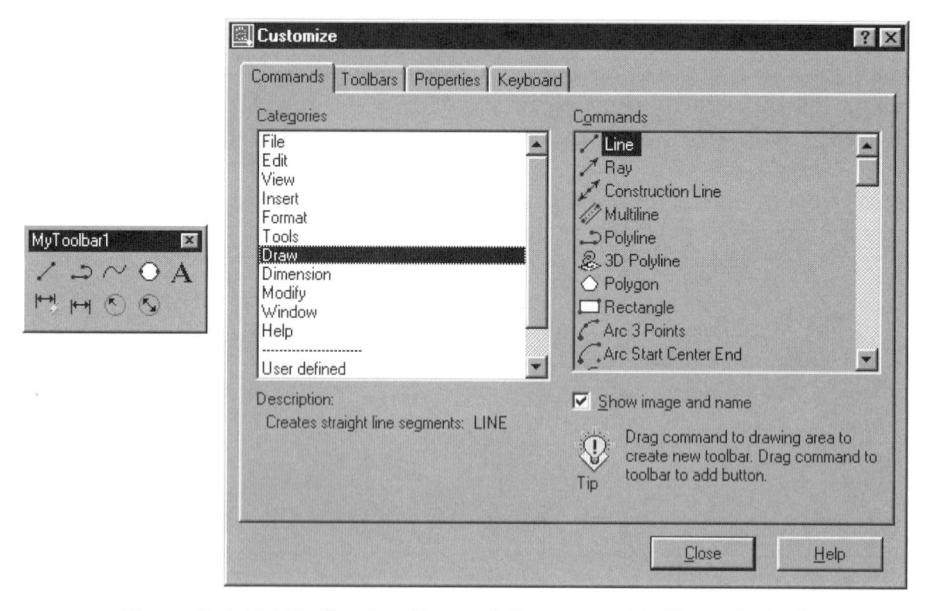

*Figure 5-25 MyToolbar1 toolbar and **Customize Toolbars** dialog box*

Step 7
Choose and drag the Line command and position it in the MyToolbar1 toolbox. Repeat the same for Polyline, Circle (Center, Radius option), Arc (Center, Start, End option), Spline, and Text (Paragraph text) buttons.

Step 8
Now, select the Dimension item in the Category list box and then select and drag some of the dimensioning commands to MyToolbar1.

Step 9
Similarly, you can open any command category and add commands to the new toolbar.

Step 10
When you are done defining the commands for the new toolbar, choose the **Close** button in the **Customize Toolbars** dialog box to return to the AutoCAD screen.

Step 11
Test the buttons in the new toolbar (MyToolbar1).

Creating a New Image and Tooltip for a Button

Step 1
Make sure the button with the image you want to edit is displayed on the screen. In this example we want to edit the image of the **Line** button of MyToolbar1.

Step 2
Right-click on any toolbar and select **Customize**; the **Customize** dialog box appears on the screen. Click on the **Button Properties** tab. Double click on the **Line** tool in the **MtTool1** toolbar; the button properties are displayed with the image of the existing **Line** tool (Figure 5-26).

Step 3
To edit the shape of the image, choose the **edit** button to access the **Button Editor dialog box**. Select the Grid to display the grid lines.

You can edit the shape by using different tools in the **Button Editor**. You can draw a Line by choosing the **Line** button and specifying two points. You can draw a circle or an ellipse by using the **Circle** button. The **Erase** button can be used to erase the image.

Step 4
In this example, we want to change the color of Line image. To accomplish this, erase the existing line, and then select the color and draw a line. Also, create the shape L in the lower right corner (Figure 5-27).

Step 5
Choose the **SaveAs** button and save the image as **MyLine** in the Tutorial directory. Choose the

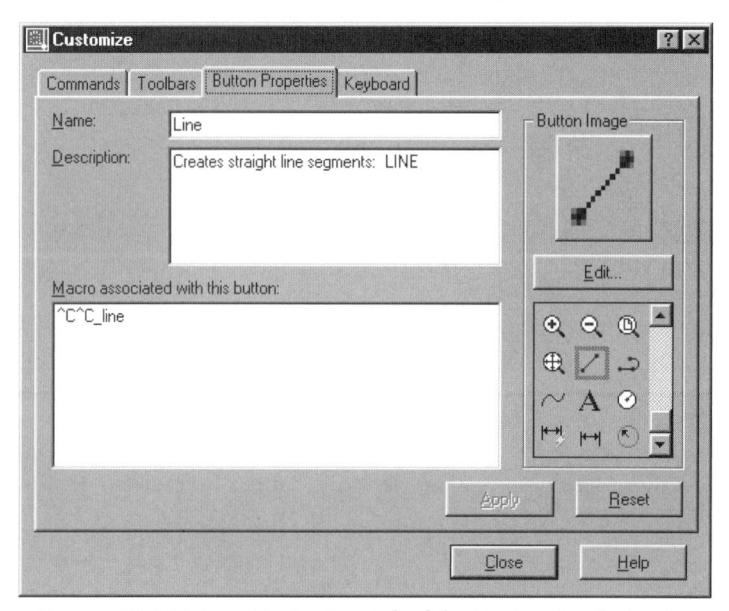

Figure 5-26 Button Properties tab of the Customize dialog box

Close button to exit the **Button Editor**. Using Save instead of SaveAs will redefine the button image for all existing toolbars

Step 6

In the **Button Properties** tab of the **Customize** dialog box, enter MyLine in the **Name** edit box. This changes the tooltip of this button. Now, choose the **Apply** button to apply the changes to the image. Close the dialog boxes to return to the AutoCAD screen. Notice the change in the button image and tooltip (Figure 5-28).

Figure 5-28 MyToollbar1 Toolbar

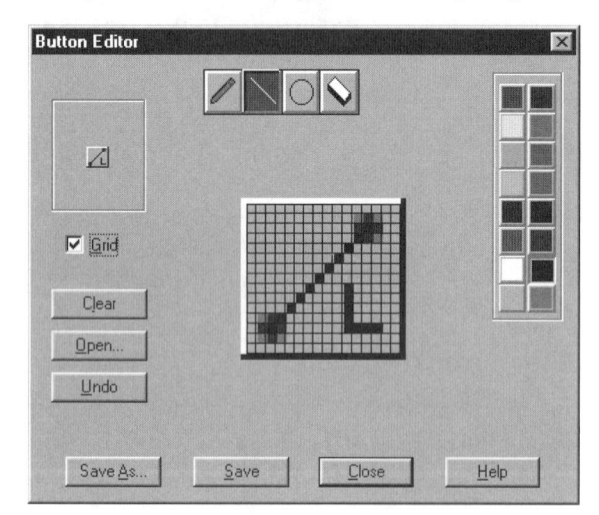

Figure 5-27 Button Editor dialog box

Deleting the Button from a Toolbar
Step 1
Right-click on any toolbar to display the shortcut menu. Select **Customize** from the shortcut menu to display the **Customize** dialog box.

Step 2
Click and drag out the **Spline** button from the **MyToolbar1** toolbox. The button will be deleted from the toolbox. Repeat the above step to delete other buttons, if needed. Close the dialog boxes to return to the screen.

Deleting a Toolbar
Select Toolbars from the **View** menu. In the Toolbar list box select the Toolbar that you want to delete and then choose the **Delete** button. The toolbar you selected is deleted.

Copying a Tool Button

Step 1
In this example we want to copy the ordinate **dimensioning** button from the **Dimensioning** toolbar to MyToolbar1. Select the Toolbars from the **View** menu or right-click on any button in any toolbar to display the shortcut menu and then select **Customize** to display the **Customize** dialog box. In the **Customize** dialog box select the **Commands** tab and then select the **Dimensions** in the **Categories** list box. The dimensioning commands are displayed in the **Commands** list box.

Step 2
Choose and drag the **Ordinate dimensioning** button from the **Commands list box** to MyToolbar1. The Ordinate dimensioning toolbar is copied to MyToolbar1.

Any changes made in the toolbars are saved in the ACAD.MNS and ACAD.MNR files. The following is the partial listing of ACAD.MNS file:

```
**MYTOOLBAR1
ID_MyToolbar1_0 [_Toolbar("MyToolbar1", _Floating, _Show, 512, 177, 1)]
ID_Line_0     [_Button("MyLine", "ICON.bmp", "ICON_24_LINE")]^C^C_line
ID_CircleCenterRadius_0 [_Button("Circle Center Radius", "ICON_16_CIRRAD",
"ICON_24_CIRRAD")]^C^C_circle
ID_Polyline_0 [_Button("Polyline", "ICON_16_PLINE", "ICON_24_PLINE")]^C^C_pline
ID_ArcCenterStartEnd_0 [_Button("Arc Center Start End", "ICON_16_ARCCSE",
"ICON_24_ARCCSE")]^C^C_arc _c
```

Creating Custom Toolbars with Flyout Icons
In this example you will create a custom toolbar with flyout buttons.

Step 1
Right-click on any toolbar, select Customize from the shortcut menu to display the **Customize**

dialog box on the screen. In the **Customize** dialog box, click on the Toolbars tab and choose the **New...** button to display the **New Toolbar** dialog box. Enter the name of the new toolbar (for example, MYToolbar2) and then choose the **OK** button to exit the box.

Step 2
In the Customize dialog box select the Commands tab and then select the Flyouts from the Categories list box. The names of the flyouts are displayed in the Commands list box. Click and drag the Draw flyout button and drop it in the MyToolbar2 toolbar. The flyout created this way contains all draw commands.

Step 3
In the **Customize** dialog box select the Commands tab and then select the **User defined** from the **Categories** list box. The names of the commands are displayed in the **Commands** list box, Figure 5-29. Click and drag the User Defined Flyout button and drop it in the MyToolbar2 toolbar. Now click on the user defined button in MyToolbar2; the AutoCAD message box appears on the screen. Click on the OK button to return to **Customize** toolbar. In the **Flyout Properties** tab select **Inquiry** and then click on the **Apply** button. The **Inquiry** toolbar gets associated with the custom flyout.

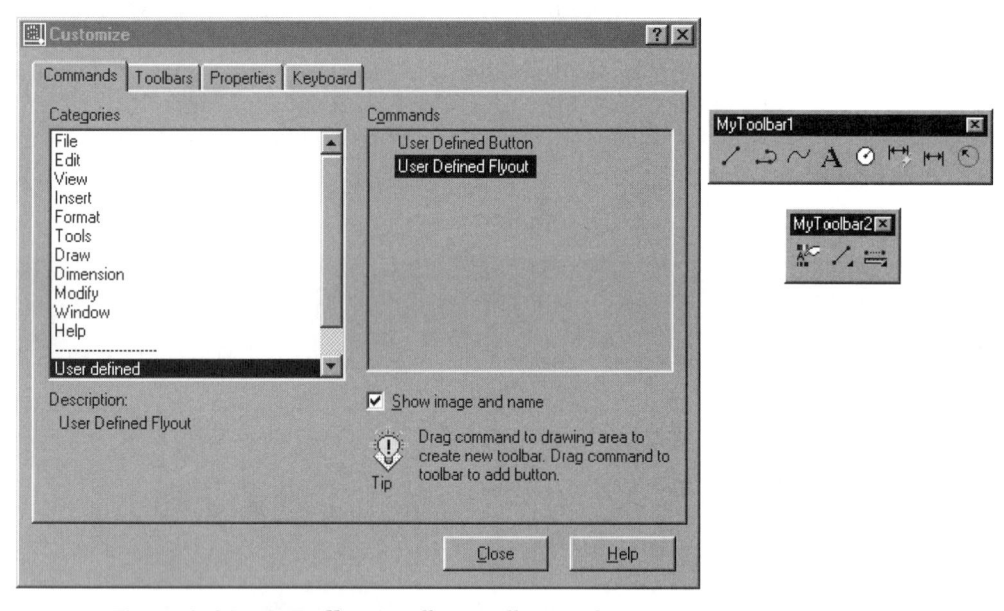

Figure 5-29 MyToolbar1 toolbar, Toolbars and Button Properties dialog box

Assigning Keyboard Shortcuts to Commands
Step 1
Select the Toolbars from the **View** menu or right-click on any button in any toolbar to display the shortcut menu and then select **Customize** to display the **Customize** dialog box. In the **Customize** dialog box select the **Keyboards** tab and then select the **Draw Menu** in the **Categories** drop-down list, Figure 5-30. The draw commands are displayed in the **Commands** list box.

Step 2

In the **Press new shortcut key** edit box, specify a key combination to be used as the keyboard shortcut for the selected menu or toolbar item. To specify a value, simultaneously press CTRL and a letter on the keyboard. You can also simultaneously press CTRL+SHIFT and a letter. In this example, hold the CTRL down and press the L key from the keyboard. You can also click on the **Show All** button to display the shortcut keys.

Step 3

Click on the Assign button to assign the command to the shortcut key and then click on the Close button to close the Customize dialog box. Now, if you hold the CTRL key down and press the L key from the keyboard, the LINE command is invoked.

Note
You cannot reassign shortcut keys that are internally assigned to Windows, for example, F10, CTRL+F4, CTRL+F6, or CTRL+ALT+DEL. If you use an invalid key combination, AutoCAD does not show the combination in the edit box. Try using another key combination.

If the keyboard shortcut you specify is already assigned to another AutoCAD command, the "Currently assigned to" message is displayed.

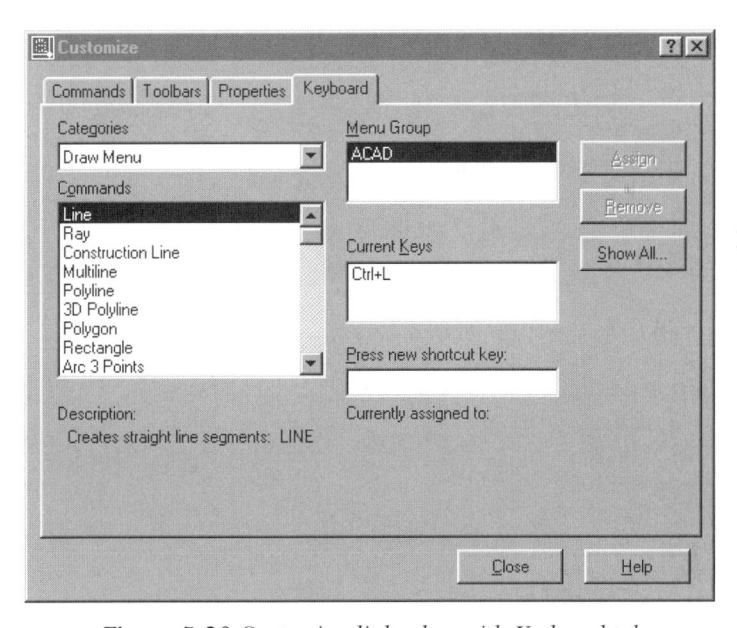

Figure 5-30 Customize dialog box with Keyboard tab

Self-Evaluation Test

Answer the following questions and then compare your answers to the correct answers at the end of this chapter.

1. The name of the standard menu file is _____ .

2. Menu files can be loaded by using the _____ command.

3. Each section in the menu file is identified by a _____.

4. The pop menu can have up to _____.

5. The tablet menus can have up to _____ different sections.

6. _____ can be given if you need to use input from a keyboard or a pointing device.

7. The sign _____ defines a cascaded submenu while _____ designates the last item in the menu.

8. All section labels in the AutoCAD menu start with _____.

9. _____ designates the last item of the pull-down or cursor menu.

10. _____ when used as a prefix, displays the item with a check mark.

11. The system variable _____s___ controls the availability of the default, edit and command mode shortcut menus.

12. A submenu definition consists of _____.

13. The Menu Customization dialog box can be invoked by either _____ or _____ command.

Review Questions

Answer the following questions.

1. A pull-down menu can have _____ sections.

2. The length of the section title should not exceed _____ characters.

3. The section titles in a pull-down menu are _____ justified.

Pull-down, Shortcut, and Partial Menus and Customizing Toolbars 5-47

4. In a pull-down menu, a line consisting of two hyphens ([--]) _____ automatically to fill the _____ of the menu.

5. If the menu item begins with a tilde (~), the items will be _____ .

6. Every cascading menu in the menu file should have a _____ name.

7. The cascading menu name can be _____ characters long.

8. The cascading menu names should not have any _____ blanks.

9. In Windows you can write partial menus, toolbars, and accelerator key definitions. (T/F)

10. A menu file can contain only one MENUGROUP. (T/F)

11. You can load the partial menu file by using the AutoCAD _____ command.

12. Once the menu is loaded, you can use the _____ (AutoLISP function) to display the partial menus.

Exercises

Exercise 3 — *General*

Write a pull-down menu for the following AutoCAD commands. (The layout of the menu is shown in Figure 5-31.)

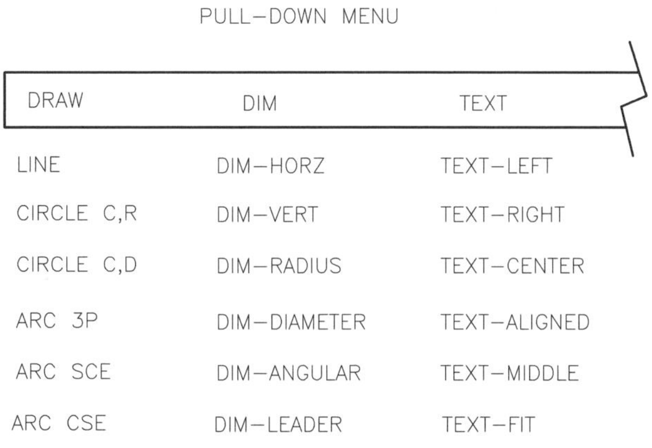

```
                    PULL-DOWN MENU

        DRAW          DIM             TEXT

        LINE          DIM-HORZ        TEXT-LEFT
        CIRCLE C,R    DIM-VERT        TEXT-RIGHT
        CIRCLE C,D    DIM-RADIUS      TEXT-CENTER
        ARC 3P        DIM-DIAMETER    TEXT-ALIGNED
        ARC SCE       DIM-ANGULAR     TEXT-MIDDLE
        ARC CSE       DIM-LEADER      TEXT-FIT
```

Figure 5-31 Layout of menu

LINE	DIMLINEAR	TEXT LEFT
CIRCLE C,R	DIMALIGNED	TEXT RIGHT
CIRCLE C,D	DIMRADIUS	TEXT CENTER
ARC 3P	DIMDIAMETER	TEXT ALIGNED
ARC SCE	DIMANGULAR	TEXT MIDDLE
ARC CSE	QLEADER	TEXT FIT

Exercise 4 — *General*

Write a pull-down menu for the following AutoCAD commands.

LINE	BLOCK
PLINE	WBLOCK
CIRCLE C,R	INSERT
CIRCLE C,D	BLOCK LIST
ELLIPSE AXIS ENDPOINT	ATTDEF
ELLIPSE CENTER	ATTEDIT

Exercise 5 — *General*

Write a partial menu for Windows. The menu file should have two menus, POP1 (MyArc) and POP2 (MyDraw). The MyArc menu should contain all Arc options and must be displayed at the sixth position. Similarly, the MyDraw menu should contain **LINE**, **CIRCLE**, **PLINE**, **TRACE**, **DTEXT**, and **MTEXT** commands and should occupy the ninth position.

Exercise 6 — *General*

Write a menu file for a toolbar with a flyout. The name of the toolbar is MyDrawX1, and it contains two buttons, **Draw** and **Modify**. When you select the **Draw** button, it should display a flyout with **all draw** buttons (Draw commands). Similarly, when you choose the **Modify** button, it should display a flyout with **modify** buttons (Modify commands).

Exercise 7 — *General*

Write a menu for the following AutoCAD commands. (The layout of the menu is shown in Figure 5-32.)

LAYER NEW	SNAP 0.25	UCS WORLD
LAYER MAKE	SNAP 0.5	UCS PREVIOUS
LAYER SET	GRID 1.0	VPORTS 2
LAYER LIST	DRID 10.0	VPORTS 4
LAYER ON	APERTURE 5	VPORTS SING.
LAYER OFF	PICKBOX 5	

```
                        PULL-DOWN MENU

        ┌─────────────────────────────────────────────┐
        │  LAYER          SETTINGS         UCS-PORT   │
        └─────────────────────────────────────────────┘

           LAYER-New      SNAP 0.25        UCS-World
           LAYER-Make     SNAP 0.5         UCS-Pre
           LAYER-Set      GRID 1.0         VPORTS-2
           LAYER-List     GRID 10.0        VPORTS-4
           LAYER-On       APERTURE 5       VPORTS-1
           LAYER-Off      PICKBOX 5
```

Figure 5-32 Design of menu for Exercise 7

Answers to the Self-Evaluation Test

1. ACAD.MNU, 2. MENU, 3. Section label, **4.** 499 sections, **5.** 4 sections, **6.** backslash, **7.** ->, <-, **8.** ***, **9.** <-<-, **10.** !, **11. SHORTCUTMENU, 12.** **, **13. TBCONFIG, Right click on any toolbar and then select Customize from the shortcut menu.**

Chapter 6

Image Tile Menus

Learning Objectives

After completing this chapter, you will be able to:
- *Write image tile menus.*
- *Reference and display submenus.*
- *Make slides for image tile menus.*

IMAGE TILE MENUS

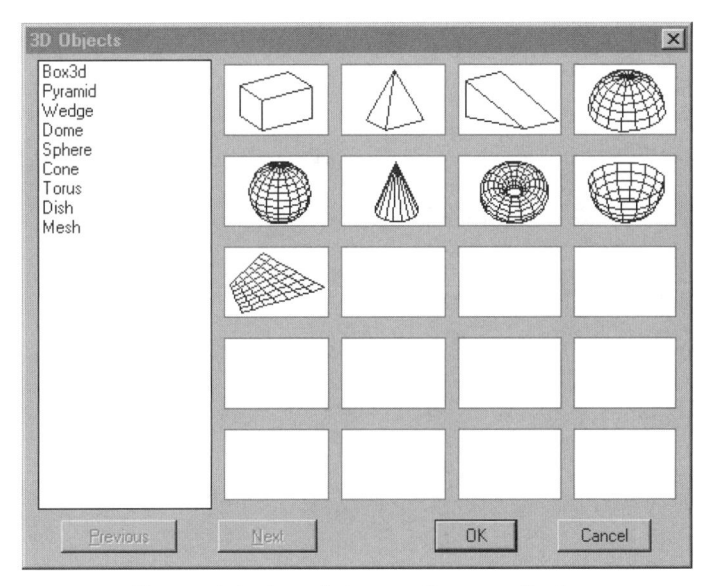

Figure 6-1 Sample image tile menu display

The image tile menus, also known as **icon menus**, are extremely useful for inserting a block, selecting a text font, or drawing a 3D object. You can also use the image tile menus to load an AutoLISP routine or a predefined macro. Thus, the image tile menu is a powerful tool for customizing AutoCAD.

The image tile menus can be accessed from the pull-down, tablet, button, or screen menu. However, the image tile menus **cannot** be loaded by entering the command. When you select an image tile, a dialog box that contains **20 image tiles** is displayed on the screen (Figure 6-1). The names of the slide files associated with image tiles appear on the left side of the dialog box with a scroll that can be used to scroll the file names. The title of the image tile menu is displayed at the top of the dialog box (Figure 6-1). When you activate the image tile menu, an arrow that can be moved to select any image tile appears on the screen. You can select an image tile by selecting the slide file name from the dialog box and then choosing the **OK** button or double-clicking on the slide file name.

When you select the slide file, AutoCAD highlights the corresponding image tile by drawing a rectangle around the image tile (Figure 6-1). You can also select an image tile by moving the arrow to the desired image tile, and then pressing the PICK button of the mouse. The corresponding slide file name will be automatically highlighted. And if you choose the **OK** button or double-click on the image tile, the command associated with that menu item will be executed. Press the **ESCAPE** key, select the **Cancel** button, or select an image to exit the image tile menu.

SUBMENUS

You can define an unlimited number of menu items in the image tile menu, but only 20 image tiles will be displayed at a time. If the number of items exceeds 20, you can use the **Next** and **Previous** buttons of the dialog box to page through different pages of image tiles. You can also define submenus that let you define smaller groups of items within an image tile menu section. When you select a submenu, the items are loaded and displayed on the screen.

Submenu Definition

A submenu label consists of two asterisks (**) followed by the name of the submenu. The image tile menu can have any number of submenus, and every submenu should have a unique name. The items that follow a submenu, up to the next section label or submenu label, belong to that submenu. The format of a submenu label is:

    **Name

        Where   ** ----------------- Two asterisks designate a submenu
                Name ------------ Name of the submenu

Note
The submenu name can be up to 31 characters long.

The submenu name can consist of letters, digits, and special characters, such as $ (dollar), - (hyphen), and _ (underscore).

Image Tile Menus

The submenu name should not have any embedded blanks.

Submenu names should be unique in a menu file.

Submenu Reference

The submenu reference is used to reference or load a submenu. It consists of a $ sign followed by a letter that specifies the menu section. The letter that specifies an image tile menu section is I. The menu section is followed by an equal sign (=) and the name of the submenu you want to activate. The submenu name should be without the **. Following is the format of a submenu reference:

$Section=Submenu
- Where **$** -------------------- "$" sign
- **Section** ---------- Menu section specifier
- **=** ------------------ "=" sign
- **Submenu** ------- Name of submenu

$I=IMAGE1
- Where **I** ------------------ I specifies image tile menu section
- **IMAGE1** -------- Name of submenu

Displaying a Submenu

When you load a submenu, the new dialog box and the image tiles are not automatically displayed. For example, if you load submenu IMAGE1, the items contained in this submenu will not be displayed. To force the display of the new image tile menu on the screen, AutoCAD uses the special command $I=*:

$I=*
- Where **I** ------------------ I for image tile menu
- ***** ------------------ Asterisk (*)

WRITING AN IMAGE TILE MENU

The image tile menu consists of the section label ***IMAGE followed by image tiles or image tile submenus. The menu file can contain only one image tile menu section (***IMAGE); therefore, all image tiles must be defined in this section.

***IMAGE
- Where ******* ---------------- Three asterisks designate a section label
- **IMAGE** ---------- Section label for an image tile

You can define any number of submenus in the image tile menu. All submenus have two asterisks followed by the name of the submenu (**PARTS or **IMAGE1):

**IMAGE1
- Where ****** ----------------- Two asterisks designate a submenu
- **IMAGE1** -------- Name of submenu

The first item in the image tile menu is the title of the image tile menu, which is also displayed at the top of the dialog box. The image tile dialog box title has to be enclosed in brackets ([PLC-SYMBOLS]) and should not contain any command definition. If it does contain a command definition, AutoCAD ignores the definition. The remaining items in the image tile menu file contain slide names in the brackets and the command definition outside the brackets.

```
***IMAGE                        Image tile menu section
**BOLTS                         Image tile submenu (BOLTS)
[HEX-HEAD BOLTS]                Image tile title
[BOLT1]^C^CINSERT;B1            BOLT1 is slide file name;
                                B1 is block name
```

SLIDES FOR IMAGE TILE MENUS

The idea behind creating slides for the image tile menus is to display graphical symbols in the image tiles. This symbol makes it easier for you to identify the operation that the image tile will perform. Any slide can be used for the image tile. However, the following guidelines should be kept in mind when creating slides for the image tile menu:

1. When you make a slide for an image tile menu, draw the object so that it fills the entire screen. The **MSLIDE** command makes a slide of the existing screen display. If the object is small, the picture in the image tile menu will be small. Use **ZOOM** Extents or **ZOOM** Window to display the object before making a slide.

2. When you use the image tile menu, it takes some time to load the slides for display in the image tiles. The more complex the slides, the more time it will take to load them. Therefore, the slides should be kept as simple as possible and at the same time give enough information about the object.

3. Do not fill the object, because it takes a long time to load and display a solid object. If there is a solid area in the slide, AutoCAD does not display the solid area in the image tile.

4. If the objects are too long or too wide, it is better to center the image with the AutoCAD **PAN** command before making a slide.

5. The space available on the screen for image tile display is limited. Make the best use of this small area by giving only the relevant information in the form of a slide.

6. The image tiles that are displayed in the dialog box have the length-to-width ratio (aspect ratio) of 1.5:1. For example, if the length of the image tile is 1.5 units, the width is 1 unit. If the drawing area of your screen has an aspect ratio of 1.5 and the slide drawing is centered in the drawing area, the slide in the image tile will also be centered.

Example 1

Write an image tile menu that will enable you to insert the block shapes from Figure 6-2 in a drawing by selecting the corresponding image tile from the dialog box. Use the menu to load the image tile menu.

Image Tile Menus 6-5

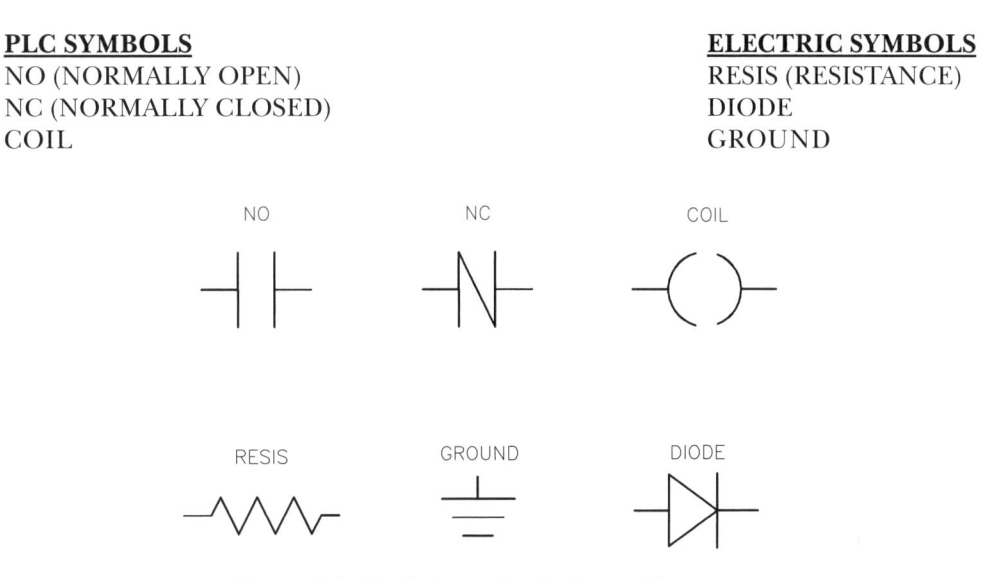

PLC SYMBOLS
NO (NORMALLY OPEN)
NC (NORMALLY CLOSED)
COIL

ELECTRIC SYMBOLS
RESIS (RESISTANCE)
DIODE
GROUND

Figure 6-2 Block shapes for the image tile menu

Step 1: Converting the given shapes in wblocks and slides
Draw the shapes shown in the example and then use the **WBLOCK** command to create wblocks. Use the MSLIDE command to make the slides that will be used to show the preview of images in the **image tile menu** dialog box. Next, use the INSERT command to insert the blocks into the drawing.

Step 2: Designing the image tile menu
Design the menu so that the commands are arranged in a desired configuration. Figure 6-3 shows one possible design for the menu and the image tile menu for Example 1.

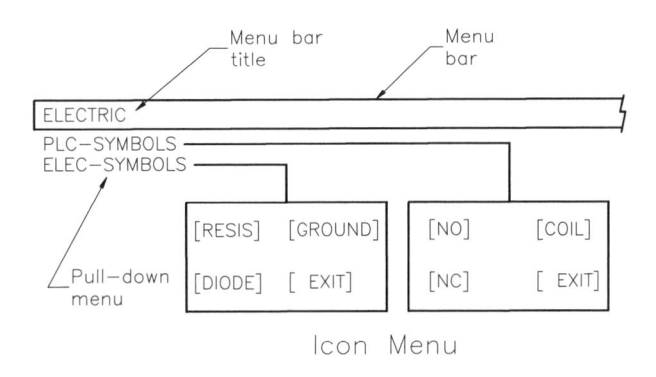

Figure 6-3 Design of the menu and image tile menu for Example 1

Step 3: Writing the image tile menu
You can use the AutoCAD **EDIT** command or any text editor like Notepad to write the file. **The line numbers in the following file are for reference and are not a part of the menu file.**

```
***POP1                                                              1
[ELECTRIC]                                                           2
[PLC-SYMBOLS]$I=IMAGE1 $I=*                                          3
[ELEC-SYMBOLS]$I=IMAGE2 $I=*                                         4
***IMAGE                                                             5
**IMAGE1                                                             6
[PLC-SYMBOLS]                                                        7
[NO]^C^CINSERT;NO;\1.0;1.0;0                                         8
[NC]^C^CINSERT;NC;\1.0;1.0;0;                                        9
[COIL]^C^CINSERT;COIL                                               10
[ No-Image]                                                         11
[blank]                                                             12
**IMAGE2                                                            13
[ELECTRICAL SYMBOLS]                                                14
[RESIS]^C^CINSERT;RESIS;\\\\                                        15
[DIODE]^C^CINSERT;DIODE;\1.0;1.0;\                                  16
[GROUND]^C^CINSERT;GRD;\1.5;1.5;0;;                                 17
```

Explanation
Line 1
***POP1
In this menu item, ***POP1 is the section label and defines the first section of the menu.

Line 2
[ELECTRIC]
In this menu item [ELECTRIC] is the menu bar label for the POP1 menu. It will be displayed in the menu bar.

Line 3
[PLC-SYMBOLS]$I=IMAGE1 $I=*
In this menu item, $I=IMAGE1 loads the submenu IMAGE1; $I=* displays the current image tile menu on the screen.

> **[PLC-SYMBOLS]$I=IMAGE1 $I=***
> Where **Image1** ---------- Loads submenu IMAGE1
> $ ------------------- Forces display of current menu

Line 5
***IMAGE
In this menu item, ***IMAGE is the section label of the image tile menu. All the image tile menus have to be defined within this section; otherwise, AutoCAD cannot locate them.

Image Tile Menus 6-7

Line 6
**IMAGE1
In this menu item, **IMAGE1 is the name of the image tile submenu.

Line 7
[PLC-SYMBOLS]
When you select line 3([PLC-SYMBOLS] $I=IMAGE1$I=*), AutoCAD loads the submenu IMAGE1 and displays the title of the image tile at the top of the dialog

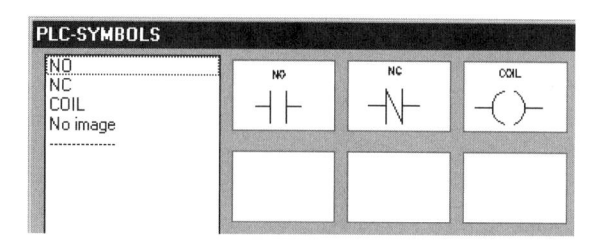

Figure 6-4 Image tile box for PLC-SYMBOLS

box (Figure 6-4). This title is defined in Line 7. If this line is missing, the next line will be displayed at the top of the dialog box. Image tile titles can be any length, as long as they fit the length of the dialog box.

Line 8
[NO]^C^CINSERT;NO;\1.0;1.0;0
In this menu item, the first NO is the name of the slide and has to be enclosed within brackets. The name should not have any trailing or leading blank spaces. If the slides are not present, AutoCAD will not display any graphical symbols in the image tiles. However, the menu items will be loaded, and if you select this item, the command associated with the image tile will be executed. The second NO is

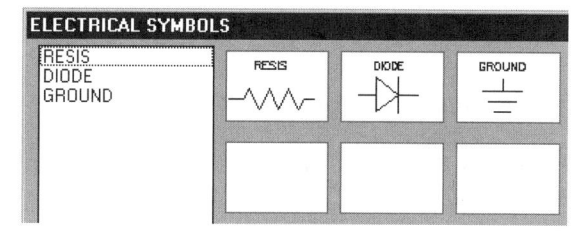

Figure 6-5 Image tile box for ELECTRICAL-SYMBOLS

the name of the block that is to be inserted. The backslash (\) pauses for user input; in this case it is the block insertion point. The first 1.0 defines the Xscale factor. The second 1.0 defines the Yscale factor, and the following 0 defines the rotation.

```
[NO]^C^CINSERT;NO;\1.0;1.0;0
    Where   NO  --------------- Block name
            \------------------- Pause for block insertion point
            1.0 ---------------- Xscale factor
            1.0 ---------------- Yscale factor
            0 ------------------ Rotation angle
```

When you select this item, it will automatically enter all the prompts of the **INSERT** command and insert the NO block at the given location. The only input you need to enter is the insertion point of the block.

Line 10
[COIL]^C^CINSERT;COIL
In this menu item the block name is given, but you need to define other parameters when inserting this block.

Line 11
[No-Image]
Notice the **blank space before No-Image**. If there is a space following the open bracket, AutoCAD does not look for a slide. AutoCAD instead displays the text, enclosed within the brackets, in the slide file list box of the dialog box.

Line 12
[blank]
Line 12 consists of **blank**; this displays a separator line in the list box and a blank image (no image) in the image box.

Line 15
[RESIS]^C^CINSERT;RESIS;
This menu item inserts the block RESIS. The first backslash (\) is for the block insertion point. The second and third backslashes are for the Xscale and Yscale factors. The fourth backslash is for the rotation angle. This menu item could also be written as:

    [RESIS]^C^CINSERT;RESIS;
           or
    [RESIS]^C^CINSERT;RESIS

If a macro runs out before the command completes then the command reverts to asking for user input.

Line 16
[DIODE]^C^CINSERT;DIODE;\1.0;1.0;
If you select this menu item, AutoCAD will prompt you to enter the block insertion point and the rotation angle. The first backslash is for the block insertion point; the second backslash is for the rotation angle.

    [DIODE]^C^CINSERT;DIODE;\1.0;1.0;
           Where \-------------------- Pause for insertion point
                    \-------------------- Pause for rotation angle

Line 17
[GROUND]^C^CINSERT;GRD;\1.5;1.5;0;;
This menu item has two semicolons (;) at the end. The first semicolon after 0 is for RETURN and completes the block insertion process. The second semicolon enters a RETURN and repeats the **INSERT** command. However, when the command is repeated you will have to respond to all of the prompts. It does not accept the values defined in the menu item.

Note
*The slide files must be in the search path. If the slide files are not in the support directory, specify the directory location in the Support File Search Path located in the **File** tab of the **Options** dialog box.*

The menu item repetition feature cannot be used with the image tile menus. For example, if the command definition starts with an asterisk ([GROUND]*^C^CINSERT;GRD;\1.5;1.5;0;;),

Image Tile Menus   6-9

the command is not automatically repeated, as is the case with a pull-down menu.

A blank line in an image tile menu terminates the menu and clears the image tiles.

The menu command $I=, which displays the current menu, cannot be entered at the keyboard.*

If you want to cancel or exit an image tile menu, press the ESC (Escape) key on the keyboard. AutoCAD ignores all other entries from the keyboard.

You can define any number of image tile menus and submenus in the image tile menu section of the menu file.

LOADING MENUS

AutoCAD automatically loads the **ACAD.MNU** file when you get into the AutoCAD drawing editor. However, you can also load a different menu file by using the AutoCAD **MENU** command.

Command: **MENU**

When you enter the **MENU** command, AutoCAD displays the **Select Menu File** dialog box (Figure 6-6) on the screen. Select the menu file that you want to load and then choose the **Open** button.

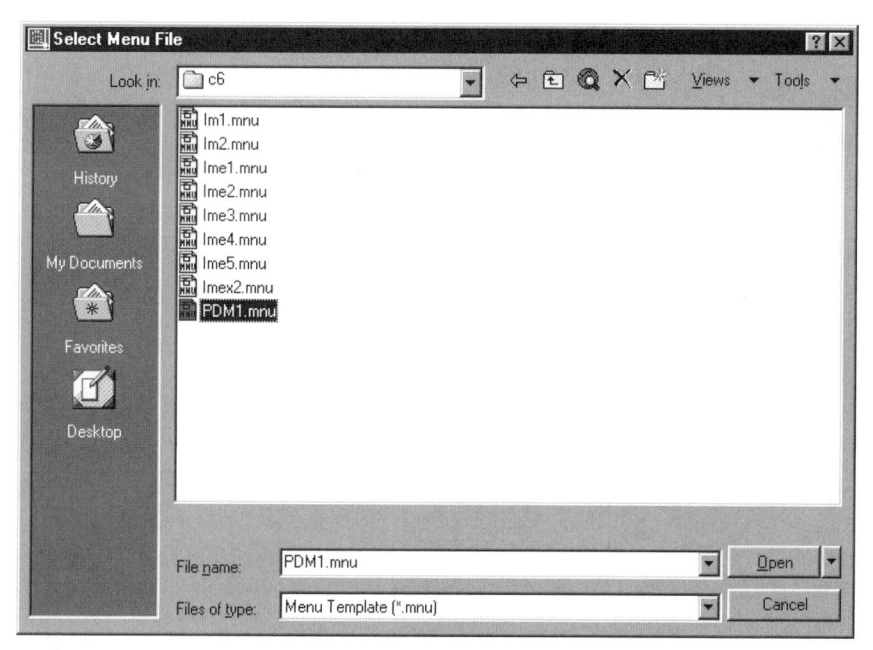

Figure 6-6 Select Menu File dialog box

You can also load the menu file from the command line.

Command: **FILEDIA**
Enter new value for FILEDIA <1>: **0**
Command: **MENU**
Menu filename <ACAD>: PDM1
    Where **PDM1** ------------ Name of menu file
               ACAD ------------ Default menu file

After you enter the **MENU** command, AutoCAD will prompt for the file name. Enter the name of the menu file without the file extension (**.MNU**), since AutoCAD assumes the extension **.MNU**. AutoCAD will automatically compile the menu file into **MNS** and **MNR** files.

Step 4: Loading the image tile menu file

Save the file as **PDM1.MNU** and then load the file with the help of the **MENU** command as described above.

Note

When you load the image tile menu, some of the commands will be displayed on the screen menu area. This happens because the screen menu area is empty. If the menu file contains a screen menu also, the pull-down menu items will not be displayed in the screen menu area.

When you load a new menu, the original menu you had on the screen prior to loading the new menu is disabled. You cannot select commands from the screen menu, pull-down menu, pointing device, or the digitizer, unless these sections are defined in the new menu file.

To load the original (**ACAD.MNU**) menu, use the **MENU** command again and enter the name of the menu file.

RESTRICTIONS

The pull-down and image tile menus are very easy to use and provide quick access to some of the frequently used AutoCAD commands. However, the menu bar, the pull-down menus, and the image tile menus are disabled during the following commands:

DTEXT Command
After you assign the text height and the rotation angle to a **DTEXT** command, the menu is automatically disabled.

SKETCH Command
The menus are disabled after you set the record increment in the **SKETCH** command.

Exercise 1   *General*

Write an image tile menu for inserting the blocks shown in Figure 6-7. Arrange the blocks in two groups so that you have two submenus in the image tile menu.

PIPE FITTINGS	**ELECTRIC SYMBOLS**
GLOBE-P	BATTERY
GLOBE	CAPACITOR
REDUCER	COUPLER
CHECK	BREAKER

Image Tile Menus 6-11

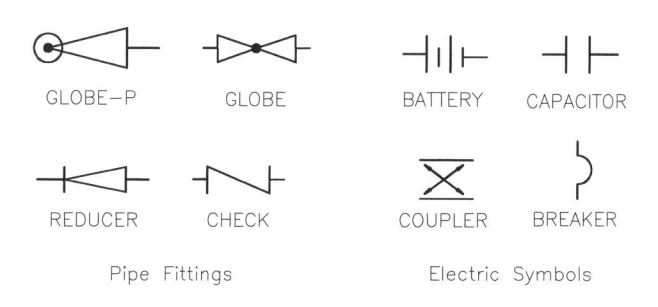

Figure 6-7 Block shapes for Exercise 1

IMAGE TILE MENU ITEM LABELS

As with screen and menus, you can use menu item labels in the image menus. However, the menu item labels in the image tile menus use different formats, and each format performs a particular function in the image tile menu. The menu item labels appear in the slide list box of the dialog box. The maximum number of characters that can be displayed in this box is 23. The characters in excess of 23 are not displayed in the list box. However, this does not affect the command that is defined with the menu item.

Menu Item Label Formats

[slidename]. In this menu item label format, **slidename** is the name of the slide displayed in the image tile. This name (slidename) is also displayed in the list box of the corresponding dialog box.

[slidename,label]. In this menu item label format, **slidename** is the name of the slide displayed in the image tile. However, unlike the previous format, the **slidename** is **not** displayed in the list box. The **label** text is displayed in the list box. For example, if the menu item label is **[BOLT1,1/2-24UNC-3LG]**, **BOLT1** is the name of the slide and **1/2-24UNC-3LG** is the label that will be displayed in the list box.

[slidelib(slidename)]. In this menu item label format, **slidename** is the name of the slide in the slide library file **slidelib**. The slide (slidename) is displayed in the image tile, and the slide file name (slidename) is also displayed in the list box of the corresponding dialog box.

[slidelib(slidename,label)]. In this menu item label format, **slidename** is the name of the slide in the slide library file **slidelib**. The slide (slidename) is displayed in the image tile, and the label text is displayed in the list box of the corresponding dialog box.

[blank]. This menu item will draw a line that extends through the width of the list box. It also displays a blank image tile in the dialog box.

[**label**]. If the **label** text is preceded by a space, AutoCAD does not look for a slide. The label text is displayed in the list box only. For example, if the menu item label is [EXIT]^C, the label text (EXIT) will be displayed in the list box. If you select this item, the cancel command (^C) defined with the item will be executed. The **label** text is **not** displayed in the image tile of the dialog box.

Example 2

Write the pull-down and image tile menus for inserting the following commands. B1 to B15 are the block names.

BLOCK
WBLOCK
ATTDEF
LIST
INSERT

BL1	BL6	BL11
BL2	BL7	BL12
BL3	BL8	BL13
BL4	BL9	BL14
BL5	BL10	BL15

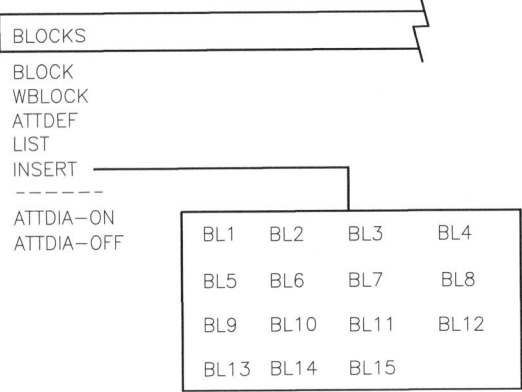

Figure 6-8 Design of screen and menus for Example 2

Step 1: Making wblocks and slides

Draw suitable figures for different blocks names shown in the example. Convert all figures into wblocks and make slides with the given name.

Step 2: Designing the image tile menu

The second step in writing a menu is to design the menu. Figure 6-8 shows the design of the menu. If you select Insert from the menu, the image tiles and block names will be displayed in the dialog box.

Image Tile Menus　　　　　　　　　　　　　　　　　　　　　　　　　　　　　6-13

Step 3: Writing the image tile menu

Use any text editor to write the menu. The following file is a listing of the menu file for Example 2. The file contains the pull-down and image tile menu sections. The line numbers are not a part of the menu file; they are given for reference only.

```
***POP1                                              1
[INSERT]                                             2
[BLOCK]^C^CBLOCK                                     3
[WBLOCK]^C^CWBLOCK                                   4
[ATTRIBUTE DEFINITION]^C^CATTDEF                     5
[LIST BLOCK NAMES]^C^CINSERT;?                       6
[INSERT]^C^C$I=IMAGE1 $I=*                           7
[--]                                                 8
[ATTDIA-ON]^C^CSETVAR ATTDIA 1                       9
[ATTDIA-OFF]^C^CSETVAR ATTDIA                       10
                                                    11
***IMAGE                                            12
**IMAGE1                                            13
[BLOCK INSERTION FOR EXAMPLE-2]                     14
[BL1]^C^CINSERT;BL1;\1.0;1.0;\                      15
[BL2]^C^CINSERT;BL2;\1.0;1.0;0                      16
[BL3]^C^CINSERT;BL3;\;;\                            17
[BL4]^C^CINSERT;BL4;\;;;                            18
[BL5]^C^CINSERT;*BL5;\1.75                          19
[BL6]^C^CINSERT;BL6;\XYZ                            20
[BL7]^C^CINSERT;BL7;\XYZ;;;\0                       21
[BL8]^C^CINSERT;BL8;\XYZ;;;;                        22
[BL9]^C^CINSERT;BL9;\XYZ;;;\                        23
[BL10]^C^CINSERT;*BL10;\XYZ;\                       24
[BL11]^C^CINSERT;BL11;\XYZ;1;1.5;2;45               25
[BL12]^C^CINSERT;BL12;\XYZ;\\;;                     26
[BL13]^C^CINSERT;*BL13;\\45                         27
[BL14]^C^CINSERT;BL14;\C;@1.0,1.0;0                 28
[BL15]^C^CINSERT;BL15;\C;@1.0,2.0;\                 29
```

Explanation

Line 1
***POP1
This is the section label of the first menu. The menu items defined on lines 2 through 10 are defined in this section.

Line 12
***IMAGE
This is the section label of the image tile menu.

Line 13
**IMAGE1
IMAGE1 is the name of the submenu; the items on lines 14 through 29 are defined in this submenu.

Line 15
[BL1]^C^CINSERT;BL1;\1.0;1.0;
In this menu item, BL1 is the name of the slide and INSERT is AutoCAD's INSERT command.

[BL1]^C^CINSERT;BL1;\1.0;1.0;\
    Where **INSERT** --------- AutoCAD **INSERT** command
                BL1 --------------- Slide file name
                1.0,1.0 ----------- X and Y scale factors

Step 4: Loading the menu file
Load the menu with the **MENU** command after saving the menu file with the .MNU extension.

Self-Evaluation Test

Answer the following questions and then compare your answers to the correct answers at the end of this chapter.

1. The image tile menu is also known as _____.

2. The image tile menu _____ be loaded by entering the command from the keyboard.

3. The submenu in an image tile menu starts with _____.

4. In order to load an image submenu, AutoCAD uses a special command _____.

5. The image tile menu consists of the section label _____.

Review Questions

Answer the following questions.

1. The maximum number of characters displayed in the menu item label is _____.

2. The image tiles are displayed in the _____ box.

3. An image tile menu can be canceled by entering _____ at the keyboard.

4. The image title dialog box can contain a maximum of _____ image tiles.

Image Tile Menus   6-15

5. A blank line in an image tile menu _____ the image tile menu.

6. The drawing for a slide should be _____ on the entire screen before making a slide.

7. You _____ fill a solid area in a slide for an image tile menu.

8. An image tile menu _____ be accessed from a tablet menu.

Exercises

Exercise 2   *General*

Write an image tile menu for inserting the following blocks.

        B1   B4   B7
        B2   B5   B8
        B3   B6   B9

Exercise 3   *General*

Write an image tile menu for the following commands. Make the slides that will graphically illustrate the function of the command.

        LINE   **CIRCLE C,R**
        PLINE   **CIRCLE C,D**
                                  CIRCLE 2P

Exercise 4   *General*

Write an image tile menu for inserting the following blocks. The B blocks and C blocks should be in separate image dialog boxes.

    B1   B2   B3   C1   C2   C3
    B4   B5   B6   C4   C5   C6
    B7   B8   B9   C7   C8   C9

Answers to the Self-Evaluation Test
1. Icon menu, **2**. cannot, **3**. **, **4**. $I=*, **5**. ***IMAGE

Chapter 7

Button and Auxiliary Menus

Learning Objectives

After completing this chapter, you will be able to:
- *Write Buttons menus.*
- *Learn special handling for button menus.*
- *Define and load submenus.*

BUTTON MENUS

The vast majority of AutoCAD installations use mice. However, you can also use a multibutton pointing device to specify points, select objects, and execute commands. These pointing devices come with different numbers of buttons, but four-button and twelve-button pointing devices are very common. In addition to selecting points and objects, the multibutton pointing devices can be used to provide access to frequently used AutoCAD commands. The commands are selected by pressing the desired button; AutoCAD automatically executes the command or the macro that is assigned to that button. Figure 7-1 shows one such pointing device with 12 buttons.

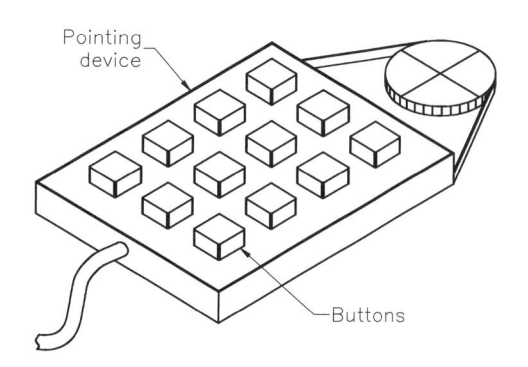

Figure 7-1 Pointing device with 12 buttons

The AutoCAD software package comes with a standard button menu that is part of the **ACAD.MNU** file. The **standard** menu is automatically loaded when you start AutoCAD and enter the drawing editor. You can write your own button menu and assign the desired commands or macros to various buttons of the pointing device.

WRITING BUTTON MENUS

In a menu file, you can have up to four button menus (BUTTONS1 through BUTTONS4) and four auxiliary menus (AUX1 through AUX4). The buttons and the auxiliary menus behave identically. However, they are OS dependent. If your system has a pointing device (digitizer puck), AutoCAD automatically assigns the commands defined in the BUTTONS sections of the menu file to the buttons of the pointing device. When you load the menu file, the commands defined in the BUTTONS1 section of the menu file are assigned to the pointing device (digitizer puck) and if your computer has a system mouse, the mouse will use the auxiliary menus. You can also access other button menus (BUTTONS2 through BUTTONS4) by using the following keyboard-and-button (buttons of the pointing device-digitizer puck) combinations.

Aux Menu	Buttons Menu	Keyboard + Button Sequence
AUX1	BUTTONS1	Press the button of the pointing device.
AUX2	BUTTONS2	Hold down the SHIFT key and press the button of the pointing device.
AUX3	BUTTONS3	Hold down the CTRL key and press the button of the pointing device.
AUX4	BUTTONS4	Hold down the SHIFT and CTRL keys and press the button of the pointing device.

One of the buttons, generally the first, is used as a pick button to specify the coordinates of the screen crosshairs and send that information to AutoCAD. This button can also be used to select commands from various other menus, such as the tablet menu, screen menu, and image tile menu. This button cannot be used to enter a command, but AutoCAD commands can be assigned to other buttons of the pointing device. Before writing a button menu, you should decide the commands and options you want to assign to different buttons, and know the prompts associated with those commands. The following example illustrates the working of the button menu and the procedure for assigning commands to different buttons.

Note
The first line after the menu section label ***AUX1 or ***BUTTONS1 is used only when the SHORTCUTMENU system variable is set to 0. If SHORTCUTMENU is set to a value other than 0, the built-in menu is used. Similarly, the second line after the ***AUX1 or ***BUTTONS1 label is used only when the MBUTTONPAN system variable is set to 0

Example 1

Write a buttons menu for the following AutoCAD commands. The pointing device has 12 buttons (Figure 7-2), and button number 1 is used as a pick button (filename **BM1.MNU**).

Button and Auxiliary Menus

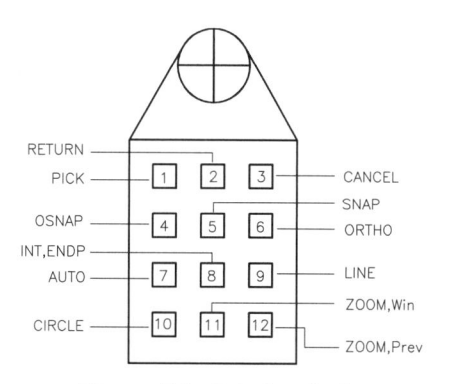

Figure 7-2 Pointing device

Button	Function	Button	Function
2	RETURN	3	CANCEL
4	CURSOR MENU	5	SNAP
6	ORTHO	7	AUTO
8	INT,END	9	LINE
10	CIRCLE	11	ZOOM Win
12	ZOOM Prev		

Step 1: Writing the menu file

You can use the AutoCAD **EDIT** command or any other text editor to write the menu file. The following file is a listing of the button menu for Example 1. **The line numbers are for reference only and are not a part of the menu file.**

```
***BUTTONS1                                                  1
;                                                            2
^C^C                                                         3
$P0=*                                                        4
^B                                                           5
^O                                                           6
AUTO                                                         7
INT,ENDP                                                     8
^C^CLINE                                                     9
^C^CCIRCLE                                                  10
'ZOOM;Win                                                   11
'ZOOM;Prev                                                  12
```

Explanation

Line 1
***BUTTONS

***BUTTONS1 is the section label for the first button menu. When the menu is loaded, AutoCAD compiles the menu file and assigns the commands to the buttons of the pointing device.

Line 2
;
This menu item assigns a semicolon (;) to button number 2. When you specify the second button on the pointing device, it enters a Return. It is like entering Return at the keyboard or the digitizer.

Line 3
^C^C
This menu item cancels the existing command twice (^C^C). This command is assigned to button number 3 of the pointing device. When you pick the third button on the pointing device, it cancels the existing command twice.

Line 4
$P0=*
This menu item loads and displays the cursor menu POP0, which contains various object snap modes. It is assumed that the POP0 menu has been defined in the menu file. This command is assigned to button number 4 of the pointing device. If you press this button, it will load and display the shortcut menu on the screen near the crosshairs location.

Line 5
^B
This menu item changes the snap mode; it is assigned to button number 5 of the pointing device. When you pick the fifth button on the pointing device, it turns the snap mode on or off. It is like holding down the CTRL key and then pressing the B key.

Line 6
^O
This menu item changes the ORTHO mode; it is assigned to button number 6. When you pick the sixth button on the pointing device, it turns the ORTHO mode on or off.

Line 7
AUTO
This menu item selects the AUTO option for creating a selection set; this command is assigned to button number 7 on the pointing device.

Line 8
INT,ENDP
In this menu item, INT is for the Intersection Osnap, and ENDP is for the Endpoint Osnap. This command is assigned to button number 8 on the pointing device. When you pick this button, AutoCAD looks for the intersection point. If it cannot find an intersection point, it then starts looking for the endpoint of the object that is within the pick box.

    INT,ENDP
        Where **INT** -------------- Intersection object snap
                    ENDP ------------ Endpoint object snap

Button and Auxiliary Menus   7-5

Line 9
^C^CLINE
This menu item defines the **LINE** command; it is assigned to button number 9. When you select this button, AutoCAD cancels the existing command and then selects the **LINE** command.

Line 10
^C^CCIRCLE
This menu item defines the **CIRCLE** command; it is assigned to button number 10. When you pick this button, AutoCAD automatically selects the **CIRCLE** command and prompts for the user input.

Line 11
'ZOOM;Win
This menu item defines a transparent **ZOOM** command with Window option; it is assigned to button number 11 of the pointing device.

> **'ZOOM;Win**
> Where   '-------------------- Single quote makes **ZOOM** command transparent
>   **ZOOM** ----------- AutoCAD **ZOOM** command
>   ; ------------------ Semicolon for RETURN
>   **Win** -------------- Window option of **ZOOM** command

Line 12
'ZOOM;Prev
This menu item defines a transparent **ZOOM** command with **previous** option; it is assigned to button number 12 of the pointing device.

Step 2: Loading the menu file
Save the file with the extension ***.MNU** and then load the menu file with the command **MENU**.

> **Note**
> *If the button menu has more menu items than the number of buttons on the pointing device, the menu items in excess of the number of buttons are ignored. This does not include the pick button. For example, if a pointing device has three buttons in addition to the pick button, the first three menu items will be assigned to the three buttons (buttons 2, 3, and 4). The remaining lines of the button menu are ignored.*
>
> *The commands are assigned to the buttons in the same order in which they appear in the file. For example, the menu item that is defined on line 3 will automatically be assigned to the fourth button of the pointing device. Similarly, the menu item that is on line 4 will be assigned to the fifth button of the pointing device. The same is true of other menu items and buttons.*
>
> *You must use the MENU command and load an entire menu. MENULOAD of partial menus ignores AUXn and BUTTONSn sections*

SPECIAL HANDLING FOR BUTTON MENUS
When you press any button on the multibutton pointing device (Figure 7-3), AutoCAD receives the following information:

1. **The button number**
2. **The coordinates of the screen crosshairs**

You can write a button menu that uses one or both pieces of information. The following example uses only the button number and ignores the coordinates of the screen crosshairs:

Example
^C^CLINE

If this command is assigned to the second button of the pointing device and you select this button, AutoCAD will receive the button number and the coordinates of the screen crosshairs. AutoCAD will execute the command that is assigned to the second button, but it will ignore the coordinates of the crosshairs. The following example uses both the button number and the coordinates of the screen crosshairs:

Example
^C^CLINE;\

In this menu item the **LINE** command is followed by a semicolon (;) and a backslash (\). The semicolon inputs an ENTER and the backslash normally causes a pause for user input. However, in the buttons menu AutoCAD will not pause for the user input. The backslash (\) in this menu item will use the coordinates of the screen crosshairs supplied by the pointing device as the starting point (From point) of the line. AutoCAD will then prompt for the second point of the line (To point).

Example 2

Write a button menu for the following AutoCAD commands. The menu items should use the information about the coordinate points of the screen crosshairs, where applicable (filename BM2.MNU).

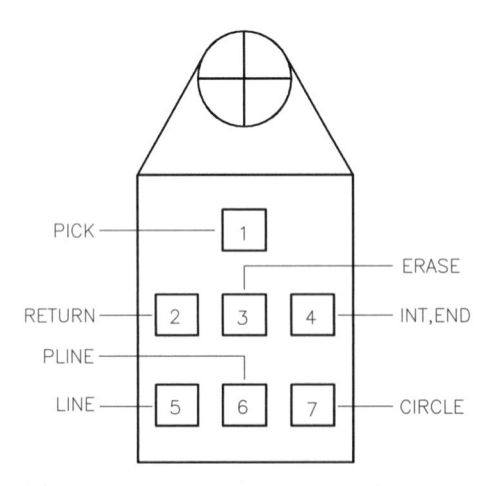

Figure 7-3 Pointing device with seven buttons

Button and Auxiliary Menus

Button	Function
1	PICK
2	ENTER
3	ERASE (with SI and NEAR options)
4	INT,END
5	LINE
6	PLINE
7	CIRCLE

Step 1: Writing the menu file

The following file is a listing of the button menu for Example 2. The line numbers are not a part of the file; they are for reference only.

```
***BUTTONS1                              1
;                                        2
^C^CERASE;SI;NEAR;\                      3
INT,ENDP;\                               4
LINE;\                                   5
PLINE;\                                  6
CIRCLE;\                                 7
```

Explanation
Line 3
^C^CERASE;SI;NEAR;

This menu item defines an **ERASE** command with single object selection option (SI) and near object snap (NEAR). The backslash is used to accept the coordinates supplied by the screen crosshairs. This command is assigned to button number 3 of the pointing device. If you point to an object and press the third button on the pointing device, the object will be erased and AutoCAD will automatically return to the command prompt.

```
^C^CERASE;SI;NEAR;\
        Where  SI ---------------- SIngle object selection mode
               NEAR ------------ NEAR object snap
               \-------------------- Accepts coordinates of screen crosshairs
```

Line 4
INT,ENDP;

In this menu item INT is for the intersection snap and ENDP is for the endpoint snap. This command is assigned to button number 8 on the pointing device. When you pick this button, AutoCAD looks for the intersection point. If it cannot find an intersection point, it starts looking for the endpoint of the object that is within the pick box. The backslash (\) is used here to accept the coordinates of the screen crosshairs. If the crosshairs are near an object within the pick box AutoCAD will snap to the intersection's point, or the endpoint of the object, when you press button number 4 on the pointing device.

INT,ENDP;\
          Where **INT** -------------- Intersection object snap
                       EBDP ------------ Endpoint object snap
                       \-------------------- Accepts the coordinates of the screen crosshairs

Line 7
CIRCLE;
This menu item generates a circle. The backslash (\) is used here to accept the coordinates of the screen crosshairs as the center of the circle. If you press button number 7 of the pointing device to draw a circle, you do not need to enter the center of the circle because the current position of the screen crosshairs automatically becomes the center of the circle. You only have to define the radius to generate the circle.

Step 2: Loading the menu file
Save the file and then load the file using the **MENU** command.

> **Note**
> *The coordinate information associated with the button can be used with the first backslash only.*
>
> *If a menu item in a button menu has more than one backslash (\), the remaining backslashes are ignored. For example, in the following menu item the first backslash uses the coordinates of screen crosshairs as the insertion point and the remaining backslashes are ignored and do not cause a pause as in other menus.*

Example
INSERT;B1\\\0
          Where \-------------------- Insertion point
                       \-------------------- X scale factor
                       \-------------------- Y scale factor
                       0 ------------------- Rotation

SUBMENUS
The facility to define submenus is not limited to screen, pull-down, and image menus only. You can also define submenus in the button menu.

Submenu Definition
A submenu label consists of two asterisk signs (**) followed by the name of the submenu. The buttons menu can have any number of submenus and every submenu should have a unique name. The items that follow a submenu, up to the next section label or submenu label, belong to that submenu. The format of a submenu label is:

    **Name
          Where ****** ----------------- Two asterisk signs (**) designate a submenu
                       Name ------------ Name of the submenu

Button and Auxiliary Menus

> **Note**
> The submenu name can be up to 31 characters long.

The submenu name can consist of letters, digits, and special characters such as: $ (dollar), - (hyphen), and _ (underscore).

The submenu name should not have any embedded blanks.

The submenu names should be unique in a menu file.

Submenu Reference

The submenu reference is used to reference or load a submenu. It consists of a "$" sign followed by a letter that specifies the menu section. The letter that specifies a button menu section is B. The menu section is followed by "=" sign and the name of the submenu that the user wants to activate. The submenu name should be without "**". Following is the format of a submenu reference:

$Section=Submenu
   Where **$** ------------------- "$" sign
          Section ---------- Menu section specifier
          = ----------------- "=" sign
          Submenu ------- Name of submenu

Example
$B=BUTTON1
   Where **B** ----------------- B specifies buttons menu section
          BUTTON1 ----- Name of submenu

LOADING MENUS

From the buttons menu, you can load any menu that is defined in the screen, pull-down, or image menu sections by using the appropriate load commands. It may not be needed in most of the applications, but if you want to, you can load the menus that are defined in other menu sections.

Loading Screen Menus

You can load any menu that is defined in the screen menu section from the button menu by using the following load commands:

$S=X $S=INSERT
   Where **S** ------------------- S specifies screen menu
          X ------------------- Submenu name defined in screen menu section
          INSERT --------- Submenu name defined in screen menu section

The first load command ($S=X) loads the submenu X that has been defined in the screen menu section of the menu file. The X submenu can contain 21 blank lines, so that when it is loaded it clears the screen menu. The second load command ($S=INSERT) loads the submenu **INSERT** that has been defined in the screen menu section of the menu file.

Loading a Menu

You can load a menu from the button menu by using the following command:

$P1=P1A $P1=*

Where **$P1=P1A** -------- Loads the submenu P1A

$P1=* ----------- Forces the display of new menu items

The first load command $P1=P1A loads the submenu P1A that has been defined in the POP1 section of the menu file. The second load command $P1=* forces the new menu items to be displayed on the screen.

Loading an Image Menu

You can also load an image menu from the button menu by using the following load command:

$I=IMAGE1 $I=*

Where **$I=IMAGE1** --- Load the submenu IMAGE1

$I=* ------------- To display the dialog box

This menu item consists of two load commands. The first load command $I=IMAGE1 loads the image submenu IMAGE1 that has been defined in the image menu section of the file. The second load command $I=* displays the new dialog box on the screen.

Example 3

Write a button menu for a pointing device that has six buttons. The functions assigned to different buttons are shown in the following table (filename BM3.MNU):

SUBMENU B1	SUBMENU B2
1. PICK	1. PICK
2. Enter	2. Enter
3. LOAD OSNAPS SUBMENU	3. LOAD IMAGE1 SUBMENU
4. LOAD ZOOM1 SUBMENU	4. EXPLODE
5. LOAD BUTTON SUBMENU B1	5. LOAD BUTTON SUBMENU B1
6. LOAD BUTTON SUBMENU B2	6. LOAD BUTTON SUBMENU B2

Note

OSNAPS submenu is defined in the POP1 section of the menu.
ZOOM1 submenu is defined in the screen section of the menu file.

IMAGE1 submenu is defined in the IMAGE section of the menu file. The IMAGE1 submenu contains four images for inserting the blocks.

In this example there are two submenus. The submenus are loaded by picking button 5 for submenu B1 and button 6 for submenu B2. When the submenu B1 is loaded, AutoCAD assigns the commands that are defined under the submenu B1 to the buttons of the pointing device. Similarly, when the submenu B2 is loaded, AutoCAD assigns the commands that are defined under the submenu B2 to the buttons of the pointing device. Figure 7-4 shows the commands that are assigned to the buttons after loading submenu B1 and submenu B2.

Button and Auxiliary Menus 7-11

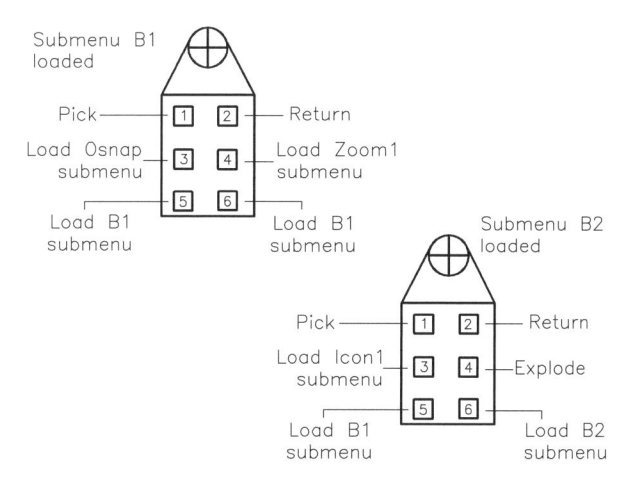

Figure 7-4 Commands assigned to different buttons of pointing device

Step 1: Writing the menu file
You can use AutoCAD's **EDIT** command to write the file. The following file is the listing of the buttons menu for Example 3. The line numbers are not a part of the file. They are shown here for reference only.

```
***BUTTON                                                   1
**B1                                                        2
;                                                           3
$P1=*                                                       4
$'ZOOM;Win                                                  5
$B=B1                                                       6
$B=B2                                                       7
**B2                                                        8
;                                                           9
^C^C$I=IMAGE1 $I=*                                         10
EXPLODE;\                                                  11
$B=B1                                                      12
$B=B2                                                      13
```

Explanation
Line 2
**B1
This menu item defines a submenu. The name of the submenu is B1.

Line 4
$P1=*
This menu item loads and displays the pull-down menu that has been defined in the POP1 section of the menu.

Line 5
'ZOOM;Win
This menu item **'ZOOM;Win** is a transparent zoom command with the window option.

Line 8
**B2
This menu item defines the submenu B2.

Line 10
^C^C$I=IMAGE1 $I=*
This menu item cancels the existing command twice and then loads the submenu IMAGE1 that has been defined in the IMAGE menu. $I=* displays the current dialog box on the screen.

    ^C^C$I=IMAGE1 $I=*
        Where **^C^$** ----------- Cancels the existing command twice
                I=IMAGE1 ---- Loads IMAGE1 submenu
                $I=* ------------- Displays the dialog box with images

Line 11
EXPLODE;
This menu item will explode an object. It utilizes the special feature of the pointing device buttons that supply the coordinates of the screen crosshair. When you select this button, it will explode the object where the screen crosshair is located.

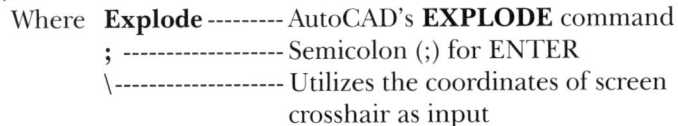

    EXPLODE;
        Where **Explode** --------- AutoCAD's **EXPLODE** command
                ; ------------------- Semicolon (;) for ENTER
                **** -------------------- Utilizes the coordinates of screen
                                    crosshair as input

Line 12
$B=B1
This menu item loads the submenu B1 and assigns the functions defined under this submenu to different buttons of the pointing device.

Line 13
$B=B2
This menu item loads the submenu B2 and assigns the functions defined under this submenu to different buttons of the pointing device.

Step 2: Loading the menu file
Save the file with the extension ***.MNU** and then load the menu file using the **MENU** command.

AUXILIARY MENUS

In a menu file, you can have up to four auxiliary menu sections (AUX1 through AUX4). The auxiliary menu sections (***AUXn) behave identically to the button menu sections. The

Button and Auxiliary Menus

difference is in the hardware. If your computer uses a system mouse, it will automatically use the auxiliary menu. BUTTONS1 menu functions identically to the AUX1 menu and so on. Generally the system mouse has two or three buttons. You can access aux menus by using the following keyboard and button combinations. The combinations of button and keyboard changes depending upon the number of buttons the system mouse has.

The following is the combination for a two-button mouse.

Aux menu	Keyboard and Button Sequence of mouse
Aux1	Right button
Aux2	Shift and right button
Aux3	Control and right button
Aux4	Shift and control and right button

The following is the combination for a three-button mouse.

Aux menu	Keyboard and Button Sequence of mouse
Aux1	Right button
	Middle button
Aux2	Shift and right button
	Shift and middle button
Aux3	Control and right button
	Control and middle button
Aux4	Shift and control and right button
	Shift and control and middle button

The following example will illustrate the use of an auxiliary menu for two-button mouse. Similarly, a three button mouse can be used for auxiliary menu. The buttons of the mouse can be assigned the most frequently used commands for the convenience of the user.

Example 4

Write an auxiliary menu for the following AutoCAD commands (filename **AUX1.MNU**).

Auxiliary menu	Function
AUX1	MOVE
AUX2	COPY
AUX3	ERASE
AUX4	SCALE

Step 1: Writing the menu file

Use any text editor or AutoCAD's **EDIT** command to write the menu file. The following file is a listing of the auxiliary menu for Example 4. The line numbers are not a part of the file; they are for reference only.

```
***AUX1                                                     1
MOVE                                                        2
                                                            3
***AUX2                                                     4
COPY                                                        5
                                                            6
***AUX3                                                     7
ERASE                                                       8
***AUX4                                                     9
SCALE                                                      10
```

Explanation
Line 1
***AUX1
***AUX1 is the section label for the first button of the mouse. When the menu is loaded, AutoCAD compiles the menu file and assigns the command to the buttons of the mouse.

Line 2
MOVE
Thw MOVE command is assigned to AUX1, the right button of the mouse. When you press the right button of the mouse, the **MOVE** command gets activated at the Command prompt.

Line 3
A space has been provided to separate two auxiliary menus. In general, each auxiliary menu must end with a blank line.

The similar description has to be followed for rest of the three auxiliary menus.

Step 2: Loading the menu file
Save the file with the extension ***.MNU** and then load the file with the command **MENU**.

Self-Evaluation Test

Answer the following questions and then compare your answers to the correct answers at the end of this chapter.

1. _____ button and _____ button pointing devices are very common.

2. In a menu file you can have up to _____ auxiliary menu sections.

3. Submenus can also be _____ in the button menu.

4. The submenu name can be up to _____ characters long.

5. A multibutton pointing device can be used to specify or select _____, or enter AutoCAD _____.

Button and Auxiliary Menus 7-15

Review Questions

Answer the following questions.

1. AutoCAD receives the button _____ and _____ of screen crosshairs when a button is activated on the pointing device.

2. If the number of menu items in the button menu is more than the number of buttons on the pointing device, the excess lines are _____.

3. Commands are assigned to the buttons of the pointing device in the _____ order in which they appear in the buttons menu.

4. The format of displaying a loaded submenu that has been defined in the image menu is _____.

5. The format of the **LOAD** command for loading a submenu that has been defined in the image menu is _____.

Exercises

Exercise 1   *General*

Write a button menu for the following AutoCAD commands. The pointing device has 10 buttons (Figure 7-5), and button number 1 is used for specifying the points. The blocks are to be inserted with a scale factor of 1.00 and a rotation of 0 degrees (file name **BME1.MNU**).

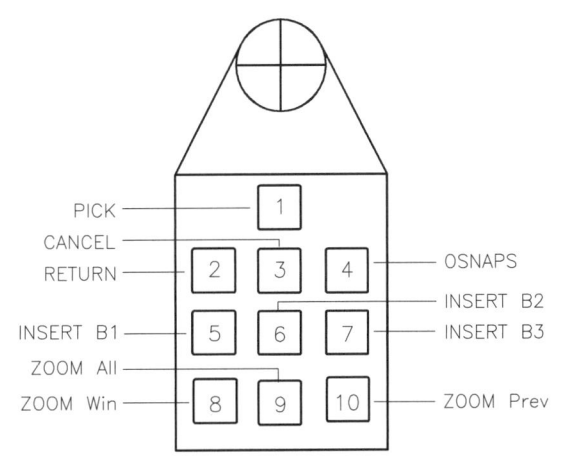

Figure 7-5 Pointing device with 10 buttons

1. PICK BUTTON   2. RETURN   3. CANCEL
4. OSNAPS   5. INSERT B1   6. INSERT B2
7. INSERT B3   8. ZOOM Window   9. ZOOM All
10. ZOOM Previous

1. B1, B2, B3 are the names of the blocks or Wblocks that have already been created.
2. Assume that the Osnap submenu has already been defined.
3. Use the transparent **ZOOM** command for **ZOOM Previous** and **ZOOM Window**.

Exercise 2 — *General*

Write a button menu for a pointing device that has 10 buttons. The functions assigned to different buttons are shown in the following table and in Figure 7-6 (filename **BME2.MNU**):

SUBMENU B1		SUBMENU B2	
1.	PICK	1.	PICK
2.	Enter	2.	Enter
3.	LINE	3.	LOAD IMAGE1
4.	CIRCLE	4.	LOAD IMAGE2
5.	LOAD OSNAPS	5.	LOAD P2 (Pull-down)
6.	ZOOM Win	6.	LOAD P3 (Pull-down)
7.	ZOOM Prev	7.	INSERT
8.	ERASE	8.	EXPLODE
9.	LOAD BUTTON MENU B1	9.	LOAD BUTTON MENU B1
10.	LOAD BUTTON MENU B2	10.	LOAD BUTTON MENU B2

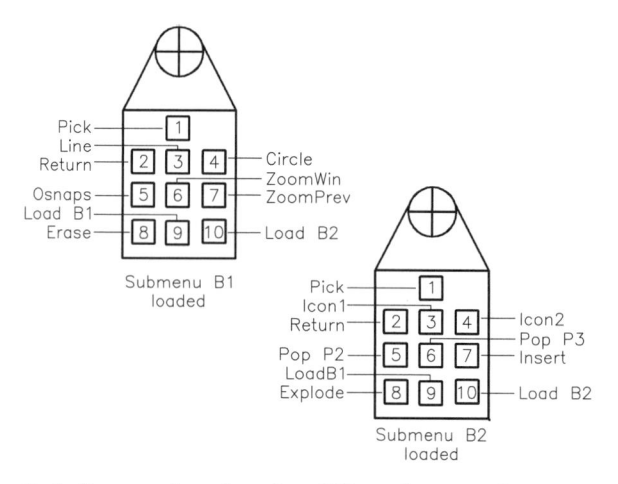

Figure 7-6 Commands assigned to different buttons of pointing device

Assume that:
1. OSNAPS submenu is defined in the POP1 or POP0 (Cursor menu) section of the menu.
2. Menus P2 and P3 are defined in the POP2 and POP3 sections of the menu.
3. IMAGE1 and IMAGE2 submenus are defined in the IMAGE section of the menu file. The IMAGE1 and IMAGE2 submenus contain four images each for inserting the blocks.

Answers to the Self-Evaluation Test

1. Four, Twelve. **2.** Four. **3.** Defined. **4.** Thirtyone. **5.** Points, Command.

Chapter 8

Tablet Menus

Learning Objectives

After completing this chapter, you will be able to:
- *Understand the features of tablet menus.*
- *Write and customize tablet menus.*
- *Load menus and configure tablet menus.*
- *Write tablet menus with different block sizes.*
- *Assign commands to tablet overlays.*

STANDARD TABLET MENU

The tablet menu provides another alternative for entering commands. In the tablet menu, the commands are selected from the template that is secured on the surface of a digitizing tablet. To use the tablet menu you need a digitizing tablet and a pointing device. You also need a tablet template (Figure 8-1) that contains AutoCAD commands arranged in various groups for easy identification.

The standard AutoCAD menu file has four tablet menu sections: TABLET1, TABLET2, TABLET3, and TABLET4. When you start AutoCAD and get into the drawing editor, the tablet menu sections TABLET1, TABLET2, TABLET3, and TABLET4 are automatically loaded. The commands defined in these four sections are then assigned to different blocks of the template.

The first tablet menu section (TABLET1) has 225 blank blocks that can be used to assign up to 225 menu items. It can be reconfigured to as many as 32,766 blocks. The remaining tablet menu sections contain AutoCAD commands, arranged in functional groups that make it easier to identify and access the commands. The commands contained in the TABLET2 section include RENDER, SOLID MODELING, UCSICON, DISPLAY, INQUIRY, LAYER, DRAW, ZOOM, and PAPER SPACE commands. The TABLET3 section contains numbers, fractions,

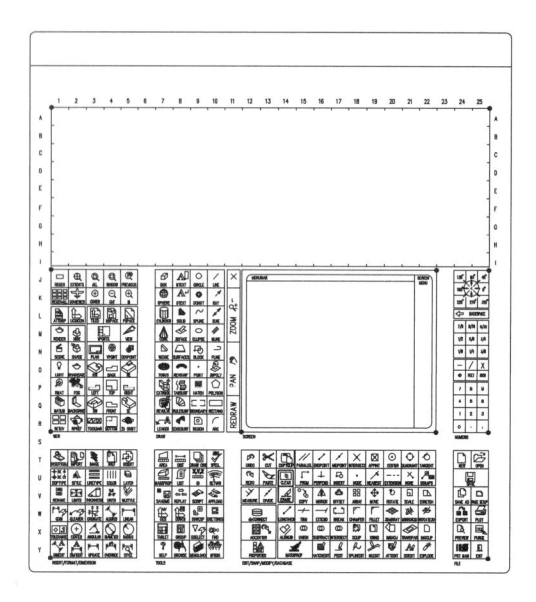

Figure 8-1 Sample tablet template

and angles. The commands contained in the TABLET4 section include **TEXT**, **DIMENSIONING**, **OBJECT SNAPS**, **EDIT**, **UTILITY**, **XREF**, and **SETTINGS**. The AutoCAD tablet template has four tablet areas (Figure 8-2), which correspond to four tablet menu sections, TABLET1, TABLET2, TABLET3, and TABLET4.

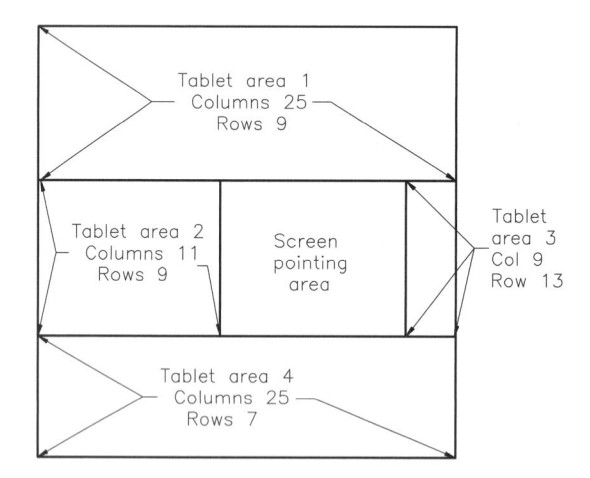

Figure 8-2 Four tablet areas of the AutoCAD tablet template

FEATURES OF A TABLET MENU
A tablet menu has the following features
1. In the tablet menu the commands can be arranged so that the most frequently used commands can be accessed directly. This can save considerable time in entering AutoCAD commands. In the screen menu or the pull-down menu some of the commands cannot be

Tablet Menus   8-3

accessed directly. For example, to generate a horizontal dimension you have to go through several steps. You first select Dimension from the root menu and then Linear. In the tablet menu you can select the Linear dimensioning command directly from the digitizer. This saves time and eliminates the distraction that takes place as you page through different screens.

2. You can have the graphical symbols of the AutoCAD commands drawn on the tablet template. This makes it much easier to recognize and select commands. For example, if you are not an expert in AutoCAD dimensioning, you may find Baseline and Continue dimensioning confusing. But if the command is supported by the graphical symbol illustrating what a command does, the chances of selecting a wrong command are minimized.

3. You can assign any number of commands to the tablet overlay. The number of commands you can assign to a tablet is limited only by the size of the digitizer and the size of the rectangular blocks.

CUSTOMIZING A TABLET MENU

As with a screen menu, you can customize the AutoCAD tablet menu. This is a powerful customizing tool to make AutoCAD more efficient.

The tablet menu can contain a maximum of four sections: TABLET1, TABLET2, TABLET3, and TABLET4. Each section represents a rectangular area on the digitizing tablet. These rectangular areas can be further divided into any number of rectangular blocks. The size of each block depends on the number of commands that are assigned to the tablet area. Also, the rectangular tablet areas can be located anywhere on the digitizer and can be arranged in any order. The AutoCAD **TABLET** command configures the tablet. The **MENU** command loads and assigns the commands to the rectangular blocks on the tablet template.

Before writing a tablet menu file, it is very important to design the layout of the design of the tablet template. A well-thought-out design can save a lot of time in the long run. The following points should be considered when designing a tablet template:

1. Learn the AutoCAD commands that you use in your profession.

2. Group the commands based on their function, use, or relationship with other commands.

3. Draw a rectangle representing a template so that it is easy for you to move the pointing device around. The size of this area should be appropriate to your application. It should not be too large or too small. Also, the size of the template depends on the active area of the digitizer.

4. Divide the remaining area into four different rectangular tablet areas for TABLET1, TABLET2, TABLET3, and TABLET4. It is not necessary to use all four areas; you can have fewer tablet areas, but four is the maximum.

5. Determine the number of commands you need to assign to a particular tablet area; then

determine the number of rows and columns you need to generate in each area. The size of the blocks does not need to be the same in every tablet area.

6. Use the **TEXT** command to print the commands on the tablet overlay, and draw the symbols of the command, if possible.

7. Plot the tablet overlay on good-quality paper or a sheet of Mylar. If you want the plotted side of the template to face the digitizer board, you can create a mirror image of the tablet overlay and then plot the mirror image.

WRITING A TABLET MENU

When writing a tablet menu you must understand the AutoCAD commands and the prompt entries required for each command. Equally important is the design of the tablet template and the placement of various commands on it. Give considerable thought to the design and layout of the template, and, if possible, invite suggestions from AutoCAD users in your trade. To understand the process involved in developing and writing a tablet menu, consider Example 1.

Example 1

Write a tablet menu for the following AutoCAD commands. The commands are to be arranged as shown in Figure 8-3. Make a tablet menu template for configuration and command selection (filename **TM1.MNU**).

LINE	CIRCLE	PLINE
CIRCLE C,D	ERASE	CIRCLE 2P

Step 1: Writing the tablet menu

Figure 8-3 represents one of the possible template designs, where the AutoCAD commands are in one row at the top of the template, and the screen pointing area is in the center. There is only one area in this template; therefore, you can place all these commands under the section label TABLET1. To write a menu file you can use any text editor like Notepad.

The name of the file is TM1, and the extension of the file is **.MNU. The line numbers are not part of the file. They are shown here for reference only.**

```
***TABLET1                                        1
^C^CLINE                                          2
^C^CPLINE                                         3
^C^CCIRCLE                                        4
^C^CCIRCLE;\D                                     5
^C^CCIRCLE;2P                                     6
^C^CERASE                                         7
```

Step 2: Explanation for writing the tablet menu
Line 1
***TABLET1

Tablet Menus

TABLET1 is the section label of the first tablet area. All the section labels are preceded by three asterisks (***).

***TABLET1
  Where ******* ---------------- Three asterisks designate the section label.
      TABLET1 ------ Section label for TABLET1

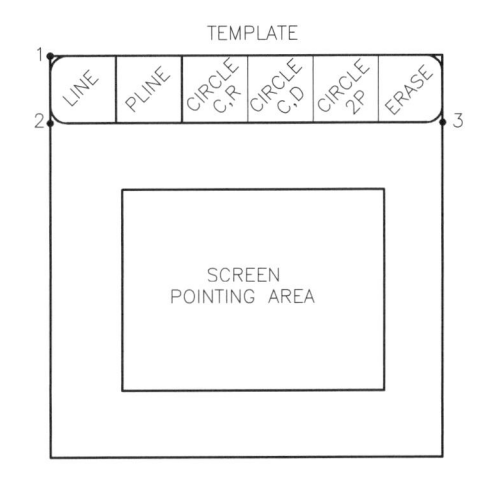

Figure 8-3 Design of tablet template

Line 2
^C^CLINE
^C^C cancels the existing command twice; **LINE** is an AutoCAD command. There is no space between the second ^C and **LINE**.

  ^C^C LINE
    Where **^C^C** ---------- Cancels the existing command twice
       Line -------------- AutoCAD **LINE** command

Line 3
^C^CPLINE
^C^C cancels the existing command twice; **PLINE** is an AutoCAD command.

Line 4
^C^CCIRCLE
^C^C cancels the existing command twice; **CIRCLE** is an AutoCAD command. The default input for the **CIRCLE** command is the center and the radius of the circle; therefore, no additional input is required for this line.

Line 5
^C^CCIRCLE;\D
^C^C cancels the existing command twice; **CIRCLE** is an AutoCAD command like the previous line. However, this command definition requires the diameter option of the circle

command. This is accomplished by using \D in the command definition. There should be no space between the backslash (\) and the D, but there should always be a space or semicolon **before** the backslash (\). The backslash (\) lets the user enter a point, and in this case it is the center point of the circle. After you enter the center point, the diameter option is selected by the letter D, which follows the backslash (\).

^C^CCIRCLE;\D

Where **CIRCLE** --------- **CIRCLE** command
; ------------------- Semicolon or Space for RETURN
\ -------------------- Pause for input
D ------------------- Diameter option

Line 6
^C^CCIRCLE;2P
^C^C cancels the existing command twice; **CIRCLE** is an AutoCAD command. Semicolon is for ENTER or RETURN. The 2P selects the two-point option of the **CIRCLE** command.

Line 7
^C^CERASE
^C^C cancels the existing command twice; **ERASE** is an AutoCAD command that erases the selected objects.

Note
In the tablet menu, the part of the menu item that is enclosed in the brackets is used for screen display only. For example, in the following menu item, T1-6 will be ignored and will have no effect on the command definition.

[T1-6]^C^CCIRCLE;2P

Where **[T1-6]** ------------ For reference only and has no effect on the command definition
T1 ----------------- Tablet area 1
6 ------------------- Item number 6

The reference information can be used to designate the tablet area and the line number.

Before you can use the commands from the new tablet menu, you need to configure the tablet and load the tablet menu.

TABLET CONFIGURATION

To use the new template to select the commands, you need to install Wintab driver and configure the tablet so that AutoCAD knows the location of the tablet template and the position of the commands assigned to each block. This is accomplished by using the AutoCAD **TABLET** command. Secure the tablet template (Figure 8-3) on the digitizer with the edges of the overlay approximately parallel to the edges of the digitizer. Enter the AutoCAD **TABLET** command, select the **Configure option**, and respond to the following prompts. Figure 8-4 shows the points you need to select to configure the tablet.

Tablet Menus

Command: **TABLET**
Enter an option [ON/OFF/CAL/CFG]:**CFG**
Enter number of tablet menus desired (0–4) <current>: **1**
Do you want to realign tablet menus? [Yes/No] <N>: **Y**
Digitize upper left corner of menu area 1; **P1**
Digitize lower left corner of menu area 1; **P2**
Digitize lower right corner of menu area 1; **P3**
Enter the number of columns for menu area 1: **6**
Enter the number of rows for menu area 1: **1**
Do you want to respecify the Fixed Screen Pointing Area? [Yes/No]<N>: **Y**
Digitize lower left corner of Fixed Screen pointing area: **P4**
Digitize upper right corner of Fixed Screen pointing area: **P5**
Do you want to specify the Floating Screen Pointing Area? [Yes/No]<N>: **N**

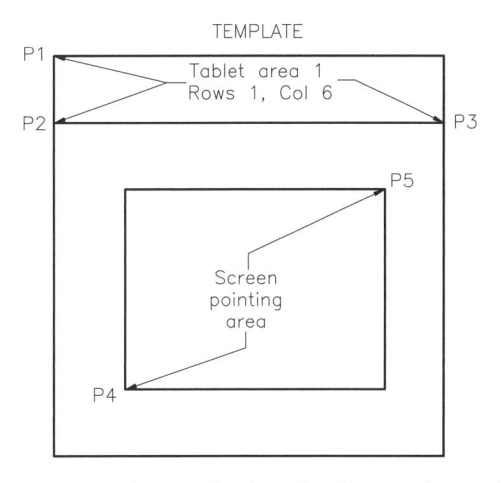

Figure 8-4 Points that need to be selected to configure the tablet

Note
The three points P1, P2, and P3 should form a 90-degree angle. If the selected points do not form a 90-degree angle, AutoCAD will prompt you to enter the points again until they do. The tablet areas should not overlap the screen pointing area.

The screen pointing area can be any size and located anywhere on the tablet as long as it is within the active area of the digitizer. The screen pointing area should not overlap other tablet areas. The screen pointing area you select will correspond to the monitor screen area. Therefore, the length-to-width ratio of the screen pointing area should be the same as that of the monitor, unless you are using the screen pointing area to digitize a drawing.

LOADING MENUS

AutoCAD automatically loads the **ACAD.MNU** file when you get into the AutoCAD drawing editor. However, you can also load a different menu file by using the AutoCAD **MENU** command.

Command: **MENU**

When you enter the **MENU** command, AutoCAD displays the **Select Menu File** dialog box on the screen (Figure 8-5). Select the menu file that you want to load and then choose the **Open** button. You can also load the menu file from the command line keeping the **FILEDIA** as 0.

Command: **FILEDIA**
Enter new value for FILEDIA <1>: **0**
Command: **MENU**
Menu filename <ACAD>: **PDM1**
            Where **ACAD** ------------ Default menu file
                      PDM1 ------------ Name of menu file

AutoCAD will automatically compile the menu file into **MNC** and **MNR** files.

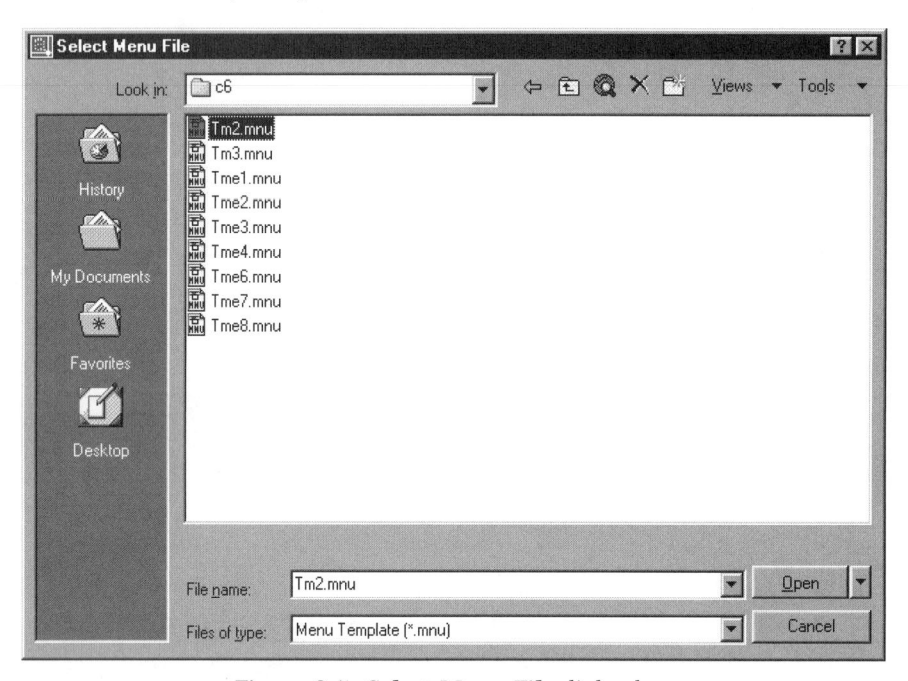

Figure 8-5 Select Menu File dialog box

Exercise 1   *General*

Write a tablet menu for the following AutoCAD commands. Make a tablet menu template for configuration and command selection (filename **TME1.MNU**).

LINE	**TEXT-Center**
CIRCLE C,R	**TEXT-Left**
ARC C.S.E	**TEXT-Right**
ELLIPSE	**TEXT-Aligned**
DONUT	

Tablet Menus   8-9

Use the template in Figure 8-6 to arrange the commands. The draw and text commands should be placed in two separate tablet areas.

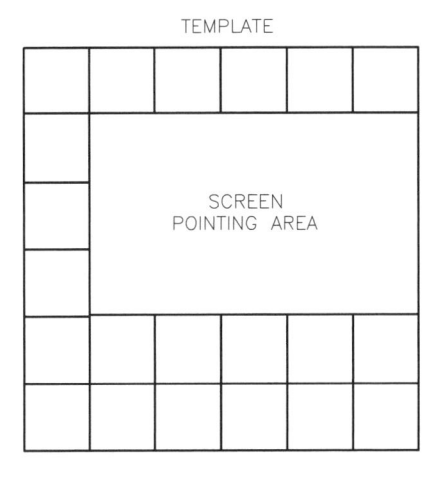

Figure 8-6 Template for Exercise 1

TABLET MENUS WITH DIFFERENT BLOCK SIZES

As mentioned earlier, the size of each tablet area can be different. The size of the blocks in these tablet areas can also be different. But the size of every block in a particular tablet area must be the same. This provides you with a lot of flexibility in designing a template. For example, you may prefer to have smaller blocks for numbers, fractions, or letters, and larger blocks for draw commands. You can also arrange these tablet areas to design a template layout with different shapes, such as L-shape and T-shape.

Tablet Area-1   **Tablet Area-2**

The following example illustrates the use of multiple tablet areas with different block sizes.

Example 2

Write a tablet menu for the tablet overlay shown in Figure 8-7(a). Figure 8-7(b) shows the number of rows and columns in different tablet areas (filename **TM2.MNU**).

Writing the tablet menu
Notice that this tablet template has four different sections in addition to the screen pointing

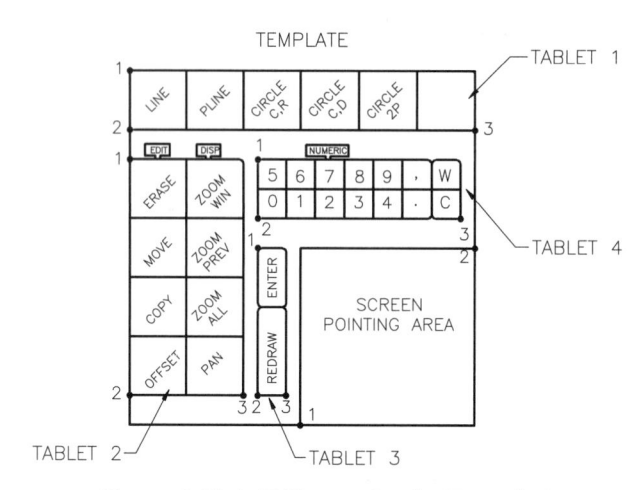

Figure 8-7(a) *Tablet overlay for Example 2*

area. Therefore, this menu will have four section labels: TABLET1, TABLET2, TABLET3, and TABLET4. You can use any text editor (Notepad) to write the file. The following file is a listing of the tablet menu of Example 2.

```
***TABLET1                                                      1
^C^CLINE                                                        2
^C^CPLINE                                                       3
^C^CCIRCLE                                                      4
^C^CCIRCLE;\D                                                   5
^C^CCIRCLE;2P                                                   6
***TABLET2                                                      7
^C^CERASE                                                       8
^C^CZOOM;W                                                      9
^C^CMOVE                                                       10
^C^CZOOM;P                                                     11
^C^CCCOPY                                                      12
^C^CZOOM;A                                                     13
^C^COFFSET                                                     14
^C^CPAN                                                        15
***TABLET3                                                     16
;                                                              17
;                                                              18
'REDRAW                                                        19
'REDRAW                                                        20
'REDRAW                                                        21
***TABLET4                                                     22
5\                                                             23
6\                                                             24
7\                                                             25
```

Tablet Menus   8-11

8\	26
9\	27
,\	28
WINDOW	29
0\	30
1\	31
2\	32
3\	33
4\	34
.\	35
CROSSING	36

Explanation

Lines 1-6
The first six lines are identical to the first six lines of the tablet menu in Example 1.

Line 9
^C^CZOOM;W
ZOOM is an AutoCAD command; W is the window option of the **ZOOM** command.

    ^C^CZOOM;W
        Where   **ZOOM** ----------- AutoCAD **ZOOM** command
                     ; ------------------- Space for Return
                     W ----------------- Window option of **ZOOM** command

This menu item could also be written as:

    ^C^CZOOM W
        Space between ZOOM and W causes a RETURN

Lines 17 and 18
The semicolon (;) is for RETURN. It has the same effect as entering RETURN at the keyboard.

Lines 19-21
'REDRAW
REDRAW is an AutoCAD command that redraws the screen. Notice that there is no ^C^C in front of the **REDRAW** command. If it had ^C^C, the existing command would be canceled before redrawing the screen. This may not be desirable in most applications because you might want to redraw the screen without canceling the existing command. The single quotation (') in front of REDRAW makes the **REDRAW** command transparent.

Line 23
5
The backslash (\) is used to introduce a pause for user input. Without the backslash you cannot enter another number or a character, because after you select the digit **5** it will automatically be followed by RETURN. For example, without the backslash (\), you will not be able to enter a number like **5.6**. Therefore, you need the backslash to enable you to enter decimal numbers

or any characters. To terminate the input, enter RETURN at the keyboard or select RETURN from the digitizer.

5\

Where ------------ \ Backslash for user input

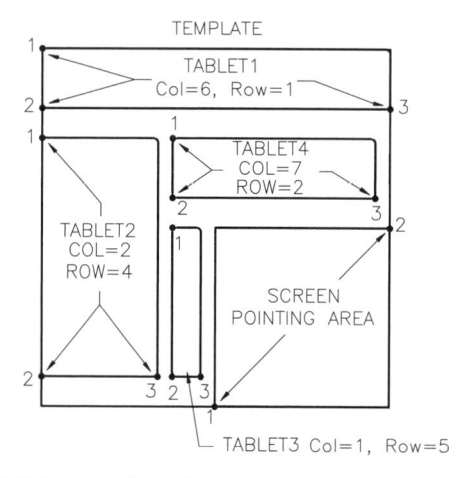

Figure 8-7(b) Number of rows and columns in different tablet

ASSIGNING COMMANDS TO A TABLET

After loading the menu by using the AutoCAD **MENU** command, you must configure the tablet. At the time of configuration, AutoCAD actually generates and stores the information about the rectangular blocks on the tablet template. When you load the menu and configure the tablet, the commands defined in the tablet menu are assigned to various blocks. For example, when you select the three points for tablet area 4, Figure 8-7(a), and enter the number of rows and columns, AutoCAD generates a grid of seven columns and two rows, as shown in the following diagram.

After Configuration

When you load the new menu, AutoCAD takes the commands under the section label TABLET4 and starts filling the blocks from left to right. That means "5", "6", "7", "8", "9", ",", and "Window" will be placed in the top row. The next seven commands will be assigned to the next row, starting from the left, as shown in the diagram on page 8-13.

Tablet Menus 8-13

After Loading Tablet Menu

5	6	7	8	9	.	Window
0	1	2	3	4	.	Cross

Similarly, tablet area 3 has been divided into five rows and one column. At first, it appears that this tablet area has only two rows and one column, as shown in Figure 8-7(a). When you configure this tablet area by specifying the three points and entering the number of rows and columns, AutoCAD divides the area into one column and five rows, as shown in the following diagrams.

After loading the menu, AutoCAD takes the commands in the TABLET3 section of the tablet menu and assigns them to the blocks. The first command (;) is placed in the first block. Since there are no more blocks in the first row, the next command (;) is placed in the second row. Similarly, the three **REDRAW** commands are placed in the next three rows. If you pick a point in the first two rows, you will select the **ENTER** command. Similarly, if you pick a point in the next three blocks, you will select the **REDRAW** command.

After Configuration

After Loading Menu

;
;
Redraw
Redraw
Redraw

This process is carried out for all the of tablet areas, and the information is stored in the AutoCAD configuration file, **ACAD2002.CFG**. If for any reason the configuration is not right, the tablet menu may not perform the desired function.

AUTOMATIC MENU SWAPPING

The screen menus can be automatically swapped by using the system variable **MENUCTL**. When the system variable **MENUCTL** is set to 1, AutoCAD automatically issues the **$S=CMDNAME** command, where **CMDNAME** is the name of the command that loads the submenu. For example, if you select the **LINE** command from the digitizer menu, or enter the **LINE** command from the keyboard, the **CMDNAME** command will load the **LINE** submenu and display it in the screen menu area. To utilize this feature of AutoCAD, the command name and the submenu name must be the same. For example, if the name of the arc submenu is **ARC** and you select the **ARC** command, the arc submenu will be automatically loaded on the screen. However, if the submenu name is different (for example MYARC), then AutoCAD will not load the arc submenu on the screen. The default value of **MENUCTL** system variable is 1. If you set the **MENUCTL** variable to 0, AutoCAD will not utilize the **$S=CMDNAME** command feature to load the submenus.

Self-Evaluation Test

Answer the following questions and then compare your answers to the correct answers at the end of this chapter.

1. To use the tablet menu you need a _____ and a _____ .

2. The first tablet menu section can be used to assign up to _____ menu items .

3. All the section labels in the tablet menu should preceded by _____ .

4. The tablet area should not _____ the screen pointing area.

5. The size of every block in a particular tablet area must be _____ .

Review Questions

Answer the following questions.

1. The maximum number of tablet menu sections is _____ .

2. A tablet menu area is _____ in shape.

3. The blocks in any tablet menu area are _____ in shape.

4. A tablet menu area can have _____ number of rectangular blocks.

5. You _____ assign the same command to more than one block on the tablet menu template.

6. The AutoCAD _____ command is used to configure the tablet menu template.

7. The AutoCAD _____ command is used to load a new menu.

Tablet Menus   8-15

Exercises

Exercise 2 *General*

Design the template and write a tablet menu to insert the following user-defined blocks:

BX1	BX5	BX9
BX2	BX6	BX10
BX3	BX7	BX11
BX4	BX8	BX12

Exercise 3 *General*

Design the template menu for the following AutoCAD commands:

LINE	**ZOOM-Win**	**DIM-Horz**
PLINE	**ZOOM-Dyn**	**DIM-Vert**
ARC	**ZOOM-All**	**DIM-Alig**
CIRCLE	**ZOOM-Pre**	**DIM-Ang**
ELLIPSE	**ZOOM-Ext**	**DIM-Rad**
POLYGON	**ZOOM-Scl**	**DIM-Cen**

Exercise 4 *General*

Write a tablet menu for the commands shown in the tablet menu template of Figure 8-8. Make a tablet menu template for configuration and command selection.

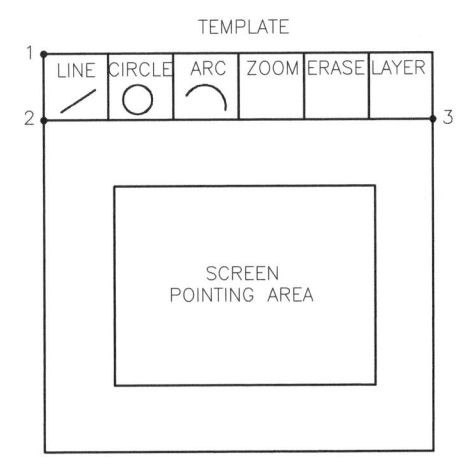

Figure 8-8 Tablet menu template for Exercise 4

Exercise 5 *General*

Write a tablet menu for the AutoCAD commands shown in Figure 8-9. Configure the tablet and then load the new menu. Make a tablet menu template for configuration and command selection.

Figure 8-9 Tablet overlay for Exercise 5

Exercise 6 *General*

Write a tablet menu file for the AutoCAD commands shown in the template of Figure 8-10. Make a tablet menu template for configuration and command selection.

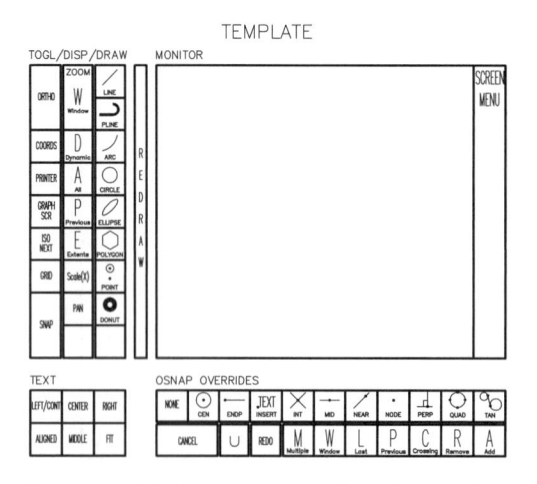

Figure 8-10 Tablet menu template for Exercise 6

Tablet Menus   8-17

Exercise 7   *General*

Write a combined pull-down and tablet menu file for the commands shown in the tablet menu template in Figure 8-11. Make a tablet menu template for configuration and command selection.

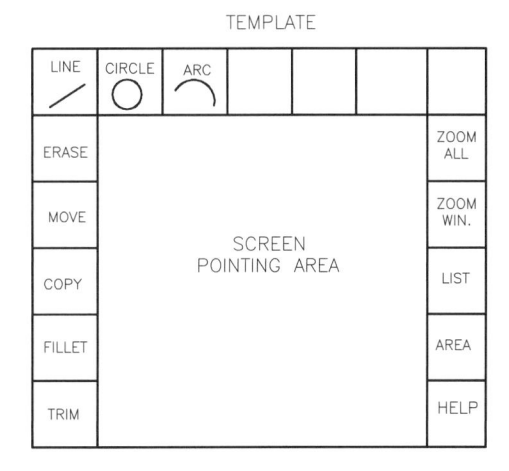

Figure 8-11 Tablet menu template for Exercise 7

Template for Example 1

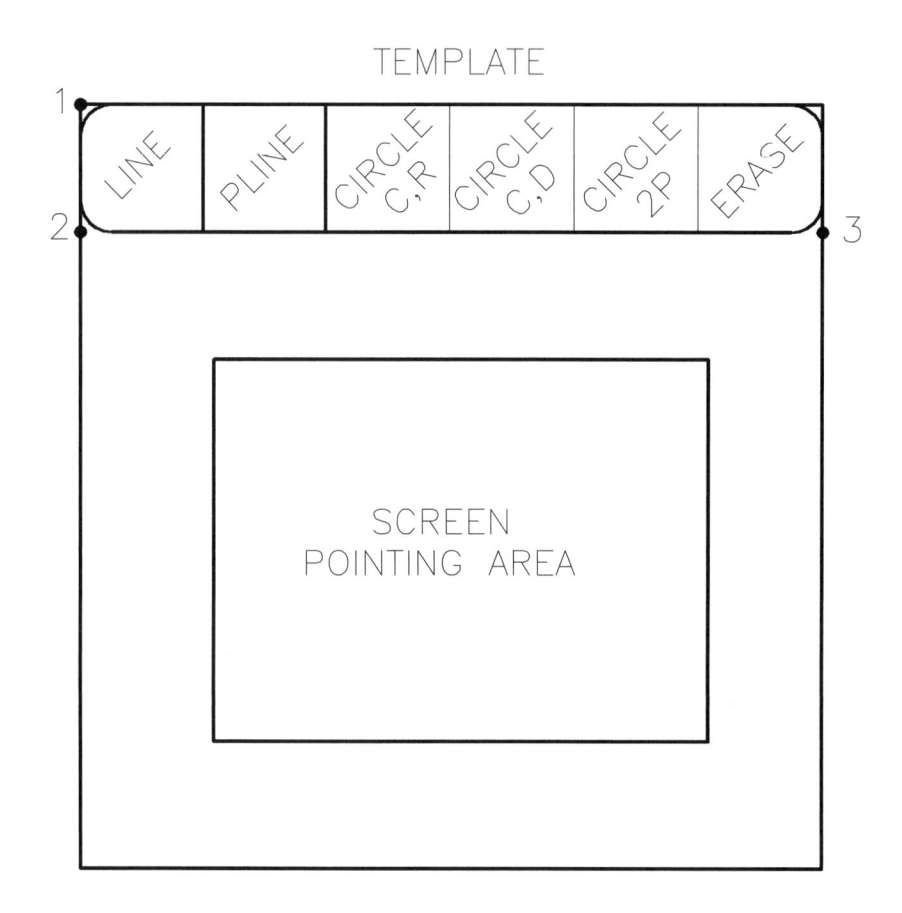

TEMPLATE

| LINE | PLINE | CIRCLE C,R | CIRCLE C,D | CIRCLE 2P | ERASE |

SCREEN POINTING AREA

Note

This template is for tablet configuration. You may make a copy of this page and then secure it on the digitizer surface for configuration.

Tablet Menus 8-19

Template for Exercise 1

TEMPLATE

Note

This template is for tablet configuration. You may make a copy of this page and then secure it on the digitizer surface for configuration.

Template for Example 2

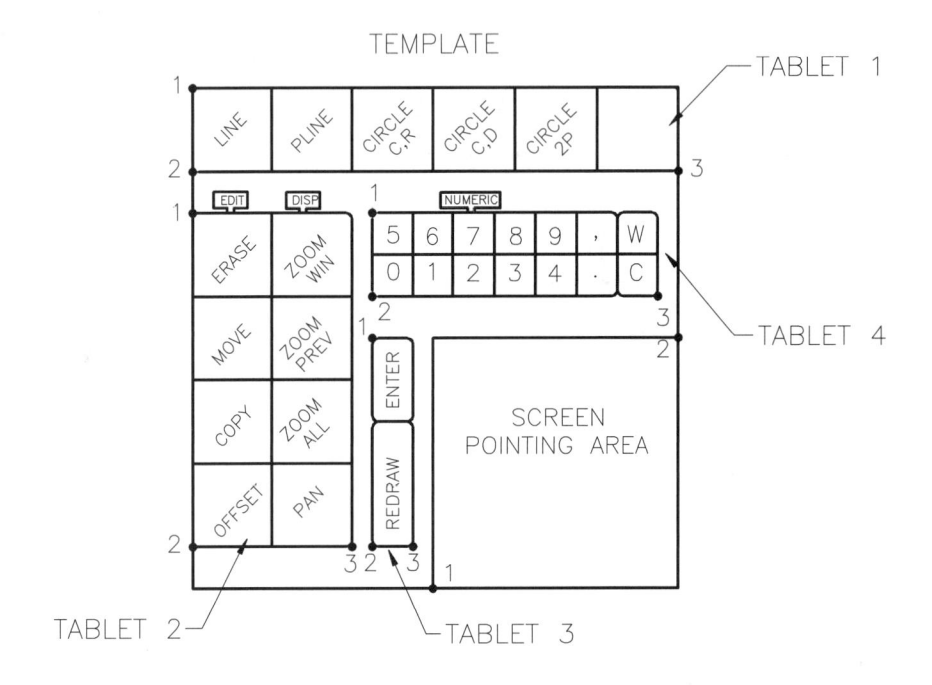

Note

This template is for tablet configuration. You may make a copy of this page and then secure it on the digitizer surface for configuration.

Template for Exercise 4

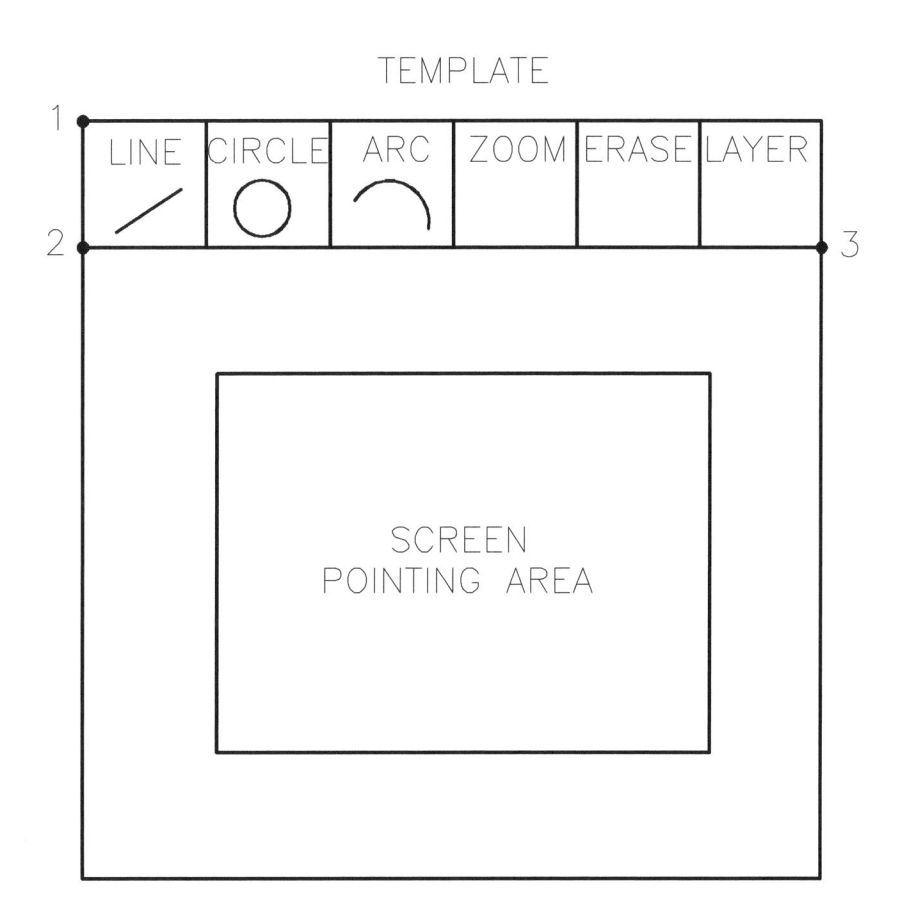

```
                    TEMPLATE
1
   | LINE | CIRCLE | ARC | ZOOM | ERASE | LAYER |
   |  /   |   O    |  )  |      |       |       |
2                                                 3

              SCREEN
           POINTING  AREA
```

Note

This template is for tablet configuration. You may make a copy of this page and then secure it on the digitizer surface for configuration.

Template for Exercise 5

TEMPLATE

SAVE	QUIT	END	SAVEAS	PLOT	@	/	,	5	6	7	8	9
					X	−	°	0	1	2	3	4
			ERASE	ZOOM WIN								
			MOVE	ZOOM PREV		REDRAW						
			COPY	ZOOM ALL		SCREEN POINTING AREA						
			OFFSET	PAN								
			TRIM	ZOOM EXTENTS		ENTER						
			CEN	ENDP	INT	LINE		PLINE		ELLIPSE		
EDIT	CHANGE		MID	NEAR	PERP	CIRCLE		CIRCLE C,D		CIRCLE 2P		

Note

This template is for tablet configuration. You may make a copy of this page and then secure it on the digitizer surface for configuration.

Tablet Menus 8-23

Template for Exercise 6

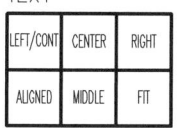

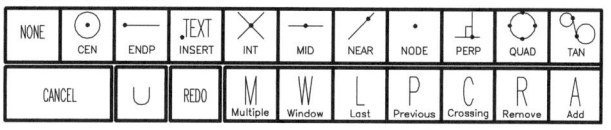

Note

This template is for tablet configuration. You may make a copy of this page and then secure it on the digitizer surface for configuration.

Template for Exercise 7

TEMPLATE

LINE	CIRCLE	ARC			
ERASE					ZOOM ALL
MOVE		SCREEN POINTING AREA			ZOOM WIN.
COPY					LIST
FILLET					AREA
TRIM					HELP

Note
This template is for tablet configuration. You may make a copy of this page and then secure it on the digitizer surface for configuration.

Answers to the Self-Evaluation Test

1. Digitizing tablet, pointing device, **2.** standard tablet 225, can be reconfigured upto 32,766, **3.** ***, **4.** overlap, **5.** same.

Chapter 9

Screen Menus

Learning Objectives

After completing this chapter, you will be able to:
- *Write screen menus.*
- *Load screen menus.*
- *Write submenus and referencing submenus.*
- *Write menu files with multiple submenus.*
- *Use menu command repetition in menus.*
- *Write menus for foreign languages.*
- *Use control and special characters in menu items.*
- *Use command definition without return or space.*
- *Use menu items with single object selection mode.*
- *Use AutoLISP and DIESEL expressions in menus.*

SCREEN MENU

When you are in the AutoCAD drawing editor, by default the screen menu is not displayed on the screen. To display it on the screen, select **Options** in the **Tools** pull-down menu to display the **Options** dialog box and then select the **Display** tab. Click on the check box located to the left of Display screen menu and then click on the **OK** button to exit the dialog box. The AutoCAD screen menu is displayed on the right of the screen with AutoCAD at the top, followed by asterisk signs (* * * *) and a list of commands (Figure 9-1).

Depending on the scope of the menu, the size of the menu file can vary from a few lines to several hundred. A menu file consists of section labels, submenus, and menu items. A menu item consists of an item label and a command definition. The menu item label is enclosed in brackets, and the command definition (menu macro) is outside the brackets.

The menu item label that is enclosed in the brackets is displayed in the screen menu area of the monitor and is not a part of the command definition. The command definition, the part of the menu item outside the bracket, is the executable part of the menu item. To understand the process of developing and writing a screen menu, consider the following example.

[LINE   ] $S=X $S=LINE ^C^CLINE

Where **[LINE   ] $S=X $S=LINE ^C^CLINE** -------- Menu item
[LINE   ] ------- Item label or menu command name
$S=X $S=LINE ^C^CLINE - Command definition

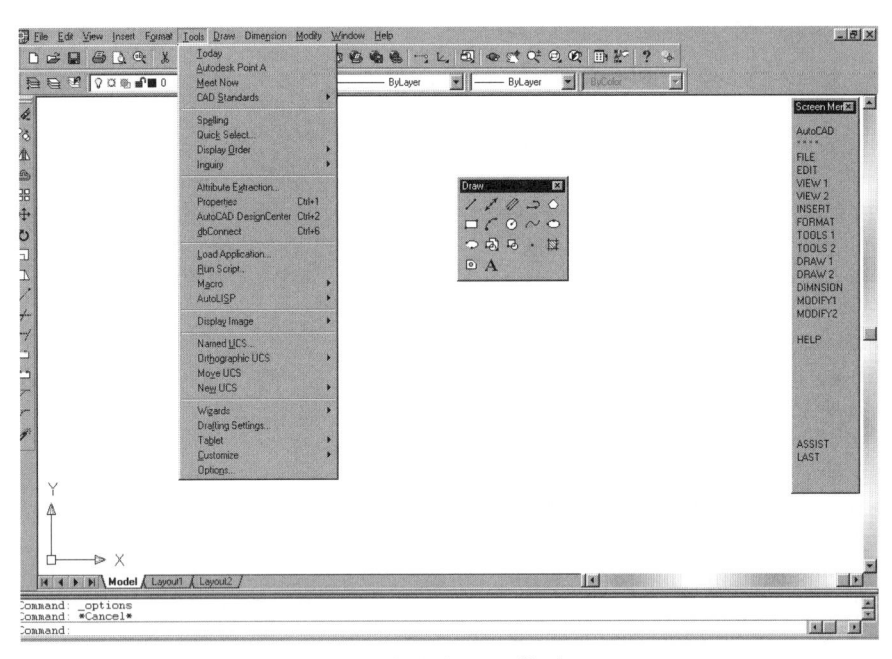

Figure 9-1 Screen display

Example 1

Write a screen menu for the following AutoCAD commands (file name **SM1.MNU**):

**LINE
CIRCLE C,R
CIRCLE C,D
CIRCLE 2P
ERASE
MOVE**

The layout of these commands is shown in Figure 9-2(a). This menu is named MENU-1 and it should be displayed at the top of the screen menu. It lets you know the menu you are using. You can use any text editor to write the file. The file extension must be **.MNU**. For Example 1, **SM1.MNU. SM1** is the name of the screen menu file, and **.MNU** is the extension of this file. All menu files have the extension **.MNU**.

Screen Menus   9-3

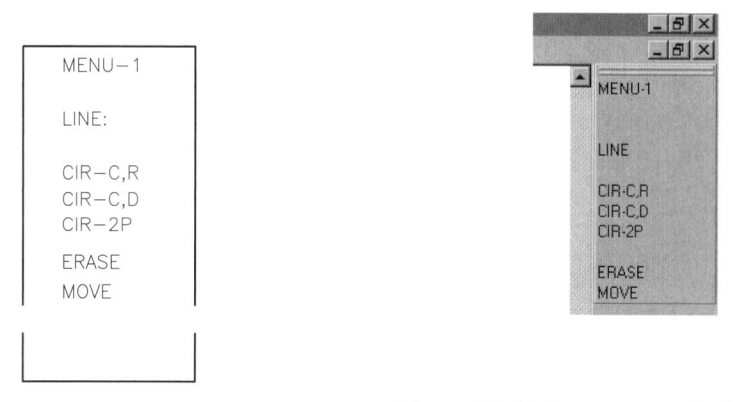

Figure 9-2(a) Layout of screen menu

Figure 9-2(b) Screen menu after loading the menu file

The following file is a listing of the screen menu for Example 1. **The line numbers on the right are not a part of the file. They are shown here for reference only.**

```
***SCREEN                                           1
[ MENU-1          ]                                 2
[                 ]                                 3
[                 ]                                 4
[LINE             ]^C^CLINE                         5
[                 ]                                 6
[CIR-C,R          ]^C^CCIRCLE                       7
[CIR-C,D          ]^C^CCIRCLE;\D                    8
[CIR- 2P          ]^C^CCIRCLE;2P                    9
[                 ]                                 10
[ERASE            ]^C^CERASE                        11
[MOVE             ]^C^CMOVE                         12
```

Explanation
Line 1
***SCREEN
***SCREEN is the section label for the screen menu. The lines that follow the screen menu are treated as a part of this menu. The screen menu definition will be terminated by another section label, such as ***TABLET1 or ***POP1.

Line 2
[MENU-1]
This menu item displays MENU-1 on the screen. Anything that is inside the brackets is for display only and has no effect on the command. The item names can be any length. However, it is recommended that the names be short and meaningful so that the screen menu does not take the valuable screen space.

Lines 3 and 4
[   ]

These menu items print a blank line on the screen menu. There are eight blank spaces inside the bracket. You could also use the brackets with no spaces between the brackets ([]). This line does not contain anything outside the bracket; therefore, no command is executed. To provide the space in the menu you can also leave a blank space in the menu file or have two brackets ([ ]). The next line, line 4, also prints a blank line.

Line 5
[LINE   ]^C^CLINE

This menu item displays LINE on the screen. The first ^C (caret C) cancels the existing command, and the second ^C cancels the command again. The two **CANCEL** (^C^C) commands are required to make sure that the existing commands are canceled before a new command is executed. Most AutoCAD commands can be canceled with just one **CANCEL** command. However, some commands, like dimensioning and pedit, need to be canceled twice to get out of the command. **LINE** will prompt the user to enter the points to draw a line. Since there is nothing after the LINE, it automatically enters a Return.

```
[LINE    ]^C^CLINE
         Where  [LINE   ] --------- For screen display only
                  ^C  --------------- ^C is the first CANCEL command
                  ^C  --------------- ^C is the second CANCEL command
                  LINE ------------- AutoCAD's LINE command
```

Line 7
[CIR-C,R]^C^CCIRCLE

The part of the menu item that is enclosed within the brackets is for screen display only. The part of the menu item that is outside the brackets is executed when this line is selected. ^C^C (caret C) cancels the existing command twice. **CIRCLE** is an AutoCAD command that generates a circle. **A space after CIRCLE automatically causes a RETURN. The space acts like pressing the SPACEBAR on the keyboard.**

```
[CIR-C,R ]^C^CCIRCLE
         Where  [CIR-C,R ] ----- For screen display only
                  ^C  --------------- ^C for first CANCEL command
                  ^C  --------------- ^C for second CANCEL command
                  CIRCLE --------- AutoCAD's CIRCLE command
```

Line 8
[CIR-C,D]^C^CCIRCLE;\D

The part of the menu item that is enclosed in the brackets is for screen display only, and the part that is outside the brackets is the executable part. ^C^C will cancel the existing command twice.

```
[CIR-C,D ]^C^CCIRCLE;\D
         Where  ^C^C ---------- ^C^C cancels the existing command
                  CIRCLE --------- CIRCLE command
```

Screen Menus 9-5

```
                          ;  ------------------ Semicolon (;) for RETURN
                          \ ------------------ AutoCAD pauses for user input
                          D ----------------- Diameter option
```

The **CIRCLE** command is followed by a semicolon, a backslash (\), and a D for diameter option. **The semicolon (;) after the CIRCLE command causes a RETURN, which has the same effect as entering RETURN at the keyboard. The backslash (\) pauses for user input.** In this case it is the center point of the circle. D is for the diameter option and it is automatically followed by RETURN. The semicolon in this example can also be replaced by a blank space, as shown in the following line. However, the semicolon is easier to spot.

[CIR-C,D]^C^CCIRCLE \D
     Where   Space between CIRCLE and \ --- Blank space for RETURN

Line 9
[CIR- 2P]^C^CCIRCLE;2P
In this menu item ^C^C cancels the existing command twice. The semicolon after **CIRCLE** enters a RETURN. 2P is for the two-point option, followed by the blank space that causes a RETURN. You will notice that the sequence of the command and input is the same as discussed earlier. Therefore, it is essential to know the exact sequence of the command and input; otherwise, the screen menu is not going to work.

[CIR- 2P]^C^CCIRCLE;2P
```
           Where  ^C^C ------------ ^C^C Cancels existing command twice
                  CIRCLE --------- CIRCLE command
                  ;  ------------------ Semicolon (;) for RETURN
                  2P ---------------- 2-Point option for CIRCLE
```

The semicolon after the **CIRCLE** command can be replaced by a blank space, as shown in the following line. The blank space causes a RETURN, like a semicolon (;).

[CIR- 2P]^C^CCIRCLE 2P
     Where Space between CIRCLE and 2P -- Blank space or semicolon for RETURN

Line 11
[ERASE]^C^CERASE
In this menu item ^C^C cancels the existing command twice, and **ERASE** is an AutoCAD command that erases the selected objects.

[ERASE]^C^CERASE
     Where **ERASE** ---------- AutoCAD **ERASE** command

Line 12
[MOVE]^C^CMOVE
In this menu item ^C^C cancels the existing command twice, and the **MOVE** command will move the selected objects.

[MOVE   ]^C^CMOVE
           Where   **Move** ------------- AutoCAD **MOVE** command

LOADING MENUS

AutoCAD automatically loads the **ACAD.MNS** file when you get into the AutoCAD drawing editor. However, you can also load a different menu file by using the AutoCAD **MENU** command.

   Command: **MENU**

When you enter the **MENU** command, AutoCAD displays the **Select Menu File** dialog box (Figure 9-3) on the screen. Select the menu file that you want to load and then choose the **Open** button.

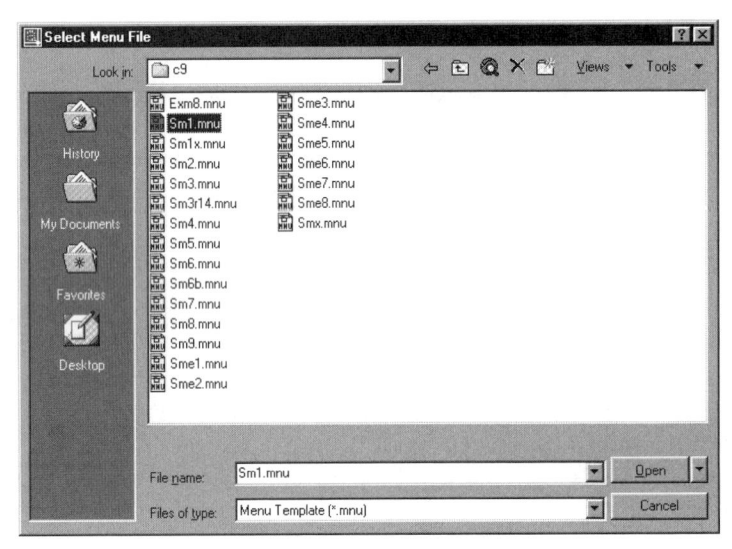

Figure 9-3 Select Menu File dialog box

You can also load the menu file from the command line.

   Command: **FILEDIA**
   Enter new value for FILEDIA <1>: **0**
   Command: **MENU**
   Enter menu file name or [. (for none)] <current>: PDM1.MNU
           Where **<current>** ------ Default menu file
                  PDM1.MNU --- Name of menu file

After you enter the **MENU** command, AutoCAD will prompt for the file name. Enter the name of the menu file with the file extension (**.MNU**).

Note

*1. After you load the new menu, you cannot use the screen menu, buttons menu, or digitizer because the original menu, **ACAD.MNU**, is not present and the new menu does not contain these menu areas.*

Screen Menus 9-7

2. To activate the original menu again, load the menu file by using the **MENU** command:

Command: **MENU**
Enter menu file name or [. (for none)]<current>: *ACAD.MNU*

3. If you need to use input from a keyboard or a pointing device in a menu macro, use the backslash (\). The system will pause for you to enter data.

4. There should be no space after the backslash (\).

5. The menu items, menu labels, and command definition can be uppercase, lowercase, or mixed.

6. You can introduce spaces between the menu items to improve the readability of the menu file.

7. The number of lines displayed on the screen depends on your system and are limited by the setting of the **SCREENBOXES** system variable. If there are more items in the menu than the number of spaces available, the excess items are not displayed on the screen. For example, if the display device limits the number of items to 28, items in excess of 28 will not be displayed on the screen and are therefore inaccessible.

Exercise 1 *General*

Design and write a screen menu for the following AutoCAD commands (file name **SME1.MNU**):

PLINE
ELLIPSE (Center)
ELLIPSE (Axis endpoint)
ROTATE
OFFSET
SCALE

SUBMENUS

The screen menu file can be so large that all items cannot be accommodated on one screen. The number of lines displayed on the screen depends on your system and are limited by the setting of the **SCREENBOXES** system variable. For example, the maximum number of items that can be displayed on your screen is 28. If the screen menu has more than 28 items, the menu items in excess of 28 are not displayed on the screen and therefore cannot be accessed. You can overcome this problem by using submenus that enable the user to define smaller groups of items within a menu section. When a submenu is selected, it loads the submenu items and displays them on screen. However, depending on your system, the number of items you can display can be higher and you may not need submenus.

Submenu Definition

A submenu definition consists of two asterisks (**) followed by the name of the submenu. A menu can have any number of submenus, and every submenu should have a unique name.

The items that follow a submenu, up to the next section label or submenu label, belong to that submenu. The format of a submenu definition is:

**Name

  Where ****** ----------------- Two asterisks (**) designate a submenu
       Name ------------ Name of the submenu

**DRAW1

  Where **DRAW1** --------- Name of the submenu
       ** ----------------- ** designates that DRAW1 is a submenu

Note

1. The submenu name can be up to 33 characters long.

2. The submenu name can consist of letters, digits, and such special characters as $ (dollar sign), - (hyphen), and _ (underscore).

3. The submenu name should not have any embedded blanks (spaces).

4. The submenu names should be unique in a menu file.

Submenu Reference

The submenu reference is used in a menu item to reference or load a submenu. It consists of a "$" sign followed by a letter that specifies the menu section. The letter that specifies a screen menu section is S. The section is followed by an "=" sign and the name of the submenu the user wants to activate. The submenu name should be without the **. The following is the format of a submenu reference:

  $Section=Submenu

    Where **$** -------------------- "$" sign
        Section ---------- Menu section specifier
        = ------------------ "=" sign
        Submenu ------- Name of submenu

  $S=EDIT

    Where **S** -------------------- S specifies screen menu section
        Edit --------------- Name of submenu

Note

$	*A special character code used to load a submenu in a menu file.*
$M=	*This is used to load a DIESEL macro from a menu item.*

The following are the section specifiers:

S	*Specifies the* **SCREEN** *menu.*
P0 - P16	*Specifies the POP menus, POP0 through POP16.*
I	*Specifies the Image Tile menu.*
B1 - B4	*Specifies the BUTTONS menu, B1 through B4.*
T1 - T4	*Specifies the TABLET menus, T1 through T4.*
A1 - A4	*Specifies the AUX menus A1 through A4.*

Nested Submenus

When a submenu is activated, the current menu is copied to a stack. If you select another submenu, the submenu that was current will be copied or pushed to the top of the stack. The maximum number of menus that can be stacked is eight. If the stack size increases to more than eight, the menu at the bottom of the stack is removed and forgotten. You can call the previous submenu by using the nested submenu call. The format of this call is:

$S=

Where $ ------------------- "$" sign
       S ------------------- Screen menu specifier
       = ------------------- "=" sign

The maximum number of nested submenu calls is eight. Each time you call a submenu (issue $S=), this pops the last item from the stack and reactivates it.

Example 2

Design a menu layout and write a screen menu for the following commands.

LINE	ERASE
PLINE	MOVE
ELLIPSE-C	ROTATE
ELLIPSE-E	OFFSET
CIR-C,R	COPY
CIR-C,D	SCALE
CIR- 2P	

As mentioned earlier, the first and the most important part of writing a menu is the design of the menu and knowing the commands and the prompts associated with those commands. You should know the way you want the menu to look and the way you want to arrange the commands for maximum efficiency. Write out the menu on a piece of paper, and check it thoroughly to make sure you have arranged the commands the way you want them. Use submenus to group the commands based on their use, function, and relationship with other submenus. Make provisions to access other frequently used commands without going through the root menu.

Figure 9-4 shows one of the possible arrangements of the commands and the design of the screen menu. It has one main menu and two submenus. One of the submenus is for draw commands, the other is for edit commands. The colon (:) at the end of the commands is not required. It is used here to distinguish the commands from those items that are not used as commands. For example, DRAW in the root menu is not a command; therefore, it has no colon at the end. On the other hand, if you select **ERASE** from the **EDIT** menu, it executes the **ERASE** command, so it has a colon (:) at the end of the command.

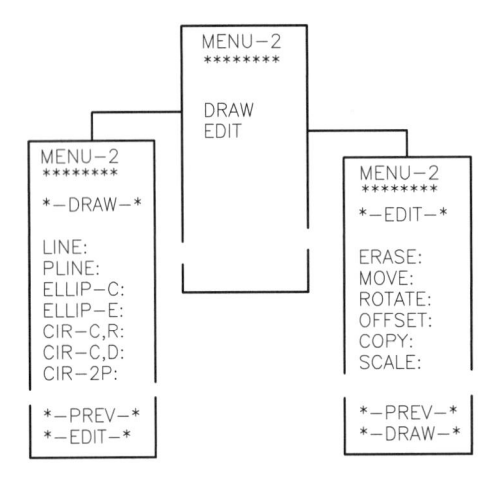

Figure 9-4 Screen menu design

The following file is a listing of the menu file of Example 1. **The line numbers on the right are not a part of the file; they are given here for reference only.**

```
***SCREEN                                           1
[ MENU-2         ]                                  2
[********        ]                                  3
[                ]                                  4
[                ]                                  5
[                ]                                  6
[                ]                                  7
[DRAW            ]^C^C$S=DRAW                       8
[EDIT            ]^C^C$S=EDIT                       9
                                                    10
**DRAW                                              11
[ MENU-2         ]^C^C$S=SCREEN                     12
[********        ]                                  13
[                                                   14
[*-DRAW-*        ]                                  15
[                ]                                  16
[LINE:           ]^C^CLINE                          17
[PLINE:          ]^C^CPLINE;\W;0.1;0.1              18
[ELLIP-C:        ]^C^CELLIPSE;C                     19
[ELLIP-E:        ]^C^CELLIPSE                       20
[CIR-C,R:        ]^C^CCIRCLE                        21
[CIR-C,D:        ]^C^CCIRCLE;\D                     22
[CIR-2P:         ]^C^CCIRCLE;2P                     23
[                ]                                  24
[                ]                                  25
[                ]                                  26
```

Screen Menus

```
    [              ]                              27
    [              ]                              28
    [              ]                              29
    [*-PREV-*      ]^C^C$S=                       30
    [*-EDIT-*      ]^C^C$S=EDIT                   31
                                                  32
        **EDIT                                    33
    [ MENU-2       ]^C^C$S=SCREEN                 34
    [********      ]                              35
    [              ]                              36
    [*-EDIT-*      ]                              37
    [              ]                              38
    [ERASE:        ]^C^CERASE                     39
    [MOVE:         ]^C^CMOVE                      40
    [ROTATE:       ]^C^CROTATE                    41
    [OFFSET:       ]^C^COFFSET                    42
    [COPY:         ]^C^CCOPY                      43
    [SCALE:        ]^C^CSCALE                     44
    [              ]                              45
    [              ]                              46
    [              ]                              47
    [              ]                              48
    [              ]                              49
    [              ]                              50
    [              ]                              51
    [*-PREV-*      ]^C^C$S=                       52
    [*-DRAW-*      ]^C^C$S=DRAW                   53
```

Explanation
Line 1
***SCREEN
***SCREEN is the section label for the screen menu.

Line 2
[MENU-2]
This menu item displays MENU-2 at the top of the screen menu.

Line 3
[********]
This menu item prints eight asterisk signs (********) on the screen menu.

Lines 4-7
[   ]
These menu items print four blank lines on the screen menu. The brackets are not required. They could be just four blank lines without brackets.
Line 8

[DRAW   ]^C^C$S=DRAW
[DRAW   ] displays DRAW on the screen, letting you know that by selecting this function you can access the draw commands. ^C^C cancels the existing command, and $S=DRAW loads the DRAW submenu on the screen.

[DRAW   ]^C^C$S=DRAW
              Where   ^C^C ---------- Cancels the existing command twice
                       = ------------------ Loads submenu DRAW
                      DRAW ---------- Name of submenu

Line 9
[EDIT   ]^C^C$S=EDIT
[EDIT   ] displays EDIT on the screen. ^C^C cancels the current command, and $S=EDIT loads the submenu **EDIT**.

Line 10
The blank lines between the submenus or menu items are not required. It just makes it easier to read the file.

Line 11
**DRAW
**DRAW is the name of the submenu; lines 12 through 31 are defined under this submenu.

Line 15
[*-DRAW-*]
This prints *-DRAW-* as a heading on the screen menu to let the user know that the commands listed on the menu are draw commands.

Line 18
[PLINE:]^C^CPLINE;\W;0.1;0.1
[PLINE:] displays PLINE: on the screen. ^C^C cancels the command, and **PLINE** is the AutoCAD polyline command. The semicolons are for RETURN. The semicolons can be replaced by a blank space. The backslash (\) is for user input. In this case it is the start point of the polyline. W selects the width option of the polyline. The first 0.1 is the starting width of the polyline, and the second 0.1 is the ending width. This command will draw a polyline of 0.1 width.

        [PLINE:]^C^CPLINE;\W;0.1;0.1
                Where **PLINE** ------------ **PLINE** (Polyline) command
                        ; ------------------ RETURN
                        \-------------------- Pause for input (start pt.)
                        W ----------------- Width
                        ; ------------------ RETURN
                        0.1 ---------------- Starting width
                        ; ------------------ RETURN
                        0.1 ---------------- Ending width

Screen Menus 9-13

Line 19
[ELLIP-C:]^C^CELLIPSE;C
[ELLIP-C:] displays ELLIP-C: on the screen menu. ^C^C cancels the existing command, and **ELLIPSE** is an AutoCAD command to generate an ellipse. The semicolon is for RETURN, and C selects the center option of the **ELLIPSE** command.

    [ELLIP-C:]^C^CELLIPSE;C
        Where **CELLIPSE** ------ **ELLIPSE** command
                ; -------------------- RETURN
                C ------------------ Center option for ellipse

Line 20
[ELLIP-E:]^C^CELLIPSE
[ELLIP-E:] displays ELLIP-E: on the screen menu. ^C^C cancels the existing command, and **ELLIPSE** is an AutoCAD command. Here the **ELLIPSE** command uses the default option, axis endpoint, instead of center.

Line 30
[*-PREV-*]^C^C$S=
[*-PREV-*] displays *-PREV-* on the screen menu, and ^C^C cancels the existing command. $S= restores the previous menu that was displayed on the screen before loading the current menu.

    [*-PREV-*]^C^C$S=
        Where **$S** ----------------- Restores the previous screen menu

Line 31
[*-EDIT-*]^C^C$S=EDIT
[*-EDIT-*] displays *-EDIT-* on the screen menu, and ^C^C cancels the existing command. $S=EDIT loads the submenu EDIT on the screen. This lets you access the **EDIT** commands without going back to the root menu and selecting **EDIT** from there.

    [*-EDIT-*]^C^C$S=EDIT
        Where **$S** ----------------- Loads the submenu EDIT
               EDIT ------------- Name of the submenu

Line 33
**EDIT
**EDIT is the name of the submenu, and lines 34 to 53 are defined under this submenu.

Line 34
[MENU-2]^C^C$S=SCREEN
[MENU-2] displays MENU-2 on the screen menu, and ^C^C cancels the existing command. $S=SCREEN loads the root menu SCREEN on the screen menu.

Line 39
[ERASE:]^C^CERASE

[ERASE:] displays ERASE: on the screen menu, and ^C^C cancels the existing command. **ERASE** is an AutoCAD command for erasing the selected objects.

Line 53
[*-DRAW-*]^C^C$S=DRAW
$S=DRAW loads the DRAW submenu on the screen. It lets you load the DRAW menu without going through the root menu. The root menu is the first menu that appears on the screen when you load a menu or start AutoCAD.

When you select DRAW from the root menu, the submenu DRAW is loaded on the screen. The menu items in the draw submenu completely replace the menu items of the root menu. If you select MENU-2 from the screen menu now, the root menu will be loaded on the screen, but some of the items are not cleared from the screen menu (Figure 9-5). This is because the root menu does not have enough menu items to completely replace the menu items of the DRAW submenu.

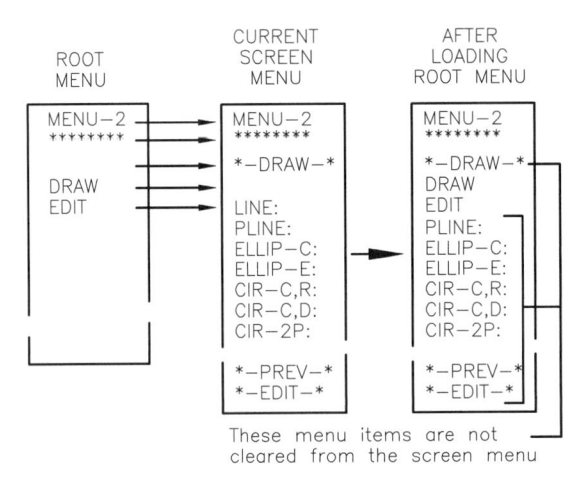

Figure 9-5 Screen menu display after loading the root menu

One of the ways to clear the screen is to define a submenu that has 21 blank lines. When this submenu is loaded it will clear the screen. If you then load another submenu, there will be no overlap, because the screen menu has already been cleared and there are no menu items left on the screen (Figure 9-6). Example 3 illustrates the use of such a submenu.

Another way to avoid overlapping the menu items is to define every submenu so that they all have the same number of menu items. The disadvantage with this approach is that the menu file will be long, because every submenu will have 21 lines.

Screen Menus

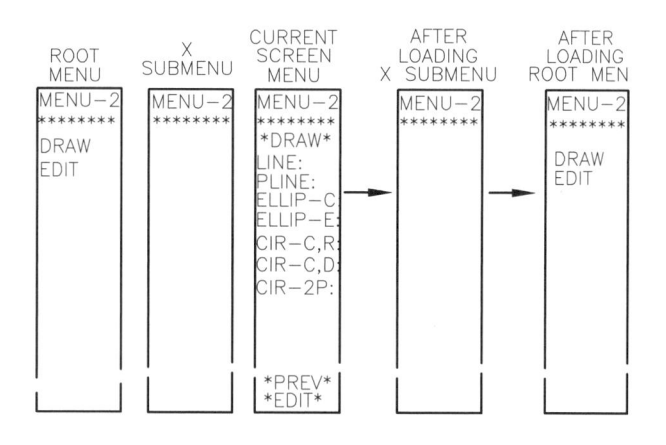

```
 ROOT       X      CURRENT    AFTER      AFTER
 MENU    SUBMENU   SCREEN    LOADING    LOADING
                    MENU    X SUBMENU  ROOT MEN
┌───────┐┌───────┐┌───────┐ ┌───────┐ ┌───────┐
│MENU-2 ││MENU-2 ││MENU-2 │ │MENU-2 │ │MENU-2 │
│*******││*******││*******│ │*******│ │*******│
│DRAW   ││       ││*DRAW* │ │       │ │DRAW   │
│EDIT   ││       ││LINE:  │ │       │ │EDIT   │
│       ││       ││PLINE: │ │       │ │       │
│       ││       ││ELLIP-C│ │       │ │       │
│       ││       ││ELLIP-E│ │       │ │       │
│       ││       ││CIR-C,R│ │       │ │       │
│       ││       ││CIR-C,D│ │       │ │       │
│       ││       ││CIR-2P:│ │       │ │       │
│       ││       ││       │ │       │ │       │
│       ││       ││*PREV* │ │       │ │       │
│       ││       ││*EDIT* │ │       │ │       │
└───────┘└───────┘└───────┘ └───────┘ └───────┘
```

Figure 9-6 Screen menu display after loading the root menu

Exercise 2 — *General*

Write a screen menu file for the following AutoCAD commands (file name **SM2.MNU**).

ARC	MIRROR
-3P	BREAK-F
-SCE	BREAK-@
-SCA	EXTEND
-SCL	STRETCH
-SEA	FILLET-0
POLYGON-C	FILLET
POLYGON-E	CHAMFER

MULTIPLE SUBMENUS

A menu file can have any number of submenus. All the submenu names have two asterisks (**) in front of them, even if there is a submenu within a submenu. Also, several submenus can be loaded in a single statement. If there is more than one submenu in a menu item, they should be separated from one another by a space. Example 3 illustrates the use of multiple submenus.

Example 3 — *General*

Design the layout of the menu, and then write the screen menu for the following AutoCAD commands.

Draw	ARC	Edit	Display
LINE	3Point	EXTEND	ZOOM
Continue	SCE	STRETCH	REGEN
Close	SCA	FILLET	SCALE
Undo	CSE		PAN

.X   CSA
.Y   CSL
.Z
.XY
.XZ
.YZ

There are several different ways of arranging the commands, depending on your requirements. The following layout is one of the possible screen menu designs for the given AutoCAD commands (Figure 9-7).

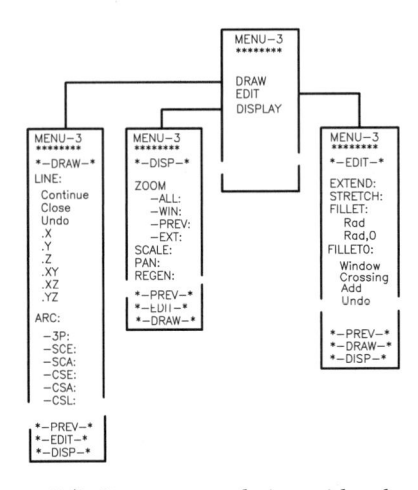

Figure 9-7 Screen menu design with submenus

The name of the following menu file is **SM3.MNU**. You can use a text editor to write the file. The following file is the listing of the menu file **SM3.MNU**. **The line numbers on the right are not a part of the file. They are shown here for reference only.**

***SCREEN		1
**S		2
[MENU-3	]^C^C$S=X $S=S	3
[********	]$S=OSNAP	4
[	]	5
[DRAW	]^C^C$S=X $S=DRAW	6
[EDIT	]^C^C$S=X $S=EDIT	7
[DISPLAY	]^C^C$S=X $S=DISP	8
**DRAW 3		9
[*-DRAW-*	]	10
[	]	11
[LINE:	]^C^CLINE	12
[Continue	]^C^CLINE;;	13
[Close	]CLOSE	14
[Undo	]UNDO	15

[.X	].X	16
[.Y	].Y	17
[.Z	].Z	18
[.XY	].XY	19
[.XZ	].XZ	20
[.YZ	].YZ	21
[	]	22
[ARC:	]^C^CArc	23
[-3P:	]^C^CARC \\DRAG	24
[-SCE:	]^C^CARC \C \DRAG	25
[-SCA:	]^C^CARC \C \A DRAG	26
[-CSE:	]^C^CARC C \\DRAG	27
[-CSA:	]^C^CARC C \\A DRAG	28
[-CSL:	]^C^CARC C \\L DRAG	29
[	]	30
[*-PREV-*	]$S= $S=	31
[*-EDIT-*	]^C^C$S=X $S=EDIT	32
[*-DISP-*	]$S=X $S=DISP	33
**EDIT 3		34
[*-EDIT-*	]	35
[	]	36
[EXTEND:	]^C^CEXTEND	37
[STRETCH:	]^C^CSTRETCH;C	38
[FILLET:	]^C^CFILLET	39
[Rad	]R;\Fillet	40
[Rad 0	]R;0;Fillet	41
[FILLET0:	]^C^CFillet;R;0::	42
Win.		43
Cross.		44
Add		45
Undo		46
[	]	47
[*-PREV-*	]$S= $S=	48
[*-DRAW-*	]^C^C$S=X $S=DRAW	49
[*-DISP-*	]$S=X $S=DISP	50
**DISP 3		51
[*-DISP-*	]	52
[	]	53
[ZOOM:	]'ZOOM	54
[-ALL	]A	55
[-WIN	]W	56
[-PREV	]P	57
[-EXT	]E	58
[	]	59
[SCALE:	]'ZOOM	60
[PAN:	]'PAN	61
[REGEN:	]^C^CREGEN	62

[	]	63
[*-PREV-*	]$S= $S=	64
[*-EDIT-*	]^C^C$S=X $S=EDIT	65
[*-DRAW-*	]$S=X $S=DRAW	66
**X 3		67
[	]	68
[	]	69
[	]	70
[	]	71
[	]	72
[	]	73
[	]	74
[	]	75
[	]	76
[	]	77
[	]	78
[	]	79
[	]	80
[	]	81
[	]	82
[	]	83
[	]	84
[	]	85
[	]	86
**OSNAP 2		87
[-OSNAPS-	]	88
[	]	89
[Center	]CEN $S=	90
[Endpoint	]END $S=	91
[Insert	]INS $S=	92
[Intersec	]INT $S=	93
[Midpoint	]MID $S=	94
[Nearest	]NEA $S=	95
[Node	]NOD $S=	96
[Perpend	]PER $S=	97
[Quadrant	]QUA $S=	98
[Tangent	]TAN $S=	99
[None	]NONE $S=	100
[	]	101
[	]	102
[	]	103
[	]	104
[	]	105
[*-PREV-*	]$S=	106

Screen Menus 9-19

Explanation
Line 3
[MENU-3]^C^C$S=X $S=S
In this menu item, [MENU-3] displays MENU-3 at the top of the screen menu. ^C^C cancels the command, and $S=X loads submenu X. Similarly, $S=S loads submenu S. Submenu X, defined on line 171, consists of 18 blank lines. Therefore, when this submenu is loaded it prints blank lines on the screen and clears the screen menu area. After loading submenu X, the system loads submenu S, and the items that are defined in this submenu are displayed on the screen menu.

[MENU-3]^C^C$S=X $S=S
   Where   ^C^C ---------- Cancels the existing command
           $S=X ------------ Loads submenu X (clears screen)
           $S=S ------------ Loads submenu S

Line 4
[********]$S=OSNAP
This menu item prints eight asterisks on the screen menu, and $S=OSNAP loads submenu OSNAP. The OSNAP submenu, defined on line 191, consists of object snap modes.

Line 6
[DRAW   ]^C^C$S=X $S=DRAW
This menu item displays DRAW on the screen menu area, cancels the existing command, and then loads submenu X and submenu DRAW. Submenu X clears the screen menu area, and submenu DRAW, as defined on line 11, loads the items defined under the DRAW submenu.

Line 9
**DRAW 3
This line is the submenu label for DRAW; 3 indicates that the first line of the DRAW submenu will be printed on line 3. Nothing will be printed on line 1 or line 2. The first line of the DRAW submenu will be on line 3, followed by the rest of the menu. All the submenus except submenus S and OSNAP have a 3 at the end of the line. Therefore, the first two lines (MENU-3) and (********) will never be cleared and will be displayed on the screen all the time. If you select MENU-3 in any menu, it will load submenu S. If you select ********, it will load submenu OSNAP.

**DRAW 3
   Where   **DRAW** ----------- Submenu name
           3 ------------------- Leaves two blank lines and prints from third line

Line 12
[LINE:]^C^CLINE
^C^C cancels the existing command, and **LINE** is the AutoCAD **LINE** command.

Line 13
[Continue]^C^CLINE;;
In this menu item, the **LINE** command is followed by two semicolons that continue the **LINE** command. To understand it, look at the **LINE** command to see how a continue option is used in this command:

Command: **LINE**
Specify first point: **RETURN (Continue)**
Specify next point or [Undo]:

Following is the command and prompt entry sequence for the **LINE** command, if you want to start a line from the last point:

LINE
RETURN
RETURN
SELECT A POINT

Therefore, in the screen menu file the **LINE** command has to be followed by two Returns to continue from the previous point.

Line 16
[.X   ].X
In this menu item, .X extracts the X coordinate from a point. The bracket, .X, and spaces inside the brackets ([.X   ]) are not needed. The line could consist of .X only. The same is true for lines 39 through 43.

Example

[.X].X	can also be written (without brackets) as	.X
[.Y].Y		.Y
[.Z].Z		.Z

Line 31
[*-PREV-*]$S= $S=
This menu item recalls the previous menu twice. AutoCAD keeps track of the submenus that were loaded on the screen. The first **$S=** will load the previously loaded menu, and the second **$S=** will load the submenu that was loaded before that. For example, in line 16 ([ARC:]$S=X $S=ARC), two submenus have been loaded: first X, and then ARC. Submenu X is stacked before loading submenu ARC, which is current and is not stacked yet. The first $S= will recall the previous menu, in this case submenu X. The second $S= will load the menu that was on the screen before selecting the item in line 16.

[*-PREV-*]$S= $S=
    Where $S= ---------------- Loads last menu
           $S= --------------- Loads next to last menu

Line 33
[*-DISP-*]$S=X $S=DISP
In this menu item, $S=X loads submenu X, and $S=DISP loads submenu DISP. Notice that there is no CANCEL (^C^C) command in the line. There are some menus you might like to load without canceling the existing command. For example, if you are drawing a line, you might want to zoom without canceling the existing command. You can select [*-DISP-*], select the appropriate zoom option, and then continue with the line command. However, if the line

Screen Menus 9-21

has a CANCEL command ([*-DISP-*]^C^C$S=X $S=DISP), you cannot continue with the **LINE** command, because the existing command will be canceled when you select *-DISP-* from the screen. In line 28 ([*-EDIT-*]^C^C$S=X $S=EDIT), ^C^C cancels the existing command, because you cannot use any editing command unless you have canceled the existing command.

Line 38
[STRETCH:]^C^CSTRETCH;C
^C^C cancels the existing command, and **STRETCH** is an AutoCAD command. The C is for the crossing option that prompts you to enter two points to select the objects.

```
[STRETCH:]^C^CSTRETCH;C
        Where  STRETCH ----- STRETCH command
               ;  ------------------ RETURN
               C  ------------------ Center O'SNAP
```

Line 40
[RAD]R;\FILLET
This menu item selects the radius option of the **FILLET** command, and then waits for you to enter the radius. After entering the radius, execute the **FILLET** command again to generate the desired fillet between the two selected objects.

```
[ RAD ]R;\FILLET
        Where  R  ------------------ Radius option of FILLET command
               ;  ------------------ Space or semicolon for RETURN
               \  ------------------ Pause for user input
               FILLET --------- AutoCAD FILLET command
```

Line 41
[RAD 0]R;0;FILLET
This menu item selects the radius option of the **FILLET** command and assigns it a 0 value. Then it executes the **FILLET** command to generate a 0 radius fillet between the two selected objects.

```
[ RAD 0]R;0;FILLET
        Where  R  ------------------ Radius option of FILLET command
               ;  ------------------ Space or semicolon for RETURN
               0  ------------------ 0 radius
               ;  ------------------ Space or semicolon for RETURN
               FILLET --------- AutoCAD FILLET command
```

Line 42
[FILLET0:]^C^CFILLET;R 0;;
This menu item defines a **FILLET** command with 0 radius, and then generates 0 fillet between the two selected objects.

[FILLET0:]^C^CFILLET;R 0;;

    Where **FILLET** --------- **FILLET** command
               ; ------------------- Semicolon or space for RETURN
               R ----------------- R for radius option
               Space ------------- Semicolon or space for RETURN
               0 ------------------- 0 radius
               ; ------------------- Semicolon for RETURN
               ; ------------------- Repeats **FILLET** command

Line 54
[ZOOM:]'ZOOM
This menu item defines a transparent **ZOOM** command.

    [ZOOM:]'ZOOM
        Where **'** -------------------- Single quote (') makes **ZOOM** transparent
                 ZOOM ----------- AutoCAD **ZOOM** command

Line 90
[Center]CEN $X=
In this menu item CEN is for center object snap, and $X= automatically recalls the previous screen menu after selecting the object.

    [Center]CEN $S=
        Where **CEN** -------------- Center object snap
                 $S= -------------- Loads previous menu

> **Note**
> *If any menu item on a screen menu includes more than one load command, the commands must be separated by a space.*

Example
[LINE:]$S=X $S=LINE
    Note that there is a space between X and $S.

Similarly, if a menu item includes both a load command and an AutoCAD command, they should be separated by a space.

Example
[LINE:]$S=LINE ^C^CLINE

Exercise 3   *General*

Write a screen menu for the AutoCAD commands shown in Figure 9-8 (file name **SME3.MNU**). The ******** are to access the object snap submenu.

Screen Menus   9-23

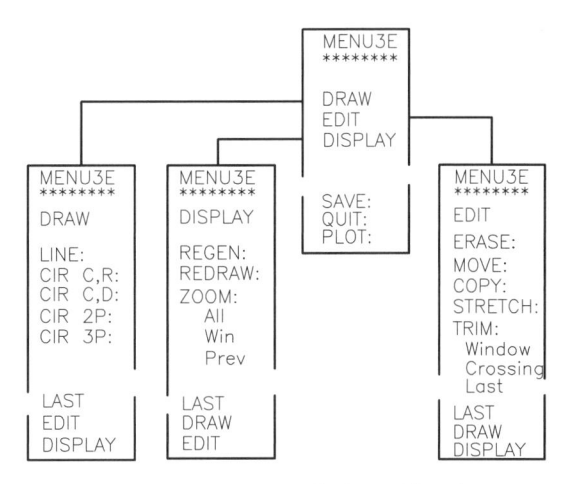

```
                    MENU3E
                    ********
                    DRAW
                    EDIT
                    DISPLAY

                    SAVE:
                    QUIT:
                    PLOT:
```

```
MENU3E        MENU3E              MENU3E
********      ********             ********
DRAW          DISPLAY              EDIT

LINE:         REGEN:               ERASE:
CIR C,R:      REDRAW:              MOVE:
CIR C,D:      ZOOM:                COPY:
CIR 2P:         All                STRETCH:
CIR 3P:         Win                TRIM:
                Prev                 Window
                                     Crossing
                                     Last
LAST          LAST                 LAST
EDIT          DRAW                 DRAW
DISPLAY       EDIT                 DISPLAY
```

Figure 9-8 Screen menu design with submenus

Example 4   *General*

Write a combined screen and tablet menu for the commands shown in the tablet menu template shown in Figure 9-9(a) and the screen menu shown in Figure 9-9(b). When the user selects a command from the digitizer template, it should automatically load the corresponding screen menu on the screen (filename TM3.MNU).

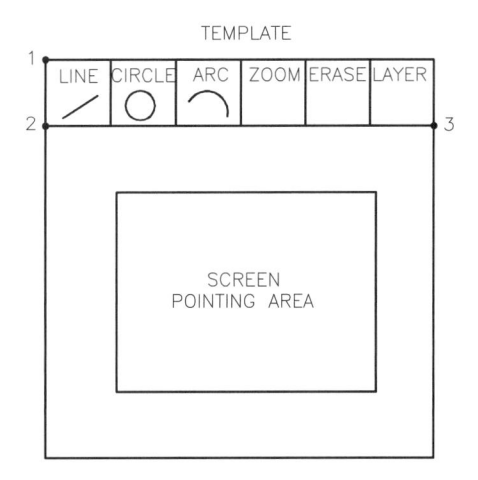

```
               TEMPLATE
    1
      | LINE | CIRCLE | ARC | ZOOM | ERASE | LAYER |
      |  /   |   O    |  )  |      |       |       |
    2                                              3


              +--------------------+
              |                    |
              |                    |
              |       SCREEN       |
              |   POINTING  AREA   |
              |                    |
              |                    |
              +--------------------+
```

Figure 9-9(a) Tablet menu template for Example 4

9-24   Customizing AutoCAD 2002

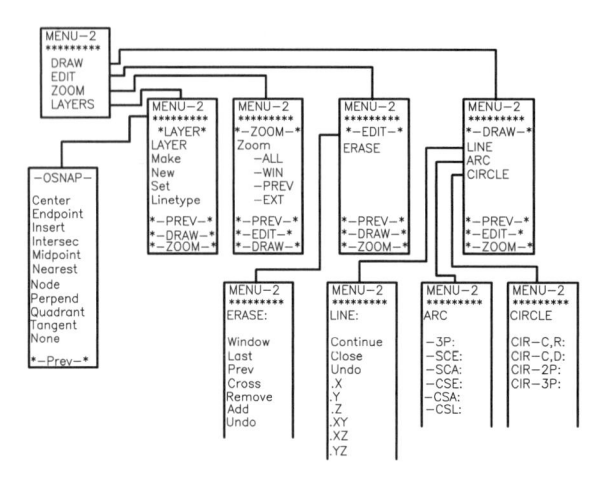

Figure 9-9(b)   *Screen menu displays*

The following file is a listing of the combined menu for Example 4. **The line numbers are not a part of the file; they are shown here for reference only.**

```
***SCREEN                                                    1
**S                                                          2
[ MENU-3    ]^C^C$S=X $S=S                                   3
[********   ]$S=OSNAP                                        4
[          ]                                                 5
[          ]                                                 6
[DRAW      ]^C^C$S=X $S=DRAW                                 7
[EDIT      ]^C^C$S=X $S=EDIT                                 8
[ZOOM      ]^C^C$S=X $S=ZOOM                                 9
[LAYER     ]^C^C$S=X $S=LAYER                               10
                                                            11
**DRAW 3                                                    12
[*-DRAW-* ]                                                 13
[          ]                                                14
[          ]                                                15
[LINE:     ]$S=X $S=LINE ^C^CLINE                           16
[ARC:      ]$S=X $S=ARC                                     17
[CIRCLE:   ]$S=X $S=CIRCLE ^C^CCIRCLE                       18
[          ]                                                19
[          ]                                                20
[          ]                                                21
[          ]                                                22
[          ]                                                23
[          ]                                                24
[          ]                                                25
[          ]                                                26
```

Screen Menus 9-25

```
[               ]                                 27
[*-PREV-*  ]$S= $S=                               28
[*-EDIT-*  ]^C^C$S=X $S=EDIT                      29
[*-ZOOM-* ]$S=X $S=ZOOM                           30
                                                  31
**LINE 3                                          32
[LINE:     ]^C^CLINE                              33
[          ]                                      34
[          ]                                      35
[Continue  ]^C^CLINE;;                            36
[Close     ]CLOSE                                 37
[Undo      ]UNDO                                  38
[.X        ].X                                    39
[.Y        ].Y                                    40
[.Z        ].Z                                    41
[.XY       ].XY                                   42
[.XZ       ].XZ                                   43
[.YZ       ].YZ                                   44
[          ]                                      45
[          ]                                      46
[          ]                                      47
[*-PREV-*  ]$S= $S=                               48
[*-EDIT-*  ]^C^C$S=X $S=EDIT                      49
[*-ZOOM-* ]$S=X $S=ZOOM                           50
                                                  51
**ARC 3                                           52
[ARC       ]                                      53
[          ]                                      54
[  -3P:    ]^C^CARC \\DRAG                        55
[  -SCE:   ]^C^CARC \C \DRAG                      56
[  -SCA:   ]^C^CARC \C \A DRAG                    57
[  -CSE:   ]^C^CARC C \\DRAG                      58
[  -CSA:   ]^C^CARC C \\A DRAG                    59
[  -CSL:   ]^C^CARC C \\L DRAG                    60
[          ]                                      61
[          ]                                      62
[          ]                                      63
[          ]                                      64
[          ]                                      65
[          ]                                      66
[          ]                                      67
[*-PREV-*  ]$S= $S=                               68
[*-EDIT-*  ]^C^C$S=X $S=EDIT                      69
[*-ZOOM-* ]$S=X $S=ZOOM                           70
                                                  71
**CIRCLE 3                                        72
[CIRCLE:  ]                                       73
```

[	]	74
[-C,R:	]^C^CCIRCLE	75
[-C,D:	]^C^CCIRCLE \D	76
[-2P:	]^C^CCIRCLE 2P	77
[-3P:	]^C^CCIRCLE 3P	78
[	]	79
[	]	80
[	]	81
[	]	82
[	]	83
[	]	84
[	]	85
[	]	86
[	]	87
[*-PREV-*	]$S= $S=	88
[*-EDIT-*	]^C^C$S=X $S=EDIT	89
[*-ZOOM-*	]$S=X $S=ZOOM	90
		91
		92
EDIT 3		93
[*-EDIT-*	]	94
[	]	95
[	]	96
[ERASE:	]$S=X $S=ERASE ^C^CERASE	97
[	]	98
[	]	99
[	]	100
[	]	101
[	]	102
[	]	103
[	]	104
[	]	105
[	]	106
[	]	107
[	]	108
[*-PREV-*	]$S= $S=	109
[*-DRAW-*	]^C^C$S=X $S=DRAW	110
[*-ZOOM-*	]$S=X $S=ZOOM	111
		112
ERASE 3		113
[ERASE:	]^C^CERASE	114
[	]	115
Window		116
Last		117
Prev		118
Cross		119
Remove		120

```
          Add                                             121
          Undo                                            122
          [              ]                                123
          [              ]                                124
          [              ]                                125
          [              ]                                126
          [              ]                                127
          [              ]                                128
          [*-PREV-*      ]$S= $S=                         129
          [*-DRAW-* ]^C^C$S=X $S=DRAW                     130
          [*-ZOOM-* ]$S=ZOOM                              131
                                                          132
          **ZOOM 3                                        133
          [*-ZOOM-* ]                                     134
          [              ]                                135
          [              ]                                136
          [ZOOM:    ]'ZOOM                                137
          [ -ALL    ]A                                    138
          [ -WIN    ]W                                    139
          [ -PREV   ]P                                    140
          [ -EXT    ]E                                    141
          [              ]                                142
          [              ]                                143
          [              ]                                144
          [              ]                                145
          [              ]                                146
          [              ]                                147
          [              ]                                148
          [*-PREV-*      ]$S= $S=                         149
          [*-EDIT-* ]^C^C$S=X $S=EDIT                     150
          [*-DRAW-* ]$S=X $S=DRAW                         151
                                                          152
          ***LAYER 3                                      153
          [*-LAYER* ]                                     154
          [              ]                                155
          [              ]                                156
          [LAYER:   ]^C^CLAYER                            157
          Make                                            158
          New                                             159
          Set                                             160
          Linetype                                        161
          Color                                           162
          [List     ]?;;                                  163
          [              ]                                164
          [              ]                                165
          [              ]                                166
          [              ]                                167
```

```
[          ]                                168
[*-PREV-*  ]$S= $S=                         169
[*-EDIT-*  ]^C^C$S=X $S=EDIT                170
[*-DRAW-*  ]$S=X $S=DRAW                    171
                                            172
**X 3                                       173
[          ]                                174
[          ]                                175
[          ]                                176
[          ]                                177
[          ]                                178
[          ]                                179
[          ]                                180
[          ]                                181
[          ]                                182
[          ]                                183
[          ]                                184
[          ]                                185
[          ]                                186
[          ]                                187
[          ]                                188
[          ]                                189
[          ]                                190
[          ]                                191
                                            192
**OSNAP 2                                   193
[-OSNAPS- ]                                 194
[          ]                                195
[Center    ]CEN $S=                         196
[Endpoint  ]END $S=                         197
[Insert    ]INS $S=                         198
[Intersec  ]INT $S=                         199
[Midpoint  ]MID $S=                         200
[Nearest   ]NEA $S=                         201
[Node      ]NOD $S=                         202
[Perpend   ]PER $S=                         203
[Quadrant  ]QUA $S=                         204
[Tangent   ]TAN $S=                         205
[None      ]NONE $S=                        206
[          ]                                207
[          ]                                208
[          ]                                209
[          ]                                210
[          ]                                211
[*-PREV-*  ]$S=                             212
                                            213
***TABLET1                                  214
```

```
          $S=X  $S=LINE   ^C^CLINE                                  215
          $S=X  $S=CIRCLE  ^C^CCIRCLE                                216
          $S=X  $S=ARC    ^C^CARC                                    217
          $S=X  $S=ZOOM   ^C^CZOOM                                   218
          $S=X  $S=ERASE  ^C^CERASE                                  219
          $S=X  $S=LAYER  ^C^CLAYER                                  220
```

Line 1
***SCREEN
This is the section label for the screen menu. The line numbers 1 through 213 are defined in this section of the menu file.

Line 3
[MENU-3]^C^C$S=X $S=S
In this menu item $S=X loads the submenu X and $S=S loads the submenu S.

Line 7
[DRAW   ]^C^C$S=X $S=DRAW
In this menu item $S=DRAW loads the submenu DRAW that is defined in the menu file.

Line 12
**DRAW 3
In this menu item DRAW is the name of the submenu. The 3 forces the submenu to be displayed on line 3. Since nothing is printed on the first two lines, [MENU-3] and [********] will be displayed on the screen all the time. If you select [MENU-3] from any menu, it will load the submenu S and display it on the screen. Similarly, if you pick [********] from any menu, it will load and display the OSNAP submenu.

Line 28
[*-PREV-*]$S= $S=
In this menu item $S= $S= loads the previous two submenus. One of them is the submenu X and the second submenu is the one that was displayed on the screen before the current menu.

Line 214
***TABLET1
TABLET1 is the section label for the tablet area 1. The line numbers 215 through 220 are defined in this section.

Line 215
$S=X $S=LINE ^C^CLINE
$S=X, defined in the screen menu section, loads the submenu X on the screen menu area. The purpose of loading the submenu X is to clear the screen so that there is no overlapping of the screen menu items. $S=LINE, defined in the screen menu section, loads the **LINE** submenu on the screen menu. ^C^CLINE cancels the existing command twice and then executes AutoCAD's **LINE** command.

When you select the LINE block from the digitizer, AutoCAD automatically clears the screen menu, loads the LINE submenu, and enters a **LINE** command. This makes it easier for the user to select the command options from the screen menu, since they are not on the digitizer template.

$S=X $S=LINE ^C^CLINE
    Where **$S=X** ------------- Loads submenu X
           LINE ------------- Loads submenu LINE
           ^C^C ------------- Cancels the existing command
           LINE ------------- AutoCAD's **LINE** command

You can also use the automatic menu swapping feature of AutoCAD to load the corresponding screen menus. If you want to utilize this feature, the TABLET1 section must be written as:

***TABLET1	214
$S=X ^C^CLINE	215
$S=X ^C^CCIRCLE	216
$S=X ^C^CARC	217
$S=X ^C^CZOOM	218
$S=X ^C^CERASE	219
$S=X ^C^CLAYER	220

LONG MENU DEFINITIONS

You can put any number of commands in one screen menu line. There is no limit to the number of commands and the order in which they appear in the line, as long as they satisfy all the requirements of the command and the sequence of prompt entries. Again, you need to know the AutoCAD commands, the options in the command, the command prompts, and the entries for various prompts. If the statement cannot fit on one line, you can put a plus (+) sign at the end of the first line and continue with a second line. The command definition that involves several commands put together in one line is also called a macro. The following example illustrates a long menu item or a macro.

Example 5   *General*

Write a screen menu command definition that performs the following functions (filename SM4.MNU):

Draw a border using polyline
           Width =   0.01
           Point-1   0,0
           Point-2   12,0
           Point-3   12,9
           Point-4   0,9
           Point-5   0,0

Initial drawing setup
           Snap   0.25
           Grid   0.5

Screen Menus

9-31

        Limits   12,9
        Zoom   All

Before writing a menu you should know the commands, options, prompts, and the prompt entries required for the commands. Therefore, first study the commands involved in the previously mentioned drawing setup.

POLYLINE
Command: **PLINE**
Specify start point: **0,0**
Specify next point or [Arc/Close/Halfwidth/Length/Undo/Width]: **W**
Specify starting width <0.0000>: **0.01**
Specify ending width <0.01>: *Press ENTER.*
Specify next point or [Arc/Close/Halfwidth/Length/Undo/Width]: **12,0**
Specify next point or [Arc/Close/Halfwidth/Length/Undo/Width]: **12,9**
Specify next point or [Arc/Close/Halfwidth/Length/Undo/Width]: **0,9**
Specify next point or [Arc/Close/Halfwidth/Length/Undo/Width]: **C**

Screen menu command definition for **PLINE**
PLINE;0,0;W;0.01;;12,0;12,9;0,9;C

SNAP
Command: **SNAP**
Specify snap spacing or [ON/OFF/Aspect/Rotate/Style/Type] <default>: **0.25**

Screen menu command definition for SNAP
SNAP;0.25

GRID
Command: **GRID**
Specify grid spacing(X) or [ON/OFF/Snap/Aspect] <default>: **0.5**

Screen menu command definition
GRID;0.5

LIMITS
Command: **LIMITS**
Specify lower left corner or [ON/OFF] <current co-ordinates>: **0,0**
Specify upper right corner <current co-ordinates>: **12,9**

Screen menu command definition
LIMITS;0,0;12,9

ZOOM
Command: **ZOOM**
Specify corner of window, enter a scale factor (nX or nXP), or
[All/Center/Dynamic/Extents/Previous/Scale/Window] <real time>: **A**

Screen menu command definition
ZOOM;A

Now you can combine these individual command definitions to form a single screen menu command definition that will perform all the functions when you select NSETUP from the new screen menu.

Combined screen menu line

[-NSETUP-]PLINE;0,0;W;0.01;;12,0;12,9;0,9;C;+
SNAP;0.25;GRID;0.5;LIMITS;+
0,0;12,9;ZOOM;A

Exercise 4 — *General*

Write a screen menu item that will set up the following parameters for the **UNITS** command (filename SME4.MNU):

System of units: Scientific
Number of digits to right of decimal point: 2
System of angle measure: Decimal
Number of fractional places for display of angles: 2
Direction for angle: 0
Angles measured counterclockwise

MENU COMMAND REPETITION

AutoCAD has made a provision for repeating a menu item until it is canceled by the user by pressing the **ESCAPE** key from the keyboard or selecting another menu item. It is particularly useful when you are editing a drawing and using the same command several times. A command can be repeated if the menu command definition starts with an asterisk sign (*).

[ERASE,W:]*^C^CERASE W
        Where   * -------------------- Asterisk for command repetition
                    ^C^C ---------- Cancels existing command
                    Erase ------------- AutoCAD's **ERASE** command
                    W ----------------- Window option

If you select this menu item, AutoCAD will prompt you to enter two points to select the objects because the window option for the **ERASE** command requires two points. When you press ENTER the objects you selected will be erased and the command will be repeated.

Example 6 — *General*

Write a screen menu for the following AutoCAD commands. Provide for automatically repeating the commands (filename SM7.MNU).

Screen Menus 9-33

LINE	**LIST**
ERASE	**INSERT**
TRIM	**DIST**

The following file is a listing of the screen menu for Example 6 that will repeat the selected command until canceled by pressing the CTRL and C keys.

```
[-REPEAT-   ]
[           ]
[           ]
[LINE:      ]*^C^CLINE
[ERASE:     ]*^C^CERASE
[TRIM:      ]*^C^CTRIM
[LIST:      ]*^C^CLIST
[INSERT:    ]*^C^CINSERT
[DIST:      ]*^C^CDIST
```

Note
*If the command definition is incorrect and you happen to select that item, the screen prompt display will be repeated indefinitely. To get out of this infinite loop, press the **ESCAPE** key. On some systems, pressing the **ESCAPE** key or selecting another command may not stop the display of the prompt. You may need to reboot the system or close the program. In Windows, press CTRL+ALT+DEL to close the program.*

*One of the major drawbacks of the menu command repetition is that it does not let you select a different command option. In the following menu item, the **ERASE** command uses the **C** (**Crossing**) option to select the objects. When the command is repeated, it does not allow you to use a different option for object selection.*

[ERASE,C:]*^C^CERASE C

AUTOMATIC MENU SWAPPING

Screen menus can be automatically swapped by using the system variable **MENUCTL**. When the system variable **MENUCTL** is set to 1, AutoCAD automatically issues the **$S=CMDNAME** command, where **CMDNAME** is the command that loads the submenu. For example, if you select the **LINE** command from the digitizer or pull-down menu, or enter the **LINE** command at the keyboard, the **CMDNAME** command will load the **LINE** submenu and display it in the screen menu area. To use this feature, the command name and the submenu name must be the same. For example, if the name of the arc submenu is ARC and you select the **ARC** command, the arc submenu will be automatically loaded on the screen. However, if the submenu name is different (for example, MYARC), AutoCAD will not load the arc submenu on the screen. The default value of the **MENUCTL** system variable is 1. If you set the **MENUCTL** variable to 0, AutoCAD will not use the **$S=CMDNAME** command feature to load the submenus.

MENUECHO SYSTEM VARIABLE

If the system variable **MENUECHO** is set to 0, the commands that you select from the digitizer, screen menu, pull-down menu, or buttons menu will be displayed in the command prompt area. For example, if you select the **CIRCLE** command from the menu, AutoCAD will display **_circle Specify center point for circle or [3P/2P/Ttr (tan tan radius)]:**. If you set the value of the **MENUECHO** variable to 1, AutoCAD will suppress the echo of the menu item and display **Specify center point for circle or [3P/2P/Ttr (tan tan radius)]:** only. You will notice that the **_circle** is not displayed when **MENUECHO** is set to 1. You can use ^P in the menu item to turn the echo on or off. The **MENUECHO** system variable can also be assigned a value of 2, 4, or 8, which controls the suppression of system prompts, disabling the ^P toggle, and debugging aid for DIESEL macros, respectively.

MENUS FOR FOREIGN LANGUAGES

Besides English, AutoCAD has several foreign language versions. If you want to write a menu that is compatible with other foreign language versions of AutoCAD, you must precede each command and keyword in the menu file with the underscore (_) character.

Examples
[New   ]^C^C_New
[Open   ]^C^C_Open
[Line   ]^C^C_Line
[Arc-SCA]^C^C_Arc;_C;_A

Any command or keyword that starts with the underscore character will be automatically translated. If you check the ACAD.MNU file, you will find that AutoCAD has made extensive use of this feature.

Example 7 *General*

Rewrite the screen menu of Example 1 so that the menu file is compatible with other foreign language versions of AutoCAD.

The following file is the listing of the menu file for Example 7:

```
***SCREEN
[ MENU-1    ]
[           ]
[           ]
[LINE       ]^C^C_LINE
[           ]
[CIR-C,R    ]^C^C_CIRCLE
[CIR-C,D    ]^C^C_CIRCLE;\_D
[CIR- 2P    ]^C^C_CIRCLE;_2P
[           ]
[ERASE      ]^C^C_ERASE
[MOVE       ]^C^C_MOVE
```

USE OF CONTROL CHARACTERS IN MENU ITEMS

You can use ASCII control characters in the command definition by using the caret sign (^) followed by the control character. For example, if you want to write a menu item that will toggle the SNAP off or on, you can use a caret sign followed by the control character B as shown in the following example.

[SNAP-TOG]^B

Where **SNAP-TOG** ----- Command label for SNAP toggle
        ^ ----------------- Caret sign (^)
        B ------------------ Control character for SNAP

Examples

^C	(Cancel)
^G	(Grid on/off)
^H	(Backspace)
^O	(Ortho on/off)
^T	(Tablet on/off)
^E	(Isoplane top/left/right)

^B is equivalent to pressing the CTRL and B keys from the keyboard that toggles the SNAP mode. SNAP-TOG is the menu item label that will be displayed on the screen menu. You can use any ASCII control character in the command definition. Following are some of the ASCII control characters:

^@	(ASCII code 0)
^[	(ASCII code 27)
^\	(ASCII code 28)
^]	(ASCII code 29)
^^	(ASCII code 30)
^_	(ASCII code 31)

SPECIAL CHARACTERS

The following is a list of the special characters that you can use in the AutoCAD menus:

Character	Description
***	Three asterisks indicate a section title
**	Two asterisks indicate a submenu
[ ]	Used to enclose a menu item label
;	Causes an ENTER
Space	Space is equivalent to pressing the spacebar
\	AutoCAD pauses for user input
_	(Underscore character) translates the AutoCAD commands and keyword that follow it into English.
+	Used to continue the menu item definition to next line
=*	Forces the display of pull-down, cursor or image tile menus

*	Menu item repetition
$M=	Special character to load a DIESEL macro expression
$S=CMDNAME	Special command that loads a screen submenu
^B	Toggles snap (On/Off)
^C	Cancels the existing command
^D	Toggles coordinate dial (On/Off)
^E	Changes the isometric plane (Left/Right/Top)
^G	Toggles grid (On/Off)
^H	Causes a backspace
^O	Toggles ortho mode (On/Off)
^P	Toggles MENUECHO on or off
^Q	Echoes all prompts to the printer
^T	Toggles tablet mode (On/Off)
^V	Change current viewport
^Z	Suppresses the automatic addition of spacebar at the end of a menu item

Example 8 — *General*

Write a screen menu for the following toggle functions (filename SM5.MNU):

ORTHO	SNAP
GRID	COORDINATE DIAL
TABLET	ISOPLANE
PRINTER	

Before writing a screen menu for these toggle functions, it is important to know the control characters that AutoCAD uses to turn these functions on or off. The following is a list of the control characters for the functions in this example:

ORTHO	CTRL O
SNAP	CTRL B
GRID	CTRL G
COORDINATE DIAL	CTRL D
TABLET	CTRL T
ISOPLANE	CTRL E
CURRENT VIEWPORT	CTRL V

The following file is a listing of the screen menu that toggles the given functions. You can use AutoCAD's **EDIT** command to write this file.

```
[-TOGGLE-    ]
[            ]
[            ]
[            ]
[ORTHO       ]^O          ———————    turns ORTHO on/off
[SNAP        ]^B          ———————    turns SNAP on/off
```

Screen Menus 9-37

[GRID	]^G	————	turns GRID on/off
[CO-ORDS	]^D	————	turns COORDINATE dial on/off
[TABLET	]^T	————	turns TABLET on/off
[	]		
[ISOPLANE	]^E	————	turns ISOPLANE on/off
[CURRENT VIEWPORT]^V		————	Changes the current viewport

COMMAND DEFINITION WITHOUT ENTER OR SPACE

All command definitions that have been discussed so far use a semicolon (;) or a space for Enter. However, sometimes a user might want to define a menu item that is not followed by an Enter or a space. This is accomplished by using the backspace ASCII control character (^H). It is especially useful when you want to write a screen menu for a numeric keypad. ^H control character can be used with any character or a group of characters. The following is the format of menu item definition without an ENTER or space:

[9]9X^H

    Where **9** ------------------- Menu item label
                  9 ------------------- Character returned by selecting this item
                  X ------------------ X is erased because of ^H (backspace)
                           (X could be substituted by any character)
              ^H -------------- ^H for backspace

^H is the ASCII control character for backspace. When you select this item, ^H erases the previous character X and therefore returns 9 only. It is not necessary to have X preceding ^H. X is a dummy character; it could be any character. You can continue selecting more characters, and when you are done and you want to press ENTER, use the keyboard or the digitizer.

Example 9 *General*

Write a screen menu for the following characters (filename SM6.MNU):

    0   5   .
    1   6   ,
    2   7   X
    3   8   Y
    4   9   Z

The following file is a listing of the screen menu of Example 9 that will enable the user to pick the characters without an ENTER.

    [-KEYPAD-   ]
    [   ]
    [   ]
    [   ]
    [0]0Y^H   ———— returns 0
    [1]1Y^H   ———— returns 1
    [2]2Y^H   ———— returns 2

[3]3Y^H
[4]4Y^H
[5]5Y^H
[6]6Y^H
[7]7Y^H
[8]8Y^H
[9]9Y^H
[.].Y^H   ———   returns period (.)
[,],Y^H   ———   returns comma (,)
[]   ———   for space in the screen menu
[X]XX^H   ———   returns X
[Y]YX^H   ———   returns Y
[Z]ZZ^H   ———   returns Z

Note

You can also use a backslash after the character to continue selecting more characters. For example, [2]2\ will return 2 and then pause for the user input, which could be another character. The following file uses a backslash (\) to let the user enter another character.

[-KEYPAD-]
[   ]
[   ]
[   ]
[0]0\
[1]1\
[2]2\
[3]3\
[4]4\
[5]5\
[6]6\
[7]7\
[8]8\
[9]9\
[.].\
[,],\
[]
[X]X\
[Y]Y\
[Z]Z\

MENU ITEMS WITH SINGLE OBJECT SELECTION MODE

The Single options for object selection, combined with the menu item repetition, can provide a powerful editing tool.

Screen Menus   9-39

[ERASE]*^C^CERASE Single

Where * -------------------- Menu item repetition
       Single ------------ Single object selection option

In this menu item the asterisk that follows the menu item label repeats the command. ^C^C cancels the existing command. **ERASE** is an AutoCAD command that erases the selected objects. The Single option enables the user to select the objects and then AutoCAD automatically terminates the object selection process. The **ERASE** command then erases the selected objects.

If the point you pick is in a blank area (if there is no object), AutoCAD automatically defaults to crossing or window option. If you pull the window to the left, it is the crossing option and if you pull the box to the right, it is the window option for selecting the objects. The asterisk in front of the menu item repeats the command.

USE OF AUTOLISP IN MENUS

You can combine AutoLISP variables and expressions with the menu items as a command definition. When you select that item from the menu, it will evaluate all the expressions and generate the necessary output. The following examples illustrate the use of AutoLISP variables and expressions. (For more information on AutoLISP, see Chapters 12 and 13.)

Example 10   *General*

Write an AutoLISP program that draws a square and then write a screen menu that utilizes the AutoLISP variables and expressions to draw a square.

The following file is the listing of the AutoLISP program that generates a square. The program prompts the user to enter the starting point and the length of the side.

```
(DEFUN C:SQR()
(SETVAR "CMDECHO" 0)
(SETQ P1 (GETPOINT "\n ENTER STARTING POINT: "))
(SETQ S (GETDIST "\n ENTER LENGTH OF SIDE: "))
(SETQ P2 (LIST (+ (CAR P1) S) (CADR P1)))
(SETQ P3 (POLAR P2 (/ PI 2) S))
(SETQ P4 (POLAR P1 (/ PI 2) S))
(COMMAND "PLINE" P1 P2 P3 P4 "C")
(SETVAR "CMDECHO" 1)
(PRINC)
)
```

The following file is the listing of the screen menu file that utilizes the AutoLISP variables and expressions to draw a square:

```
[-SQUARE- ]
[         ]
[         ]
```

```
[SQUARE: ](SETQ P1(GETPOINT "ENTER STARTING POINT:+
"));\+
(SETQ S (GETDIST "ENTER LENGTH OF SIDE: "));\+
(SETQ P2 (LIST (+ (CAR P1) S) (CADR P1)))+
(SETQ P3 (POLAR P2 (/ pi 2) S))+
(SETQ P4 (POLAR P1 (/ pi 2) S));+
PLINE !P1 !P2 !P3 !P4 C
```

If you compare the AutoLISP program with the screen menu, you will notice that the statements that prompt the user to enter the information and the statements that do actual calculations are identical. Therefore, you can utilize AutoLISP variables and expressions in the menu file to customize AutoCAD.

DIESEL EXPRESSION IN MENUS

You can also define a DIESEL expression in the screen, tablet, pull-down, or button menu. When you select the menu item, it will automatically assign the value to the **MODEMACRO** system variable and then display the new status line. The following example illustrates the use of DIESEL expression in the screen menu.

Example 11 *General*

Write a screen menu that will load the DIESEL macro to display the following information in the status line:

Macro-1	**Macro-2**	**Macro-3**
Project name	Pline width	Dimtad
Drawing name	Fillet radius	Dimtix
Current layer	Offset distance	Dimscale

The following file is a listing of the screen menu that contains the definition of three DIESEL macros of Example 11. This menu can be loaded by using AutoCAD's MENU command and then entering the name of the menu file. If you select the first item, DIESEL1, it will automatically display the corresponding status line (Figure 9-10).

```
***screen
[*DIESEL*]
[DIESEL1:]^C^CMODEMACRO;$M=Cust-Acad,N:$(GETVAR,DWGNAME)+
,L:$(GETVAR,CLAYER);
[DIESEL2:]^C^CMODEMACRO;$M=PLWID:$(GETVAR,PLINEWID),+
FRAD:$(GETVAR,FILLETRAD),OFFSET:$(GETVAR,OFFSETDIST),+
LTSCALE:$(GETVAR,LTSCALE);
[DIESEL3:]^C^CMODEMACRO;$M=DTAD:$(GETVAR,DIMTAD),+
DTIX:$(GETVAR,DIMTIX),DSCALE:$(GETVAR,DIMSCALE);
```

You can also use the **$M=** command to load a DIESEL macro. AutoCAD evaluates the DIESEL string expression that follows the equal sign (=) and the value returned by the DIESEL expression becomes a part of the menu item.

Screen Menus 9-41

Figure 9-10 Screen menu and status line for Example 11

Example
[SCR-MODE]SCREENMODE $M=$(-,1,$(GETVAR,SCREENMODE))

In this example, AutoCAD evaluates the DIESEL string expression $(-,1,$(GETVAR, SCREENMODE)). The **GETVAR** command retrieves the value of the system variable SCREENMODE and subtracts the value from 1. If the value of the system variable **SCREENMODE** is 1, then the DIESEL expression returns 0, and if the value of **SCREENMODE** is 0, then the DIESEL expression returns 1. The value returned by the DIESEL expression is assigned to the system variable **SCREENMODE**. Therefore, this menu item can be used to change the screen display from the text to graphics mode or from the graphics to text mode.

Self-Evaluation Test

Answer the following questions and then compare your answers to the correct answers at the end of this chapter.

1. The command definition that involves several commands put together in one line is also called a _____.

2. A command can be repeated if the menu command definition starts with an _____.

3. The screen menus can be automatically swapped by using the system variable _____.

4. _____ command will load the submenu.

5. _____ system variable is used to display the new status line.

6. You can use _____ in the menu item to turn the echo on or off.

7. If the statement can't fit on one line, you can put a _____ at the end of the first line and then continue with a second line.

8. The _____ is used in a menu item to refer to or load a submenu.

9. Command definition without enter or space can be accomplished by using _____.

Review Questions

Answer the following questions.

1. The AutoCAD menu file can have up to _____ main sections.

2. A submenu is designated by _____.

3. The part of the menu item that is inside the brackets is for _____ only.

4. Only the first _____ characters can be displayed on the screen menu.

5. If the number of characters inside the bracket exceeds eight, the screen menu _____ work.

6. In a menu file you can use _____ to cancel the existing command.

7. _____ is used to pause for input in a screen menu definition.

8. The menu item label _____ be a combination of uppercase and lowercase characters.

9. Submenu names can be _____ characters long.

10. The maximum number of accessible menu items in a screen menu depends on the _____.

Screen Menus

Exercises

Exercise 5 — *General*

Design and write a screen menu for the following AutoCAD commands (file name SME5.MNU).

POLYGON (Center)
POLYGON (Edge)
ELLIPSE (Center)
ELLIPSE (Axis End point)
CHAMFER
EXPLODE
COPY

Exercise 6 — *General*

Write a screen menu file for the following AutoCAD commands. Use submenus if required (file name **SME6.MNU**).

ARC	ROTATE
-3P	ARRAY
-SCE	DIVIDE
-CSE	MEASURE
BLOCK	
INSERT	LAYER
WBLOCK	SET
MINSERT	LIST

Exercise 7 — *General*

Write a screen menu item that will set up the following layers, linetypes, and colors (file name SME7.MNU).

Layer Name	Color	Linetype
0	WHITE	CONTINUOUS
OBJECT	RED	CONTINUOUS
HIDDEN	YELLOW	HIDDEN
CENTER	BLUE	CENTER
DIM	GREEN	CONTINUOUS

Exercise 8 — *General*

Write a screen menu for the commands in Figure 9-11. Use the menu item ******** to load Osnaps, and use the menu item MENU-8 to load the root menu (file name SME8.MNU).

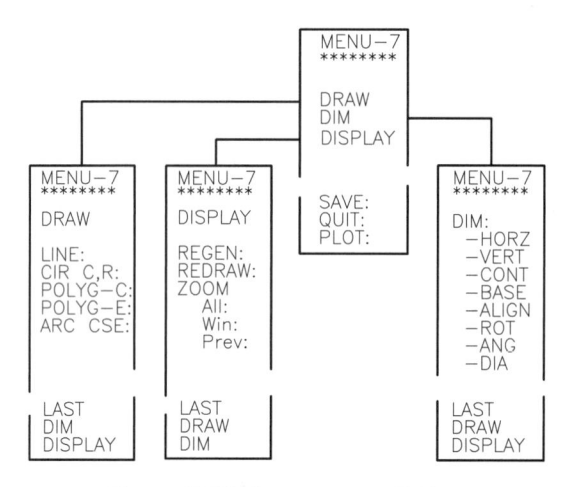

```
              MENU-7
              ********
              DRAW
              DIM
              DISPLAY

 MENU-7       MENU-7              MENU-7
 ********     ********             ********
 DRAW         DISPLAY    SAVE:     DIM:
                         QUIT:     -HORZ
 LINE:        REGEN:     PLOT:     -VERT
 CIR C,R:     REDRAW:              -CONT
 POLYG-C:     ZOOM                 -BASE
 POLYG-E:      All:                -ALIGN
 ARC CSE:      Win:                -ROT
               Prev:               -ANG
                                   -DIA

 LAST         LAST                 LAST
 DIM          DRAW                 DRAW
 DISPLAY      DIM                  DISPLAY
```

Figure 9-11 Screen menu displays

Exercise 9   *General*

Write a coordinated pull-down and screen menu for the following AutoCAD commands. Use a cascading menu for the **LINE** command options in the menu. When the user selects an item from the menu, the corresponding screen menu should be automatically loaded on the screen. The layout of the screen menu and the pull-down menu are shown in Figures 9-12(a) and 9-12(b).

LINE	ZOOM All	TIME
Continue	ZOOM Win	LIST
Close	ZOOM Pre	DISTANCE
Undo	PAN	AREA
.X	DBLIST	
.Y	STATUS	
.Z		
CIRCLE		
ELLIPSE		

Screen Menus

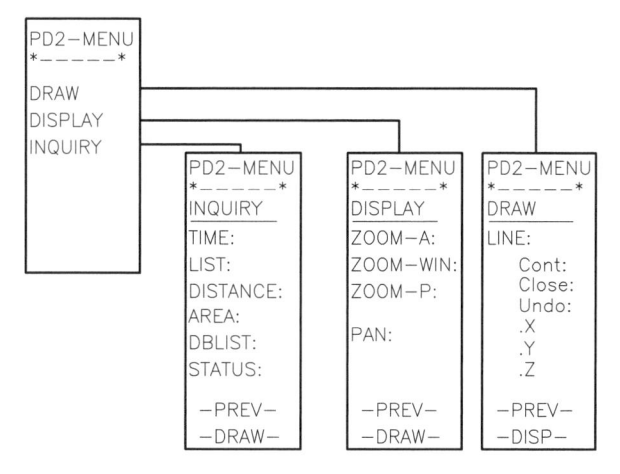

9-45

Figure 9-12(a) Design of screen menu for Exercise 9

Figure 9-12(b) Design of menu for Exercise 9

Exercise 10   *General*

Write a menu, a screen menu, and a tablet menu for the following AutoCAD commands. When you select a command from the template or the menu, the corresponding screen menu should be automatically loaded on the screen. The layout of the screen is shown in Figure 9-13.

LINE	BLOCK
PLINE	WBLOCK
CIRCLE C,R	INSERT
CIRCLE C,D	BLOCK LIST

ELLIPSE AXIS ENDPOINT   DDATTDEF
ELLIPSE CENTER   DDATTE

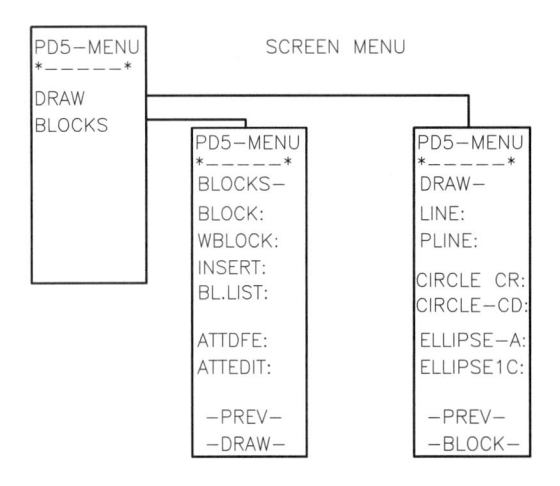

```
PD5-MENU                SCREEN MENU
*-----*
DRAW
BLOCKS
            PD5-MENU              PD5-MENU
            *-----*               *-----*
            BLOCKS-               DRAW-
            BLOCK:                LINE:
            WBLOCK:               PLINE:
            INSERT:               CIRCLE CR:
            BL.LIST:              CIRCLE-CD:
            ATTDFE:               ELLIPSE-A:
            ATTEDIT:              ELLIPSE1C:
            -PREV-                -PREV-
            -DRAW-                -BLOCK-
```

Figure 9-13 Design of screen menu for Exercise 10

Exercise 11   *General*

Write a screen, pull-down, and image tile menu for the commands shown in Figure 9-14. When the user selects a command from the menu, it should automatically load the corresponding menu on the screen.

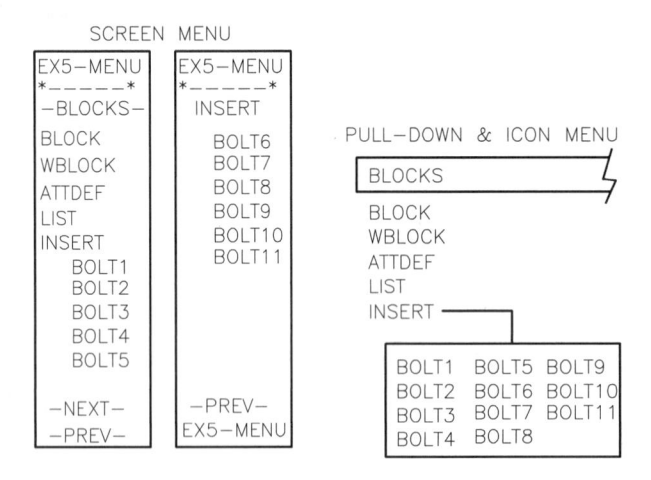

```
         SCREEN MENU
EX5-MENU    EX5-MENU
*-----*     *-----*
-BLOCKS-    INSERT
BLOCK       BOLT6         PULL-DOWN & ICON MENU
WBLOCK      BOLT7         ┌──────────────┐
ATTDEF      BOLT8         │ BLOCKS       │
LIST        BOLT9         └──────────────┘
INSERT      BOLT10        BLOCK
  BOLT1     BOLT11        WBLOCK
  BOLT2                   ATTDEF
  BOLT3                   LIST
  BOLT4                   INSERT ──┐
  BOLT5                            │
                          ┌────────┴──────────────┐
 -NEXT-     -PREV-        │ BOLT1 BOLT5 BOLT9     │
 -PREV-     EX5-MENU      │ BOLT2 BOLT6 BOLT10    │
                          │ BOLT3 BOLT7 BOLT11    │
                          │ BOLT4 BOLT8           │
                          └───────────────────────┘
```

Figure 9-14 Design of screen, pull-down, and image tile menus for Exercise 11

Screen Menus

Exercise 12 — *General*

Write a screen, tablet, pull-down, and image tile menu for inserting the following commands. B1 to B15 are the block names.

BLOCK
WBLOCK
ATTDEF
LIST
INSERT

BL1	BL6	BL11
BL2	BL7	BL12
BL3	BL8	BL13
BL4	BL9	BL14
BL5	BL10	BL15

Exercise 13 — *General*

Write a pull-down and screen menu for the following AutoCAD commands. Use submenus for the **ARC** and **CIRCLE** commands. When the user selects an item from the menu, the corresponding screen menu should be automatically loaded on the screen.

LINE	**BLOCK**	**QUIT**
PLINE	**INSERT**	**SAVE**
ARC	**WBLOCK**	———
ARC 3P		**PLOT**
ARC SCE		
ARC SCA		
ARC CSE		
ARC CSA		
ARC CSL		
CIRCLE		
CIRCLE C,R		
CIRCLE C,D		
CIRCLE 2P		

The layout shown in Figure 9-15(a) is one of the possible designs for this menu. The **ARC** and **CIRCLE** commands are in separate groups that will be defined as submenus in the menu file. The layout of the screen menu is shown in Figure 9-15(b).

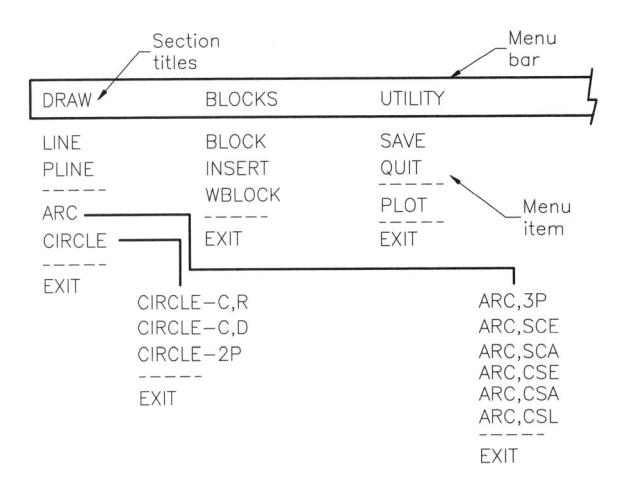

Figure 9-15(a) Design of pull-down menu for Exercise 13

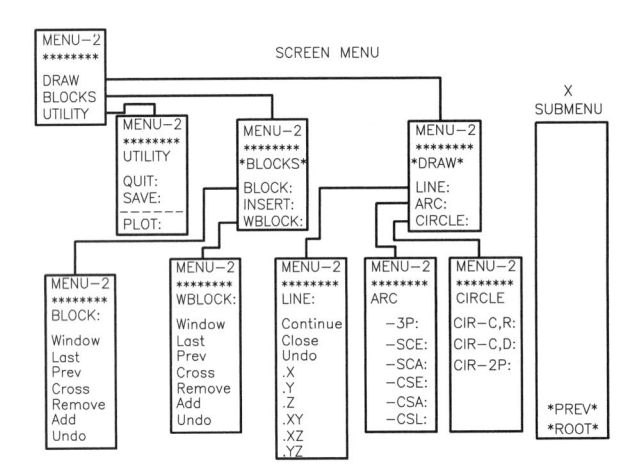

Figure 9-15(b) Design of screen menu

Answers to the Self-Evaluation Test

1. MACRO, **2.** asterisk(*), **3.** MENUCTL, **4.** CMDNAME, **5.** MODEMACRO, **6.** ^p, **7.** plus(+)
8. Submenu Reference, **9.** ^H.

Chapter 10

Customizing the Standard AutoCAD Menu

Learning Objectives

After completing this chapter, you will be able to:
- *Edit the standard AutoCAD menu file, ACAD.MNU.*
- *Load menus and submenus.*
- *Customize tablet areas.*
- *Customize buttons menus.*
- *Customize pull-down and cursor menus.*
- *Customize image tile menus and screen menus.*

THE STANDARD AUTOCAD MENU

The AutoCAD software package comes with a standard menu file: ACAD.MNU. AutoCAD stores the name of the menu that was used last in the system registry. When you start AutoCAD, the last used menu file is automatically loaded (Figure 10-1). The menu sections of a menu file are identified by section labels. The general format of the section label is ***section_name**. For example, if it is the screen menu section, the section label is ***SCREEN. The following is the list of the section labels.

***SCREEN		Screen menu
***TABLET(n)	**n** is from 1 to 4	Tablet menu
***IMAGE		Image tile menu
***POP(n)	**n** is from 1 to 499	Pull-down menu
	n is 0 and from 500 to 999	For shortcut menus
***BUTTONS(n)	**n** is from 1 to 4	Pointing device menu

***AUX(n)	n is from 1 to 4	System pointing device menu
***MENUGROUP		Menu file group name
***TOOLBARS		Toolbar definition
***HELPSTRING		Text displayed in the status bar
***ACCELERATORS		Accelerator key definitions

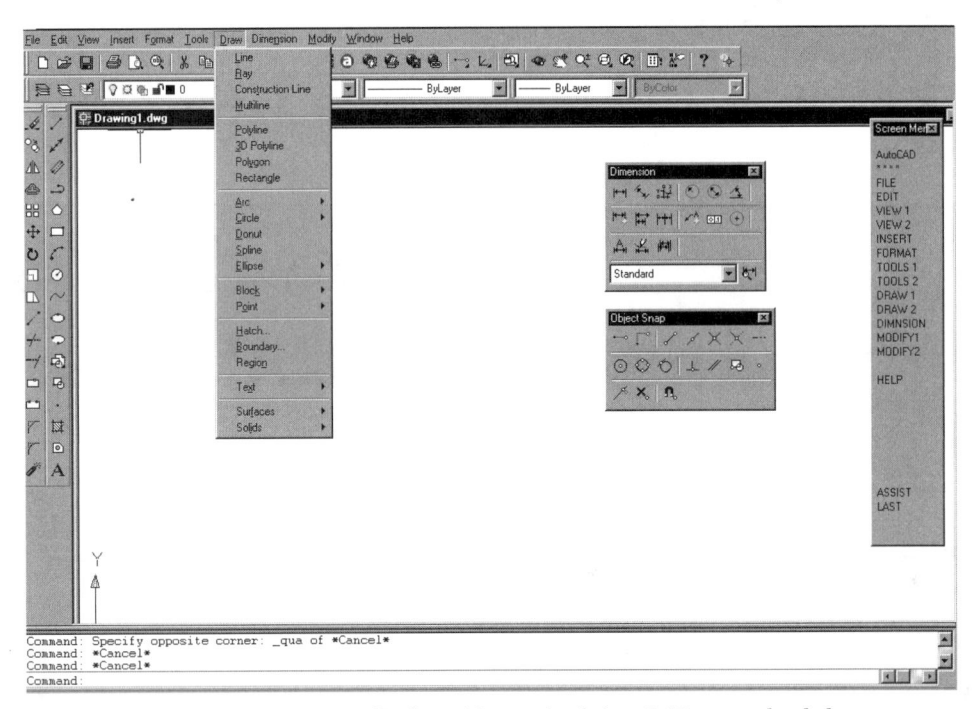

Figure 10-1 Screen display with standard AutoCAD menu loaded

The following file is a partial listing of the ACAD.MNU file that shows some of the section labels:

```
///     AutoCAD 2002 Menu
//      Dec. 15, 2000
//      Copyright (C) 1986, 1987, 1988, 1989, 1990, 1991, 1992, 1994, 1996, 1997,
//      1998, 1999, 2000, 2001 by Autodesk, Inc.
```

***MENUGROUP=ACAD
// Begin AutoCAD Digitizer Button Menus

***BUTTONS1
// Simple + button
// if a grip is hot bring up the Grips Cursor Menu (POP 500), else send a carriage return
// If the SHORTCUTMENU sysvar is not 0 the first item (for button 1) is NOT USED.
$M=$(if,$(eq,$(substr,$(getvar,cmdnames),1,5),GRIP_),$P0=ACAD.GRIPS $P0=*);
$P0=SNAP $p0=*

Customizing the Standard AutoCAD Menu

```
^C^C
^B
^O
^G
^D
^E
^T
```

***BUTTONS2**
// Shift + button
$P0=SNAP $p0=*

***BUTTONS3**
// Control + button

***BUTTONS4**
// Control + shift + button
//   Begin System Pointing Device Menus

***AUX1**
// Simple button
// if a grip is hot bring up the Grips Cursor Menu (POP 500), else send a carriage return
// If the SHORTCUTMENU sysvar is not 0 the first item (for button 1, the "right button")
// is NOT USED.
$M=$(if,$(eq,$(substr,$(getvar,cmdnames),1,5),GRIP_),$P0=ACAD.GRIPS $P0=*);
$P0=SNAP $p0=*
```
^C^C
^B
^O
^G
^D
^E
^T
```

***AUX2**
// Shift + button
$P0=SNAP $p0=*
$P0=SNAP $p0=*

***AUX3**
// Control + button
$P0=SNAP $p0=*

***AUX4**
// Control + shift + button
$P0=SNAP $p0=*

***POP0
**SNAP
// Shift-right-click if using the default AUX2 and/or BUTTONS2
// menus. [&Object Snap Cursor Menu]
ID_Tracking [Temporary trac&k point]_tt
ID_From   [&From]_from
ID_MnPointFi [->Poin&t Filters]
ID_PointFilx [.X].X
ID_PointFily [.Y].Y
ID_PointFilz [.Z].Z

***POP1
**FILE
ID_MnFile   [&File]
ID_New   [&New...\tCtrl+N]^C^C_new
ID_Open   [&Open...\tCtrl+O]^C^C_open
ID_DWG_CLOSE [&Close]^C^C_close
ID_PartialOp [$(if,$(eq,$(getvar,fullopen),0),,~)Partia&l Load]^C^C_partiaload
ID_Save   [&Save\tCtrl+S]^C^C_qsave
ID_Saveas   [Save &As...]^C^C_saveas
ID_ETransmit [e&Transmit...]^C^C_etransmit
ID_Publish   [Publish to &Web...]^C^C_publishtoweb
ID_Export   [&Export...]^C^C_export

***POP2
**EDIT
ID_MnEdit   [&Edit]
ID_U   [&Undo\tCtrl+Z]^C^C_u
ID_Redo   [&Redo\tCtrl+Y]^C^C_redo
             [--]
ID_Cutclip   [Cu&t\tCtrl+X]^C^C_cutclip
ID_Copyclip [&Copy\tCtrl+C]^C^C_copyclip
ID_Copybase [Copy with &Base Point]^C^C_copybase
ID_Copylink [Copy &Link]^C^C_copylink

***POP3
**VIEW
ID_MnView   [&View]
ID_Redrawall [&Redraw]'_redrawall
ID_Regen   [Re&gen]^C^C_regen
ID_Regenall [Regen &All]^C^C_regenall

Customizing the Standard AutoCAD Menu   10-5

```
ID_ZoomRealt  [&Realtime]'_zoom ;
ID_ZoomPrevi  [&Previous]'_zoom _p
ID_ZoomWindo  [&Window]'_zoom _w
ID_ZoomDynam  [&Dynamic]'_zoom _d
ID_ZoomScale  [&Scale]'_zoom _s
ID_ZoomCente  [&Center]'_zoom _c
```

***POP4
**INSERT
```
ID_MnInsert  [&Insert]
ID_Ddinsert  [&Block...]^C^C_insert
ID_Xattach   [E&xternal Reference...]^C^C_xattach
ID_Imageatta [Raster &Image...]^C^C_imageattach
```

***POP5
**FORMAT
```
ID_MnFormat [F&ormat]
ID_Layer    [&Layer...]'_layer
ID_Ddcolor  [&Color...]'_color
ID_Linetype [Li&netype...]'_linetype
ID_Linewt   [Line&weight...]'_lweight
```

***POP6
**TOOLS
```
ID_MnTools   [&Tools]
ID_Today     [&Today]^C^C_Today
ID_PointA    [&Autodesk Point A]^C^C_ ^C^C_browser http://pointa.autodesk.com
ID_MeetNow   [&Meet Now]^C^C_MeetNow
ID_MnCadStd  [->CAD &Standards]
ID_Standards [&Configure...]^C^C_standards
```

***POP7
**DRAW
```
ID_MnDraw  [&Draw]
ID_Line    [&Line]^C^C_line
ID_Ray     [&Ray]^C^C_ray
ID_Xline   [Cons&truction Line]^C^C_xline
ID_Mline   [&Multiline]^C^C_mline
ID_Pline   [&Polyline]^C^C_plineID_3dpoly  [&3D Polyline]^C^C_3dpoly
```

```
ID_Polygon   [Pol&ygon]^C^C_polygon
ID_Rectang   [Rectan&gle]^C^C_rectang
```

***POP8
**DIMENSION
```
ID_MnDimensi [Dime&nsion]
ID_QDim      [&QDIM]^C^C_qdim
        [--]
ID_Dimlinear [&Linear]^C^C_dimlinear
ID_Dimaligne [Ali&gned]^C^C_dimaligned
ID_Dimordina [&Ordinate]^C^C_dimordinate
        [--]
ID_Dimradius [&Radius]^C^C_dimradius
ID_Dimdiamet [&Diameter]^C^C_dimdiameter
ID_Dimangula [&Angular]^C^C_dimangular
```

***POP9
**MODIFY
```
ID_MnModify  [&Modify]
ID_Ai_propch [&Properties]^C^C_properties
ID_Matchprop [&Match Properties]'_matchprop
ID_MnObject  [->&Object]
ID_MnExterna [->&External Reference]
ID_Xbind     [&Bind...]^C^C_xbind
ID_Xclipfram                                                    [<-
$(if,$(eq,$(getvar;xclipframe),1),!.)&Frame]$M=$(if,$(eq,$(getvar;xclipframe),1),^C^C_xclipframe
0,^C^C_xclipframe 1)
ID_MnImage   [->&Image]
ID_Imageadju [&Adjust...]^C^C_imageadjust
ID_Imagequal [&Quality]^C^C_imagequality
ID_Transpare [&Transparency]^C^C_transparency
```

***POP10
**WINDOW
```
ID_MnWindow  [&Window]
ID_DWG_CLOSE       [Cl&ose]^C^C_close
ID_WINDOW_CLOSEALL [C&lose All]^C^C_closeall
ID_WINDOW_CASCADE  [&Cascade]^C^C_syswindows;_cascade
ID_WINDOW_TILE_HORZ [Tile &Horizontally]^C^C_syswindows;_hor
```

Customizing the Standard AutoCAD Menu 10-7

***POP11
**HELP
ID_MnHelp   [&Help]
ID_Help   [&Help\tF1]'_help
ID_ASSIST   [&Active Assistance]'_ASSIST
ID_DevHelp   [&Developer Help]^C^C^P(help "acad_dev" "") ^P
ID_Support   [&Support Assistance]^C^C^P(help "acad" "asa") ^P
ID_WhatsNew [&What's New]^C^C^P(progn (help "acadwnew")(princ)) ^P
ID_LAssist   [&Learning Assistance]^C^C^P(help "acad" "elearning") ^P
ID_AUGI   [Autodesk &User Group International]^C^C^P(command "_browser" (findfile "augi.htm")) ^P

***POP500
**GRIPS
// When a grip is hot, then display the following shortcut menu for grips. See also AUX1 menu. [&Grips Cursor Menu]
ID_Enter   [&Enter];
ID_GripMove [&Move]_move
ID_GripMirro [M&irror]_mirror
ID_GripRotat [&Rotate]_rotate
ID_GripScale [Sca&le]_scale

***POP501
**CMDEFAULT [Context menu for default mode]
ID_CMNonLast [&Repeat %s]^C^C;
 ID_Cutclip   [Cu&t]^C^C_cutclip
ID_Copyclip [&Copy]^C^C_copyclip
ID_Copybase [Copy with &Base Point]^C^C_copybase
ID_Pasteclip [&Paste]^C^C_pasteclip
ID_Pastebloc [Paste as Bloc&k]^C^C_pasteblock
ID_Pasteorig [Paste to Original Coor&dinates]^C^C_pasteorig

***POP502
**CMEDIT [Context menu for edit mode]
ID_CMSelLast [&Repeat %s]^C^C;
ID_Cutclip   [Cu&t]^C^C_cutclip
ID_Copyclip [&Copy]^C^C_copyclip
ID_Copybase [Copy with &Base Point]^C^C_copybase
ID_Pasteclip [&Paste]^C^C_pasteclip
ID_Pastebloc [Paste as Bloc&k]^C^C_pasteblock

ID_Pasteorig [Paste to Original Coor&dinates]^C^C_pasteorig

***POP503
**CMCOMMAND   [Context menu for command mode]
ID_Enter   [&Enter];
ID_Cancel   [&Cancel]^C

***POP504
**OBJECTS_DIMENSION   [Context Menu for Dimension Objects]
ID_DimText   [->Dim Te&xt position]
ID_DimAbove   [&Above dim line]^C^C_ai_dim_textabove
ID_DimTxtCen   [&Centered]^C^C_ai_dim_textcenter
ID_DimHome   [&Home text]^C^C_ai_dim_texthome
ID_DimTxtMove2 [&Move text alone]^C^C_aidimtextmove _2

***POP505
**OBJECT_VIEWPORT   [Context menu for Viewport Object]
ID_VpClip   [&Viewport Clip]^C^C_vpclip
ID_Vport_disp   [->&Display Viewport Objects]
ID_Vport_dispon [&Yes]^C^C_-vports _on _p;;
ID_Vport_dispoff [<-&No]^C^C_-vports _off _p;;
ID_Vport_lock   [->Display &Locked]

***POP506
**OBJECTS_XREF   [Context menu for XREF Objects]
ID_Xclip   [Xref Cl&ip]^C^C_xclip
ID_XRef   [Xref Ma&nager...]^C^C_xref

***POP507
**OBJECT_MTEXT [Context menu for MTEXT Object]
ID_Mtedit   [Mtext Ed&it...]^C^C_mtedit

***POP508
**OBJECT_TEXT [Context menu for TEXT object]
ID_Ddedit   [Text Ed&it...]^C^C_ddedit

Customizing the Standard AutoCAD Menu

***POP509
**OBJECT_HATCH [Context menu for HATCH object]
ID_Hatchedit   [&Hatch Edit...]^C^C_hatchedit

***POP510
**OBJECT_LWPOLYLINE [Context menu for PLINE object]
ID_Pedit   [Polyline Ed&it]^C^C_pedit

***POP511
**OBJECT_SPLINE [Context menu for SPLINE object]
ID_Splinedit [Spline Ed&it]^C^C_splinedit

***TOOLBARS
**TB_DIMENSION
ID_TbDimensi [_Toolbar("Dimension", _Floating, _Hide, 100, 130, 1)]
ID_Dimlinear [_Button("Linear Dimension", ICON_16_DIMLIN, ICON_16_DIMLIN)]^C^C_dimlinear
ID_Dimaligne [_Button("Aligned Dimension", ICON_16_DIMALI, ICON_16_DIMALI)]^C^C_dimaligned
ID_Dimordina [_Button("Ordinate Dimension", ICON_16_DIMORD, ICON_16_DIMORD)]^C^C_dimordinate
ID_Dimradius [_Button("Radius Dimension", ICON_16_DIMRAD, ICON_16_DIMRAD)]^C^C_dimradius
ID_Dimdiamet [_Button("Diameter Dimension", ICON_16_DIMDIA, ICON_16_DIMDIA)]^C^C_dimdiameter
ID_Dimangula [_Button("Angular Dimension", ICON_16_DIMANG, ICON_16_DIMANG)]^C^C_dimangular

**TB_DRAW
ID_TbDraw   [_Toolbar("Draw", _Left, _Show, 0, 0, 1)]
ID_Line   [_Button("Line", ICON_16_LINE, ICON_16_LINE)]^C^C_line
ID_Xline   [_Button("Construction Line", ICON_16_XLINE, ICON_16_XLINE)]^C^C_xline
ID_Mline   [_Button("Multiline", ICON_16_MLINE, ICON_16_MLINE)]^C^C_mline

**TB_INQUIRY
ID_TbInquiry [_Toolbar("Inquiry", _Floating, _Hide, 100, 170, 1)]

ID_Dist   [_Button("Distance", ICON_16_DIST, ICON_16_DIST)]'_dist
ID_Area   [_Button("Area", ICON_16_AREA, ICON_16_AREA)]^C^C_area
ID_Massprop [_Button("Mass Properties", ICON_16_MASSPR, ICON_16_MASSPR)]^C^C_massprop
ID_List   [_Button("List", ICON_16_LIST, ICON_16_LIST)]^C^C_list
ID_Id   [_Button("Locate Point", ICON_16_ID, ICON_16_ID)]'_id

**TB_INSERT
ID_TbInsert [_Toolbar("Insert", _Floating, _Hide, 100, 190, 1)]
ID_Ddinsert [_Button("Insert Block", ICON_16_DINSER, ICON_16_DINSER)]^C^C_insert
ID_Xref   [_Button("External Reference", ICON_16_XREATT, ICON_16_XREATT)]^C^C_xref

**TB_LAYOUTS
ID_TbLayouts [_Toolbar("Layouts", _Floating, _Hide, 100, 350, 1)]
ID_LayNew   [_Button("New Layout", ICON_16_LAYNEW, ICON_16_LAYNEW)]^C^C_-layout _n
ID_LayTemp   [_Button("Layout from Template", ICON_16_LAYTEM, ICON_16_LAYTEM)]^C^C_-layout _t
ID_PlotSetup [_Button("Page Setup", ICON_16_PLTSET, ICON_16_PLTSET)]^C^C_pagesetup
ID_VpDialog [_Button("Display Viewports Dialog", ICON_16_VPDLG, ICON_16_VPDLG)]^C^C_vports

**TB_MODIFY
ID_TbModify [_Toolbar("Modify", _Left, _Show, 1, 0, 1)]
ID_Erase   [_Button("Erase", ICON_16_ERASE, ICON_16_ERASE)]^C^C_erase
ID_Copy   [_Button("Copy Object", ICON_16_COPYOB, ICON_16_COPYOB)]$M=$(if,$(eq,+
$(substr,$(getvar,cmdnames),1,4),grip),_copy,^C^C_copy)
ID_Mirror   [_Button("Mirror", ICON_16_MIRROR, ICON_16_MIRROR)]$M=$(if,$(eq,+

**TB_MODIFY_II
ID_TbModifII [_Toolbar("Modify II", _Floating, _Hide, 100, 270, 1)]
ID_Draworder
[_Button("Draworder",ICON_16_DRWORD,ICON_16_DRWORD)]^C^C_draworder
ID_Hatchedit [_Button("Edit Hatch", ICON_16_HATEDI,

Customizing the Standard AutoCAD Menu

ICON_16_HATEDI)]^C^C_hatchedit

**TB_OBJECT_PROPERTIES
ID_TbObjectP [_Toolbar("Object Properties", _Top, _Show, 0, 1, 1)]
ID_Ai_molc [_Button("Make Object's Layer Current", ICON_16_MOLC, ICON_16_MOLC)]^C^C_ai_molc
ID_Layer [_Button("Layers", ICON_16_LAYERS, ICON_16_LAYERS)]'_layer
ID_CtrlLayer [_Control(_Layer)]
ID_CtrlColor [_Control(_Color)]

**TB_OBJECT_SNAP
ID_TbOsnap [_Toolbar("Object Snap", _Floating, _Hide, 100, 210, 1)]
ID_Tracking [_Button("Temporary Tracking Point", ICON_16_OSNTMP, ICON_16_OSNTMP)]_tt
ID_From [_Button("Snap From", ICON_16_OSNFRO, ICON_16_OSNFRO)]_from
ID_OsnapEndp [_Button("Snap to Endpoint", ICON_16_OSNEND, ICON_16_OSNEND)]_endp

**TB_ORBIT
ID_TbOrbit [_Toolbar("3D Orbit", _Floating, _Hide, 100, 350, 1)]
ID_3dpan [_Button("3D Pan", ICON_16_3DPAN, ICON_16_3DPAN)]^C^C_3dpan
ID_3dzoom [_Button("3D Zoom", ICON_16_3DZOOM, ICON_16_3DZOOM)]^C^C_3dzoom
ID_3dorbit [_Button("3D Orbit", ICON_16_3DORBIT,

**TB_SHADING
ID_TbShading [_Toolbar("Shade", _Floating, _Hide, 100, 390, 1)]
ID_2doptim [_Button("2D Wireframe", ICON_16_2DOPTIM,

**TB_REFEDIT
ID_TbRefedit [_Toolbar("Refedit", _Floating, _Hide, 100, 170, 1)]
ID_RefEditor [_Button("Edit block or Xref", ICON_16_REFED_EDIT, ICON_16_REFED_EDIT)]^C^C_refedit;

TB_REFERENCE
ID_TbReferen [_Toolbar("Reference", _Floating, _Hide, 100, 370, 1)]
ID_Xref   [_Button("External Reference", ICON_16_XREATT, ICON_16_XREATT)]^C^C_xref
ID_Xattach [_Button("External Reference Attach", ICON_16_XATTAC, ICON_16_XATTAC)]^C^C_xattach
ID_Xclip   [_Button("External Reference Clip",ICON_16_XRECLI,ICON_16_XRECLI)]^C^C_xclip

TB_RENDER
ID_TbRender [_Toolbar("Render", _Floating, _Hide, 100, 230, 1)]
ID_Hide   [_Button("Hide", ICON_16_HIDE, ICON_16_HIDE)]^C^C_hide
ID_Render   [_Button("Render", ICON_16_RENDER,

TB_SOLIDS
ID_TbSolids [_Toolbar("Solids", _Floating, _Hide, 100, 250, 1)]
ID_Box   [_Button("Box", ICON_16_BOX, ICON_16_BOX)]^C^C_box
ID_Sphere   [_Button("Sphere", ICON_16_SPHERE, ICON_16_SPHERE)]^C^C_sphere
ID_Cylinder [_Button("Cylinder", ICON_16_CYLIND, ICON_16_CYLIND)]^C^C_cylinder

TB_SOLIDS2
ID_TbSolids2 [_Toolbar("Solids Editing", _Floating, _Hide, 100, 250, 1)]
ID_Union   [_Button("Union", ICON_16_UNION, ICON_16_UNION)]^C^C_union
ID_Subtract [_Button("Subtract", ICON_16_SUBTRA, ICON_16_SUBTRA)]^C^C_subtract
ID_Intersect [_Button("Intersect", ICON_16_INTERS,

TB_STANDARD
ID_TbStandar [_Toolbar("Standard Toolbar", _Top, _Show, 0, 0, 1)]
ID_New   [_Button("New", ICON_16_NEW, ICON_16_NEW)]^C^C_new
ID_Open   [_Button("Open", ICON_16_OPEN, ICON_16_OPEN)]^C^C_open
ID_Save   [_Button("Save", ICON_16_SAVE, ICON_16_SAVE)]^C^C_qsave

TB_SURFACES
ID_TbSurface [_Toolbar("Surfaces", _Floating, _Hide, 100, 290, 1)]
ID_Solid   [_Button("2D Solid", ICON_16_SOLID, ICON_16_SOLID)]^C^C_solid

Customizing the Standard AutoCAD Menu

ID_3dface   [_Button("3D Face", ICON_16_3DFACE,

**TB_UCS
ID_TbUcs   [_Toolbar("UCS", _Floating, _Hide, 100, 310, 1)]
ID_Ucs   [_Button("UCS", ICON_16_UCS, ICON_16_UCS)]^C^C_ucs
ID_Dducs   [_Button("Display UCS Dialog", ICON_16_DDUCS,
ICON_16_DDUCS)]^C^C_+ucsman 0

**TB_UCSII
ID_TbUcsName [_Toolbar("UCS II", _Floating, _Hide, 100, 300, 1)]
ID_Dducs   [_Button("Display UCS Dialog", ICON_16_DDUCS,
ICON_16_DDUCS)]^C^C_+ucsman 0
ID_UcsMove [_Button("Move UCS Origin", ICON_16_UCSMOVE,
ICON_16_UCSMOVE)]^C^C_ucs _move
ID_UcsCombo [_Control(_UCSManager)]

**TB_VIEWPOINT
ID_TbViewpoi [_Toolbar("View", _Floating, _Hide, 100, 330, 1)]
ID_Ddview   [_Button("Named Views", ICON_16_DDVIEW,
ICON_16_DDVIEW)]^C^C_view
ID_VpointTop [_Button("Top View", ICON_16_VIETOP,
ICON_16_VIETOP)]^C^C_-view
ID_VpointBot [_Button("Bottom View", ICON_16_VIEBOT,

ID_TbVpCreat [_Toolbar("Viewports", _Floating, _Hide, 100, 350, 1)]
ID_VpDialog [_Button("Display Viewports Dialog", ICON_16_VPDLG,
ICON_16_VPDLG)]^C^C_vports
ID_VpSingle [_Button("Single Viewport", ICON_16_VPONE,
ICON_16_VPONE)]$M=$(if,$(eq,$(getvar,tilemode),1),^C^C_-vports

**TB_WEB
ID_TbWeb   [_Toolbar("Web", _Floating, _Hide, 100, 380, 1)]
ID_HlnkBack [_Button("Go Back", ICON_16_HLNK_BACK,
ICON_16_HLNK_BACK)]'_hyperlinkBack
ID_HlnkFwd [_Button("Go Forward", ICON_16_HLNK_FWD,
ICON_16_HLNK_FWD)]'_hyperlinkFwd
ID_HlnkStop   [_Button("Stop Navigation", ICON_16_HLNK_STOP,

ICON_16_HLNK_STOP)]'_hyperlinkStop
ID_Browser [_Button("Browse the Web", ICON_16_WEB,

***image
**image_3DObjects
[3D Objects]
[acad(Box3d,Box3d)]^C^C_ai_box
[acad(Pyramid,Pyramid)]^C^C_ai_pyramid
[acad(Wedge,Wedge)]^C^C_ai_wedge

**image_poly
[Set Spline Fit Variables]
[acad(pm-quad,Quadric Fit Mesh)]'_surftype 5
[acad(pm-cubic,Cubic Fit Mesh)]'_surftype 6
[acad(pm-bezr,Bezier Fit Mesh)]'_surftype 8
[acad(pl-quad,Quadric Fit Pline)]'_splinetype 5
[acad(pl-cubic,Cubic Fit Pline)]'_splinetype 6

**image_vporti
[Tiled Viewport Layout]
[acad(vport-1,Single)]^C^C(ai_tiledvp 1 nil)
[acad(vport-3v,Three: Vertical)]^C^C(ai_tiledvp 3 "_V")
[acad(vport-3h,Three: Horizontal)]^C^C(ai_tiledvp 3 "_H")
[acad(vport-4,Four: Equal)]^C^C(ai_tiledvp 4 nil)

***SCREEN
**S
[AutoCAD]^C^C^P(ai_rootmenus) ^P
[* * * *]$S=ACAD.OSNAP
[FILE   ]$S=ACAD.01_FILE
[EDIT   ]$S=ACAD.02_EDIT
[VIEW 1]$S=ACAD.03_VIEW1
[VIEW 2]$S=ACAD.04_VIEW2
[INSERT]$S=ACAD.05_INSERT
[FORMAT]$S=ACAD.06_FORMAT
[TOOLS 1]$S=ACAD.07_TOOLS1
[TOOLS 2]$S=ACAD.08_TOOLS2
[DRAW 1]$S=ACAD.09_DRAW1
[DRAW 2]$S=ACAD.10_DRAW2

Customizing the Standard AutoCAD Menu

[DIMNSION]$S=ACAD.11_DIMENSION
[MODIFY1]$S=ACAD.12_MODIFY1

// The SNAP_TO menu can be called on high resolution displays to paint the snap
// options on the bottom of the screen menu.
// To add lines to the bottom of the Root menu delete the next line.

**SNAP_TO 28
[Endpoint]_endp
[Midpoint]_mid
[Intersec]_int
[App Int]_appint

**ASSIST 3
[Last   ]_l
[Previous]_p
[All   ]_all
[Cpolygon]_cp
[Wpolygon]_wp

**02_EDIT 3
[Undo   ]^C^C_u
[Redo   ]^C^C_redo
[Cut   ]^C^C_cutclip
[Copy   ]^C^C_copyclip
[CopyBase]^C^C_copybase
[CopyLink]^C^C_copylink
[Paste   ]^C^C_pasteclip
[PasteBlk]^C^C_pasteblock
[PasteOri]^C^C_pasteorig
[PasteSpe]^C^C_pastespec
[OLElinks]^C^C_olelinks

**03_VIEW1 3
[Redraw]'_redraw
[Redrawal]'_redrawall
[Regen   ]^C^C_regen
[Regenall]^C^C_regenall
[Zoom   ]'_zoom

[Pan   ]'_pan
[Dsviewer]'_dsviewer
[Tilemode]^C^C_tilemode
[Mspace]^C^C_mspace
[Pspace]^C^C_pspace

**04_VIEW2 3
[Hide   ]^C^C_hide
[Shade]^C^C_shademode

[Render]^C^C_render
[Scene]^C^C_scene
[Light]^C^C_light
[Mapping]^C^C_setuv
[Backgrnd]^C^C_background
[Fog   ]^C^C_fog

**06_FORMAT 3
[Layer]'_layer
[Color]'_color
[Linetype]'_linetype

[Style]'_style
[Ddim   ]'_dimstyle
[Ddptype]'_ddptype
[Mlstyle]^C^C_mlstyle

[Units]'_units
[Thicknes]'_thickness
[Limits]'_limits

[Rename]^C^C_rename
**ENDSCREEN

***TABLET1
**TABLET1STD
[A-1]\
[A-2]\
[A-3]\
[A-4]\
[A-5]\

Customizing the Standard AutoCAD Menu

10-17

```
          [A-6]\
          [A-7]\
          [A-8]\
          [A-9]\
          [A-10]\


***TABLET2
**TABLET2STD
//  Row J View
^C^C_regen
'_zoom _e
'_zoom _a
'_zoom _w
'_zoom _p
[Draw]\


***TABLET3
**TABLET3STD
//  Row 1
\
\
\
<<135
<<135
<<90
<<90
<<45
<<45


***TABLET4
**TABLET4STD
//  Row S


***HELPSTRINGS
ID_2doptim   [Set viewport to 2D wireframe:  SHADEMODE 2]
ID_3darray   [Creates a three-dimensional array:  3DARRAY]
ID_3dclip    [Starts 3DORBIT and opens the Adjust Clipping Planes window:  3DCLIP]
ID_3dclipbk  [Toggles the back clipping plane on or off in the 3D Orbit Adjust Clipping
Planes window:  DVIEW ALL CL B]
```

ID_3dclipfr [Toggles the front clipping plane on or off in the 3D Orbit Adjust Clipping Planes window: DVIEW ALL CL F]
ID_3dcorbit [Starts the 3DORBIT command with continuous orbit active in the 3D view: 3DCORBIT]
ID_3ddistanc [Starts the 3DORBIT command and makes objects appear closer or farther away: 3DDISTANCE]
ID_3dface   [Creates a three-dimensional face: 3DFACE]

***ACCELERATORS
// Bring up hyperlink dialog
ID_Hyperlink [CONTROL+"K"]
// Toggle Orthomode
[CONTROL+"L"]^O
// Next Viewport
[CONTROL+"R"]^V
ID_Copyclip [CONTROL+"C"]
ID_New   [CONTROL+"N"]
ID_Open   [CONTROL+"O"]
ID_Print   [CONTROL+"P"]
ID_Save   [CONTROL+"S"]
ID_Pasteclip [CONTROL+"V"]
ID_Cutclip   [CONTROL+"X"]
ID_Redo   [CONTROL+"Y"]
ID_U   [CONTROL+"Z"]
ID_Modify   [CONTROL+"1"]
ID_Content [CONTROL+"2"]
ID_dbConnect [CONTROL+"6"]
ID_VBARun   [ALT+"F8"]
ID_VBAIDE   [ALT+"F11"]

SUBMENUS

The number of items in a submenu section can be so large that they cannot be accommodated on one screen. For example, the maximum number of items that can be displayed on some of the display devices is 21. If the menu or the screen menu has more than 21 items, the menu items in excess of 21 are not displayed on the screen and cannot be accessed. Similarly, the maximum number of assignable blocks in tablet area 1 is 225. If the number of items in the TABLET1 section is more than 225, the menu items in excess of 225 are not assigned to any block on the template and are therefore inaccessible. The user can overcome this problem by using submenus that let the user define smaller groups of items within a menu section. When a submenu is selected, it loads the submenu items and displays them on the screen. In the case of a tablet menu the commands are assigned to different blocks on the template.

Customizing the Standard AutoCAD Menu 10-19

Submenu Definition

A submenu definition consists of two asterisk signs (**) followed by the name of the submenu. A menu can have any number of submenus and each submenu should have a unique name. The items that follow a submenu, up to the next section label or submenu label, belong to that submenu. The following is the format of a submenu label:

**Name

Where ** ----------------- Two asterisk signs (**) designate a submenu
Name ------------ Name of the submenu

Note
The submenu name can be up to 31 characters long.

The submenu name can consist of letters, digits, and special characters such as: $ (dollar), - (hyphen), and _ (underscore).

The submenu name should not have any embedded blanks.

The submenu names should be unique in a menu file.

Submenu Reference

The submenu reference is used to reference or load a submenu. It consists of a "$" sign followed by a letter that specifies the menu section. For example, the letter S specifies a screen menu, B specifies a button menu, I specifies an image tile menu, and Pn specifies a menu where n designates the number of the menu. Similarly, the letter that specifies a tablet menu section is Tn, where n designates the number of the tablet menu section. The menu section is followed by an "=" sign and the name of the submenu that the user wants to activate. The submenu name should be without "**". The following is the format of a submenu reference:

$Section=Submenu

Where $ ------------------- "$" sign
Section ---------- Menu section specifier
= ------------------ "=" sign
Submenu ------- Name of submenu

Example
$S=BLOCKS

Where S ------------------- S specifies screen menu
BLOCKS -------- Name of submenu

Loading Screen Menus

You can load a menu that is defined in the screen menu section from any menu by using the following load command:

$S=(name1) $S=(name2)

Where "name1" and "name2" are the names of the submenus

Example
$S=X $S=INSERT
    Where S ------------------- S specifies screen menu
           X ----------------- Submenu name defined in screen menu section
           INSERT --------- Submenu name defined in screen menu section

The first load command ($S=X) loads the submenu X that has been defined in the screen menu section of the menu file. The X submenu, defined in the screen menu section, contains 21 blank lines. When it is loaded, it clears the screen menu. The second load command ($S=INSERT) loads the submenu INSERT that has also been defined in the screen menu section of the menu file.

Loading Pull-down Menus

You can load a pull-down menu from any menu by using the following command:

$P(n)=(name) $P(n)=*
        Where "n" ranges from 1 to 10 (POP1 — POP10) and "name" is the name of the submenu defined in the pull-down menu.

Example
$P1=P1A $P1=*
        Where $P1=P1A ------- Loads the submenu P1A
               $P1=* ---------- Forces the display of new menu items

The first load command $P1=P1A loads the submenu P1A that has been defined in the POP1 section of the menu file. The second load command $P1=* is a special AutoCAD command that forces the new menu items to be displayed on the screen.

Loading Image Tile Menus

You can also load an image tile menu from any menu by using the following load command:

$I=(name) $I=*
        Where "name" is the name of the submenu defined in the image submenu

Example
$I=IMAGE1 $I=*
        Where $I=IMAGE1 --- Load the submenu IMAGE1
               $I=* ------------- To display the dialog box

This menu item consists of two load commands. The first load command $I=IMAGE1 loads the image submenu IMAGE1 that has been defined in the image tile menu section of the file. The second load command $I=* displays the new dialog box on the screen.

CUSTOMIZING TABLET AREA-1

The tablet menu has four sections: TABLET1, TABLET2, TABLET3, and TABLET4 (Figure 10-2). Tablet areab 1 of the template has 25 columns numbered 1 to 25 and 9 rows designated by letters A through I. The total number of blocks for tablet area 1 is 225 (25 x 9 = 225). You can utilize this area to customize AutoCAD's tablet menu by assigning commands or macros to different blocks. Figure 10-3 shows tablet area-1 of the standard AutoCAD template. Before making any changes or additions to the tablet menu it is very important to figure out the commands that you want to add to the tablet menu and determine their location on the template. A well thought-out tablet design will save a lot of revision time in the long run. Figure 10-3 shows a drawing that has 25 columns and 9 rows. You may draw a similar drawing and make several copies of it to arrange the commands and design the template.

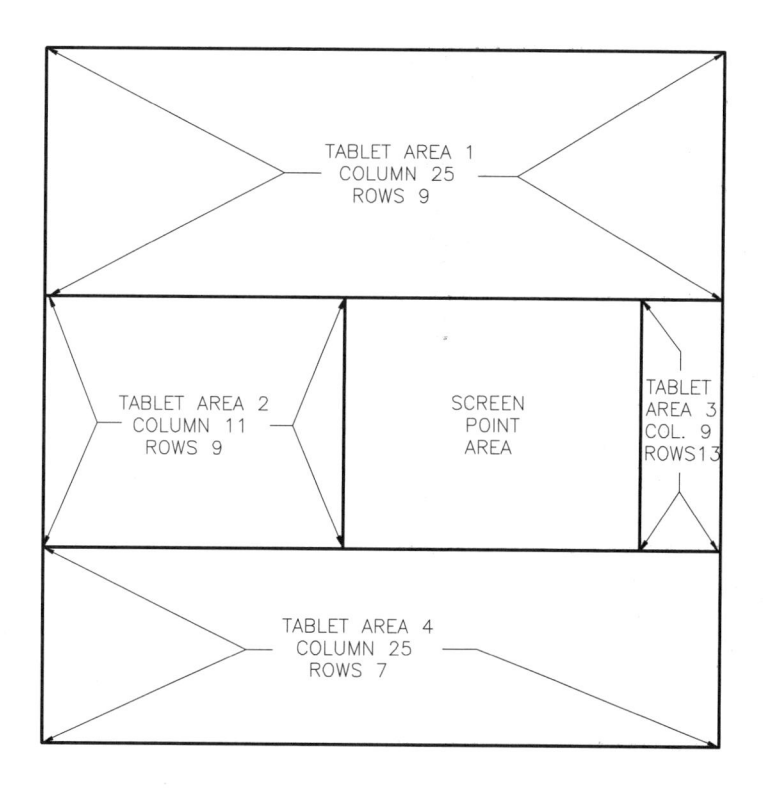

Figure 10-2 Four tablet areas of the AutoCAD template

Note
Before making any changes to the menu file make sure that the original file has been properly saved. You may also make a copy of the ACAD.MNU file and then make the changes in the new file.

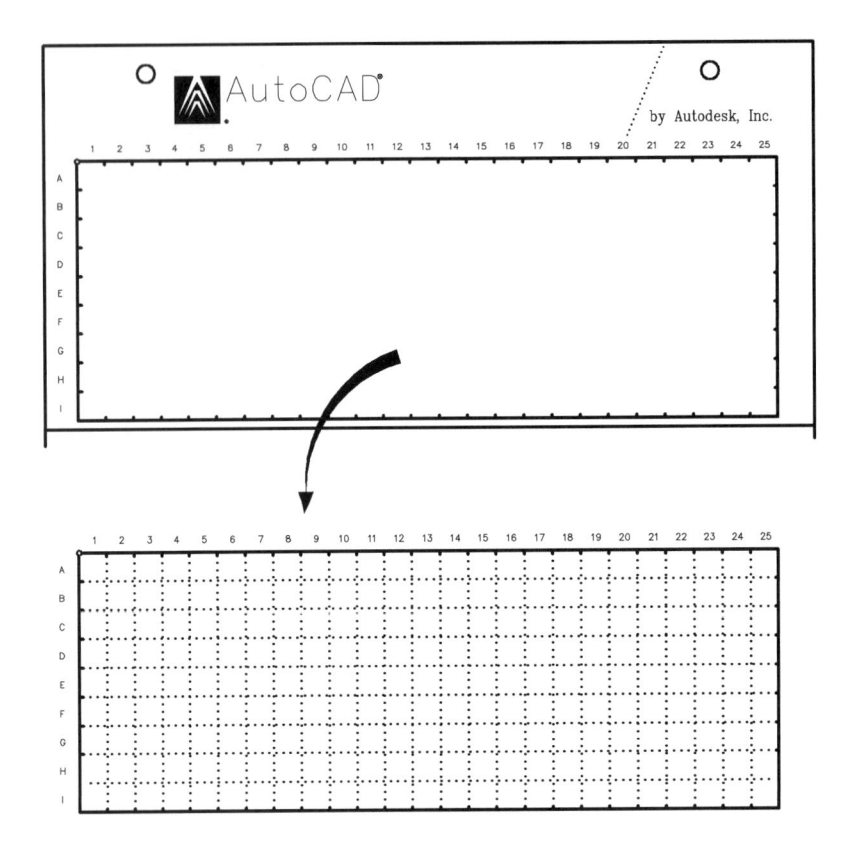

Figure 10-3 Tablet area-1 with 25 columns and 9 rows

Example 1

Add the commands shown in Figure 10-4 to the TABLET1 section of CUSTOM.MNU file. B10 through B25 are the names of the blocks. The user should be able to insert the blocks with X-scale and Y-scale factors of 1.0 and a rotation angle of 0.

Before making any changes in the file, you need to determine the location of each block on the template. For example, the WBlock B10 is to be assigned to a block that is in row B and column number 22. Similarly, WBlock B17 is in row C and column 25. The following table shows the location of WBlocks in tablet area 1 of the template:

WBLOCK IN NAME	ROW	COLUMN	LOCATION TABLET1
B10	B	22	B-22
B11	B	23	B-23

Customizing the Standard AutoCAD Menu

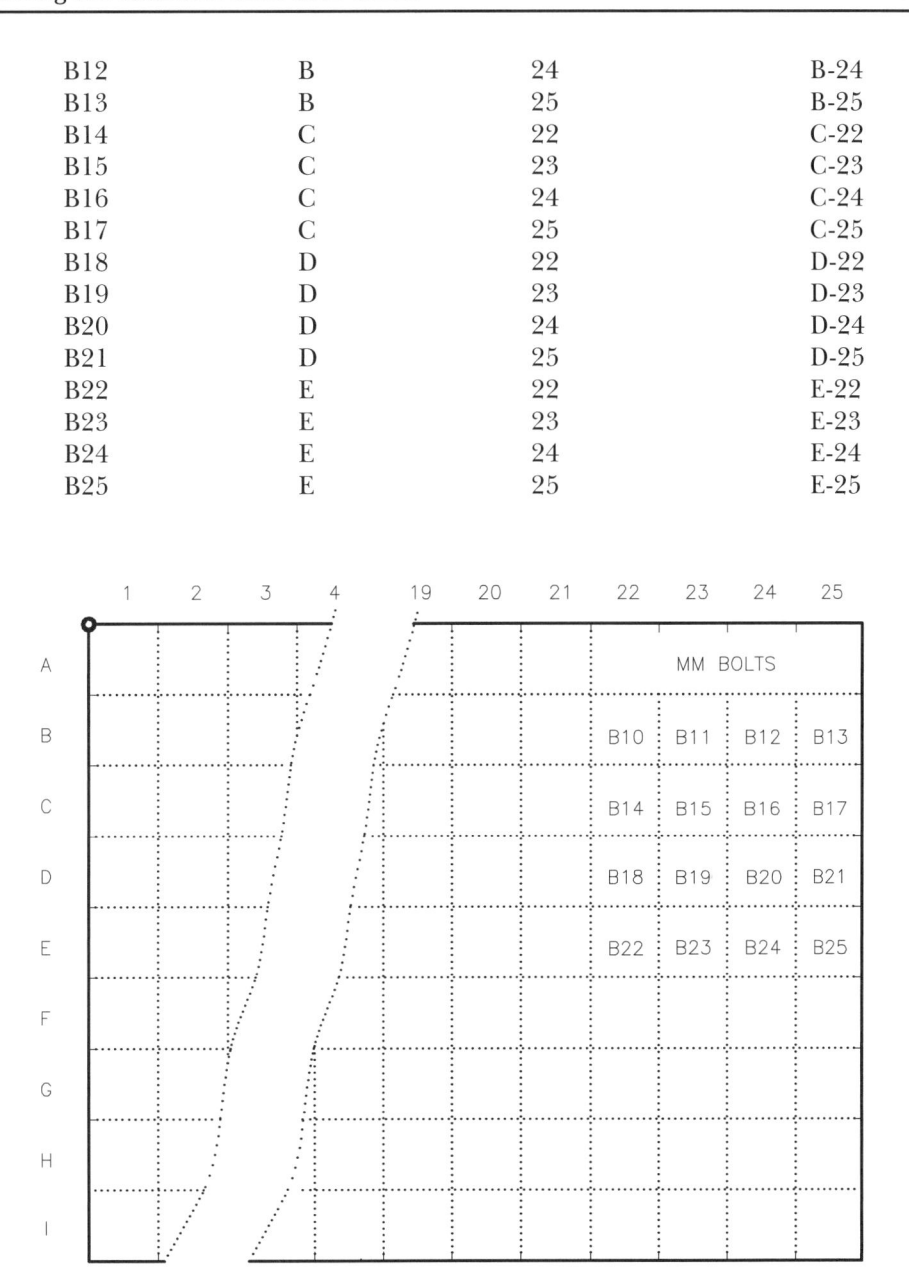

B12	B	24	B-24
B13	B	25	B-25
B14	C	22	C-22
B15	C	23	C-23
B16	C	24	C-24
B17	C	25	C-25
B18	D	22	D-22
B19	D	23	D-23
B20	D	24	D-24
B21	D	25	D-25
B22	E	22	E-22
B23	E	23	E-23
B24	E	24	E-24
B25	E	25	E-25

Figure 10-4 Commands assigned to tablet area 1

You can use any word processor or text editor to load and edit the menu file. After the CUSTOM.MNU file is loaded, search for ***TABLET1 section label. The letters and the numbers inside the brackets indicate the row and column. For example, in [A-1], A is for row number and 1 is for column number in the tablet area-1.

[A-1]

        Where **A** ------------------ Row
                    1 ------------------- Column number

The location of the first WBlock B10 in the menu file is B-22. Locate the menu item B-22 in the menu file and add the **INSERT** command to the menu item command definition. The following file is a partial listing of the TABLET1 section of the menu file after editing:

***TABLET1
**TABLET1STD
[A-1]
[A-2]
[A-3]
[A-4]
[A-5]
[A-6]
[A-7]
[A-8]
[A-9]
[A-10]
[A-11]
[A-12]
[A-13]
[A-14]
[A-15]
[A-16]
[A-17]
[A-18]
[A-19]
[A-20]
[A-21]
[A-22]
[A-23]
[A-24]
[A-25]
[B-1]
[B-2]
[B-3]
[B-4]
[B-5]
[B-6]
[B-7]
[B-8]
[B-9]
[B-10

Customizing the Standard AutoCAD Menu　　　　　　　　　　　　　　　　　10-25

```
[B-22]^C^CINSERT;B10;\1.0;1.0;0
[B-23]^C^CINSERT;B11;\1.0;1.0;0
[B-24]^C^CINSERT;B12;\1.0;1.0;0
[B-25]^C^CINSERT;B13;\1.0;1.0;0
[C-1]
   |
   |
   |
[C-21]
[C-22]^C^CINSERT;B14;\1.0;1.0;0
[C-23]^C^CINSERT;B15;\1.0;1.0;0
[C-24]^C^CINSERT;B16;\1.0;1.0;0
[C-25]^C^CINSERT;B17;\1.0;1.0;0
[D-1]
   |
   |
   |
[D-20]
[D-21]
[D-22]^C^CINSERT;B18;\1.0;1.0;0
[D-23]^C^CINSERT;B19;\1.0;1.0;0
[D-24]^C^CINSERT;B20;\1.0;1.0;0
[D-25]^C^CINSERT;B21;\1.0;1.0;0
   |
   |
   |
[E-21]
[E-22]^C^CINSERT;B22;\1.0;1.0;0
[E-23]^C^CINSERT;B23;\1.0;1.0;0
[E-24]^C^CINSERT;B24;\1.0;1.0;0
[E-25]^C^CINSERT;B25;\1.0;1.0;0
```

Note
To load the new menu CUSTOM.MNU, use AutoCAD's MENU command.

　　　　Command: **MENU**
　　　　Menu file name <ACAD>: **CUSTOM**

When you select the block insert command from the template, AutoCAD will prompt you to enter an insertion point. The X-scale and Y-scale factors and the rotation angle are already defined in the command definition.

^C^CINSERT;B22;\1.0;1.0;0
　　　　Where　**B22** --------------- Block or Wblock name
　　　　　　　　\--------------------- Pause for user input
　　　　　　　　1.0 ---------------- X-scale factor
　　　　　　　　; ------------------- Enter

     1.0 ---------------- Y-scale factor
     ; ------------------ Enter
     0 ------------------ Rotation angle

SUBMENUS

The number of items in the TABLET1STD section of the menu file can be so large that they may not be accommodated on one template. For example, the maximum number of assignable blocks in the standard AutoCAD tablet area-1 is 225. It can be re-configured up to 32,766 blocks. If the number of items in the standard TABLET1STD section is more than 225, the menu items in excess of 225 are not assigned to any block on the template and are therefore inaccessible. The user can overcome this problem by using submenus that let the user define any number of menu items in the tablet area-1 of the template. When the user selects the submenu, AutoCAD automatically loads the new submenu and assigns the commands to different blocks in template area-1. The format of the submenu reference is:

  $Section=Submenu
     Where **$** -------------------- "$" sign
         Section ---------- Menu section specifier
         = ------------------- "=" sign
         Submenu ------- Name of submenu

  Example
  $T1=TAB1
     Where **$T1** --------------- T1 specifies TABLET1 menu section
         Tab1 -------------- Name of submenu

Example 2

Edit the CUSTOM.MNU file to add the commands shown in Figure 10-5. Use the submenus and make a provision for swapping the submenus. When the user selects a block insert command from the tablet menu, AutoCAD should automatically load the corresponding screen menu.

In this example it is assumed that there is no room for adding more commands to tablet area-1; therefore, submenus have to be created to make room for the commands in excess of 225. Two submenus TABA and TABB have been defined in the TABLET1 section of the menu file CUSTOM.MNU. When you load the menu file CUSTOM.MNU, the first submenu TABA is automatically loaded and you can select the block insert commands from the template. You can load the submenu TABB by selecting the **Load TABB** block from the template. AutoCAD loads the submenu TABB and now you can select the commands from the new submenu. If you want to load the submenu TABA back, select the **Load TABA** from the template. The following file is a partial listing of the TABLET1 section of the menu file after inserting the new command definitions:

  ***TABLET1
  **TABLET1STD
  **TABA

Customizing the Standard AutoCAD Menu 10-27

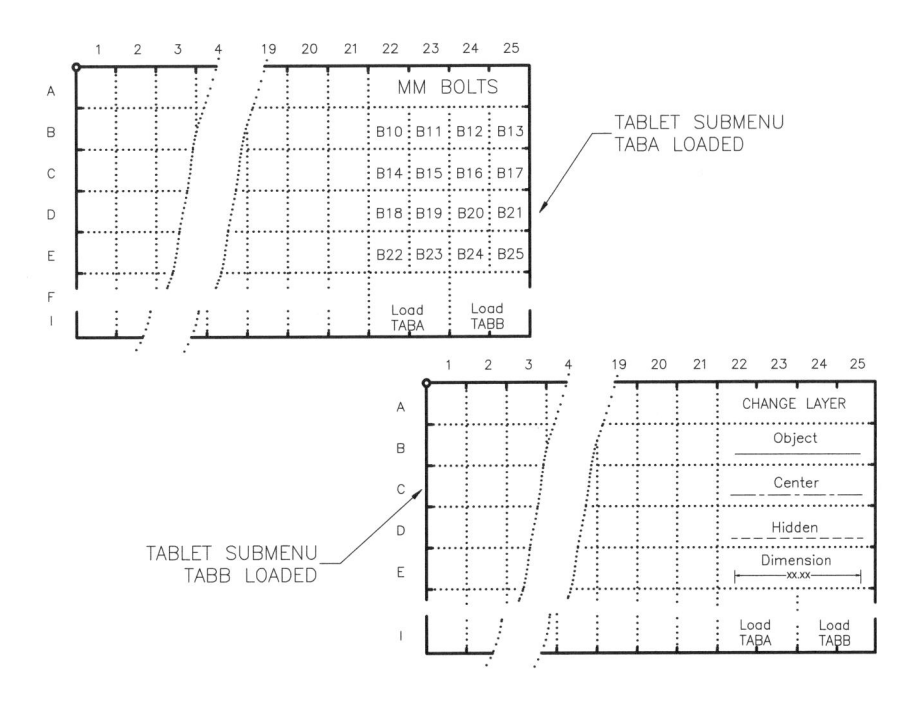

Figure 10-5 Commands assigned to tablet area-1

```
[A-1]
[A-2]
[A-3]
[A-4]
[A-5]
[A-6]
[A-7]


[B-18]
[B-19]
[B-20]
[B-21]
[B-22] ^C^C$S=X $S=INSBLK INSERT;B10;\1.0;1.0;0
[B-23] ^C^C$S=X $S=INSBLK INSERT;B11;\1.0;1.0;0
[B-24] ^C^C$S=X $S=INSBLK INSERT;B12;\1.0;1.0;0
[B-25] ^C^C$S=X $S=INSBLK INSERT;B13;\1.0;1.0;0
[C-1]
```

[C-21]
[C-22] ^C^C$S=X $S=INSBLK INSERT;B14;\1.0;1.0;0
[C-23] ^C^C$S=X $S=INSBLK INSERT;B15;\1.0;1.0;0
[C-24] ^C^C$S=X $S=INSBLK INSERT;B16;\1.0;1.0;0
[C-25] ^C^C$S=X S=INSBLK INSERT;B17;\1.0;1.0;0
[D-1]

[D-20]
[D-21]
[D-22] ^C^C$S=X $S=INSBLK INSERT;B18;\1.0;1.0;0
[D-23] ^C^C$S=X $S=INSBLK INSERT;B19;\1.0;1.0;0
[D-24] ^C^C$S=X $S=INSBLK INSERT;B20;\1.0;1.0;0
[D-25] ^C^C$S=X $S=INSBLK INSERT;B21;\1.0;1.0;0
[E-1]

[E-21]
[E-22] ^C^C$S=X $S=INSBLK INSERT;B22;\1.0;1.0;0
[E-23] ^C^C$S=X $S=INSBLK INSERT;B23;\1.0;1.0;0
[E-24] ^C^C$S=X $S=INSBLK INSERT;B24;\1.0;1.0;0
[E-25] ^C^C$S=X $S=INSBLK INSERT;B25;\1.0;1.0;0
[F-1]

[H-19]
[H-20]
[H-21]
[H-22] ^C^C$T1=TABA
[H-23] ^C^C$T1=TABA
[H-24] ^C^C$T1=TABB
[H-25] ^C^C$T1=TABB

**TABB
[A-1]
[A-2]
[A-3]
[A-4]
[A-5]
[A-6]
[A-7]
[A-8]

Customizing the Standard AutoCAD Menu

[B-18]
[B-19]
[B-20]
[B-21]
[B-22] ^C^CLAYER;SET;OBJECT;;
[B-23] ^C^CLAYER;SET;OBJECT;;
[B-24] ^C^CLAYER;SET;OBJECT;;
[B-25] ^C^CLAYER;SET;OBJECT;;
[C-1]

|

[C-21]
[C-22] ^C^CLAYER;SET;CENTER;;
[C-23] ^C^CLAYER;SET;CENTER;;
[C-24] ^C^CLAYER;SET;CENTER;;
[C-25] ^C^CLAYER;SET;CENTER;;
[D-1]

|

[D-20]
[D-21]
[D-22] ^C^CLAYER;SET;HIDDEN;;
[D-23] ^C^CLAYER;SET;HIDDEN;;
[D-24] ^C^CLAYER;SET;HIDDEN;;
[D-25] ^C^CLAYER;SET;HIDDEN;;
[E-1]

|

[E-21]
[E-22] ^C^CLAYER;SET;DIM;;
[E-23] ^C^CLAYER;SET;DIM;;
[E-24] ^C^CLAYER;SET;DIM;;
[E-25] ^C^CLAYER;SET;DIM;;
[F-1]

|

[H-19]
[H-20]
[H-21]
[H-22] ^C^C$T1=TABA
[H-23] ^C^C$T1=TABA
[H-24] ^C^C$T1=TABB
[H-25] ^C^C$T1=TABB

> **Note**
> *When you load the submenu TABB, the template overlay for tablet area-1 must be changed.*
>
> *The menu items defined in H-22, H-23, H-24, and H-25 load the submenus TABA and TABB respectively.*
>
> /^C^C$T1=TABA
>   Where ^C^C ---------- Cancels the existing command twice
> $T1=TABA ---- Loads tablet submenu TABA

CUSTOMIZING TABLET AREA-2

Tablet area-2 has 11 columns and 9 rows. The columns are numbered from 1 to 11 and the rows are designated by the letters J through R as shown in Figure 10-6. The total number of blocks in tablet area-2 is 99 (11 x 9 = 99). There are no empty blocks in this area like tablet area-1 and the commands that have been assigned to the blocks are defined in the TABLET2 menu sections of the standard menu file. You can change or delete the command definition assigned to these blocks. You can even delete the entire TABLET2 menu sections and write your own menu to best fit your needs. However, you must be careful in developing a new menu and you may need quite some time to come up with a reliable menu.

Exercise 1   *General*

Assign the following commands to the blocks in the tenth column of tablet area-2 as shown in Figure 10-6.

    LINE
    PLINE
    ARC,CSE
    ARC,SCE
    ARC,CSA
    CIRCLE-C,R
    CIRCLE-C,D
    CIRCLE-2P

CUSTOMIZING TABLET AREA-3

Tablet area-3 has 9 columns and 13 rows. In Figure 10-7 the columns are numbered from 1 to 9, and the rows are numbered from 1 to 13. The total number of blocks in tablet area-3 are 117 (13 x 9 = 117). The size of the blocks in this area is smaller than the blocks in other sections of the tablet template. Also, the blocks are rectangular in shape, whereas in other tablet areas the blocks are square. You can modify or delete the command definitions assigned to these blocks. You can even delete the entire TABLET3 menu sections and write a menu that best fits your needs. However, you have to be careful when you develop a new menu. The following example illustrates the editing process for adding commands to the TABLET3 section of the CUSTOM.MNU file.

Customizing the Standard AutoCAD Menu   10-31

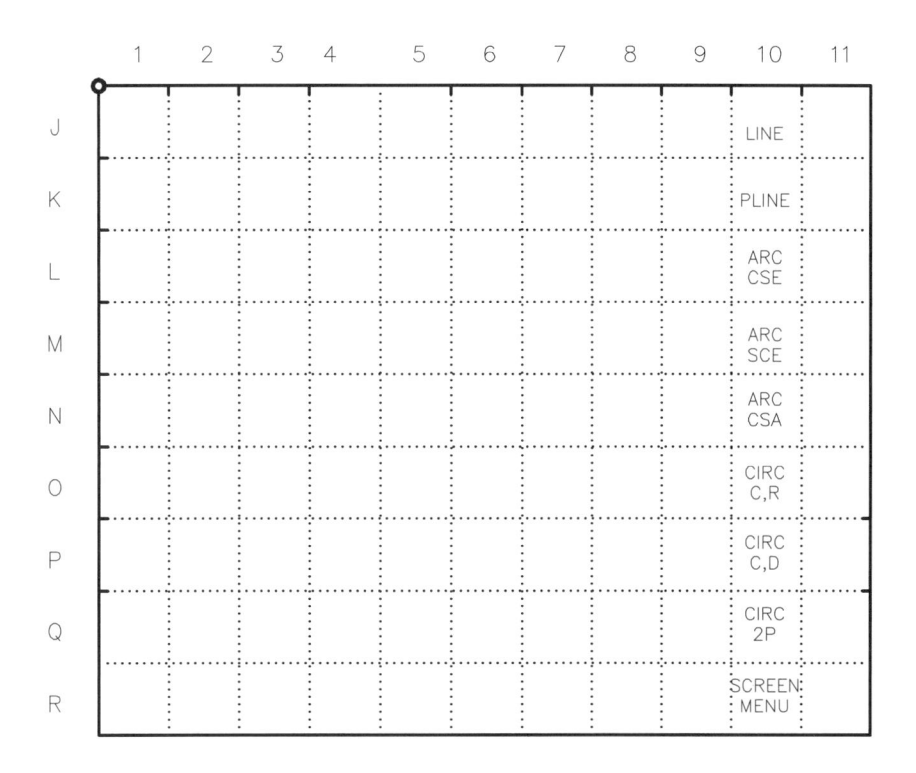

	1	2	3	4	5	6	7	8	9	10	11
J										LINE	
K										PLINE	
L										ARC CSE	
M										ARC SCE	
N										ARC CSA	
O										CIRC C,R	
P										CIRC C,D	
Q										CIRC 2P	
R										SCREEN MENU	

Figure 10-6 Commands assigned to tablet area 2

Example 3

Add the following angles to the TABLET3 section of CUSTOM.MNU file. The layout of tablet area 3 is shown in Figure 10-7.

ANGLES
30
120
210
330

Use your word processor or text editor to load CUSTOM.MNU file and search for the **TABLET3 section of the menu file. Locate the lines that you want to edit and then assign the required command definitions to these lines. The following file is a partial listing of the TABLET3 section of the CUSTOM.MNU file after editing:

**TABLET3
;

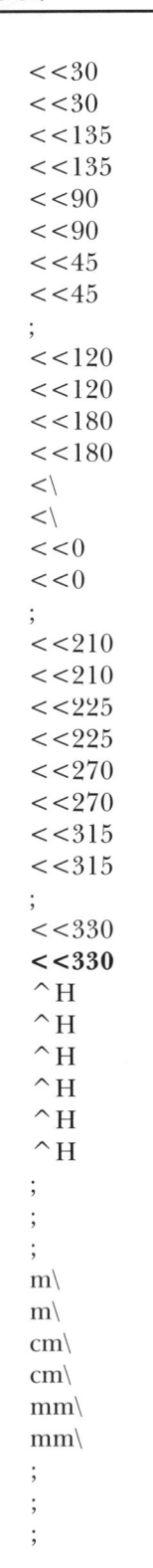

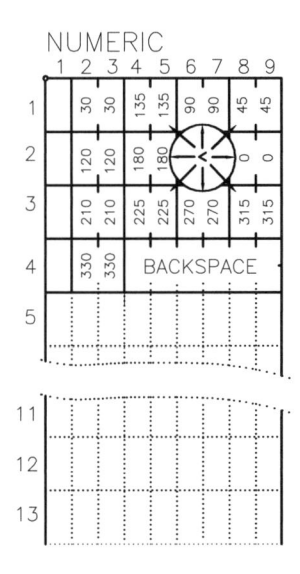

```
<<30
<<30
<<135
<<135
<<90
<<90
<<45
<<45
;
<<120
<<120
<<180
<<180
<\
<\
<<0
<<0
;
<<210
<<210
<<225
<<225
<<270
<<270
<<315
<<315
;
<<330
<<330
^H
^H
^H
^H
^H
^H
;
;
;
m\
m\
cm\
cm\
mm\
mm\
;
;
;
```

Figure 10-7 Angles assigned to tablet area-3

Customizing the Standard AutoCAD Menu

10-33

```
.\
.\
+\
+\
%%d\
%%d\
;
;
;
,\
,\
%%p\
%%p\
%%c\
%%c\
```

CUSTOMIZING TABLET AREA-4

Tablet area-4 has 25 columns and 7 rows. As shown in Figure 10-8, the columns are numbered from 1 to 25 and the rows are designated by the letters S through Y. The total number of blocks in tablet area-4 is 175 (7 * 25 = 175). You can modify or delete the command definition assigned to these blocks. You can even delete the entire TABLET4 menu sections and write a menu that best fits your needs. However, you have to be careful when you develop a new menu.

Exercise 2 *General*

Add the following commands to the TABLET4 section of CUSTOM.MNU file. The partial layout of tablet area-4 is shown in Figure 10-8.

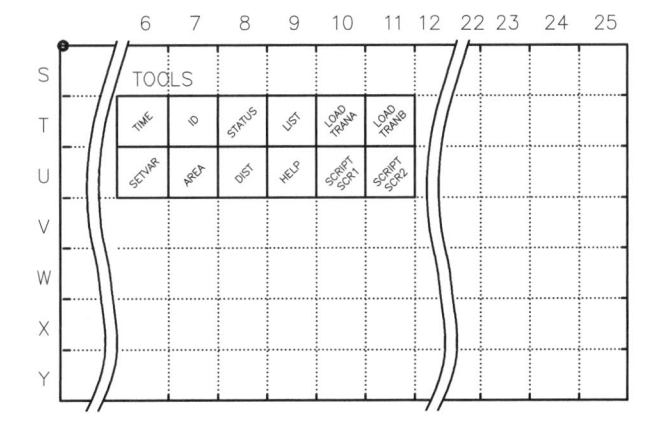

Figure 10-8 Commands assigned to tablet area-4

1. Load and run the AutoLISP routines TRANA.LSP and TRANB.LSP. (It is assumed that the files TRANA.LSP and TRANB.LSP are predefined.)

2. Run the script files SCR1 and SCR2.

Note
^C^C(LOAD "TRANA");TRANA   *In this menu, item (LOAD "TRANA") loads the AutoLISP file TRANA. The TRANA that is outside the parenthesis, after the semicolon, executes the function TRANA. Refer to Chapter 12 for detailed information.*

^C^C(LOAD "TRANA");TRANA
    Where   **LOAD** ------------ Loads AutoLISP program TRANA
                TRANA --------- Name of the AutoLISP function

^C^CSCRIPT;SCR1   *In this menu item SCR1 is the name of the script file, and **SCRIPT** is an AutoCAD command for running a script file.*

^C^CSCRIPT;SCR1
    Where   **SCRIPT** --------- AutoCAD's **SCRIPT** command
                SCR1 ------------- Name of the script file

CUSTOMIZING BUTTONS AND AUXILIARY MENUS

The standard AutoCAD menu file ACAD.MNU contains buttons and auxiliary menu sections. The buttons and auxiliary menu sections are identical and the first section has nine menu items. The following file is a listing of the buttons and auxiliary menu sections of the standard AutoCAD menu file, ACAD.MNU:

```
***BUTTONS1
$M=$(if,$(eq,$(substr,$(getvar,cmdnames),1,5),GRIP_),$P0=ACAD.GRIPS $P0=*);
$P0=SNAP $p0=*
^C^C
^B
^O
^G
^D
^E
^T


***BUTTONS2
$P0=SNAP $p0=*

***AUX1
$M=$(if,$(eq,$(substr,$(getvar,cmdnames),1,5),GRIP_),$P0=ACAD.GRIPS $P0=*);
$P0=SNAP $p0=*
^C^C
^B
```

Customizing the Standard AutoCAD Menu 10-35

```
^O
^G
^D
^E
^T

***AUX2
$p0=SNAP $p0=*
```

> **Note**
> *The following table shows the function of the menu items that are defined in the BUTTONS section of the CUSTOM.MNU file.*

MENU ITEM	FUNCTION
***BUTTONS	Section label
$p0=*	Displays the cursor menu
^C^C	Cancels the existing command twice
^B	Snap On/off (CTRL B)
^O	Ortho On/off (CTRL O)
^G	Grid On/off (CTRL G)
^D	Coordinate Dial On/off (CTRL D)
^E	Isoplane (CTRL E)
^T	Tablet On/off (CTRL T)

Like any other menu you can make changes in the buttons menu to assign different commands to the buttons of the pointing device. The pointing devices generally come in a 4-button or 10-button configuration. The first button is always the pick button and cannot be used for any other purpose. AutoCAD commands can be assigned to the remaining buttons. The following example describes the editing procedure for the BUTTONS section of CUSTOM.MNU file (CUSTOM.MNU file is a copy of ACAD.MNU file).

Example 4

Change the buttons sections of CUSTOM.MNU to assign the following commands to the four buttons of the pointing device (Figure 10-9):

1. PICK
2. Enter
3. ZOOM Win
4. ZOOM Prev

The ZOOM Window command assigned to button number 3 should automatically zoom in an area that is 2 units from the selected point.

Before making any changes it is very important to know the commands and the prompt entries that are associated with the commands. Use your word processor to load the CUSTOM.MNU file and search for the ***BUTTONS1 section. The lines that follow the ***BUTTONS1 section are the lines that need to be edited to assign new commands to the buttons of the pointing device. The following file is a listing of the button menu section of the CUSTOM.MNU file after making the required changes:

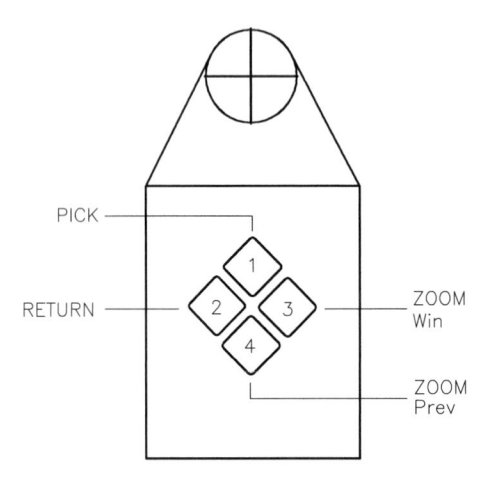

Figure 10-9 Commands assigned to 4 buttons of a pointing device

***BUTTONS1	1
;	2
'ZOOM;WIN;\@2,2	3
'ZOOM;PRE	4
^B	5
^O	6
^G	7
^D	8
^E	9
^T	10

Line 1
***BUTTONS1
This is the section label for the BUTTONS1 menu section.

Line 2
;
The semicolon (;) causes an ENTER. It is like pressing the Enter key from the keyboard or template.

Line 3
'ZOOM;WIN;\@2,2
When you select this menu item, it will zoom as shown in Figure 10-10. The first point you select becomes the first corner of the zoom window and the other corner of the zoom window is 2.0,2.0 units from the selected point.

'ZOOM;WIN;\@2,2
Where ' -------------------- Makes the ZOOM command transparent
\ -------------------- Selected point becomes first corner
2,2 ---------------- Other corner

Customizing the Standard AutoCAD Menu

Before selecting the third button of the pointing device for this command, move the screen crosshairs to the point where you want to zoom and then click the button. AutoCAD will zoom in the area since the two corners of the window are defined in the menu item. The single quote (') in front of **ZOOM** makes the **ZOOM Window** command transparent.

Line 4
'ZOOM;PRE
This menu item defines the **ZOOM** command with the Previous option. The single quote (') in front of the **ZOOM** command makes the **ZOOM Previous** command transparent.

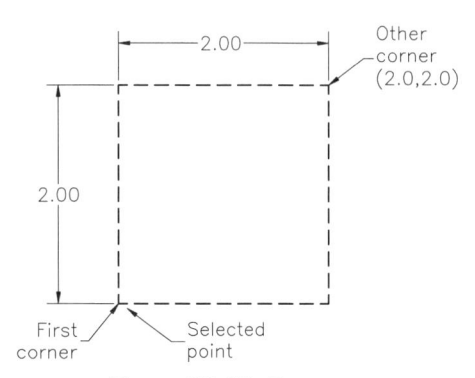

Figure 10-10 Zoom area

Note
*The commands on the first three lines (2 - 4), excluding the ***BUTTONS1 line, will be assigned to the second, third, and fourth button of the four-button pointing device. The remaining items will be ignored and do not affect the commands that are assigned to other buttons. Therefore, you can leave them in the file.*

Example 5

Edit the buttons and the auxiliary menu sections of the menu file CUSTOM.MNU to add the following AutoCAD commands. The pointing device has 10 buttons (Figure 10-11) and button number 1 is used as a pick button. The blocks should be inserted with a scale factor of 1.00 and a rotation angle of 0 degrees. (The CUSTOM.MNU file is a copy of the ACAD.MNU menu file.)

1. PICK BUTTON
2. ENTER
3. CANCEL
4. OSNAPS
5. INSERT B1
6. INSERT B2
7. INSERT B3
8. **ZOOM Window**
9. **ZOOM All**
10. **ZOOM Previous**

Note
B1, B2, B3 are the names of the blocks or wblocks that have already been created.

It is assumed that the cursor menu POP0 has already been defined in the menu file.

*Use the transparent **ZOOM** command for **ZOOM Previous** and **ZOOM Window**.*

The following file is a listing of the buttons menu section of the CUSTOM.MNU file after making the required changes:

```
***BUTTONS                                                          1
;                                                                   2
```

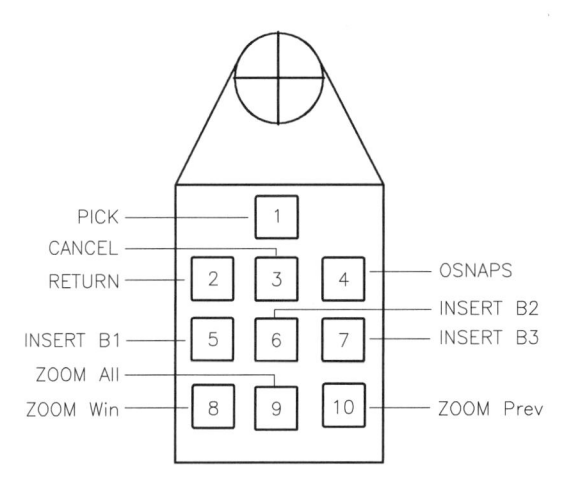

Figure 10-11 *Commands assigned to different buttons of a 10-button pointing device*

```
  ^C^C                                   3
  $P0=*                                  4
  ^C^CINSERT;B1;\1.0;1.0;0               5
  ^C^CINSERT;B2;\1.0;1.0;0               6
  ^C^CINSERT;B3;\1.0;1.0;0               7
  'ZOOM;Win                              8
  ^C^CZOOM;All                           9
  'ZOOM;Prev                            10
  ***AUX1                               11
  ;                                     12
  ^C^C                                  13
  $P0=*                                 14
  ^C^CINSERT;B1;\1.0;1.0;0              15
  ^C^CINSERT;B2;\1.0;1.0;01             16
  ^C^CINSERT;B3;\1.0;1.0;0              17
  'ZOOM;Win                             18
  ^C^CZOOM;All                          19
  'ZOOM;Prev                            20
```

Line 3
^C^C
This command definition is assigned to button number 3. It cancels the existing command twice.

Line 4
$P0=*
In this menu item $P0=* is a special command that you can use to access the cursor menu. When you select this item, AutoCAD will display the cursor menu on the screen near the cursor location. The cursor menu that contains the object snap mode commands is defined in the POP0 menu section of ACAD.MNU file. This command definition is assigned to button

Customizing the Standard AutoCAD Menu 10-39

number 4 of the pointing device.

> **$S=X $S=OSNAPS1**
> Where = ------------------ Loads submenu X defined in screen menu section
> **OSNAPS1** ------ Loads submenu OSNAPS1 defined in screen menu section

Line 5
^C^CINSERT;B1;\1.0;1.0;0
In this menu item ^C^C cancels the existing command twice. **INSERT** is an AutoCAD command that can be used to insert a Block or Wblock. B1 is the name of the block and the backslash (\) pauses for the user input. In this menu item, it is the insertion point of the block. The first 1.0 is the X-scale factor and the second 1.0 is the Y-scale factor of the block. The 0 at the end is for the rotation angle of the block.

> **^C^CINSERT;B1;\1.0;1.0;0**
> Where **INSERT** --------- AutoCAD command
> **B1** ------------------ Block name
> **** -------------------- Pause for block insertion point
> **1.0** ---------------- X-scale factor
> **1.0** ---------------- Y-scale factor
> **;** ------------------- ENTER
> **;** ------------------- ENTER
> **0** ------------------- Rotation angle

Line 9
^C^CZOOM;All
The command definition of this menu item is assigned to button number 9 of the pointing device. If you select this key, it will enter a **ZOOM All** command.

> **^C^CZOOM;All**
> Where **ZOOM** ----------- AutoCAD's **ZOOM** command
> **;** ------------------- ENTER
> **All** ---------------- All option of **ZOOM** command

Line 10
'ZOOM;Prev
This menu item defines the transparent **ZOOM Previous** command. It is assigned to button number 10 of the pointing device.

> **'ZOOM;Prev**
> Where **'** -------------------- Single quote makes **ZOOM Previous** command transparent
> **Prev** -------------------- **Previous** option of **ZOOM** command

Line 11
***AUX1
In this line AUX1 is the section label for auxiliary menu section. Lines 12 through 20 are defined in this section.

CUSTOMIZING PULL-DOWN AND SHORTCUT MENUS

The pull-down and shortcut menus are a part of the standard AutoCAD menu file (ACAD.MNU) that comes with the AutoCAD software package. The ACAD.MNU file is automatically loaded when you get into the Drawing Editor, provided the standard configuration of AutoCAD has not been changed. In addition to POP0, the menu can have a maximum of 499 sections defined as POP1, POP2, POP3 — — — POP499. The standard AutoCAD menu uses POP0 and POP500 through POP999 for shortcut menus. The shortcut menu can be loaded and displayed by using the command $P0=*.

CASCADING SUBMENUS IN MENUS

The cascading feature of AutoCAD allows the pull-down and cursor menus to be displayed in a hierarchical order that makes it easier to select submenus. To use the cascading feature in the pull-down and cursor menus, AutoCAD has provided some special characters. For example, the characters -> defines a cascaded submenu and <- designates the last item in the menu. The following table lists some of the frequently used characters that can be used with the pull-down or cursor menus:

Character	Character Description
--	The item label consisting of two hyphens automatically expands to fill the entire width of the pull-down menu. Example: [--]
+	Used to continue the menu item to the next line. This character has to be the last character of the menu item. Example: [Triang:]^C^CLine;1,1;+ 3,1;2,2;
->	This label character defines a cascaded submenu and it must precede the name of the submenu. Example: [->Draw]
<-	This label character designates the last item of the cascaded pull-down or cursor menu. The character must precede the label item. Example: [<-CIRCLE 3P]^C^CCIRCLE;3P
<-<-...	This label character designates the last item of the pull-down or cursor menu and also terminates the parent menu. The character must precede the label item. Example: [<-<-Center Mark]^C^C_dim;_center
$(	This label character can be used with the pull-down and cursor menus to evaluate a DIESEL expression. The character must precede the label item. Example: "$(if,$(getvar,orthomode),Ortho)"
~	This item grays-out the label item. The character must precede the item. Example: [~—]

Customizing the Standard AutoCAD Menu   10-41

The length of each menu bar depends on the Windows settings. For eample, a display devices that provides space for a maximum of 80 characters, if there are 16 menus, the length of each menu title should be up to 5 characters long. If the combined length of all menu bar titles exceeds 80 characters, AutoCAD will automatically truncate the characters from the longest menu title until it fits all menu titles in the menu bar. The following is a list of some additional features of the menu:

1. The section labels of the menus are ***POP1 through ***POP16. The menu bar titles are displayed in the menu bar.

2. The menus can be accessed by selecting the menu title from the menu bar at the top of the screen.

3. A maximum of 999 menu items can be defined in the menu. This includes the items that are defined in the submenus. The menu items in excess of 999 are ignored.

4. The number of menu items that can be displayed on the screen depends on the display device that you are using. If the cursor or the menu contains more items than what can be accommodated on the screen, the excess items are truncated. For example, if your system can display 21 menu items, then the menu items in excess of 21 are automatically truncated.

5. If AutoCAD is not configured to show the status line, the menus, cursor menus, and the menu bar are automatically disabled.

Example 6

Edit the POP4 section of the menu to add a new insert command with the label NEW-INSERT (Figure 10-12). When the user selects NEW-INSERT from the POP4 menu, it should load and display a cascading submenu that contains the commands for inserting the following blocks:

      INSERT-BLOCKS ———- Cascading submenu title
      DOOR1
      DOOR2
      - - - - - - - -
      WINDOW1
      WINDOW2

Note
The doors and windows are saved as WBLOCKS or individual drawings in the SYMBOLS subdirectory on the D drive.

The user should be able to insert the block at the selected point with X-scale and Y-scale factors of 1.25 and a rotation angle of 0.

Do not edit the ACAD.MNU file. Make a copy of the ACAD.MNU file and then edit the new file (CUSTOM.MNU).

Use your word processor to load the menu file CUSTOM.MNU and search for the **INSERT** command in the menu section ***POP4. Now, insert a line that contains the definition of the INSERT-BLOCKS cascading submenu as shown in Figure 10-12.

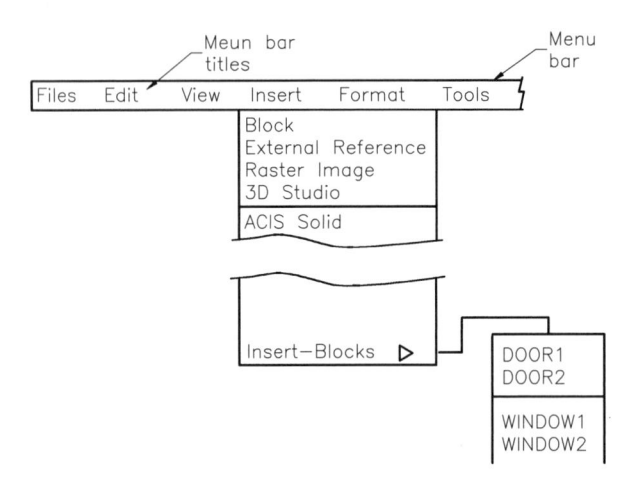

Figure 10-12 POP4 section of menu

The following file is a partial listing of CUSTOM.MNU file after editing the POP4 section of the menu and adding a new submenu:

```
***POP1
**FILE
ID_MnFile    [&File]
ID_New       [&New...\tCtrl+N]^C^C_new
ID_Open      [&Open...\tCtrl+O]^C^C_open
ID_FILE_CLOSE [&Close]
ID_PartialOp [$(if,$(eq,$(getvar,fullopen),0),,~)Pa&rtial Load]^C^C_partiaload
ID_Save      [&Save\tCtrl+S]^C^C_qsave
ID_Saveas    [Save &As...]^C^C_saveas
ID_Export    [&Export...]^C^C_export
ID_PlotSetup [Pa&ge Setup...]^C^C_pagesetup



***POP2
**EDIT
ID_MnEdit    [&Edit]
ID_U         [&Undo\tCtrl+Z]^C^C_u
ID_Redo      [&Redo\tCtrl+Y]^C^C_redo
ID_Cutclip   [Cu&t\tCtrl+X]^C^C_cutclip
ID_Copyclip  [&Copy\tCtrl+C]^C^C_copyclip
ID_Copybase  [Copy with &Base Point]^C^C_copybase
ID_Copylink  [Copy &Link]^C^C_copylink
```

Customizing the Standard AutoCAD Menu 10-43

***POP3
**VIEW
ID_MnView   [&View]
ID_Redrawall [&Redraw]'_redrawall
ID_Regen   [Re&gen]^C^C_regen
ID_Regenall [Regen &All]^C^C_regenall
ID_MnZoom   [->&Zoom]
ID_ZoomRealt   [&Realtime]'_zoom ;
ID_ZoomPrevi   [&Previous]'_zoom _p
ID_ZoomWindo   [&Window]'_zoom _w
ID_ZoomDynam   [&Dynamic]'_zoom _d
ID_ZoomScale   [&Scale]'_zoom _s
ID_ZoomCente   [&Center]'_zoom _c

***POP4
**INSERT
ID_MnInsert [&Insert]
ID_Ddinsert [&Block...]^C^C_insert
ID_Xattach [E&xternal Reference...]^C^C_xattach
ID_Imageatta [Raster &Image...]^C^C_imageattach

[->Insert-Blocks]
 [Door1]^C^C_insert;d:/symbols/door1;\1.25;1.25;0
 [Door2]^C^C_insert;d:/symbols/door2;\1.25;1.25;0
 [--]
[Window1]^C^C_inserd;d:/symbols/window1;\1.25;1.25;0
[<-Window2]^C^C_insert;d:/symbols/window2;\1.25;1.25;0

Note
[DOOR1]^C^CINSERT;D:/SYMBOLS/DOOR1;\1.25;1.25;

When you define the search path in the menu file, replace the backslashes (\) with forward slashes (/). To load the file DOOR1 from the SYMBOLS subdirectory in the D drive, the search path normally will be defined as D:\SYMBOLS\DOOR1. But the same statement in a menu file will be defined as D:/SYMBOLS/DOOR1. In the menu file, backslash is used for user input.

[->Insert-Blocks]

In this menu item, the character -> defines a cascading submenu. When you select this item from the POP3 menu, AutoCAD will display the menu on the side of the menu.

[<-Window2]^C^C_insert;d:/symbols/window2;\1.25;1.25;0
In this menu item, the character <- defines the last menu item in the cascading submenu.

SHORTCUT MENUS

The shortcut menus are similar to the pull-down menus, except that the shortcut menu can contain only 499 items compared to 999 items in the pull-down menu. The section label of the cursor menu must be ***POP0 or POP500 through POP999. The shortcut menus are displayed near or at the cursor location. Therefore, it can be used to provide a convenient and quick access to some of the frequently used commands. The following is a list of some of the features of the cursor menu:

1. The section label of the cursor menu is ***POP0 and POP500 through POP999. The menu bar title defined under these section labels are not displayed in the menu bar.

2. On most systems, the menu bar title is not displayed at the top of the cursor menu. However, for compatibility reasons it is recommended to give a dummy menu bar title.

3. The POP0 menu can be accessed through the **$P0=*** menu command. The shortcut menus POP500 through POP999 must be referenced by their alias names. The reserved alias names for AutoCAD use are GRIPS, CMDEFAULT, CMEDIT, and CMCOMMAND. For example, to reference POP500 for grips, use the **GRIPS command line under POP5000. This command can be issued by a menu item in another menu, such as the button menu, auxiliary menu, or the screen menu. The command can also be issued from an AutoLISP or ADS program.

4. A maximum of 499 menu items can be defined in the cursor menu. This includes the items that are defined in the cursor submenus. The menu items in excess of 499 are ignored.

5. The number of menu items that can be displayed on the screen depends on the system that you are using. If the cursor or the menu contains more items than what can be accommodated on the screen, the excess items are truncated. For example, if your system can display 21 menu items, then the menu items in excess of 21 are automatically truncated.

SUBMENUS

The number of items in a menu file can be so large that the items cannot be accommodated on one screen. For example, the maximum number of items that can be displayed on some of the display devices is 21. If the menu has more than 21 items, the menu items in excess of 21 are not displayed on the screen and cannot be accessed. The user can overcome this problem by using cascading menus or by swapping the submenus that let the user define smaller groups of items within a menu section. When a submenu is selected, it loads the submenu items and displays them on the screen.

Swapping Menus

The menus that use AutoCAD's cascading feature are the most efficient and easy to write. The submenus follow a logical pattern that are easy to load and use without causing any confusion. It is strongly recommended to use the cascading menus whenever you need to write the pull-down or cursor menus. However, AutoCAD provides the option to swap the submenus in the menus. These menus can sometimes cause distraction, because the original menu is completely replaced by the submenu when swapping the menus.

$Section=Submenu

Where **$** ------------------ "$" sign
Section ---------- Menu section specifier
= ------------------ "=" sign
Submenu ------- Name of submenu

Example
$P1=P1A

Where **$P1** --------------- P1 specifies menu section 1
P1A --------------- Name of submenu

CUSTOMIZING IMAGE TILE MENUS

The image tile menus are extremely useful for inserting blocks and selecting a hatch pattern or a text font. You can also use the image tile menus to load an AutoLISP routine or a pre-defined macro. Therefore, the image tile menu is a powerful tool for customizing AutoCAD.

The image tile menus can be accessed from the pull-down, tablet, button, or screen menu. However, the image tile menus cannot be loaded by entering the command from the keyboard.

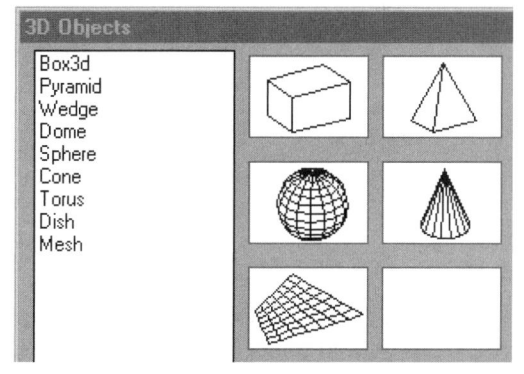

When the user selects an image, a dialog box is displayed on the screen that contains **twenty images**. The names of the slide files associated with the images appear on the left side of the dialog box with the scrolling bar that can be used to scroll the filenames. The title of the image tile menu is displayed at the top of the dialog box. When you activate the image tile menu, an arrow appears on the screen that can be moved to select any image. You can select an image by selecting the slide file name from the dialog box and then choosing the **OK** button from the dialog box or double-clicking on the slide file name.

When you select the slide file, AutoCAD highlights the corresponding image by drawing a rectangle around the image. You can also select an image by moving the arrow in the desired image and then pressing the pick button of the pointing device. The corresponding slide file name will be automatically highlighted and if you choose the **OK** button or double click on the image, the command associated with that menu item will be executed. You can cancel an image tile menu by pressing the CTRL and C keys or the ESCAPE key on the keyboard, or selecting an image from the dialog box.

An image tile menu will work only if the system is configured so that the status line is not disabled. Otherwise, the image tile menus or the menus cannot be used. The image tile menu consists of a section label ***IMAGE. There is only one image tile menu section in a file and all the image tile menus are defined in this section.

***IMAGE
>   Where *** ---------------- Three asterisks (***) designate a section label
>   IMAGE ---------- Section label for an image

SUBMENUS

You can define an unlimited number of menu items in the image tile menu, but only 20 images will be displayed at a time. If the number of items exceeds 20, you can use the **Next** or **Previous buttons** of the dialog box to page through different pages of images. You can also define submenus that let the user define smaller groups of items within an image tile menu section. When a submenu is selected, it loads the submenu items and displays them on the screen.

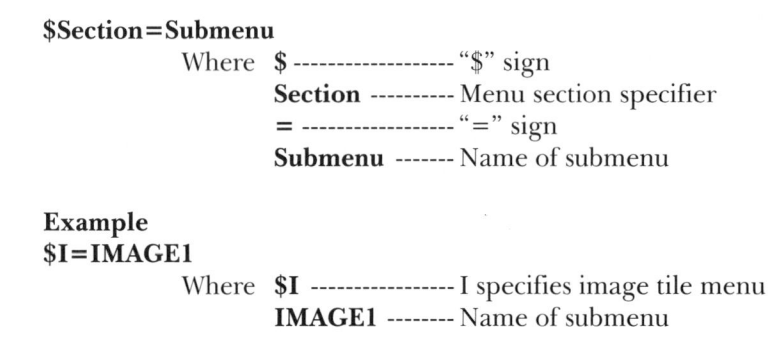

> $Section=Submenu
>   Where $ ------------------- "$" sign
>   Section ---------- Menu section specifier
>   = ------------------ "=" sign
>   Submenu ------- Name of submenu
>
> Example
> $I=IMAGE1
>   Where $I ----------------- I specifies image tile menu
>   IMAGE1 -------- Name of submenu

IMAGE TILE MENU ITEM LABELS

Like screen and pull-down menus, you can use the menu item labels in the pull-down menus. However, the menu item labels in the image tile menus use different formats and each format performs a particular function in the image tile menu. The menu item labels appear in the slide list box of the dialog box. The maximum number of characters that can be displayed in this box is 17. The characters in excess of 17 are not displayed in the list box. However, it does not affect the command that is defined with the menu item. The following are the different formats of the menu item labels:

[slidename]
In this menu item label format, **slidename** is the name of the slide that is displayed in the image. This name (slidename) is also displayed in the list box of the corresponding dialog box.

[slidename,label]
In this menu item label format, **slidename** is the name of the slide that is displayed in the image. However, unlike the previous format, the **slidename** is not displayed in the list box. It is the label text that is displayed in the list box. For example, if the menu item label is **[BOLT1,1/2-24UNC-3LG]**, **BOLT1** is the name of the slide and **1/2-24UNC-3LG** is the label that will be displayed in the list box.

[slidelib(slidename)]
In this menu item label format, **slidename** is the name of the slide in the slide library file

Customizing the Standard AutoCAD Menu   10-47

slidelib. The slide (slidename) is displayed in the image and the slide filename (slidename) is also displayed in the list box of the corresponding dialog box.

[slidelib(slidename,label)]
In this menu item label format, **slidename** is the name of the slide in the slide library file **slidelib**. The slide (slidename) is displayed in the image and the **label** text is displayed in the list box of the corresponding dialog box.

[blank]
This menu item will draw a line that extends through the width of the list box. It also displays a blank image in the dialog box.

[label]
If the **label** text is preceded by a space, AutoCAD does not look for a slide. The **label** text is displayed in the list box only. For example, if the menu item label is [EXIT]^C, the label text (EXIT) will be displayed in the list box. If you select this item, the cancel command (^C) defined with the item will be executed. The **label** text is not displayed in the image of the dialog box.

Example 7

Write an image tile menu that can be accessed from POP12 for inserting the following blocks (see Figure 10-13):

    BL1   BL4
    BL2   BL5
    BL3   BL6

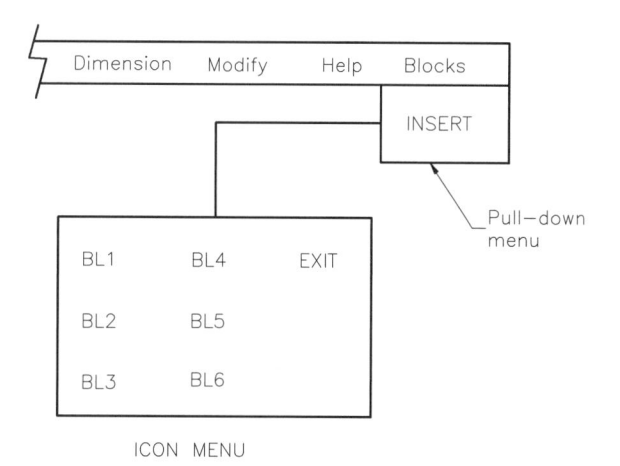

Figure 10-13 Commands defined in the image tile menu

Note
It is assumed that the slides have already been created and the names of the slides are the same as blocks.

Do not edit the ACAD.MNU file. Make a copy of the ACAD.MNU file and then edit the new file (CUSTOM.MNU).

Use your word processor to load the CUSTOM.MNU file and search for the ***IMAGE section label. The standard ACAD.MNU file does not have the POP12 section in the pull-down menu. Therefore, you can define POP12 just before the ***IMAGE section label. Similarly, you can define the INSTBLK image tile menu after the image section label ***IMAGE. The following file is a partial listing of the CUSTOM.MNU file after editing the file.

```
***POP12
[BLOCKS]
[INSERT]^C^C$I=INSTBLK $I=*

***image
**INSTBLK
[INSERT CUSTOMIZED BLOCKS]
[BL1]^C^CINSERT;BL1;\1.0;1.0;0
[BL2]^C^CINSERT;BL2;\1.0;1.0;0
[BL3]^C^CINSERT;BL3;\1.0;1.0;0
[BL4]^C^CINSERT;BL4;\1.0;1.0;0
[BL5]^C^CINSERT;BL5;\1.0;1.0;0
[BL6]^C^CINSERT;BL6;\1.0;1.0;0
[ EXIT]^C^C
```

Note
[INSERT]^C^C$I=INSTBLK $I=*

In this menu item, $I=INSTBLK loads the image submenu INSTBLK that has been defined in the image tile menu section. $I= forces the display of the dialog box with the images.*

[INSERT]^C^C**$I=INSTBLK $I=***
    Where **$I=INSTBLK** - Loads the submenu INSTBLK that
                    has been defined in the image tile menu section
           $I=* -------------- Displays dialog box with images

***IMAGE
This is the section label of the Image tile menu. All image tile menus have to be defined in this section (***IMAGE) of the menu file.

**INSTBLK
This is the submenu label. The name of the submenu is INSTBLK.

[INSERT CUSTOMIZED BLOCKS]

Customizing the Standard AutoCAD Menu 10-49

This menu item describes the contents of the image tile menu and it is displayed at the top of the dialog box.

[BL1]^C^CINSERT;BL1;\1.0;1.0;0
The BL1 that is inside the brackets is the name of the slide that is displayed in one of the images of the dialog box. ^C^C cancels the existing command twice and the **INSERT** *command inserts the block BL1 with the X,Y scale factor of 1.0 and the rotation angle of 0 degrees.*

[BL1]^C^CINSERT;BL1;\1.0;1.0;0
Where [BL1] ------------ Name of the slide
       BL1 -------------- Block name
       \ ------------------ User input (Insertion point)
       1.0 --------------- X-scale factor
       1.0 --------------- Y-scale factor
       0 ------------------ Rotation angle

CUSTOMIZING THE SCREEN MENU

Like other menu sections, the screen menu section can be edited to modify or delete the existing commands or add new commands and submenus. Before writing a menu, it is very important to design a menu, know the exact sequence of AutoCAD commands, and know the prompts associated with the commands. You can edit the standard AutoCAD menu and arrange the commands in a way that provides the user an easy and quick access to most frequently used commands. A careful design will save quite some time in the long run. Therefore, it is strongly recommended to consider several possible alternatives and then select the one that is best suited for the job.

SUBMENUS

The number of items in the screen menu file can be so large that the items cannot be accommodated on one screen. For example, the maximum number of items that can be displayed on some of the display devices is 21. If the screen menu has more than 21 items, the menu items in excess of 21 are not displayed on the screen and therefore cannot be accessed. The user can overcome this problem by using submenus that let the user define smaller groups of items within a menu section. When a submenu is selected, it loads the submenu items and displays them on the screen.

$Section=Submenu
       Where $ ------------------ "$" sign
              Section ---------- Menu section specifier
              = ------------------ "=" sign
              Submenu ------- Name of submenu
Example
$S=EDIT
       Where S ------------------ S specifies screen menu section
              EDIT ------------- Name of submenu

Nested Submenus

When a submenu is activated, the current menu is copied to a stack. If you select another submenu, the submenu that was current will be copied or pushed to the top of the stack. The maximum number of menus that can be stacked is eight. If the stack size increases to more than eight, the menu that is at the bottom of the stack is removed and forgotten. You can call the previous submenu by using the nested submenu call. The format of the call is:

$S=

    Where **$** ------------------- "$" sign
                    S ------------------- Screen menu specifier
                    = ------------------- "=" sign

The maximum number of nested submenu calls that AutoCAD can have is eight. Each time you call a submenu, this pops the last item off the stack and reactivates it.

> **Note**
> *To load the original menu (ACAD.MNU), load the menu file by using the MENU command.*
>
> Command: **MENU**
> Enter menu file name or [. (for none)]<SM1>: **ACAD.MNU**
>
> *If you need to use input from a keyboard or a pointing device use backslash "\". The system will pause for the user to enter data.*
>
> *There should be no space after the backslash "\".*
>
> *The menu items, menu labels, and the command definition can be in uppercase, lowercase, or mixed.*
>
> *You can introduce spaces between the menu items to improve the readability of the menu file.*
>
> *If there are more items in the menu than the number of spaces available, the excess items are not displayed on the screen. For example, in the screen menu if the display device limits the number of items to 21, the items in excess of 21 will not be displayed on the screen, and are therefore inaccessible.*

Example 8

Edit the standard AutoCAD menu to add the commands as shown in Figure 10-14.

> **Note**
> *It is assumed that the image submenu **INSTBLK** has already been defined in the image tile menu section of the menu file. Do not edit the ACAD.MNU file. Make a copy of the ACAD.MNU file and then edit the new file (CUSTOM.MNU).*

You can use your word processor to load the menu file CUSTOM.MNU and search for

Customizing the Standard AutoCAD Menu   10-51

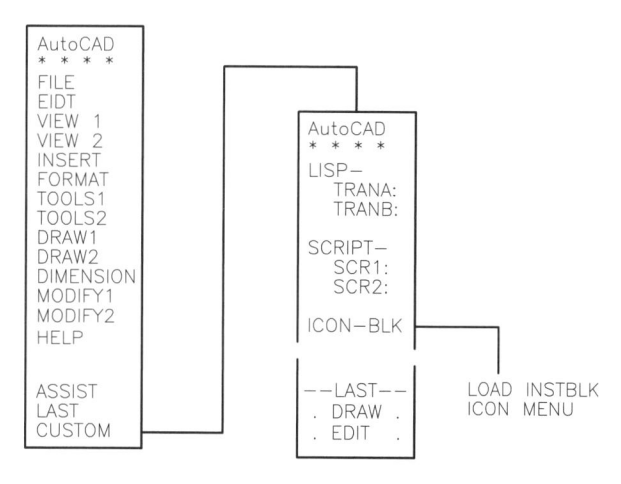

Figure 10-14 Modified screen menu

***SCREEN** section label. Add the new menu item at the end of the submenu ****S** and then define the submenu ****CUSTOM** and the menu items as shown in Figure 10-14. The following file is a partial listing of the **CUSTOM.MNU** file after editing:

***IMAGE
**INSTBLK
[INSERT CUSTOMIZED BLOCKS]
[BL1]^C^CINSERT;BL1;\1.0;1.0;0
[BL2]^C^CINSERT;BL2;\1.0;1.0;0
[BL3]^C^CINSERT;BL3;\1.0;1.0;0
[BL4]^C^CINSERT;BL4;\1.0;1.0;0

[BL5]^C^CINSERT;BL5;\1.0;1.0;0
[BL6]^C^CINSERT;BL6;\1.0;1.0;0
[EXIT]^C^C

***SCREEN
**S
[AutoCAD]^C^C^P(ai_rootmenus) ^P
[* * * *]$S=ACAD.OSNAP
[FILE   ]$S=ACAD.01_FILE
[EDIT   ]$S=ACAD.02_EDIT
[VIEW 1]$S=ACAD.03_VIEW1
[VIEW 2]$S=ACAD.04_VIEW2
[INSERT]$S=ACAD.05_INSERT

```
**ASSIST 3
[Last   ]_l
[Previous]_p
[All    ]_all
[Cpolygon]_cp
[Wpolygon]_wp



[CUSTOM]^C^C$S=X $S=CUSTOM

**CUSTOM 3
[LISP-]
[ TRANA:]^C^C(LOAD "TRANA");TRANA
[ TRANB:]^C^C(LOAD "TRANB");TRANB
[ ]
[SCRIPT-]
[ SCR1:]^C^CSCRIPT;SCR1
[ SCR2:]^C^CSCRIPT;SCR2
[ ]
[IMAGE-BLK]^C^C$I=INSTBLK $I=*
```

Note

[CUSTOM]^C^C$S=X $S=CUSTOM *In this menu item, $S=X loads the submenu X that has been defined in the screen menu section. $S=CUSTOM loads the submenu CUSTOM that has also been defined in the screen menu section of the menu file CUSTOM.MNU.*

[CUSTOM]^C^C**$S=X $S=CUSTOM**
  Where **$S=X** ------------ Loads submenu X
      $S=Custom ---- Loads submenu CUSTOM

**CUSTOM 3 *In this menu item, CUSTOM is the name of the submenu. The 3 that follows the submenu name prints the menu items defined in the submenu CUSTOM from the line number 3. Nothing is printed on the first two lines; therefore, the first two lines AutoCAD and * * * * stay on the screen.*

**CUSTOM 3
  Where **Custom** ---------- Submenu name
      3 ------------------- Start printing from line number 3

[IMAGE-BLK]^C^C$I=INSTBLK $I=* *In this menu item, $I=INSTBLK loads the image submenu INSTBLK that has been defined in the image section of the menu file. $I=* forces the display of the new image tile menu on the screen.*

[IMAGE-BLK]^C^C$I=INSTBLK $I=*
  Where **$1=INSTBLK** Loads the image submenu INSTBLK
      $SI=* ------------ Forces display of new image tile menu on screen

Customizing the Standard AutoCAD Menu   10-53

Review Questions

Answer the following questions and then compare your answers to the correct answers at the end of this chapter.

1. When you start AutoCAD, the last used menu file is _____ loaded.

2. Tablet area 1 of standard ACAD.MNU has _____ columns and _____ rows.

3. The blocks are _____ in shape in the tablet areas.

4. The shortcut menu can be loaded and displayed by using the command _____.

5. Tablet area 2 of the standard ACAD.MNU has _____ columns and _____ rows.

6. An image tile menu will work only if the status line is not _____.

7. Swapping of submenus sometimes causes _____.

8. The maximum number of nested submenu calls that AutoCAD can have is _____.

9. Tablet area 3 of the standard ACAD.MNU has _____ columns and _____ rows.

10. The size of the blocks in tablet area 3 is _____ than the blocks in other sections of the tablet template.

11. Tablet area 4 of standard ACAD.MNU has _____ columns and _____ rows.

Review Questions

Answer the following questions.

1. The AutoCAD menu file can have up to _____ sections.

2. A tablet menu can have up to _____ sections.

3. The section label is designated by _____.

4. The submenu label is designated by _____.

5. In a menu file you can use _____ to cancel the existing command.

6. Submenu names can be _____ characters long.

7. You _____ assign the same command to more than one block on the template.

8. AutoCAD's _____ command is used to configure the tablet menu template.

9. AutoCAD's _____ command is used to load a new menu.

10. You need to enter _____ points that are at _____ degrees to configure different tablet areas.

11. The commands are assigned to the buttons of the pointing device in the _____ order in which they appear in the buttons menu.

12. The format of the command used for loading a submenu that has been defined in the screen menu section is _____.

13. The format of the command used for loading a submenu that has been defined in the pull-down menu section is _____.

14. The format of the command used for loading a submenu that has been defined in the image tile menu section is _____.

15. The command that is used to force the display of the current pull-down menu is _____.

Exercises

Exercise 3   *General*

Add the following commands to the TABLET1 section of the standard AutoCAD menu file ACAD.MNU. Figure 10-15 shows the layout of tablet area-1 of the template.

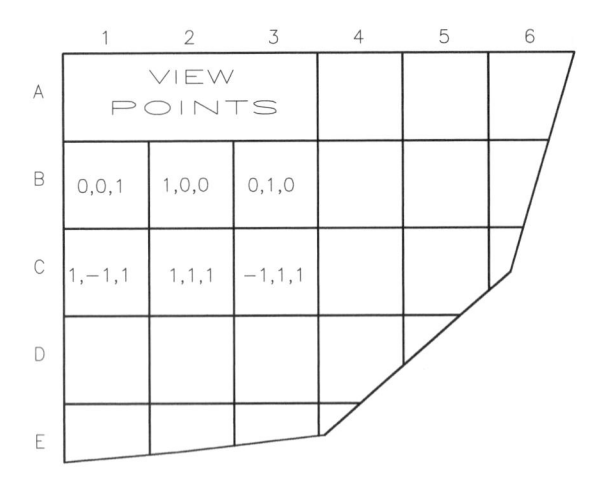

Figure 10-15 Commands assigned to tablet area-1

Customizing the Standard AutoCAD Menu   10-55

VIEW POINTS

0,0,1	1,0,0	0,1,0
1,-1,1	1,1,1	-1,1,1

Exercise 4   *General*

Add the following commands to the TABLET1 section of the standard AutoCAD menu file ACAD.MNU. The layout of tablet area-1 is shown in Figure 10-16.

INSERT NO	PLOT 12x18	SETLAYER OBJ
INSERT NC	PLOT 18x24	SETLAYER HID
INSERT COIL	PLOT 24x36	SETLAYER CEN
INSERT RESIS	PRPLOT	SETLAYER DIM

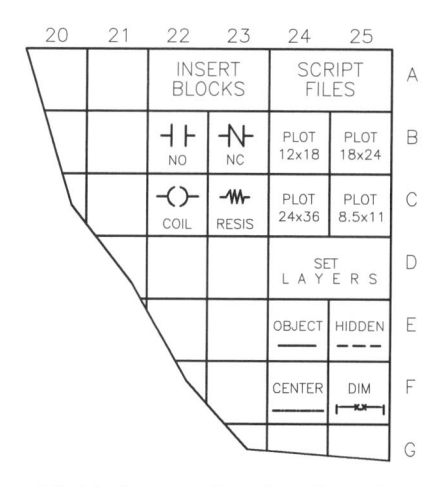

Figure 10-16 Commands assigned to tablet area-1

Exercise 5   *General*

Write a button menu for the following AutoCAD commands (Figure 10-17). Add the commands to the BUTTONS2 section of ACAD.MNU. The pointing device has 10 buttons; button number 1 is used for picking the points. The blocks are to be inserted with a scale factor of 1.00 and a rotation of 0 degrees (Filename BME1.MNU).

1. PICK BUTTON   2. Enter   3. CANCEL
4. OSNAPS   5. END PT   6. CENTER
7. NEAR   8. ZOOM Window   9. ZOOM Prev
10. PAN

Exercise 6   *General*

Add the commands shown in Figure 10-18 to the POP12 section of the standard AutoCAD menu ACAD.MNU

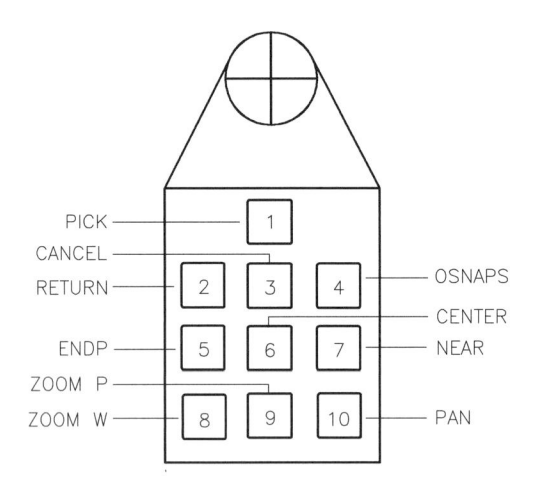

Figure 10-17 Commands assigned to different buttons of a pointing device

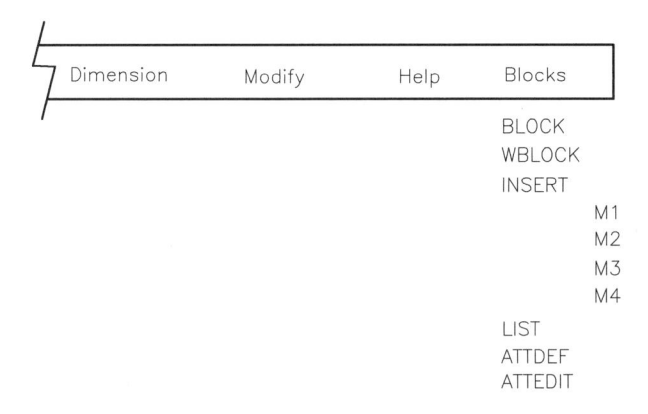

Figure 10-18 POP11 section of menu

Exercise 7   *General*

Write an image tile menu for inserting the following blocks that can be accessed through the POP13 section. It is assumed that the blocks have already been created.

SYMBOL-X	SYMBOL-Y	SYMBOL-Z
LOGO-1	LOGO-2	LOGO-3
TBLOCK-1	TBLOCK-2	TBLOCK-3

Answers to the Self Evaluation Test
1. Automatically, **2.** 25 columns, 9 rows **3.** rectangular, **4.** $PO=*, **5.** 11,9 **6.** displayed, **7.** distraction, **8.** eight, **9.** 9,13 **10.** smaller, **11.** 25,4.

Chapter 11

Shapes and Text Fonts

Learning Objectives
After completing this chapter, you will be able to:
• *Write shape files.*
• *Use vector length and direction encoding to write shape files.*
• *Compile and load shape/font files.*
• *Use special codes to define a shape.*
• *Write text font files.*

SHAPE FILES
AutoCAD provides a facility to define shapes and text fonts. These files are ASCII files with the extension **.SHP**. You can write these files using any text editor like Notepad.

Shape files contain information about the individual elements that constitute the shape of an object. The basic objects that are used in these files are lines and arcs. You can define any shape using these basic objects, and then insert them anywhere in a drawing. The shapes are easy to insert, and they take less disk space than blocks. However, there are some disadvantages to using shapes. For example, you cannot edit a shape or change it. Blocks, on the other hand, can be edited after exploding them with the AutoCAD **EXPLODE** command.

SHAPE DESCRIPTION
Shape description consists of the following two parts: a header and a shape specification.

Header
The header line has the following format:

   *SHAPE NUMBER, DEFBYTES, SHAPE NAME

```
*201,21,HEXBOLT
         Where *201 -------------- Shape number
                21 ---------------- Number of data bytes in shape specification
           HEXBOLT ----- Shape name
```

Every header line starts with an asterisk (*), followed by the **SHAPE NUMBER**. The shape number is any number between 1 and 255 in a particular file, and these numbers cannot be repeated within the same file. However, the numbers can be repeated in another shape file with a different name. **DEFBYTES** is the number of data bytes used by the shape specification and includes the terminating zero. **SHAPE NAME** is the name of a shape, in uppercase letters. The name is ignored if the letters are lowercase. The file must not contain two shapes with the same name.

Shape Specification

The shape specification line contains the complete definition of the shape of an object. The shape is described with special codes, hexadecimal numbers, and decimal numbers. A hexadecimal number is designated by a leading zero (012), and a decimal number is a regular number without a leading zero (12). The data bytes are separated by a comma (,). The maximum number of data bytes is 2,000 bytes per shape, and in a particular shape file there can be more than one shape. The shape specification can have multiple lines. You should define the shape in logical blocks and enter each block on a separate line. This makes it easier to edit and debug the files. The number of characters on any line must not exceed 80. The shape specification is terminated with a zero.

VECTOR LENGTH AND DIRECTION ENCODING

Figure 11-1 shows the vector direction codes. All vectors in this figure have the same length specification. The diagonal vectors have been extended to match the closest orthogonal vector. Let us assume that the endpoint of vector 0 is two grid units from the intersection point of vectors. The endpoint of vector 1 is one grid directly above the endpoint of vector 0. Therefore,

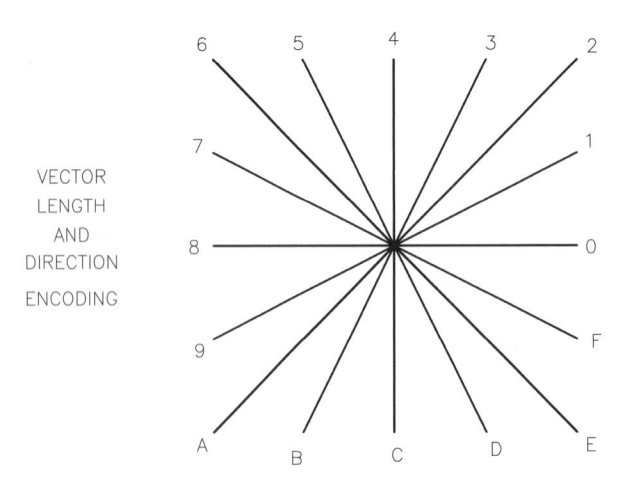

Figure 11-1 Vector length and direction encoding

Shapes and Text Fonts

the angle of vector 1 is 26.565 degrees (Tan-1 1/2 = Tan-1 0.5 = 26.565). Similarly, vector 2 is at 45 degrees (Tan-1 2/2 = Tan-1 1 = 45).

All these vectors have the same magnitude or length specification. In other words, although their actual lengths vary, they are all considered one unit in definition. To define a vector you need its magnitude and direction. That means each shape specification byte contains vector length and a direction code. The maximum length of the vector is 15 units. Example 1 illustrates the use of vectors.

Example 1

Write a shape file for the resistor shown in Figure 11-2. The name of the file is **SH1.SHP**, and the shape name is RESIS.

Step 1: Writing the shape file

Use any text editor to write the shape file. The following two lines define the shape file for the given resistor:

*201,8,RESIS
020,023,04D,043,04D,023,020,0

The first line is the **header line**; the second line is the **shape specification**.

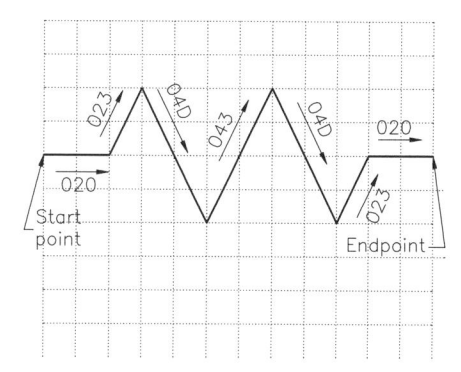

Figure 11-2 Resistor

Explanation

Header Line

*201,8,RESIS

*201 is the shape number, and 8 is the number of data bytes contained in the shape specification line. RESIS is the name of the shape.

Shape Specification

<u>020</u>,023,04D,043,04D,023,020,0

        Where   **0** ------------------- Hexadecimal notation
                     2 ------------------- Vector length
                     0 ------------------- Direction code

Each data byte in this line, except the terminating zero, has three elements. The first element (0) is the hexadecimal, the second is the length of the vector, and the third element is the direction code. For the first data byte, 020, the length of the vector is 2, and the direction is along the direction vector 0. Similarly, for the second data byte, 023, the first element, 0, is for hexadecimal; the second element, 2, is the length of the vector; and the third element, 3, is the direction code for the vector.

Step 2: Compiling the shape file

Save the file with the extension **.SHP**. Now you can compile the shape file or the font file by using the **COMPILE** command. Enter the **COMPILE** command at the Command prompt to display the **Select Shape or Font File** dialog box (Figure 11-3). From this dialog box, select the shape file that you want to compile. AutoCAD will compile the file, and if the compilation process is successful, the following prompt will be displayed at the Command prompt line.

    Compilation successful
    Output file name.shx contains nn bytes

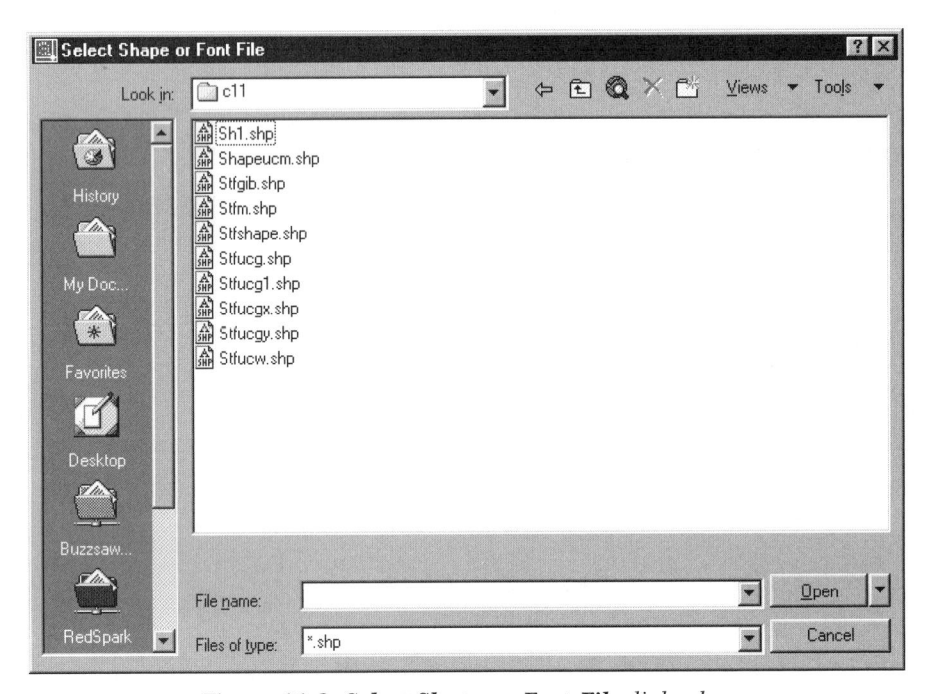

Figure 11-3 Select Shape or Font File dialog box

Compilation translates a file with extension **.SHP** into **.SHX**. For Example 1, the name of the compiled output file is **SH1.SHX** and the number of bytes is 49. If AutoCAD encounters an error in compiling a shape file, an error message will be displayed, indicating the type of error and the line number where the error occurred. You can also enter the options at the Command prompt keeping the **FILEDIA** as 0.

Step 3: Loading the shape file

Enter the **LOAD** command at the Command prompt to display the **Select Shape File** dialog box. Select the file you want to load and then choose the **Open** button (Figure 11-4) or enter the command **LOAD** at the Command prompt keeping the **FILEDIA** as 0.

    Command: **LOAD**
    Name of shape file to load (or ?): *Name of file*. **SH1**

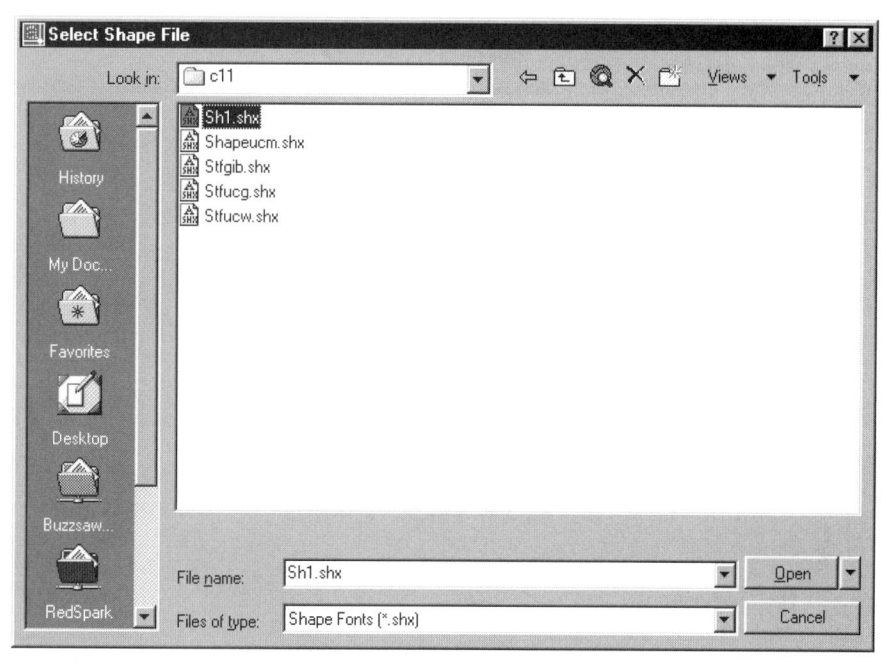

Figure 11-4 Select Shape File dialog box

SH1 is the name of the shape file for Example 1. Do not include the extension **.SHX** with the name because AutoCAD automatically assumes the extension. If the shape file is present, AutoCAD will display the shape names that are loaded.

To insert the loaded shapes, use the AutoCAD **SHAPE** command:

Command: **SHAPE**
    Shape name (or ?)<default>: *Shape name.*
    Start point: *Shape origin.*
    Height<1.0>: *Number or point.*
    Rotation angle<0.0>: *Number or point.*

For Example 1, the shape name is RESIS. After you enter the information about the start point, height, and rotation, the shape will be displayed on the screen (Figure 11-5).

SPECIAL CODES

Generating shapes with the direction vectors has certain limitations. For example, you cannot draw an arc or a line that is not along the standard direction vectors. These limitations can be overcome by using special codes that add flexibility and give you better control over the shapes you want to create.

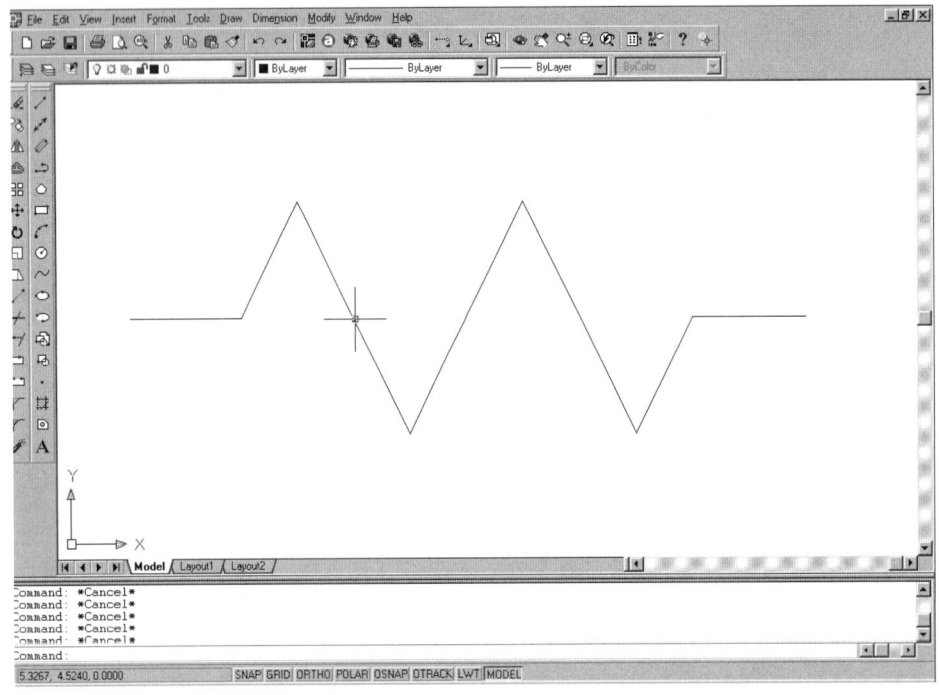

Figure 11-5 The shape (RESIS) inserted in the drawing

Standard Codes

000	End of shape definition
001	Activate draw mode (pen down)
002	Deactivate draw mode (pen up)
003	Divide vector lengths by next byte
004	Multiply vector lengths by next byte
005	Push current location from stack (saving a location)
006	Pop current location from stack (restoring the location)
007	Draw subshape numbers given by next byte (calling a subshape)
008	X-Y displacement given by the next two bytes (non uniform lines)
009	Multiply X-Y displacement, terminated by (0,0) (continuous use of non uniform line)
00A or 10	Octant arc defined by next two bytes
00B or 11	Fractional arc defined by next five bytes
00C or 12	Arc defined by X-Y displacement and bulge
00D or 13	Multiple bulge-specified arcs (continuous use of bulge arc)
00E or 14	Process next command only if vertical text style

Code 000: End of Shape Definition

This code marks the end of a shape definition.

Shapes and Text Fonts

Code 001: Activate Draw Mode
This code turns the draw mode on. When you start a shape, the draw mode is on, so you do not need to use this code. However, if the draw mode has been turned off, you can use code 001 to turn it on.

Code 002: Deactivate Draw Mode
This code turns the draw mode off. It is used when you want to move the pen without drawing a line.

```
       1         2         3         4
```

Let us say the distance from point 1 to point 2, from point 2 to point 3, and from point 3 to point 4 is 2 units each. The shape specification for this line is:

020,002,020,001,020,0

The first data byte, 020, generates a line 2 units long along direction vector 0. The second data byte, 002, deactivates the draw mode; and the third byte, 020, generates a blank line 2 units long. The fourth data byte, 001, activates the draw mode; and the next byte, 020, generates a line that is 2 units long along direction vector 0. The last byte, 0, terminates the shape description.

Example 2

Write a shape file to generate the character "G" as shown in Figure 11-6.

Step 1: Writing the shape file
You can use any text editor like notepad to write a shape file. The name of the file is **CHRGEE** and the shape name is GEE. **In the following file, the line numbers are not a part of the file; they are for reference only.**

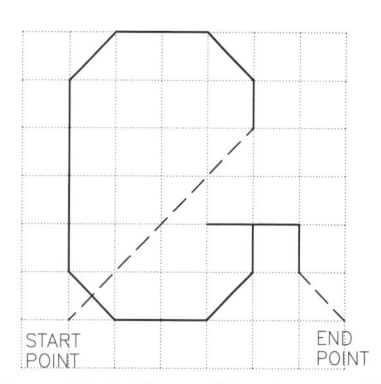

Figure 11-6 *Shape of the character "G"*

```
*215,20,GEE                                    1
002,042,                                       2
001,014,016,028,01A,                           3
04C,01E,020,012,014,                           4
002,018,                                       5
001,020,01C,                                   6
002,01E,0                                      7
```

Explanation

Line 1
*215,20,GEE
The first data byte contains an asterisk (*) and shape number 215. The second data byte is the number of data bytes contained in the shape specification, including the terminating 0. GEE is the name of the shape.

Line 2
002,042,
The data byte 002 deactivates the draw mode (pen up), and the next data byte defines a vector 4 units long along direction vector 2.

Line 3
001,014,016,028,01A,
The data byte 001 activates the draw mode (pen down), and 014 defines a vector that is 1 unit long at 90 degrees (direction vector 4). The data byte 016 defines a vector that is 1 unit long along direction vector 6. The data byte 028 defines a vector that is 2 units long along direction vector 8 (180 degrees). The data byte 01A defines a unit vector along direction vector A.

Line 4
04C,01E,020,012,014,
The data byte 04C defines a vector that is 4 units long along direction vector C. The data byte 01E defines a direction vector that is 1 unit along direction vector E. The data byte 020 defines a direction vector that is 2 units long along direction vector 0 (0 degrees). The data byte 012 defines a direction vector that is 1 unit long along direction vector 2. Similarly, 014 defines a vector that is 1 unit long along direction vector 4.

Line 5
002,018,
The data byte 002 deactivates the pen (pen up), and 018 defines a vector that is 1 unit long along direction vector 8.

Line 6
001,020,01C,
The data byte 001 activates the pen (pen down), and 020 defines a vector that is 2 units long along direction vector 0. The data byte 01C defines a vector that is 1 unit long along direction vector C.

Line 7
002,01E,0
The data byte 002 deactivates the pen, and the next data byte, 01E, defines a vector that is 1 unit long along direction vector E. The data byte 0 terminates the shape specification.

Tip
The shape definition file does NOT need to be recompiled; this only needs to be done once. It will reload automatically when the drawing is reopened. If you change the shape file in anyway and after loading and compiling it doesn't reflects the changes then open a new

Shapes and Text Fonts

drawing and then compile and load the shape again. The existing drawing file does not incorporate the changes made in the shape file.

Code 003: Divide Vector Lengths by Next Byte

This code is used if you want to divide a vector by a certain number. In Example 2, if you want to divide the vectors by 2, the shape description can be written as:

003,2,020,002,020,001,020,0

The first byte, 003, is the division code, and the next byte, 2, is the number by which all the remaining vectors are divided. The length of the lines and the gap between the lines will be equal to 1 unit now:

Also, the scale factors are cumulative within a shape. For example, if we insert another code, 003, in the preceding shape description, the length of the last vector, 020, will be divided by 4 (2 x 2):

003,2,020,002,020,001,003,2,020,0
Where **003,2** ------------- All the vectors are divided by 2
003,2 ------------- All the remaining vectors are divided by 4 (2 x 2)

Here is the output of this shape file:

Code 004: Multiply Vector Lengths by Next Byte

This code is used if you want to multiply the vectors by a certain number. It can also be used to reverse the effect of code 003. All multiplying and dividing vectors must be integers, 1-255.

003,2,020,002,020,001,004,2,020,0
Where **003,2** ------------- Divides all the vectors on the right by 2
004,2 ------------- Multiplies all the vectors on the right by 2

In this example, the code 003 divides all the vectors to the right by 2. Therefore, a vector that was 1 unit long will be 0.5 units long now. The second code, 004, multiplies the vectors to the right by 2. We know the scale factors are cumulative; therefore, the vectors that were divided by 2 earlier will be multiplied by 2 now. Because of this cumulative effect, the length of the last vector remains unchanged. This file will produce the following shape:

Codes 005 and 006: Location Save/Restore

Code 005 lets you save the current location of the pen, and code 006 restores the saved location.

Example 3

The following example illustrates the use of codes 005 and 006.

Step 1: Writing the shape file

Figure 11-7(a) shows three lines that are unit vectors and intersect at one point. After drawing the first line, the pen has to return to the origin to start a second vector. This is done using code 005, which saves the starting point (origin) of the first vector, and code 006, which restores the origin. Now, if you draw another vector, it will start from the origin. Since there are three lines, you need three code 005s and three code 006s. The following file shows the header line and the shape specification for generating three lines as shown in Figure 11-7(a):

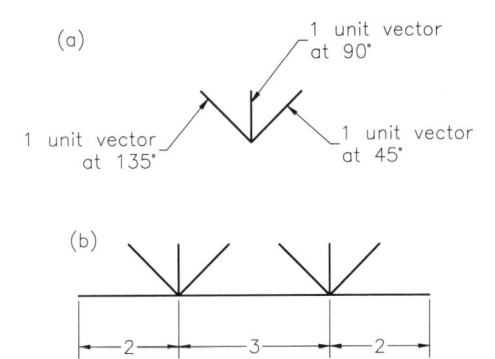

Figure 11-7 (a) Three unit vectors intersecting at a point; (b) repeating predefined subshapes

```
*210,10,POP1
005,005,005,012,006,014,006,016,006,0
```
         Where   005 --------------- Saves origin three times
                 012 --------------- Generates first vector
                 006 --------------- Restores origin
                 014 --------------- Generates second vector
                 006 --------------- Restores origin

The number of saves (code 005) has to equal the number of restores (code 006). If the number of saves (code 005) is more than the number of restores (code 006), AutoCAD will display the following message when the shape is drawn:

Position stack overflow in shape (shape number)

Similarly, if the number of restores (code 006) is more than the number of saves (code 005), the following message will be displayed:

Position stack underflow in shape (shape number)

The maximum number of saves and restores you can use in a particular shape definition is four.

Step 2: Compiling and loading the shape

Use the **COMPILE and LOAD** command for compiling and loading the shape file. Next use the **SHAPE** command to insert the shape as described in Example 2.

> **Note**
> *The name of the shape must be uppercase. Names with lowercase characters are ignored and are used to label font shape definitions. One shape file can contain 255 different shapes. Each shape is recognized by the unique name assigned to it.*

Shapes and Text Fonts

Code 007: Subshape

You can define a subshape like a subroutine in a program. To reference a subshape, the subshape code, 007, has to be followed by the shape number of the subshape. The subshape has to be defined in the same shape file, and the shape number has to be from 1 to 255.

*210,10,POP1
005,005,005,012,006,014,006,016,006,0
*211,8,SUB1
020,007,210,030,007,210,020,0
        Where   ***210** -------------- Shape number
                     007 --------------- Subshape reference
                       210 --------------- Shape number

The shape that this example generates is shown in Figure 11-7(b).

Code 008: X-Y Displacement

In the previous examples, you might have noticed some limitations with the vectors. As mentioned earlier, you can draw vectors only in the 16 predefined directions, and the length of a vector cannot exceed 15 units. These restrictions make the shape files easier and more efficient, but at the same time, they are limiting. Therefore, codes 008 and 009 allow you to generate nonstandard vectors by entering the displacements along the X and Y directions. The general format for Code 008 is:

008, XDISPLACEMENT, YDISPLACEMENT
or
008, (XDISPLACEMENT, YDISPLACEMENT)

X and Y displacements can range from +127 to -128. Also, a positive displacement is designated by a positive (+) number, and a negative displacement is designated by a negative (-) number. The leading positive sign (+) is optional in a positive number. The parentheses are used to improve readability, but they have no effect on the shape specification.

Code 009: Multiple X-Y Displacements

Whereas code 008 allows you to generate nonstandard vectors by entering a single X and Y displacement, code 009 allows you to enter multiple X and Y displacements. It is terminated by a pair of 0 displacements (0,0). The general format is:

009,(XDISPL, YDISPL), (XDISP, YDISPL), . . ., (0,0)

Code 00A or 10: Octant Arc

If you divide 360 degrees into eight equal parts, each angle will be 45 degrees. Each 45-degree angle segment is called an octant, and the two lines that contain an octant are called an **octant boundary**. The octant boundaries are numbered from 0 to 7, as shown in Figure 11-8. The general format is:

10,(R,+/-0SN)

```
Where  R ----------------- Radius of arc
       +/- --------------- Defines direction, + Counterclockwise, - Clockwise
       0 ----------------- Hexadecimal notation
       S ----------------- Starting octant boundary
       N ----------------- Number of octants
```

10,(3,-043)

The first number, 10, is the code 00A for the octant arc. The second number, 3, is the radius of the octant arc. The negative sign indicates that the arc is to be generated in a clockwise direction. If it is positive (+) or if there is no sign, the arc will be generated in a counterclockwise direction. Zero is the hexadecimal notation, and the following number, 4, is the number of the octant boundary where the octant arc will start. The next element, 3, is the number of octants that this arc will extend. This example will generate the arc in Figure 11-9. The following is the listing of the shape file that will generate the shape shown in Figure 11-9:

```
*214,5,FOCT1
001,10,(3,-043),0
```

Figure 11-8 Octant boundaries

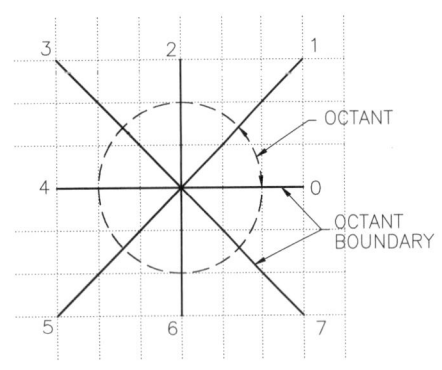

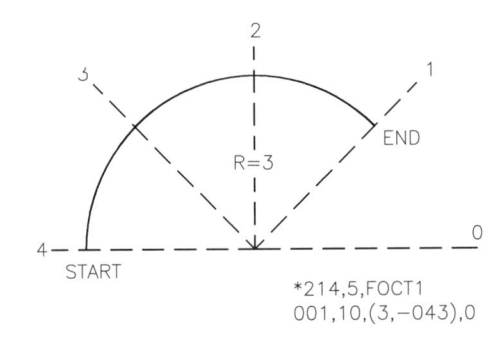

Figure 11-9 Octant arc

Code 00B or 11: Fractional Arc

You can generate a nonstandard fractional arc by using code 00B or 11. This code will allow you to start and end an arc at any angle. The definition uses five bytes, and the general format is:

11,(START OFFSET, END OFFSET, HIGHRADIUS, LOWRADIUS, +/-0SN)

The START OFFSET represents how far from an octant boundary the arc starts, and the END OFFSET represents how far from an octant boundary the arc ends. The HIGHRADIUS is zero if the radius is equal to or less than 255 units, and the LOWRADIUS is the radius of the arc. The positive (+) or negative (-) sign indicates, respectively, whether the arc is drawn counterclockwise or clockwise. The next element, S, is the number of the octant where the arc starts, and element N is the number of octants the arc goes through. The following example illustrates the fractional arc concept.

Shapes and Text Fonts

Start offset = (starting angle of the arc - nearest angle of the octant whose sum is less than the starting angle) x 256 / 45

End offset = (end angle of the arc - nearest angle of the octant whose sum is less than the end angle) x 256 / 45

Example 4

Draw a fractional arc of radius 3 units that starts at a 20-degree angle and ends at a 140-degree angle (counterclockwise).

The solution involves the following steps:

Step 1: Calculating the parameters

Find the nearest octant boundary whose angle is less than 140 degrees. The nearest octant boundary is the number 4 octant boundary, whose angle is 135 degrees (3 * 45 = 135). Calculate the end offset to the nearest whole number (integer):

End offset = (140 - 135) * 256/45 = 28.44 = 28

Find the nearest octant boundary whose angle is less than 20 degrees. The nearest octant boundary is 0 and its angle is 0 degrees. Calculate the start offset to the nearest whole number:

Start offset = (20 - 0) * 256/45 = 113.7 = 114

Find the number of octants the arc passes through. In this example, the arc starts in the first octant and ends in the fourth octant. Therefore, the number of octants the arc passes through is four (counterclockwise).

Find the octant where the arc starts. In this example, it starts in the 0 octant.

Radius of the arc is 3. Its high radius is 0 as radius is less than 255.

Substitute the values in the general format of the fractional arc:

11,(114,28,0,3,004)

Step 2: Writing the shape file

Use any text editor or word processor to write down the shape file. The following shape file will generate the fractional arc shown in Figure 11-10:

*221,8,FOCT2
001,11,(114,28,0,3,004),0

*221,8,FOCT2
001,11,(114,28,0,3,004),0

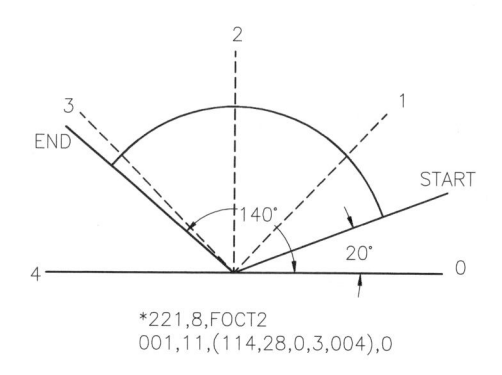

Figure 11-10 *Fractional arc*

Step 3: Saving and loading the shape file

Save the file with the extension **.SH.** Use the **COMPILE and LOAD** command for compiling and loading the shape file. Next use the **SHAPE** command to insert the shape in the drawing editor.

Code 00C or 12: Arc Definition by Displacement and Bulge

Code C can be used to define an arc by specifying the displacement of the endpoint of an arc and the bulge factor. X and Y displacements may range from -127 to +127, and the bulge factor can also range from -127 to +127. A semicircle will have a bulge factor of 127, and a straight line will have a bulge factor of 0. If the bulge factor has a negative sign, the arc is drawn clockwise.

Bulge factor = ((2 * H)/D) * 127
Where H ------------------ Height of arc
       D ------------------ Displacement

For a semicircle, 2H   = D
Therefore, bulge   = (D/D) * 127 = 127
For a straight line, H   = 0
Therefore, bulge   = (0/D) * 127 = 0

In Figure 11-11, the distance between the start point and the endpoint of an arc is 4 units, and the height is 1 unit. Therefore, the bulge can be calculated by substituting the values in the previously mentioned relation:

Bulge = (2 * 1/4) * 127 = 63.5 = 63
(integer)

The following shape description will generate the arc shown in Figure 11-11.

*213,5,BULGE1
12,(4,0,-63),0
           Where 4 -------------------- X displacement
                  0 -------------------- Y displacement
                  - -------------------- Negative (-), generates clockwise arc
                  63 ----------------- Bulge factor

Code 00D or 13: Multiple Bulge-Specified Arcs

Code 00D or 13 can be used to generate multiple arcs with different bulge factors. It is terminated by a (0,0). The following shape description defines the arc configuration of Figure 11-12:

*214,16,BULGE2
13,(4,0,-111),
(0,4,63),

Shapes and Text Fonts

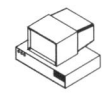

(-4,0,-111),
(0,-4,63),(0,0),0

Tip
The information about different arcs can be given in one line. For the convenience of the users the above example is divided into different lines.

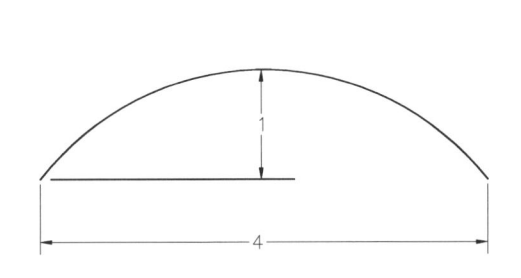

Figure 11-11 Calculating bulge

Figure 11-12 Different arc configuration

Code 00E or 14: Flag Vertical Text

Code 00E or 14 is used when the same text font description is to be used in both the horizontal and vertical orientations. If the text is drawn in a horizontal direction, the vector next to code 14 is ignored. If the text is drawn in a vertical position, the vector next to code 14 is not ignored. This lets you generate text in a vertical or horizontal direction with the same shape file.

For horizontal text, the start point is the lower left point, and the endpoint is on the lower right. In vertical text, the start point is at the top center, and the endpoint is at the bottom center of the text, as shown in Figure 11-13. At first, it appears that you need two separate shape files to define the shape of a horizontal and a vertical text. However, with code 14 you can avoid the dual shape definition.

Figure 11-13 shows the pen movements for generating the text character "G." If the text is horizontal, the line that is next to code 14

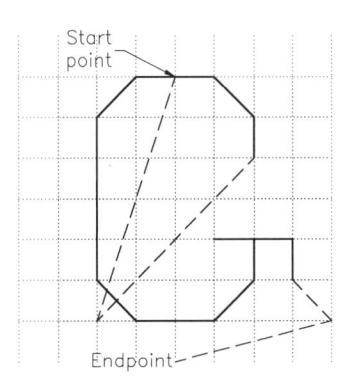

Figure 11-13 Pen movement for generating the character "G"

is automatically ignored. However, if the text is vertical, the line is not ignored, resulting in resetting the start and endpoints appropriately for vertically aligned text.

```
1*15,28,FLAG
002,14,                          ┐ If text is horizontal, code 14
008,(-2,-6),                     │ automatically ignores next line.
042,001,                         ┘ 008, (-2,-6),
014,016,028,01A,
04C,01E,020,012,014,
002,018,
001,020,01C,
002,01E,
14,                              ┐ If text is horizontal, code 14
008,(-4,-1),                     │ automatically ignores next line.
0                                ┘ 008,(-4,-1),
```

Example 5

Write a shape file for a hammer as shown in Figure 11-14(a). (The name of the shape file is HMR.SHP and the shape name is HAMMER.) The following file defines the shape of a hammer. The line numbers are not a part of the file; they are for reference only.

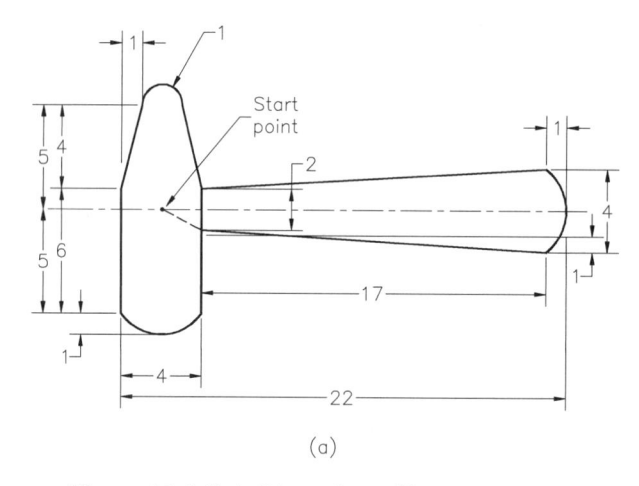

Figure 11-14(a) Dimensions of hammer in units

Step 1: Writing the shape file
Use any text editor or word processor to write the following shape file.

```
*204,34,HAMMER                                                     1
003,22,                                                            2
002,8,(2,-1),                                                      3
001,024,                                                           4
```

Shapes and Text Fonts

8,(-1,4),	5
00A,(1,004),	6
8,(-1,-4),06C,	7
00C,(4,0,63),	8
044,8,(17,-1),	9
00C,(0,4,63),	10
8,(-17,-1),0	11

Explanation
Line 1
*204,34,HAMMER
This is the header line that consists of the shape number (204), the number of data bytes in the shape specification (34), and the name of the shape (HAMMER).

Line 2
003,22,
Data byte 003 has been used to divide the vectors by the next data byte, 22. This reduces the hammer shape to a unit size that facilitates the scaling operation when you insert the shape in a drawing.

Line 3
002,8,(2,-1),
Data byte 002 deactivates the pen (pen up), and the next data byte (code 008) defines a vector that has X-displacement of 2 units and Y-displacement of 1 unit. No line is drawn because the pen is deactivated.

Line 4
001,024,
Data byte 001 activates the pen (pen down) and the next data byte defines a vector that is 2 units long along the direction vector 2.

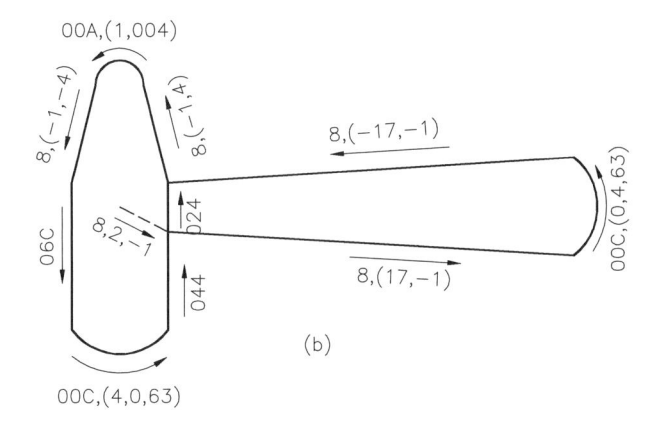

Figure 11-14(b) Pen movements for generating the shape of a hammer

Line 5
8,(-1,-4),
The first data byte, 8 (code 008), defines a vector whose X-displacement and Y-displacement is given by the next two data bytes. The X-displacement of this vector is -1 and the Y-displacement is -4 units.

Line 6
00A,(1,004),
Data byte 00A defines an octant arc that has a radius of 1 unit as defined by the next data byte. The first element (0) of data byte 004 is a hexadecimal notation. The second element (0) defines the starting octant of the arc, and the third element (4) defines the ending octant of the arc.

Line 7
8,(-1,-4),06C,
Data byte 8 (code 008) defines a vector that has X-displacement of -1 unit, and Y-displacement of -4 units. The next data byte defines a vector that is 6 units long along the direction vector C.

Line 8
00C,(4,0,63),
The first data byte (00C) defines an arc that has X-displacement of 4 units, Y-displacement of 0 units, and a bulge factor of 63.

$$\begin{aligned} \text{Bulge factor} &= (2 * H)/D * 127 \\ &= (2 * 1)/4 * 127 \\ &= 63.5 \\ &= 63 \text{ (integer)} \end{aligned}$$

Line 9
044,8,(17,-1),
The first data byte defines a vector that is 4 units long along the direction vector 4. The second data byte (8) defines a vector that has X-displacement of 17 units and Y-displacement of -1 unit.

Line 10
00C,(0,4,63),
The first data byte (00C) defines an arc that has X-displacement of 0 units, Y-displacement of 4 units, and a bulge factor of 63.

$$\begin{aligned} \text{Bulge factor} &= (2 * H)/D * 127 \\ &= (2 * 1)/2 * 127 \\ &= 63.5 \\ &= 63 \text{ (integer)} \end{aligned}$$

Line 11
8,(-17,-1),0
The first data byte (8) defines a vector that has X-displacement of -17 units and Y-displacement of -1 unit. The data byte 0 terminates the shape definition.

Step 3: Loading the shape file
Save the file with the extension **.SHP** and then use **COMPILE**, **LOAD**, and **SHAPE** command to load the shape as described in Example 1.

TEXT FONT FILES

In addition to shape files, AutoCAD provides a facility to create new text fonts. After you have created and compiled a text font file, text can be inserted in a drawing like regular text and using the new font. These text files are regular shape files with some additional information describing the text font and the line feed. The following is the general layout of the text font file:

Text font description
Line feed
Shape definition

Text Font Description
The text font description consists of two lines:

*0,4,font name
ABOVE, BELOW, MODES, 0

Where ***0** ----------------- Special shape number for text font
4 ------------------- Number of data bytes
font name ------- Name of font in lowercase
ABOVE ---------- Upper distance
BELOW --------- Lower distance
MODES --------- 0 for horizontal text, 2 for dual orientation
0 ------------------- Terminating zero

For example, if you are writing a shape definition for an uppercase M, the text font description would be:

*0,4,ucm
10,4,2,0

In the first line, the first data byte (0) is a special shape number for the text font, and every text font file will have this shape number. The next data byte (4) is the number of data bytes in the next line, and ucm is the shape name (name of font). The shape names in all text font files should be lowercase so that the computer does not have to save the names in memory. You can still reference the shape names for editing.

In the second line, the first data byte (10) specifies the height of an uppercase letter above the baseline. For example, in Figure 11-15 (page 11-22), the height of the letter M above the baseline is 10 units. The next data byte (4) specifies the distance of lowercase letters below the baseline. AutoCAD uses this information to scale the text automatically. For example, if you enter the height of text as 1 unit, the text will be 1 unit, although it was drawn 10 units high in the text font definition. The third data byte (2) defines the mode. It can have one of only two values, 0 or 2. If the text is horizontal, the mode is 0; if the text has dual orientation (horizontal and vertical), the mode is 2. The fourth data byte (0) is the zero that terminates the definition.

Line Feed

The line feed is used to space the lines so that characters do not overlap and so that a desired distance is maintained between the lines. AutoCAD has reserved the ASCII number 10 to define the line feed.

*10,5,lf
2,8,(0,-14),0

Where	*10	(10) ASCII number reserved for line feed
	5	(5) Number of data bytes in shape specification
	lf	(lf) Shape name
	2	Deactivate pen (pen up)
	(0,-14)	Line feed of 14 units
	0	Terminating zero

In the first line, the first data byte (10) is the shape number reserved for line feed, and the next data byte (5) is the number of characters in the shape specification. The data byte lf is the name of the shape.

In the second line, the first data byte (2) deactivates the pen. The next data byte (8) is a special code, 008, that defines a vector by X displacement and Y displacement. The third and fourth data bytes (0,-14) are the X displacement and Y displacement of the displacement vector; they produce a line feed that is 14 units below the baseline. The fifth data byte (0) is the zero that terminates the shape definition.

Shape Definition

The shape number in the shape definition of the text font corresponds to the ASCII code for that character. For example, if you are writing a shape definition for an uppercase M, the shape number is 77.

*77,50,ucm

Where	*77	Shape number - ASCII code of uppercase "M"
	50	Number of data bytes
	ucm	Shape name

The ASCII codes can be obtained from the ASCII character table, which gives the ASCII codes for all characters, numbers, and punctuation marks.

32	space	56	8	80	P	104	h
33	!	57	9	81	Q	105	i
34	"	58	:	82	R	106	j
35	#	59	;	83	S	107	k
36	$	60	<	84	T	108	l
37	%	61	=	85	U	109	m

Shapes and Text Fonts   11-21

38	&	62	>	86	V	110	n
39	'	63	?	87	W	111	o
40	(	64	@	88	X	112	p
41	)	65	A	89	Y	113	q
42	*	66	B	90	Z	114	r
43	+	67	C	91	[	115	s
44	,	68	D	92	\	116	t
45	-	69	E	93	]	117	u
46	.	70	F	94	^	118	v
47	/	71	G	95	_	119	w
48	0	72	H	96	`	120	x
49	1	73	I	97	a	121	y
50	2	74	J	98	b	122	z
51	3	75	K	99	c	123	{
52	4	76	L	100	d	124	³
53	5	77	M	101	e	125	}
54	6	78	N	102	f	126	~
55	7	79	O	103	g		

Example 6   *General*

Write a text font shape file (UCM) for an uppercase M as shown in Figure 11-15. The font file should be able to generate horizontal and vertical text. Each grid is 1 unit, and the directions of vectors are designated with leader lines.

Step 1: Writing the shape file
In the following file, the line numbers at the right are not a part of the file; they are for reference only. Use any text editor to write down the following shape file.

```
*0,4,uppercase m                                        1
10,0,2,0                                                2
*10,13,lf                                               3
002,8,(0,-14),14,9,(0,14),(14,0),(0,0),0                4
*77,51,ucm                                              5
2,14,8,(-5,-10),                                        6
001,009,(0,10),(1,0),(4,-6),(4,6),(1,0),                7
(0,-10),(-1,0),(0,0),                                   8
003,2,                                                  9
009,(0,17),(-7,-11),(-2,0),(-7,11),                     10
(0,-17),(-2,0),(0,0),                                   11
002,8,(28,0),                                           12
004,2,                                                  13
14,8,(-9,-4),0                                          14
```

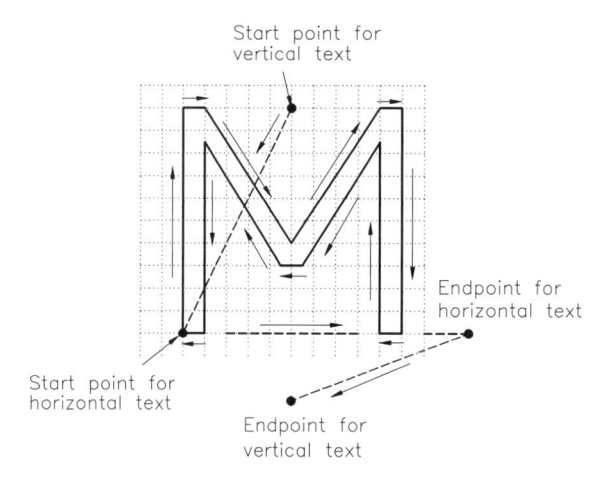

Figure 11-15 Shape and pen movement of uppercase "M"

Explanation
Line 1
***0,4,uppercase m**
The first data byte (0) is the special shape number for the text font file. The next data byte (4) is the number of data bytes, and the third data byte is the name of the shape.

Line 2
10,0,2,0
The first data byte (10) represents the total height of the character M, and the second data byte (0) represents the length of the lowercase letters that extend below the base line. Data byte 2 is the text mode for dual orientation of the text (horizontal and vertical). If the text was required in the horizontal direction only, the mode would be 0. The fourth data byte (0) is the zero that terminates the definition of this particular shape.

Line 3
*10,13,lf
The first data byte (10) is the code reserved for line feed, and the second data byte (13) is the number of data bytes in the shape specification. The third data byte (lf) is the name of the shape.

Line 4
002,8,(0,-14),14,9,(0,14),(14,0),(0,0),0
The first data byte (002 or 2) is the code to deactivate the pen (pen up). The next three data bytes [8,(0,-14)] define a displacement vector whose X displacement and Y displacement are 0 and -14 units, respectively. This will cause a carriage return that is 14 units below the text insertion point of the first text line. This will work fine if the text is drawn in the horizontal direction only. However, if the text is vertical, the carriage return should produce a displacement to the right of the existing line. This is accomplished by the next seven data bytes. Data byte

Shapes and Text Fonts

14 ignores the next code if the text is horizontal. If the text is vertical, the next code is processed. The next set of data bytes (0,14) defines a displacement vector that is 14 units below the previous point, D1 in Figure 11-16.

Data bytes (14,0) define a displacement vector that is 14 units to the right, D2 in Figure 11-16. These four data bytes combined will result in a carriage return that is 4 units to the right of the existing line. The next set of data bytes (0,0) terminates the code 9, and the last data byte (0) terminates the shape specification.

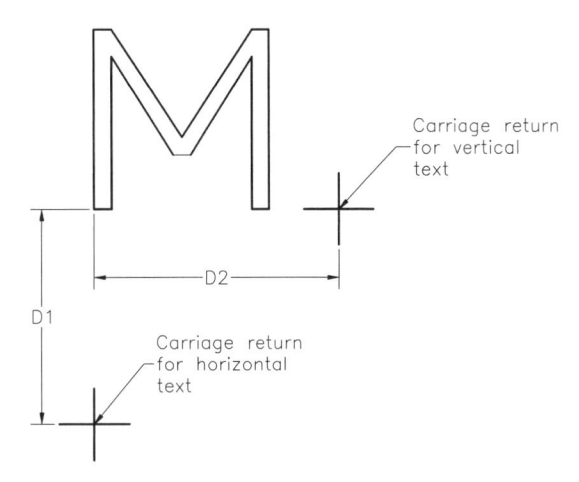

Figure 11-16 Carriage return for vertical and horizontal texts

Line 5
*77,51,ucm
The first data byte (77) is the ASCII code of the uppercase M. The second data byte (51) is the number of data bytes in the shape specification. The next data byte (ucm) is the name of the shape file in lowercase letters.

Line 6
2,14,8,(-5,-10),
The first data byte code (2) deactivates the pen (pen up), and the next data byte code (14) will cause the next code to be ignored if the text is horizontal. In the horizontal text, the insertion point of the text is the starting point of that text line (Figure 11-15). However, if the text is vertical, the starting point of the text is the upper middle point of the character M. This is accomplished by the next three data bytes [8,(-5,-10)], which displace the starting point of the text 5 units to the left (width of character M is 10) and 10 units down (height of character M is 10).

Lines 7,8
001,009,(0,10),(1,0),(4,-6),(4,6),(1,0),
(0,-10),(-1,0),(0,0),
The first byte (001) activates the draw mode (pen down), and the remaining bytes define the

next seven vectors.
Lines 9,10,11
003,2,
009,(0,17),(-7,-11),(-2,0),(-7,11),
(0,-17),(-2,0),(0,0),
The inner vertical line of the right leg of the character M is 8.5 units long, and you cannot define a vector that is not an integer. However, you can define a vector that is 2 x 8.5 = 17 units long and then divide that vector by 2 to get a vector 8.5 units long. This is accomplished by code 003 and the next data byte, 2. All the vectors defined in the next two lines will be divided by 2.

Line 12
002,8,(28,0),
The first data byte (002) deactivates the draw mode, and the next three data bytes define a vector that is 28/2 = 14 units to the right. This means that the next character will start 14 - 10 = 4 units to the right of the existing character so that it will produce a horizontal text.

Line 13
004,2,
The code 004 multiplies the vectors that follow it by 2; therefore, it nullifies the effect of code 003,2.

Line 14
14,8,(-9,-4),0
If the text is vertical, the next letter should start below the previous letter. This is accomplished by data bytes 8,(-9,-4), which define a vector that is -9 units along the X axis and -4 units along the Y axis. The data byte 0 terminates the definition of the shape.

Step 2: Loading the shape file
Save the file as **ucm.shp**. To compile the shape file, use the **COMPILE** command. Now, you can create text using this shape. To define a text style, select **Text Style** in the **Format** pull-down menu and then create a new style. You can also use the **-STYLE** command at the Command prompt to create a style corresponding to the compiled **.SHX** file. The prompt responses to the **-STYLE** command is as follows. Let us say **MYUCM1** is the current new text style.

> Command: **-STYLE**
> Enter name of text style or [?] <Standard>: **MYUCM1**
>
> New style.
> Specify full font name or font filename (TTF or SHX) <txt>: **ucm.shx**
> Specify height of text <0.0000>:**1**
> Specify width factor <1.0000>: *Press ENTER.*
> Specify obliquing angle <0>: *Press ENTER.*
> Display text backwards? [Yes/No] <N>: *Press ENTER.*
> Display text upside-down? [Yes/No] <N>: *Press ENTER.*

Shapes and Text Fonts

11-25

Vertical? <N> *Press ENTER.*
"**MYUCM1**" is now the current text style.

Now use the **DTEXT** command to write the text as **M** with the text style **s1** shown in

Figure 11-17 Using the defined text font

Example 7

Figure 11-17.
Write a text font shape file for the lowercase m as shown in Figure 11-18. The font file should be able to generate horizontal and vertical text. Each grid is 1 unit, and the direction of the

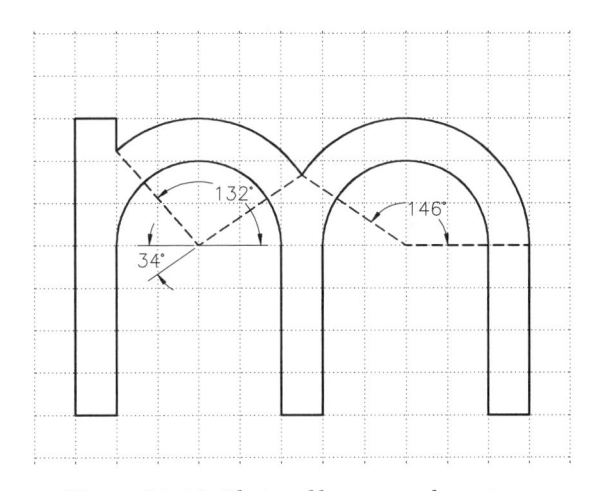

Figure 11-18 Shape of lowercase character, m

vectors is designated with the leader lines.
Step 1: Writing the shape file
The following file is a listing of the text font shape file for Example 7. The line numbers are

not a part of the file; they are shown here for reference only. Use any text editor to write down the following shape file.

*0,4,lower-case m	1
14,3,2,0	2
*10,13,lf	3
002,8,(0,-18),14,9,(0,18),(27,0),(0,0),0	4
*109,57,lcm	5
2,14,8,(-11,-14),	6
005,005,001,020,084,	7
00A,(4,-044),	8
08C,020,084,	9
00A,(4,-044),	10
08C,020,084,	11
00B,(0,62,0,6,004),	12
00B,(193,239,0,6,003),	13
006,9,(0,14),(2,0),(0,0),	14
003,5,07C,004,5	15
006,2,8,(27,0),	16
14,8,(-16,-5),0	17

Explanation

The format of most of the lines is the same as in the previous example, except for the following lines, which use save/restore origin, octant, and fractional arcs.

Line 7
005,005,001,020,084,
The first and second data bytes (005) are used to save the location of the point twice. The remaining data bytes activate the draw mode and define the vectors.

Line 8
00A,(4,-044),
The first data byte code (00A) is the code for the octant arc and the second data byte (4) defines the radius of the arc. The negative sign (-) in the third data byte generates an arc in a clockwise direction. The first element (0) is a hexadecimal notation. The second element defines the starting octant and the third element (4) defines the number of octants that the arc passes through.

Line 12
00B,(0,62,0,6,004),
The first data byte (00B) is the code for the fractional arc that is defined by the next five data bytes. The second data byte (0) is the starting offset of the first arc, as shown in the following calculations:

<u>**1st Arc**</u>
Starting angle = 0 Ending angle = 146
Starting octant = 0 Ending octant = 4
Starting offset = (0-0)*256/45

Shapes and Text Fonts   11-27

$$\text{Ending offset} \quad \begin{aligned} &= 0 \\ &= (146-135)*256/45 \\ &= 62.57 \\ &= 62 \text{ (integer)} \end{aligned}$$

The third data byte (62) is the ending offset of the arc and the fourth data byte (0) is the high radius. The fifth data byte (6) defines the radius of the arc. The second element (0) of the next data byte is the starting octant and the third element (4) is the number of octants the arc goes through.

Line 13
00B,(193,239,0,6,003),
The first data byte (00B) is the code for the fractional arc. The remaining data bytes define various parameters of the fractional arc as explained earlier. The offset angles have been obtained from the following calculations:

2nd Arc

$$\begin{aligned} \text{Starting angle} &= 34 \quad \text{Ending angle} = 132 \\ \text{Starting octant} &= 0 \quad \text{Ending octant} = 3 \\ \text{Starting offset} &= (34-0)*256/45 \\ &= 193.4 \\ &= 193 \text{ (integer)} \\ \text{Ending offset} &= (132-90)*256/45 \\ &= 238.9 \\ &= 239 \text{ (integer)} \end{aligned}$$

Note
Since the offset values have been rounded, it is not possible to describe an arc that is very accurate. Therefore, in this example the origin has been restored after two arcs were drawn. This origin was then used to draw the remaining lines.

Line 14
006,9,(0,14),(2,0),(0,0),
The first data byte (006) restores the previously saved point, and the remaining data bytes define the vectors using code 009.

Step 2: loading the shape file
Follow the procedure described in Example 6 to load the shape file and then use this shape to write the text.

Note
The compiled font files (.SHX), like shape files, must be present whenever the drawing is opened. If they send a drawing to someone else, they also must send the shape and font files.

Self-Evaluation Test

Answer the following questions and then compare your answers with the answers given at the end of the chapter.

1. The basic objects used in shape files are _____ and _____.

2. Shapes are easy to insert and take less disk space than _____. However, there are certain disadvantages to using shapes. For example, you cannot _____ a shape.

3. The shape number could be any number between 1 and _____ in a particular file, and these numbers cannot be repeated within the same file.

4. The shape file may not contain two _____ with the same name.

5. A hexadecimal number is designated by a leading _____.

6. The maximum number of data bytes is _____ bytes per shape.

7. To define a vector, you need its magnitude and _____.

8. To load the shape file, use the AutoCAD _____ command.

9. Generating shapes with direction vectors has some limitations. For example, you cannot draw an arc or a line that is not along the _____ vectors. These limitations can be overcome by using the _____, which add a lot of flexibility and give you better control over the shapes you want to create.

10. Code 001 activates the _____ mode, and code _____ deactivates the draw mode.

11. The byte that follows the division code divides the _____ vectors.

12. Code 004 is used if you want to multiply the vectors by a certain number. It can also be used to _____ the effect of code 003.

Review Questions

Answer the following questions.

1. Scale factors are _____ within the shape.

2. The number of saves (code 005) must be equal to the number of _____ code _____.

3. The maximum number of saves and restores you can use in a particular shape definition

Shapes and Text Fonts   11-29

is _____.

4. You can define a subshape like a subroutine in a program. To reference the subshape, use code _____.

5. Vectors can be drawn in the 16 predefined directions only, and the length of the vector cannot exceed _____ units.

6. A nonstandard fractional arc can be generated by using code 00B or _____.

7. Code _____ can be used to define an arc by specifying the displacement of the endpoint of an arc and the bulge factor.

8. The bulge factor can range from -127 to _____.

9. Code 00E or _____ is used when the same text font description is to be used in both horizontal and vertical orientation.

10. The text files are regular _____ files with some additional information about the text font description and the line feed.

11. The shape names in all text font files should be lowercase so that the computer does not have to save the names in its _____.

12. The line feed is used to space the lines so that the characters do not _____.

13. The shape number in the shape definition of the text font corresponds to the _____ code for that character.

14. To recall a subshape, the subshape has to be _____ within the same shape file.

15. Each 45 degree angle segment is called an _____.

16. AutoCAD has reserved the ASCII code no. 10 to define the _____.

17. Shapes cannot be _____ after insertion.

18. The shape name is the name of the shape in _____ letter.

19. During the definition of the shape, the first line is the _____ line and the second line is the _____.

Exercises

Exercise 1 — General

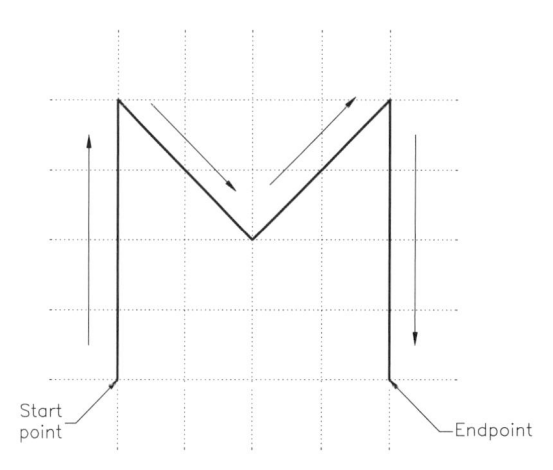

Figure 11-19 Uppercase letter M

Exercise 2 — General

Write a shape file for the uppercase M shown in Figure 11-19.
Write a shape file for generating the tapered gib-head key shown in Figure 11-20.

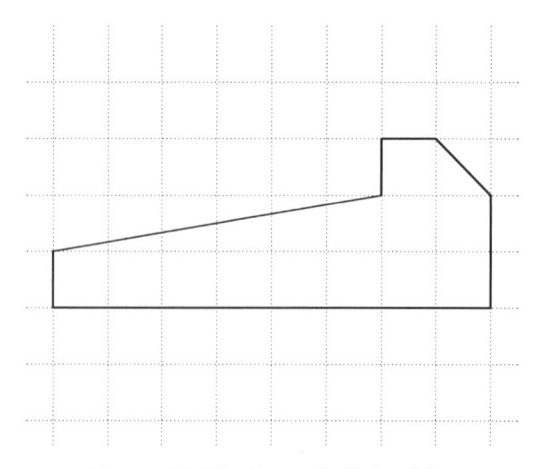

Figure 11-20 Tapered gib-head key

Shapes and Text Fonts 11-31

Exercise 3 *General*

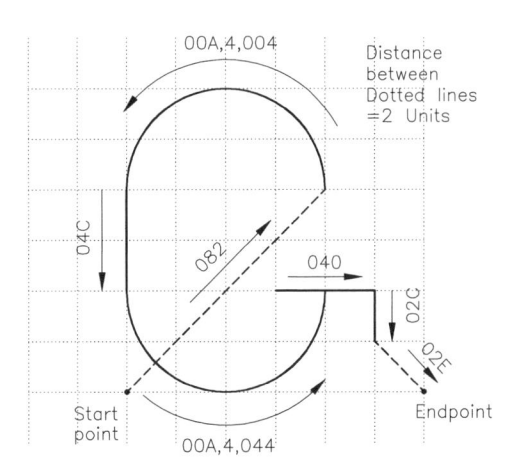

Figure 11-21 Uppercase letter G

Exercise 4 *General*

Write a text font shape file for the uppercase G shown in Figure 11-21.
Write a text font shape file for the uppercase W shown in Figure 11-22. The font file should be

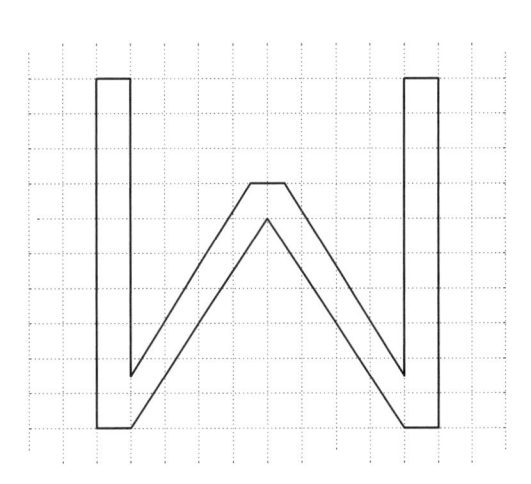

Figure 11-22 Uppercase letter W

able to generate horizontal and vertical text.

Answers to the Self-Evaluation Test
1. line and Arc, **2.** blocks, edit, **3.** 255, **4.** shapes, **5.** zero, **6.** 2000, **7.** direction, **8. LOAD**, **9.** direction, special codes, **10.** draw mode, 002, **11.** whole, **12.** nullify

Chapter 12

AutoLISP

Learning Objectives
After completing this chapter, you will be able to:
• *Perform mathematical operations using AutoLISP.*
• *Use trigonometrical functions in AutoLISP.*
• *Understand the basic AutoLISP functions and their applications.*
• *Load and run AutoLISP programs.*
• *Use the **Load/Unload Applications** dialog box.*
• *Use flowcharts to analyze problems.*
• *Test a condition using conditional functions.*

ABOUT AutoLISP

Developed by Autodesk, Inc., **AutoLISP** is an implementation of the **LISP** programming language (LISP is an acronym for **LISt Processor**.) The first reference to LISP was made by John McCarthy in the April 1960 issue of *The Communications of the ACM*.

Except for **FORTRAN** and **COBOL**, most of the languages developed in the early 1960s have become obsolete. But LISP has survived and has become a leading programming language for artificial intelligence (AI). Some of the dialects of the LISP programming language are Common LISP, BYSCO LISP, ExperLISP, GCLISP, IQLISP, LISP/80, LISP/88, MuLISP, TLCLISP, UO-LISP, Waltz LISP, and XLISP. XLISP is a public-domain LISP interpreter. The LISP dialect that resembles AutoLISP is Common LISP. The AutoLISP interpreter is embedded within the AutoCAD software package. However, AutoCAD LT and AutoCAD Versions 2.17 and lower lack the AutoLISP interpreter; therefore, you can use the AutoLISP programming language only with AutoCAD Release 2.18 and up.

The AutoCAD software package contains most of the commands used to generate a drawing. However, some commands are not provided in AutoCAD. For example, AutoCAD has no command to make global changes in the drawing text objects. With AutoLISP you can write a program in the AutoLISP programming language that will make global or selective changes in the drawing text objects. As a matter of fact, you can use AutoLISP to write any program or embed it in the menu and thus customize your system to make it more efficient.

The AutoLISP programming language has been used by hundreds of third-party software developers to write software packages for various applications. For example, the author of this text has developed a software package, **SMLayout**, that generates flat layouts of various geometrical shapes like transitions, intersection of pipes and cylinders, elbows, cones, and tank heads. There is demand for AutoLISP programmers as consultants for developing application software and custom menus.

This chapter assumes that you are familiar with AutoCAD commands and AutoCAD system variables. However, you need not be an AutoCAD or programming expert to begin learning AutoLISP. This chapter also assumes that you have no prior programming knowledge. If you are familiar with any other programming language, learning AutoLISP may be easy. A thorough discussion of various functions and a step-by-step explanation of the examples should make it fun to learn. This chapter discusses the most frequently used AutoLISP functions and their application in writing a program. For those functions not discussed in this chapter, refer to the **AutoLISP Programmers Reference Manual** from Autodesk. AutoLISP does not require any special hardware. If your system runs AutoCAD, it will also run AutoLISP. To write AutoLISP programs you can use any text editor.

MATHEMATICAL OPERATIONS

A mathematical function constitutes an important feature of any programming language. Most of the mathematical functions commonly used in programming and mathematical calculations are available in AutoLISP. You can use AutoLISP to add, subtract, multiply, and divide numbers. You can also use it to find the sine, cosine, and arctangent of angles expressed in radians. There is a host of other calculations you can do with AutoLISP. This section discusses the most frequently used mathematical functions supported by the AutoLISP programming language.

Addition
Format **(+ num1 num2 num3 - - -)**

This function (+) calculates the sum of all the numbers to the right of the plus (+) sign (num1 + num2 + num3 + . . .). The numbers can be integers or real. If the numbers are integers, then the sum is an integer. If the numbers are real, the sum is real. However, if some numbers are real and some are integers, the sum is real. In the following display, all numbers in the first two examples are integers, so the result is an integer. In the third example, one number is a real number (50.0), so the sum is a real number.

Examples

Command: (+ 2 5)	returns 7
Command: (+ 2 30 4 50)	returns 86
Command: (+ 2 30 4 50.0)	returns 86.0

Subtraction

Format **(- num1 num2 num3 - - -)**

This function (-) subtracts the second number from the first number (num1 - num2). If there are more than two numbers, the second and subsequent numbers are added and the sum is subtracted from the first number [num1 - (num2 + num3 + . . .)]. In the first of the following examples, 14 is subtracted from 28 and returns 14. Since both numbers are integers, the result is an integer. In the third example, 20 and 10.0 are added, and the sum of these two numbers (30.0) is subtracted from 50, returning a real number, 20.0.

Examples
Command: (- 28 14)   returns 14
Command: (- 25 7 11)   returns 7
Command: (- 50 20 10.0)   returns 20.0
Command: (- 20 30)   returns -10
Command: (- 20.0 30.0)   returns -10.0

Multiplication

Format **(* num1 num2 num3 - - -)**

This function (*) calculates the product of the numbers to the right of the asterisk (num1 x num2 x num3 x . . .). If the numbers are integers, the product of these numbers is an integer. If one of the numbers is a real number, the product is a real number.

Examples
Command: (* 2 5)   returns 10
Command: (* 2 5 3)   returns 30
Command: (* 2 5 3 2.0)   returns 60.0
Command: (* 2 -5.5)   returns -11.0
Command: (* 2.0 -5.5 -2)   returns 22.0

Division

Format **(/ num1 num2 num3 - - -)**

This function (/) divides the first number by the second number (num1/num2). If there are more than two numbers, the first number is divided by the product of the second and subsequent numbers [num1 / (num2 x num3 x . . .)]. In the fourth of the following examples, 200 is divided by the product of 5.0 and 4 [200 / (5.0 * 4)].

Examples
Command: (/ 30)   returns 30
Command: (/ 3 2)   returns 1
Command: (/ 3.0 2)   returns 1.5
Command: (/ 200.0 5.0 4)   returns 10.0
Command: (/ 200 -5)   returns -40
Command: (/ -200 -5.0)   returns 40.0

INCREMENTED, DECREMENTED, AND ABSOLUTE NUMBERS

Incremented Number

Format **(1+ number)**

This function **(1+)** adds 1 (integer) to **number** and returns a number that is incremented by 1. In the second example below, 1 is added to -10.5 and returns -9.5.

Examples
(1+ 20)	returns 21
(1+ -10.5)	returns -9.5

Decremented Number

Format **(1- number)**

This function **(1-)** subtracts 1 (integer) from the **number** and returns a number that is decremented by 1. In the second example below, 1 is subtracted from -10.5 and returns -11.5

Examples
(1- 10)	returns 9
(1- -10.5)	returns -11.5

Absolute Number

Format **(abs num)**

The **abs** function returns the absolute value of a number. The number may be an integer number or a real number. In the second example below, the function returns 20 because the absolute value of -20 is 20.

Examples
(abs 20)	returns 20
(abs -20)	returns 20
(abs -20.5)	returns 20.5

TRIGONOMETRIC FUNCTIONS

sin

Format **(sin angle)**

The **sin** function calculates the sine of an angle, where the angle is expressed in radians. In the second of the following examples, the **sin** function calculates the sine of pi (180 degrees) and returns 0.

Examples

Command: (sin 0)   returns 0.0
Command: (sin pi)   returns 0.0
Command: (sin 1.0472)   returns 0.866027

cos

Format **(cos angle)**

The **cos** function calculates the cosine of an angle, where the angle is expressed in radians. In the third of the following examples, the **cos** function calculates the cosine of pi (180 degrees) and returns -1.0.

Examples

Command: (cos 0)   returns 1.0
Command: (cos 0.0)   returns 1.0
Command: (cos pi)   returns -1.0
Command: (cos 1.0)   returns 0.540302

atan

Format **(atan num1)**

The **atan** function calculates the arctangent of **num1**, and the calculated angle is expressed in radians. In the second of the following examples, the **atan** function calculates the arctangent of 1.0 and returns 0.785398 (radians).

Examples

Command: (atan 0.5)   returns 0.463648
Command: (atan 1.0)   returns 0.785398
Command: (atan -1.0)   returns -0.785398

You can also specify a second number in the atan function:

Format **(atan num1 num2)**

If the second number is specified, the function returns the arctangent of (num1/num2) in radians. In the first of the following examples the first number (0.5) is divided by the second number (1.0), and the **atan** function calculates the arctangent of the dividend (0.5/1.0 = 0.5).

Examples

Command: (atan 0.5 1.0)   returns 0.463648 radians
Command: (atan 2.0 3.0)   returns 0.588003 radians
Command: (atan 2.0 -3.0)   returns 2.55359 radians
Command: (atan -2.0 3.00)   returns -0.588003 radians
Command: (atan -2.0 -3.0)   returns -2.55359 radians
Command: (atan 1.0 0.0)   returns 1.5708 radians
Command: (atan -0.5 0.0)   returns -1.5708 radians

angtos

Format **(angtos angle [mode [precision]])**

The **angtos** function returns the angle expressed in radians in a string format. The format of the string is controlled by the **mode** and **precision** settings.

Examples

(angtos 0.588003 0 4)	returns "33.6901"
(angtos 2.55359 0 4)	returns "146.3099"
(angtos 1.5708 0 4)	returns "90.0000"
(angtos -1.5708 0 2)	returns "270.00"

Note

In *(angtos angle [mode [precision]])*
angle is angle in radians
mode is the angtos mode that corresponds to the AutoCAD system variable AUNITS
The following modes are available in AutoCAD:

ANGTOS MODE	EDITING FORMAT
0	Decimal degrees
1	Degrees/minutes/seconds
2	Grads
3	Radians
4	Surveyor's units

*Precision is an integer number that controls the number of decimal places. Precision corresponds to the AutoCAD system variable **AUPREC**. The minimum value of **precision** is zero and the maximum is four.*

In the first example above, angle is 0.588003 radians, mode is 0 (angle in degrees), and precision is 4 (four places after decimal). The function will return 33.6901.

RELATIONAL STATEMENTS

Programs generally involve features that test a particular condition. If the condition is true, the program performs certain functions, and if the condition is not true, then the program performs other functions. For example, the relational statement (if (< x 5)) tests true if the value of the variable x is less than 5. This type of test condition is frequently used in programming. The following section discusses various relational statements used in AutoLISP programming.

Equal to

Format **(= atom1 atom2 - - - -)**

This function (=) checks whether the two atoms are equal. If they are equal, the condition is true and the function will return T. Similarly, if the specified atoms are not equal, the condition is false and the function will return nil.

AutoLISP

Examples

(= 5 5)	returns T
(= 5 4.9)	returns nil
(= 5.5 5.5 5.5)	returns T
(= "yes" "yes")	returns T
(= "yes" "yes" "no")	returns nil

Not equal to

Format (/= atom1 atom2 - - - -)

This function (/=) checks whether the two atoms are not equal. If they are not equal, the condition is true and the function will return T. Similarly, if the specified atoms are equal, the condition is false and the function will return nil.

Examples

(/= 50 4)	returns T
(/= 50 50)	returns nil
(/= 50 -50)	returns T
(/= "yes" "no")	returns T

Less than

Format (< atom1 atom2 - - - -)

This function (<) checks whether the first atom (**atom1**) is less than the second atom (**atom2**). If it is true, then the function will return **T**. If it is not, the function will return **nil**.

Examples

(< 3 5)	returns T
(< 5 3 4 2)	returns nil
(< "x" "y")	returns T

Less than or equal to

Format (<= atom1 atom2 - - - -)

This function (<=) checks whether the first atom (**atom1**) is less than or equal to the second atom (atom2). If it is, the function will return **T**. If it is not, the function will return nil.

Examples

(<= 10 15)	returns T
(<= "c" "b")	returns nil
(<= -2.0 0)	returns T

Greater than

Format (> atom1 atom2 - - - -)

This function (>) checks whether the first atom (**atom1**) is greater than the second atom

(atom2). If it is, the function will return **T**. If it is not, then the function will return **nil**. In the first example below, 15 is greater than 10. Therefore, this relational function is true and the function will return T. In the second example, 10 is greater than 9, but this number is not greater than the second 9; therefore, this function will return **nil**.

Examples

(> 15 10)	returns T
(> 10 9 9)	returns nil
(> "c" "b")	returns T

Greater than or equal to

Format **(>= atom1 atom2 - - - -)**

This function (>=) checks whether the first atom **(atom1)** is greater than or equal to the second atom (atom2). If it is, the function returns **T**; otherwise, it will return nil. In the first example below, 78 is greater than 50, but 78 is not equal to 50; therefore, it will return **nil**.

Examples

(>= 78 50)	returns T
(>= "x" "y")	returns T

defun, setq, getpoint, and Command Functions

defun

The **defun** function is used to define a function in an AutoLISP program. The format of the **defun** function is:

(defun name [argument])
    Where **Name** ------------ Name of the function
           Argument ------- Argument list

Examples
(defun ADNUM ( )
Defines a function ADNUM with no arguments or local symbols. This means that all variables used in the program are global variables. A global variable does not lose its value after the programs ends.

(defun ADNUM (a b c)
Defines a function ADNUM that has three arguments: **a, b,** and **c**. The variables **a, b,** and **c** receive their value from outside the program.

(defun ADNUM (/ a b)
Defines a function ADNUM that has two local variables: **a** and **b**. A local variable is one that retains its value during program execution and can be used within that program only.

(defun C:ADNUM ( )

AutoLISP

With **C:** in front of the function name, the function can be executed by entering the name of the function at the AutoCAD Command: prompt. If **C:** is not used, the function name has to be enclosed in parentheses.

> **Note**
> *AutoLISP contains some built-in functions. Do not use any of those names for function or variable names. The following is a list of some of the names reserved for AutoLISP built-in functions. (Refer to the AutoLISP Programmer's Reference manual for a complete list of AutoLISP built-in functions.)*

abs	ads	alloc
and	angle	angtos
append	apply	atom
ascii	assoc	atan
atof	atoi	distance
equal	fix	float
if	length	list
load	member	nil
not	nth	null
open	or	pi
read	repeat	reverse
set	type	while

setq

The **setq** function is used to assign a value to a variable. The format of the **setq** function is:

(setq Name Value [Name Value].......)
  Where **Name** ------------ Name of variable
    Value ------------- Value assigned to variable

The value assigned to a variable can be any expression (numeric, string, or alphanumeric).

  Command: **(setq X 12)**
  Command: **(setq X 6.5)**
  Command: **(setq X 8.5 Y 12)**

In this last expression, the number 8.5 is assigned to the variable **X**, and the number 12 is assigned to the variable **Y**.

  Command: **(setq answer "YES")**

In this expression the string value "YES" is assigned to the variable **answer**.

The **setq** function can also be used in conjunction with other expressions to assign a value to a variable. In the following examples the **setq** function has been used to assign values to different variables.

  (setq pt1 (getpoint "Enter start point: "))

(setq ang1 (getangle "Enter included angle:"))
(setq answer (getstring "Enter YES or NO: "))

Note
*AutoLISP uses some built-in function names and symbols. Do not assign values to any of those functions. The following functions are valid ones, but must never be used because the **pi** and **angle** functions that are reserved functions will be redefined.*

(setq pi 3.0)
(setq angle (. . .))

getpoint
The **getpoint** function pauses to enable you to enter the X, Y coordinates or X, Y, Z coordinates of a point. The coordinates of the point can be entered from the keyboard or by using the screen cursor. The format of the **getpoint** function is:

(getpoint [point] [prompt])
    Where **Point** ------------- Enter a point, or select a point
                Prompt ---------- Prompt to be displayed on screen

Example
(setq pt1 (getpoint))
(setq pt1 (getpoint "Enter starting point"))

Note
You cannot enter the name of another AutoLISP routine in response to the getpoint function.

A 2D or a 3D point is always defined with respect to the current user coordinate system (UCS).

Command
The **Command** function is used to execute standard AutoCAD commands from within an AutoLISP program. The AutoCAD command name and the command options have to be enclosed in double quotation marks. The format of the **Command** function is:

(Command "commandname")
    Where **Command** ----------------- AutoLISP function
               commandname ---------- AutoCAD command

Example
(Command "line" pt1 pt2 "")
    Where **"Line"** ----------- AutoCAD LINE Command
              Pt1 ---------------- First point
              Pt2 ---------------- Second point
              "" ----------------- for Return (two double quotes with no space between)

AutoLISP

Note

*Prior to AutoCAD Release 12, the **Command** function **could not be used** to execute the AutoCAD PLOT command. For example: (Command "plot" . . .) was not a valid statement. In AutoCAD Release 13 and later, you can use plot with the Command function (Command "plot" . . .).*

*The **Command** function cannot be used to enter data with the AutoCAD DTEXT or TEXT command. (You can issue the DTEXT and TEXT command with the **Command** function. You can also enter text height and text rotation, but you cannot enter the text when DTEXT or TEXT prompts for text entry.)*

*You cannot use the input functions of AutoLISP with the **Command** function. The input functions are **getpoint, getangle, getstring**, and **getint**. For example, (Command "getpoint" . . .) or (Command "getangle" . . .) are not valid functions. If the program contains such a function, it will display an error message when the program is loaded. However, you can enclose them in parentheses (Command "text" (getpoint) (getdist) (getangle) "hello, sailer")*

Example 1

Write a program that will prompt you to select three points of a triangle and then draw lines through those points to generate the triangle shown in Figure 12-1.

Step 1: Understanding the lisp program

Most programs consist of essentially three parts: **input, output,** and **process**. **Process** includes what is involved in generating the desired output from the given input (Figure 12-2). Before writing a program, you must identify these three parts. In this example, the **input** to the program is the coordinates of the three points. The desired **output** is a triangle. The **process** needed to generate a triangle is to draw three lines from P1 to P2, P2 to P3, and P3 to P1. Identifying these three sections makes the programming process less confusing.

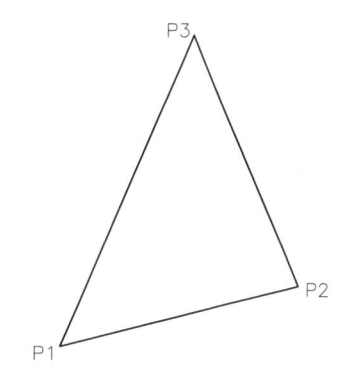

Figure 12-1 Triangle P1, P2, P3

The process section of the program is vital to the success of the program. Sometimes it is simple, but sometimes it involves complicated calculations. If the program involves many calculations, divide them into sections (and perhaps subsections) that are laid out in a logical and systematic order. Also, remember that programs need to be edited from time to time, perhaps by other programmers. Therefore, it is wise to document the programs as clearly and unambiguously as

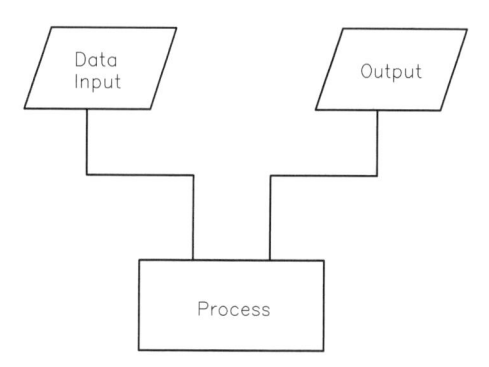

Figure 12-2 Three elements of a Program

possible so that other programmers can understand what the program is doing at different stages of its execution. Give sketches, and identify points where possible.

Input
Location of point P1
Location of point P2
Location of point P3

Output

Triangle P1, P2, P3

Process
Line from P1 to P2
Line from P2 to P3
Line from P3 to P1

Step 2: Writing the lisp program
Use any text editor to write the following lisp program. The following file is a listing of the AutoLISP program for Example 1. **The line numbers at the right are not a part of the program; they are shown here for reference only.**

```
;This program will prompt you to enter three points              1
;of a triangle from the keyboard, or select three points         2
;by using the screen cursor. P1, P2, P3 are triangle corners.    3
                                                                 4
(defun c:TRIANG1()                                               5
   (setq P1 (getpoint "\n Enter first point of Triangle: "))     6
   (setq P2 (getpoint "\n Enter second point of Triangle: "))    7
   (setq P3 (getpoint "\n Enter third point of Triangle: "))     8
   (Command "LINE" P1 P2 P3 "C")                                 9
)                                                                10
```

Explanation
Lines 1-3
The first three lines are comment lines describing the function of the program. These lines are important because they make it easier to edit a program. Comments should be used when needed. All comment lines must start with a semicolon (;). These lines are ignored when the program is loaded.

Line 4
This is a blank line that separates the comment section from the program. Blank lines can be used to separate different modules of a program. This makes it easier to identify different sections that constitute a program. The blank lines have no effect on the program.

Line 5
(defun c:TRIANG1( )
In this line, **defun** is an AutoLISP function that defines the function **TRIANG1**. **TRIANG1** is the name of the function. The c: in front of the function name, **TRIANG1**, enables it to be executed like an AutoCAD command. If the c: is missing, the **TRIANG1** command can be executed only by enclosing it in parentheses (TRIANG1). The TRIANG1 function has three

global variables (P1, P2, and P3). When you first write an AutoLISP program, it is a good practice to keep the variables global, because after you load and run the program, you can check the values of these variables by entering an exclamation point (!) followed by the variable name at the AutoCAD Command prompt (Command: !P1). Once the program is tested and it works, you should make the variable local, (defun c:TRIANG1(/ P1 P2 P3)).

Line 6
(setq P1 (getpoint "\n Enter first point of Triangle: "))
In this line, the **getpoint** function pauses for you to enter the first point of the triangle. The prompt, **Enter first point of Triangle,** is displayed in the prompt area of the screen. You can enter the coordinates of this point at the keyboard or select a point by using the screen cursor. The setq function then assigns these coordinates to the variable **P1**. **\n** is used for the carriage return so that the statement that follows \n is printed on the next line ("n" stands for "newline".)

Lines 7 and 8
(setq P2 (getpoint "\n Enter second point of Triangle: "))
(setq P3 (getpoint "\n Enter third point of Triangle: "))
These two lines prompt you to enter the second and third corners of the triangle. These coordinates are then assigned to the variables **P2** and **P3**. **\n** causes a carriage return so that the input prompts are displayed on the next line.

Line 9
(Command "LINE" P1 P2 P3 "C")
In this line, the **Command** function is used to enter the AutoCAD LINE command and then draw a line from P1 to P2 and from P2 to P3. "C" (for "close" option) joins the last point, P3, with the first point, P1. All AutoCAD commands and options, when used in an AutoLISP program, have to be enclosed in double quotation marks. The variables P1, P2, P3 are separated by a blank space.

Line 10
This line consists of a closing parenthesis that completes the definition of the function, TRIANG1. This parenthesis could have been combined with the previous line. It is good practice to keep it on a separate line so that any programmer can easily identify the end of a definition. In this program there is only one function defined, so it is easy to locate the end of a definition. But in some programs a number of definitions or modules within the same program might need to be clearly identified. The parentheses and blank lines help to identify the start and end of a definition or a section in the program.

LOADING AN AUTOLISP PROGRAM

Save the file with the extension **.LSP**. There are generally two names associated with an AutoLISP program: the program file name and the function name. For example, **TRIANG.LSP** is the name of the file, not a function name. All AutoLISP file names have the extension **.LSP**. An AutoLISP file can have one or several functions defined within the same file. For example, TRIANG1 in Example 1 is the name of a function. To execute a function, the AutoLISP program file that defines that function must be loaded. Use the following procedure to load an AutoLISP file when you are in the drawing editor.

Loading of an AutoLISP program can be either achieved through the dialog box or by entering a command at the Command prompt. Choose the **Tools > AutoLISP > Load** menu or choose **Load Application** from the **Tools** menu to display the **Load/Unload Application** dialog box (Figure 12-3). This dialog box can be used to load LSP, VLX, FAS, VBA, DBX, and ObjectARX applications. VBA, DBX, and ObjectARX files are loaded immediately when you select a file. LSP, VLX, and FAS files are qued and loaded when you close the **Load/Unload Application** dialog box. The top portion of the dialog box lists the files in the selected directory. The file type can be changed by entering the file type (*.lsp) in the **Files of type:** edit box or by selecting the file type in the pop-down list. A file can be loaded by selecting the file and then choosing the **Load** button. The following is the description of other features of the **Load/Unload Application** dialog box:

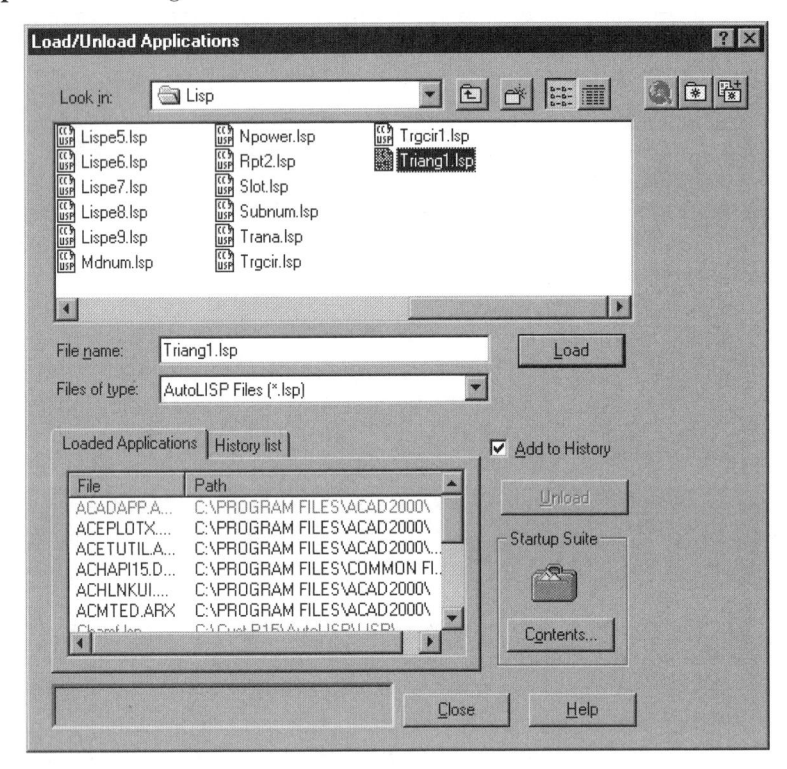

*Figure 12-3 Loading AutoLISP files using the **Load/Unload Applications** dialog box*

Load
The **Load** button can be used to load and reload the selected files. The files can be selected from the file list box, Load Application tab, or History List tab. The Objecr ARX files cannot be reloaded. You must first unload the ObjectARX file and then reload it.

Load Application Tab
When you choose the **Load Application** tab, AutoCAD displays the applications that are currently loaded. You can add files to this list by dragging the file names from the file list box and then dropping them in the Load Application list.

History List Tab

When you choose the **History List** tab, AutoCAD displays the list of files that have been previously loaded with Add to History selected. If the Add to History is not selected and you drag and drop the files in the History list, the files are loaded but not added to the History List.

Add to History

When Add to History is selected, it automatically adds the files to the History List when you drag and drop the files in the History List.

Unload

The **Unload** button appears when you choose the **Loaded Application** tab. To unload an application, select the file name in the Loaded Application file list and then choose the **Unload** button. LISP files and the ObjectARX files that are not registered for unloading cannot be unloaded.

Remove

The **Remove** button appears when you choose the **History List** tab. To remove a file from the History List, select the file in the History List and then choose the **Remove** button.

Startup Suit

The files in the Startup Suit are automatically loaded each time you start AutoCAD. When you choose the Startup Suit, AutoCAD displays the **Startup Suit** dialog box that contains a list of files. You can add files to the list by choosing the **Add** button. You can also drag the files from the file list box and drop them in the Startup Suit. To add files from the History List, right click on the file.

The AutoLISP program can also be loaded by entering the LOAD command in the following format.

```
Command: (load "[path]file name")
        Where  Command ------ AutoCAD command prompt
               load    -------------- Loads an AutoLISP program file
               file name -------- Path and name of AutoLISP program file
```

The AutoLISP file name and the optional path name must be enclosed in double quotes. The **load** and **file name** must be enclosed in parentheses. If the parentheses are missing, AutoCAD will try to load a shape or a text font file, not an AutoLISP file. The space between **load** and **file name** is not required. If AutoCAD is successful in loading the file, it will display the name of the function in the Command: prompt area of the screen.

To run the program, type the name of the function at the AutoCAD Command: prompt, and press ENTER (Command: **TRIANG1**). If the function name does not contain **C:** in the program, you can run the program by enclosing the function name in parentheses:

Command: **TRIANG1** or Command: **(TRIANG1)**

> **Note**
> *Use a forward slash when defining the path for loading an AutoLISP program. For example, if the AutoLISP file TRIANG is in the LISP subdirectory on the C drive, use the following command to load the file. You can also use a double backslash (\\) in place of the forward slash.*
> Command: **(load "c:/lisp/triang")** or Command: **(load "c:\\lisp\\triang")**

> **Tip**
> *You can also load an application by using the standard Windows drag and drop technique. To load a LISP program, select the file in the Windows Explorer and then drag and drop it in the graphics window of AutoCAD. The selected program will be automatically loaded.*

Exercise 1 — General

Write an AutoLISP program that will draw a line between two points (Figure 12-4). The program must prompt the user to enter the X and Y coordinates of the points.

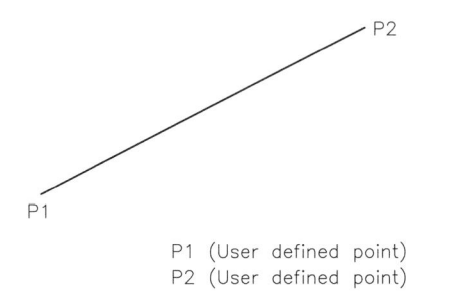

P1 (User defined point)
P2 (User defined point)

Figure 12-4 Draw line from point P1 to P2

getcorner, getdist, AND setvar FUNCTIONS

getcorner

The **getcorner** function pauses for you to enter the coordinates of a point. The coordinates of the point can be entered at the keyboard or by using the screen crosshairs. This function requires a base point, and it displays a rectangle with respect to the base point as you move the screen crosshairs around the screen. The format of the **getcorner** function is:

(getcorner point [prompt])
    Where **point** ------------------------ Base point
                  prompt -------------------- Prompt displayed on screen

Examples
(getcorner pt1)
(setq pt2 (getcorner pt1))
(setq pt2 (getcorner pt1 "Enter second point: "))

> **Note**
> *The base point and the point that you select in response to the getcorner function are located with respect to the current UCS.*

AutoLISP   12-17

If the point you select is a 3D point with X, Y, and Z coordinates, the Z coordinate is ignored. The point assumes current elevation as its Z coordinate.

getdist

The **getdist** function pauses for you to enter distance, and it then returns the distance as a real number. The format of the **getdist** function is:

(getdist [point] [prompt])
Where **Point** ------------- First point for distance
Prompt ---------- Any prompt that needs to be displayed on screen

Examples
(getdist)
(setq dist (getdist))
(setq dist (getdist pt1))
(setq dist (getdist "Enter distance"))
(setq dist (getdist pt1 "Enter second point for distance"))

The distance can be entered by selecting two points on the screen. For example, if the assignment is **(setq dist (getdist))**, you can enter a number or select two points. If the assignment is **(setq dist (getdist pt1))**, where the first point (pt1) is already defined, you need to select the second point only. The getdist function will always return the distance as a real number. For example, if the current setting is architectural and the distance is entered in architectural units, the getdist function will return the distance as a real number.

setvar

The **setvar** function assigns a value to an AutoCAD system variable. The name of the system variable must be enclosed in double quotes. The format of the **setvar** function is:

(setvar "variable-name" value)
Where **Variable-name** - AutoCAD system variable
Value ------------- Value to be assigned to the system variable

Examples
(setvar "cmdecho" 0)
(setvar "dimscale" 1.5)
(setvar "ltscale" 0.5)
(setvar "dimcen" -0.25)

Example 2

Write an AutoLISP program that will generate a chamfer between two given lines by entering the chamfer angle and the chamfer distance. To generate a chamfer, AutoCAD uses the values assigned to system variables **CHAMFERA** and **CHAMFERB**. When you select the AutoCAD CHAMFER command, the first and second chamfer distances are automatically assigned to the system variables **CHAMFERA** and **CHAMFERB**. The CHAMFER command then uses these assigned values to generate a chamfer. However, in most engineering drawings, the preferred way to generate the chamfer is to enter the chamfer length and the chamfer angle, as shown in Figure 12-5.

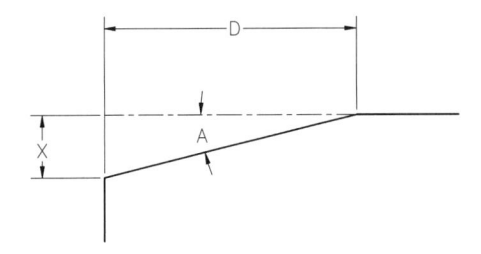

Figure 12-5 Chamfer with angle A and distance D

Input
First chamfer distance (D)
Chamfer angle (A)

Output
Chamfer between any two selected lines

Step 1: Understanding the program algorithm

Process
1. Calculate second chamfer distance.
2. Assign these values to the system variables CHAMFERA and CHAMFERB.
3. Use AutoCAD CHAMFER command. to generate chamfer,

Calculations
x/d = tan a
x =   d * (tan a)
  =   d * [(sin a) / (cos a)]

Step 2: Writing the lisp prgram

Use any text editor to write down the lisp program. The following file is a listing of the program for Example 3. **The line numbers on the right are not a part of the file; they are for reference only.**

```
;This program generates a chamfer by entering                    1
;the chamfer angle and the chamfer distance                      2
;                                                                3
(defun c:chamf (/ d a)                                           4
(setvar "cmdecho" 0)                                             5
   (graphscr)                                                    6
   (setq d (getdist "\n Enter chamfer distance: "))              7
   (setq a (getangle "\n Enter chamfer angle: "))                8
```

```
        (setvar "chamfera" d)                                    9
        (setvar "chamferb" (* d (/ (sin a) (cos a))))           10
        (command "chamfer")                                     11
      (setvar "cmdecho" 1)                                      12
      (princ)                                                   13
    )                                                           14
```

Explanation

Line 7
(setq d (getdist "\n Enter chamfer distance: "))
The **getdist** function pauses for you to enter the chamfer distance, then the **setq** function assigns that value to variable d.

Line 8
(setq a (getangle "\n Enter chamfer angle: "))
The **getangle** pauses for you to enter the chamfer angle, then the **setq** function assigns that value to variable a.

Line 9
(setvar "chamfera" d)
The **setvar** function assigns the value of variable d to AutoCAD system variable **chamfera**.

Line 10
(setvar "chamferb" (* d (/ (sin a) (cos a))))
The **setvar** function assigns the value obtained from the expression (* d (/ (sin a) (cos a))) to the AutoCAD system variable **chamferb**.

Line 11
(command "chamfer")
The **Command** function uses the AutoCAD **CHAMFER** command to generate a chamfer.

Step 3: Loading the lisp program
Save the file with the extension **.LSP** and then follow the procedure described in Example 1 to load the lisp file.

Exercise 2   *General*

Write an AutoLISP program that will generate the drawing shown in Figure 12-6. The program should prompt the user to enter points P1 and P2 and diameters D1 and D2.

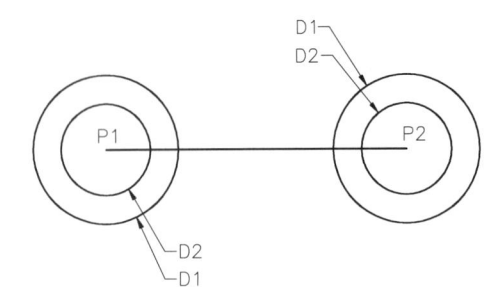

Figure 12-6 Concentric circles with connecting line

list FUNCTION

In AutoLISP the **list** function is used to define a 2D or 3D point. The list function can also be designated by using the single quote character ('), if the expression does not contain any variables or undefined items.

Examples

(Setq x (list 2.5 3 56))	returns (2.5 3 56)
(Setq x '(2.5 3 56))	returns (2.5 3 56)

Tip
A list when included with integers or real numbers can be enclosed within a single paretheses. A list when represented in the form of a single quote can be given outside the parentheses.

car, cdr, and cadr functions

car

The **car** function returns the first element of a list. If the list does not contain any elements, the function will return **nil**. The format of the **car** function is:

(car list)

        Where **car** ---------------- Returns the first element
                    list ---------------- List of elements

Examples

(car '(2.5 3 56))	returns 2.5
(car '(x y z))	returns X
(car '((15 20) 56))	returns (15 20)
(car '( ))	returns nil
(car (list 2 3.0 4))	returns 2
(car '(A B C))	

The single quotation mark signifies a list.

cdr

The **cdr** function returns a list with the first element removed from the list. The format of the **cdr** function is:

(cdr list)

    Where **cdr** ---------------- Returns a list with the first element removed
          list ---------------- List of elements

Examples
(cdr '(2.5 3 56)	returns (3 56)
(cdr '(x y z))	returns (Y Z)
(cdr '((15 20) 56))	returns (56)
(cdr '( ))	returns nil

cadr

The **cadr** function performs two operations, **cdr** and **car**, to return the second element of the list. The **cdr** function removes the first element, and the **car** function returns the first element of the new list. The format of the **cadr** function is:

(cadr list)

    Where **cadr** -------------- Performs two operations (car (cdr '(x y z))
          list ---------------- List of elements

Examples
(cadr '(2 3))	returns 3
(cadr '(2 3 56))	returns 3
(cadr '(x y z))	returns y
(cadr '((15 20) 56 24))	returns 56

In these examples, **cadr** performs two functions:

(cadr '(x y z))   = (car (cdr '(x y z)))
                = (car (y z)) returns y

Note

In addition to the functions car, cdr, and cadr, several other functions can be used to extract different elements of a list. Following is a list of these functions, where the function f consists of a list '((x y) z w)).

(setq f '((x y) z w))
(caar f) = (car (car f))	*returns x*
(cdar f) = (cdr (car f))	*returns y*
(cadar) = (car (cdr (car f)))	*returns y*
(cddr f) = (cdr (cdr f))	*returns w*
(caddr f) = (car (cdr (cdr f)))	*returns w*
(last f)	*returns w*

graphscr, textscr, princ, and terpri functions

graphscr
The **graphscr** function switches from the text window to the graphics window, provided the system has only one screen. If the system has two screens, this function is ignored.

textscr
The **textscr** function switches from the graphics window to the text window, provided the system has only one screen. If the system has two screens, this function is ignored.

princ
The **princ** function prints (or displays) the value of the variable. If the variable is enclosed in double quotes, the function prints (or displays) the expression that is enclosed in the quotes. The format of the **princ** function is:

(princ [variable or expression])

Examples
(princ)	prints a blank on screen
(princ a)	prints the value of variable a on screen
(princ "Welcome")	prints Welcome on screen

terpri
The **terpri** function prints a new line on the screen, just as **\n** does. This function is used to print the line that follows the **terpri** function.

Examples
(setq p1 (getpoint "Enter first point: "))(terpri)
(setq p2 (getpoint "Enter second point: "))

The first line (Enter first point:) will be displayed in the screen's command prompt area. The **terpri** function causes a carriage return; therefore, the second line (Enter second point:) will be displayed on a new line, just below the first line. If the terpri function is missing, the two lines will be displayed on the same line (Enter first point: Enter second point:).

Example 3

Write a program that will prompt you to enter two opposite corners of a rectangle and then draw the rectangle on the screen as shown in Figure 12-7.

Step 1: Understanding the program algorithm

Input	Output
Coordinates of point P1	Rectangle
Coordinates of point P3	

Process
1. Calculate the coordinates of the points P2 and P4.
2. Draw the following lines.
   Line from P1 to P2
   Line from P2 to P3
   Line from P3 to P4
   Line from P4 to P1

The X and Y coordinates of points P2 and P4 can be calculated using the **car** and **cadr** functions. The **car** function extracts the X coordinate of a given list, and the **cadr** function extracts the Y coordinate.

X coordinate of point p2
x2 = x3
x2 = car (x3 y3)
x2 = car p3

Y coordinate of point p2
y2 = y1
y2 = cadr (x1 y1)
y2 = cadr p1

X coordinate of point p4
x4 = x1
x4 = car(x1 y1)
x4 = car p1

Y coordinate of point p4
y4 = y3
y4 = cadr (x3 y3)
y4 = cadr p3

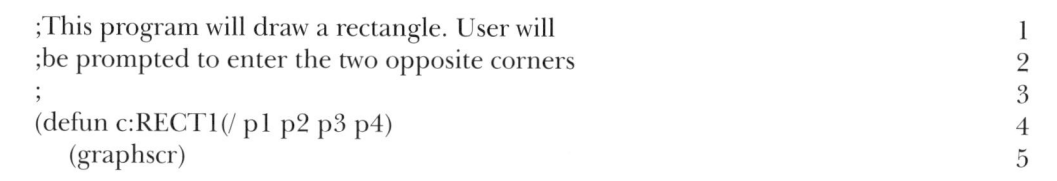

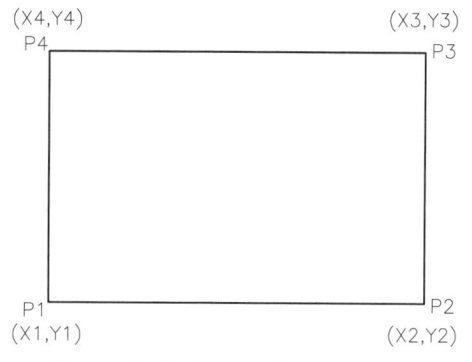

Figure 12-7 Rectangle P1 P2 P3 P4

Therefore, points p2 and p4 are:
p2 = (list (car p3) (cadr p1))
p4 = (list (car p1) (cadr p3))

Step 2: Writing the lisp program
Use any text editor to write the lisp program. The following file is a listing of the program for Example 3. **The line numbers at the right are for reference only; they are not a part of the program.**

```
;This program will draw a rectangle. User will        1
;be prompted to enter the two opposite corners        2
;                                                     3
(defun c:RECT1(/ p1 p2 p3 p4)                         4
   (graphscr)                                         5
```

```
        (setvar "cmdecho" 0)                                    6
        (prompt "RECT1 command draws a rectangle")(terpri)      7
        (setq p1 (getpoint "Enter first corner"))(terpri)       8
        (setq p3 (getpoint "Enter opposite corner"))(terpri)    9
        (setq p2 (list (car p3) (cadr p1)))                    10
        (setq p4 (list (car p1) (cadr p3)))                    11
    (command "line" p1 p2 p3 p4 "c")                           12
    (setvar "cmdecho" 1)                                       13
    (princ)                                                    14
)                                                              15
```

Explanation

Lines 1-3

The first three lines are comment lines that describe the function of the program. All comment lines that start with a semicolon are ignored when the program is loaded.

Line 4

(defun c:RECT1(/ p1 p2 p3 p4)

The **defun** function defines the function **RECT1**.

Line 5

(graphscr)

This function switches the text screen to the graphics screen, if the current screen happens to be a text screen. Otherwise, this function has no effect on the display screen.

Line 6

(setvar "cmdecho" 0)

The **setvar** function assigns the value 0 to the AutoCAD system variable cmdecho, which turns the echo off. When **cmdecho** is off, AutoCAD command prompts are not displayed in the command prompt area of the screen.

Line 7

(prompt "RECT1 command draws a rectangle")(terpri)

The **prompt** function will display the information in double quotes ("RECT1 command draws a rectangle"). The function **terpri** causes a carriage return so that the next text is printed on a separate line.

Line 8

(setq p1 (getpoint "Enter first corner"))(terpri)

The **getpoint** function pauses for you to enter a point (the first corner of the rectangle), and the **setq** function assigns that value to variable p1.

Line 9

(setq p3 (getpoint "Enter opposite corner"))(terpri)

The **getpoint** function pauses for you to enter a point (the opposite corner of the rectangle), and the **setq** function assigns that value to variable p3.

AutoLISP   12-25

Line 10
(setq p2 (list (car p3) (cadr p1)))
The **cadr** function extracts the Y coordinate of point p1, and the car function extracts the X coordinate of point p3. These two values form a list, and the setq function assigns that value to variable p2.

Line 11
(setq p4 (list (car p1) (cadr p3)))
The **cadr** function extracts the Y coordinate of point p3, and the **car** function extracts the X coordinate of point p1. These two values form a list, and the **setq** function assigns that value to variable p4.

Line 12
(command "line" p1 p2 p3 p4 "c")
The command function uses the AutoCAD **LINE** command to draw lines between points p1, p2, p3, and p4. The c (close) joins the last point, p4, with the first point, p1.

Line 13
(setvar "cmdecho" 1)
The **setvar** function assigns a value of 1 to the AutoCAD system variable **cmdecho**, which turns the echo on.

Line 14
(princ)
The princ function prints a blank on the screen. If this line is missing, AutoCAD will print the value of the last expression. This value does not affect the program in any way. However, it might be confusing at times. The **princ** function is used to prevent display of the last expression in the command prompt area.

Line 15
The closing parenthesis completes the definition of the function **RECT1** and ends the program.

Note
In this program the rectangle is generated after you define the two corners of the rectangle. The rectangle is not dragged as you move the screen crosshairs to enter the second corner. However, the rectangle can be dragged by using the getcorner function, as shown in the following program listing:

```
;This program will draw a rectangle with the
;drag mode on and using getcorner function
;
(defun c:RECT2(/ p1 p2 p3 p4)
   (graphscr)
   (setvar "cmdecho" 0)
   (prompt "RECT2 command draws a rectangle")(terpri)
   (setq p1 (getpoint "enter first corner"))(terpri)
   (setq p3 (getcorner p1 "Enter opposite corner" ))(terpri)
```

```
            (setq p2 (list (car p3) (cadr p1)))
            (setq p4 (list (car p1) (cadr p3)))
       (command "line" p1 p2 p3 p4 "c")
       (setvar "cmdecho" 1)
       (princ)
)
```

Step 3: Loading the lisp program
Save the file with the extension ***.LSP** and then load the file using the **APPLOAD** command as described in Example 1.

Getangle and getorient functions

getangle
The **getangle** function pauses for you to enter the angle. Then it returns the value of that angle in radians. The format of the **getangle** function is:

(getangle [point] [prompt])
         Where **point** ------------- First point of the angle
                        prompt ---------- Any prompt that needs to be displayed on screen

Examples
(getangle)
(setq ang (getangle))
(setq ang (getangle pt1))———————— pt1 is a predefined point
(setq ang (getangle "Enter taper angle"))
(setq ang (getangle pt1 "Enter second point of angle"))

The angle you enter is affected by the angle setting. The angle settings can be changed using the AutoCAD **UNITS** command or by changing the value of the AutoCAD system variables, **ANGBASE** and **ANGDIR**. Following are the default settings for measuring an angle:

The angle is measured with respect to the positive X-axis (3 o'clock position). The value of this setting is saved in the AutoCAD system variable **ANGBASE**.

The angle is positive if it is measured in the counterclockwise direction and negative if it is measured in the clockwise direction. The value of this setting is saved in the AutoCAD system variable **ANGDIR**.

If the angle has a default setting [Figure 12-8(a)], the getangle function will return 2.35619 radians for an angle of 135.

Examples
(setq ang (getangle "Enter angle"))   returns 2.35619 for an angle of 135 degrees

Figure 12-8(b) shows the new settings of the angle, where the Y-axis is 0 degrees and the angles measured clockwise are positive. The **getangle** function will return 3.92699 for an

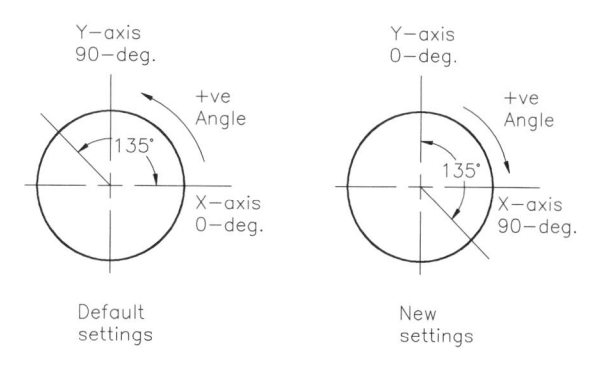

AutoLISP

Figure 12-8(a)   *Figure 12-8(b)*

angle of 135 degrees. The getangle function calculates the angle in the counterclockwise direction, **ignoring the direction set in the system variable ANGDIR**, with respect to the angle base as set in the system variable ANGBASE [Figure 12-9(b)].

Examples
(setq ang (getangle "Enter angle"))   returns 3.92699

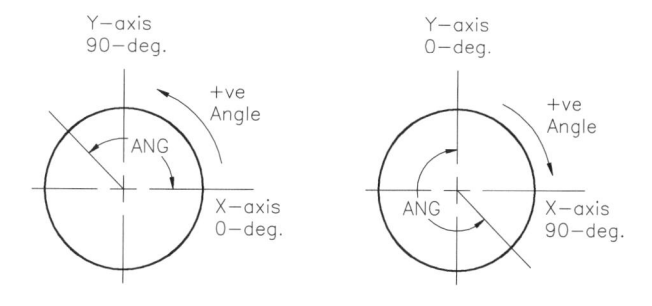

Figure 12-9(a)   *Figure 12-9(b)*

getorient

The **getorient** function pauses for you to enter the **angle**, then it returns the value of that angle in radians. The format of the **getorient** function is:

    (getorient [point] [prompt])
              Where   **point** ------------- First point of the angle
                          prompt ---------- Any prompt that needs to be displayed on the screen

Examples
(getorient)
(setq ang (getorient))
(setq ang (getorient pt1))

(setq ang (getorient "Enter taper angle"))
(setq ang (getorient pt1 "Enter second point of angle"))

The **getorient** function is just like the **getangle** function. Both return the value of the angle in radians. However, the **getorient** function always measures the angle with a positive X-axis (3 o'clock position) and in a counterclockwise direction. **It ignores the ANGBASE and ANGDIR settings.** If the settings have not been changed, as shown in Figure 12-10(a) (default settings for ANGDIR and ANGBASE), for an angle of 135 degrees the **getorient** function will return 2.35619 radians. If the settings are changed, as shown in Figure 12-10(b), for an angle of 135 degrees the **getorient** function will return 5.49778 radians. Although the settings have been changed where the angle is measured with the positive Y-axis and in a clockwise direction, the **getorient** function ignores the new settings and measures the angle from positive X-axis and in a counterclockwise direction.

Note
For the getangle and getorient functions you can enter the angle by typing the angle at the keyboard or by selecting two points on the screen. If the assignment is (setq ang (getorient pt1)), where the first point pt1 is already defined, you will be prompted to enter the second point. You can enter this point by selecting a point on the screen or by entering the coordinates of the second point.

180 degrees is equal to pi (3.14159) radians. To calculate an angle in radians, use the following relation:

Angle in radians = (pi x angle)/180

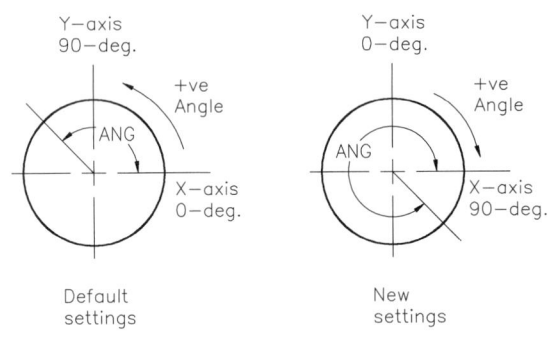

Figure 12-10(a)   *Figure 12-10(b)*

Getint, getreal, getstring, and getvar Functions

getint
The **getint** function pauses for you to enter an integer. The function always returns an integer, even if the number you enter is a real number. The format of the **getint** function is:

(getint [prompt])
     Where **prompt** ---------- Optional prompt that you want to display on screen

Examples
(getint)
(setq numx (getint))
(setq numx (getint "Enter number of rows: "))
(setq numx (getint "\n Enter number of rows: "))

getreal

The **getreal** function pauses for you to enter a real number and it returns a real number. If you happen to enter an integer, it returns a interger. The format of the **getreal** function is:

(getreal [prompt])
        Where **prompt** ---------- Optional prompt that is displayed on screen

Examples
(getreal)
(setq realnumx (getreal))
(setq realnumx (getreal "Enter num1: "))
(setq realnumx (getreal "\n Enter num2: "))

getstring

The **getstring** function pauses for you to enter a string value, and it always returns a string, even if the string you enter contains numbers only. The format of the **getstring** function is:

(getstring [prompt])
        Where **prompt** ---------- Optional prompt that is displayed on screen

Examples
(getstring)
(setq answer (getstring))
(setq answer (getstring "Enter Y for yes, N for no:))
(setq answer (getstring "\n Enter Y for yes, N for no:))

> **Note**
> *The maximum length of the string is 256 characters. If the string exceeds 256 characters, the excess is ignored.*

getvar

The **getvar** function lets you retrieve the value of an AutoCAD system variable. The format of the **getvar** function is:

(getvar "variable")
        Where **variable** --------- AutoCAD system variable name

Examples
(getvar)
(getvar "dimcen")   returns 0.09

(getvar "ltscale")	returns 1.0
(getvar "limmax")	returns 12.00,9.00
(getvar "limmin")	returns 0.00,0.00

Note
The system variable name should always be enclosed in double quotes.

You can retrieve only one variable value in one assignment. To retrieve the values of several system variables, use a separate assignment for each variable.

polar and sqrt functions

Polar

The **polar** function defines a point at a given angle and distance from the given point (Figure 12-11). The angle is expressed in radians, measured positive in the counterclockwise direction (assuming default settings for **ANGBASE** and **ANGDIR**). The format of the **polar** function is:

(polar point angle distance)

Where **point** ------------- Reference point
angle ------------- Angle the point makes with the referenced point
distance --------- Distance of the point from the referenced point

Examples
(polar pt1 ang dis)
(setq pt2 (polar pt1 ang dis))
(setq pt2 (polar '(2.0 3.25) ang dis))

sqrt

The **sqrt** function calculates the square root of a number, and the value this function returns is always a real number, Figure 12-12. The format of the **sqrt** function is:

(sqrt number)

Where **number** ---------- Number you want to find the square root of (real or integer)

Examples
(sqrt 144)	returns 12.0
(sqrt 144.0)	returns 12.0
(setq x (sqrt 57.25))	returns 7.566373
(setq x (sqrt (* 25 36.5)))	returns 30.207615
(setq x (sqrt (/ 7.5 (cos 0.75))))	returns 3.2016035
(setq hyp (sqrt (+ (* base base) (* ht ht))))	

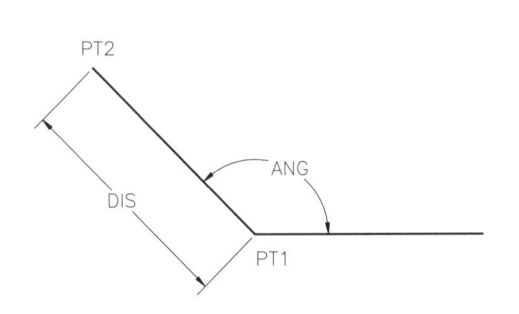

AutoLISP 12-31

Figure 12-11 *Using the **polar** function to define a point*

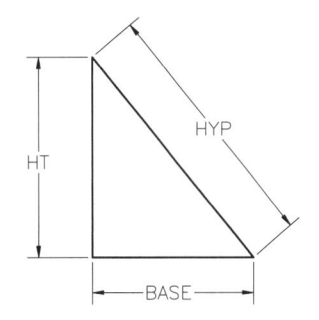

(setq hyp (sqrt (+(*base base) (*ht ht))))

Figure 12-12 *Application of the **sqrt** function*

Example 4

Write an AutoLISP program that will draw an equilateral triangle outside a circle (Figure 12-13). The sides of the triangle are tangent to the circle. The program should prompt you to enter the radius and the center point of the circle.

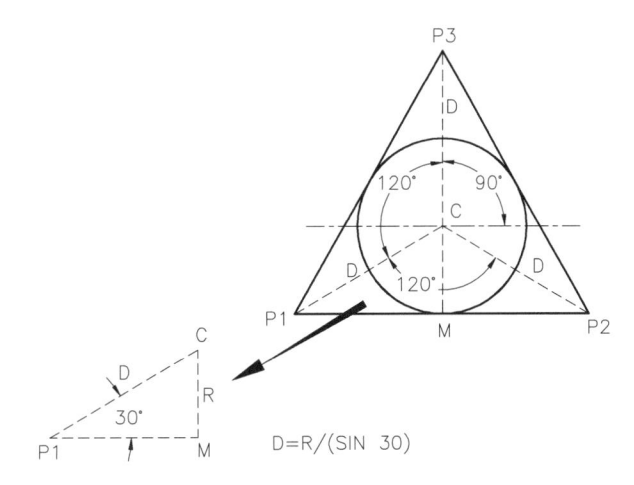

Figure 12-13 *Equilateral triangle outside a circle*

Step 1: Writing the lisp program
Use any text editor to write the lisp program. The following file is the listing of the AutoLISP program for Example 4.

;This program will draw a triangle outside
;the circle with the lines tangent to circle
:
(defun dtr (a)

```
    (* a (/ pi 180.0))
)
(defun c:trgcir(/ r c d p1 p2 p3)
(setvar "cmdecho" 0)
(graphscr)
    (setq r(getdist "\n Enter circle radius: "))
    (setq c(getpoint "\n Enter center of circle: "))
    (setq d(/ r (sin(dtr 30))))
    (setq p1(polar c (dtr 210) d))
    (setq p2(polar c (dtr 330) d))
    (setq p3(polar c (dtr 90) d))
(command "circle" c r)
(command "line" p1 p2 p3 "c")
(setvar "cmdecho" 1)
(princ)
)
```

Step 2: Loading the lisp program
Save the lisp file and then load it using **APPLOAD** command as described in Example 1.

Exercise 3 — *General*

Write an AutoLISP program that will draw an isosceles triangle P1,P2,P3. The base of the triangle (P1,P2) makes an angle B with the positive X-axis (Figure 12-14). The program should prompt you to enter the starting point, P1, length L1, and angles A and B.

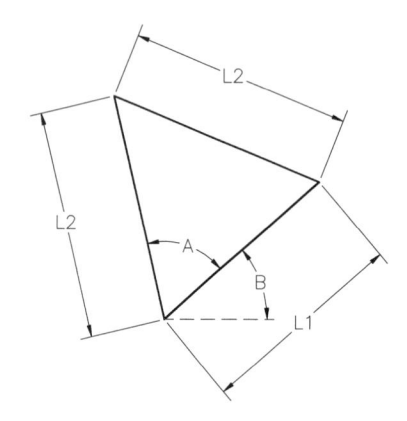

Figure 12-14 Isosceles triangle at an angle

Exercise 4 — *General*

Write a program that will draw a slot with centerlines. The program should prompt you to enter slot length, slot width, and the layer name for the centerlines (Figure 12-15).

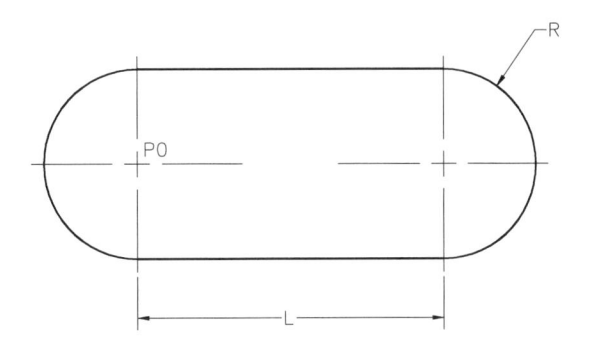

Figure 12-15 Slot of length L and radius R

itoa, rtos, strcase, and prompt functions

itoa

The **itoa** function changes an integer into a string and returns the integer as a string. The format of the **itoa** function is:

(itoa number)

        Where **Number** --------- The integer number that you want to convert into a string

Examples

(itoa 89)	returns "89"
(itoa -356)	returns "-356"

(setq intnum 7)
(itoa intnum)   returns "7"

(setq intnum 345)
(setq intstrg (itoa intnum))   returns "345"

rtos

The **rtos** function changes a real number into a string and the function returns the real number as a string. The format of the **rtos** function is:

(rtos realnum)

        Where **Realnum** -------- The real number that you want to convert into a string

Examples
(rtos 50.6)   returns "50.6"
(rtos -30.0)   returns "-30.0"
(setq realstrg (rtos 5.25))   returns "5.25"

(setq realnum 75.25)
(setq realstrg (rtos realnum))   returns "75.25"

The **rtos** function can also include mode and precision. If mode and precision are not provided, then the current AutoCAD settings are used. The format of the **rtos** function with mode and precision is:

(rtos realnum [mode] [precision])
Where **realnum** --------- Real number
mode ------------- Unit mode, like decimal, scientific
precision -------- Number of decimal places or denominator of fractional units

strcase

The **strcase** function converts the characters of a string into uppercase or lowercase. The format of the **strcase** function is:

(strcase string [true])
Where **String** ------------ String that needs to be converted to uppercase or lowercase
True -------------- If it is not nil, all characters are converted to lowercase

The **true** is optional. If it is missing or if the value of **true** is nil, the string is converted to uppercase. If the value of true is not nil, the string is converted to lowercase.

Examples
(strcase "Welcome Home")   returns "WELCOME HOME"

(setq t 0)
(strcase "Welcome Home" t)   returns "welcome home"
(strcase "Welcome Home" a)   returns "WELCOME HOME"

(setq answer (strcase (getstring "Enter Yes or No: ")))

prompt

The **prompt** function is used to display a message on the screen in the command prompt area. The contents of the message must be enclosed in double quotes. The format of the **prompt** function is:

(prompt message)
Where **message** --------- Message that you want to display on the screen

AutoLISP

12-35

Examples
(prompt "Enter circle diameter: ")
(setq d (getdist (prompt "Enter circle diameter: ")))

Note
On a two-screen system, the prompt function displays the message on both screens.

Example 5

Write a program that will draw two circles of radii r1 and r2, representing two pulleys that are separated by a distance d. The line joining the centers of the two circles makes an angle a with the X-axis, as shown in Figure 12-16.

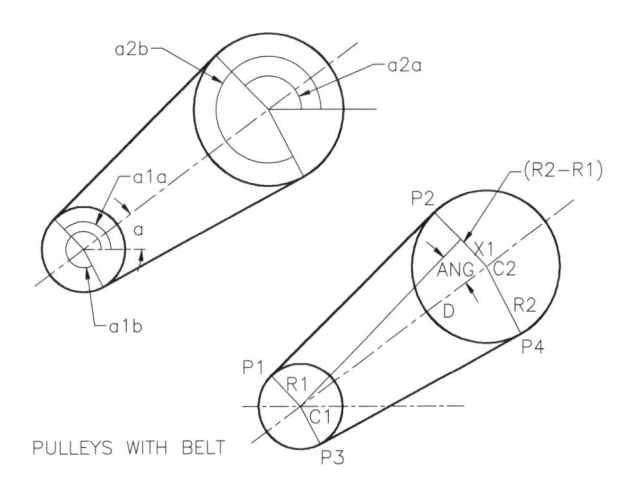

Figure 12-16 Two circles with tangent lines

Step 1: Understanding the program algorithm

Input
Radius of small circle - r1
Radius of large circle - r2
Distance between circles - d
Angle of center line - a
Center of small circle - c1

Output
Small circle of radius - r1
Large circle of radius - r2
Lines tangent to circles

Process
1. Calculate distance x1, x2.
2. Calculate angle ang.
3. Locate point c2 with respect to point c1.
4. Locate points p1, p2, p3, p4.
5. Draw small circle with radius r1 and center c1.

6. Draw large circle with radius r2 and center c2.
7. Draw lines p1 to p2 and p3 to p4.

Calculations

x1 = r2 - r1
x2 = SQRT [d^2 - (r2 - r1)^2]
tan ang = x1 / x2
ang = atan (x1 / x2)
a1a = 90 + a + ang
a1b = 270 + a - ang
a2a = 90 + a + ang
a2b = 270 + a - ang

Step 2: Writing the lisp program

The following file is a listing of the AutoLISP program for Example 5. **The line numbers on the right are not a part of the file. These numbers are for reference only.** Use any text editor to write down the following lisp file.

```
;This program draws a tangent (belt) over two                                    1
;pulleys that are separated by a given distance.                                 2
                                                                                 3
;This function changes degrees into radians                                      4
(defun dtr (a)                                                                   5
  (* a (/ pi 180.0))                                                             6
  )                                                                              7
;End of dtr function                                                             8
;The belt function draws lines that are tangent to circles                       9
(defun c:belt(/ r1 r2 d a c1 x1 x2 c2 p1 p2 p3 p4)                              10
    (setvar "cmdecho" 0)                                                        11
    (graphscr)                                                                  12
    (setq r1(getdist "\n Enter radius of small pulley: "))                      13
    (setq r2(getdist "\n Enter radius of larger pulley: "))                     14
    (setq d(getdist "\n Enter distance between pulleys: "))                     15
    (setq a(getangle "\n Enter angle of pulleys: "))                            16
    (setq c1(getpoint "\n Enter center of small pulley: "))                     17
    (setq x1 (- r2 r1))                                                         18
    (setq x2 (sqrt (- (* d d) (* (- r2 r1) (- r2 r1)))))                        19
    (setq ang (atan (/ x1 x2)))                                                 20
    (setq c2 (polar c1 a d))                                                    21
    (setq p1 (polar c1 (+ ang a (dtr 90)) r1))                                  22
    (setq p3 (polar c1 (- (+ a (dtr 270)) ang) r1))                             23
    (setq p2 (polar c2 (+ ang a (dtr 90)) r2))                                  24
    (setq p4 (polar c2 (- (+ a (dtr 270)) ang) r2))                             25
                                                                                26
;The following lines draw circles and lines                                     27
(command "circle" c1 p3)                                                        28
```

(command "circle" c2 p2)	29
(command "line" p1 p2 "")	30
(command "line" p3 p4 "")	31
(setvar "cmdecho" 1)	32
(princ))	33

Explanation

Line 5
(defun dtr (a)
In this line, the **defun** function defines a function, **dtr (a)**, that converts degrees into radians.

Line 6
(* a (/ pi 180.0))
(/ pi 180) divides the value of **pi** by 180, and the product is then multiplied by angle a (180 degrees is equal to **pi** radians).

Line 10
(defun c:belt(/ r1 r2 d a c1 x1 x2 c2 p1 p2 p3 p4)
In this line, the function **defun** defines a function, c:belt, that generates two circles with tangent lines.

Line 18
(setq x1 (- r2 r1))
In this line, the function **setq** assigns a value of r2 - r1 to variable x1.

Line 19
(setq x2 (sqrt (- (* d d) (* (- r2 r1) (- r2 r1)))))
In this line, **(- r2 r1)** subtracts the value of r1 from r2 and **(* (- r2 r1) (- r2 r1))** calculates the square of (- r2 r1). **(sqrt (- (* d d) (* (- r2 r1) (- r2 r1))))** calculates the square root of the difference, and **setq x2** assigns the product of this expression to variable x2.

Line 20
(setq ang (atan (/ x1 x2)))
In this line, **(atan (/ x1 x2))** calculates the arctangent of the product of **(/ x1 x2)**. The function **setq ang** assigns the value of the angle in radians to variable **ang**.

Line 21
(setq c2 (polar c1 a d))
In this line, **(polar c1 a d)** uses the **polar** function to locate point c2 with respect to c1 at a distance d and making an angle a with the positive X-axis.

Line 22
(setq p1 (polar c1 (+ ang a (dtr 90)) r1))
In this line, **(polar c1 (+ ang a (dtr 90)) r1))** locates point p1 with respect to c1 at a distance r1 and making an angle **(+ ang a (dtr 90))** with the positive X-axis.

Line 28
(command "circle" c1 p3)
In this line, the **Command** function uses the AutoCAD **CIRCLE** command to draw a circle with center c1 and a radius defined by the point p3.

Line 30
(command "line" p1 p2 "")
In this line, the **Command** function uses the AutoCAD **LINE** command to draw a line from p1 to p2. The pair of double quotes ("") at the end introduces a Return, which terminates the LINE command.

Step 3: Loading the lisp program
Save the program and then load the program as desribed in the Example 1.

Exercise 5   *General*

Write an AutoLISP program that will draw two lines tangent to two circles, as shown in Figure 12-17. The program should prompt you to enter the circle diameters and the center distance between the circles.

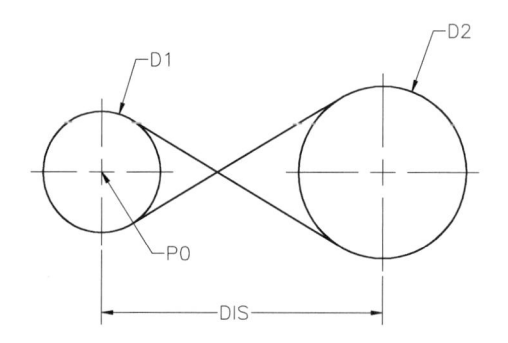

Figure 12-17 *Circles with two tangent lines*

FLOWCHARTS

A **flowchart** is a graphical representation of an algorithm. It can be used to analyze a problem systematically. It gives a better understanding of the problem, especially if the problem involves some conditional statements. It consists of standard symbols that represent a certain function in the program. For example, a rectangle is used to represent a process that takes place when the program is executed. The blocks are connected by lines indicating the sequence of operations. Figure 12-18 gives the standard symbols that can be used in a flowchart.

CONDITIONAL FUNCTIONS

The relational functions discussed earlier in the chapter establish a relationship between two atoms. For example, (< x y) describes a test condition for an operation. To use such functions in a meaningful way, a conditional function is required. For example, (if (< x y) (setq z (- y x)) (setq z (- x y))) describes the action to be taken when the condition is true (T) and when it is false (nil). If the condition is true, then z = y - x. If the condition is not true, then z = x - y. Therefore, conditional functions are very important for any programming language, including AutoLISP.

AutoLISP

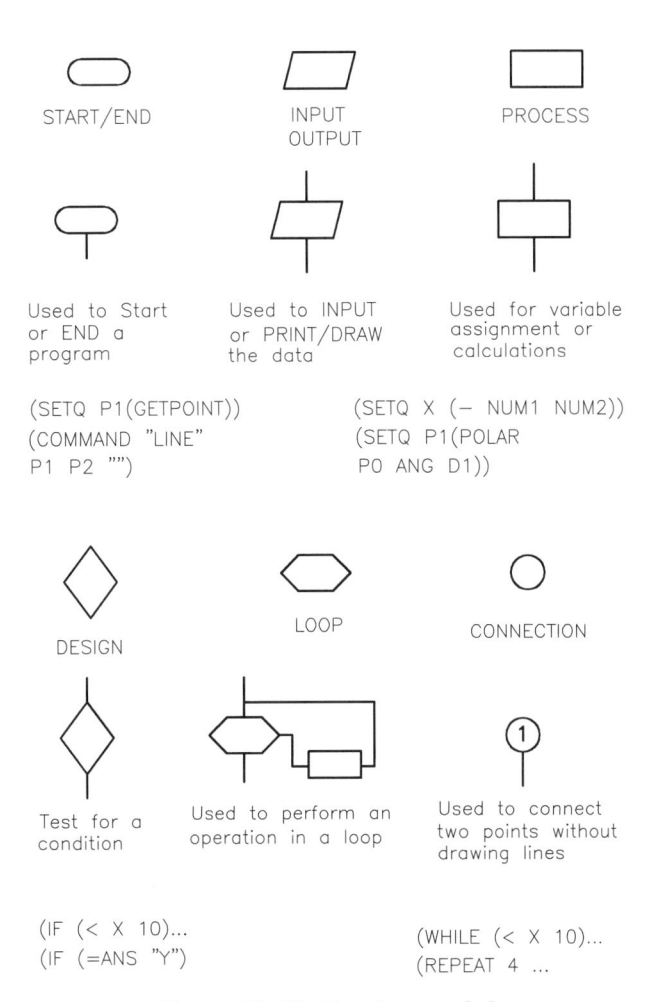

Figure 12-18 Flowchart symbols

if

The **if** function (Figure 12-19) evaluates the first expression (then) if the specified condition returns "true," it evaluates the second expression (else) if the specified condition returns "nil." The format of the **if** function is:

(if condition then [else])
       Where **condition** ------- Specified conditional statement
                      then -------------- Expression evaluated if the condition returns T
                      else -------------- Expression evaluated if the condition returns nil

Examples
(if (= 7 7) ("true"))   returns "true"
(if (= 5 7) ("true") ("false"))   returns "false"

```
(setq ans "yes")
(if (= ans "yes") ("Yes") ("No"))              returns "Yes"
(setq num1 8)
(setq num2 10)
(if (> num1 num2)
   (setq x (- num1 num2))
   (setq x (- num2 num1))
)                                               returns 2
```

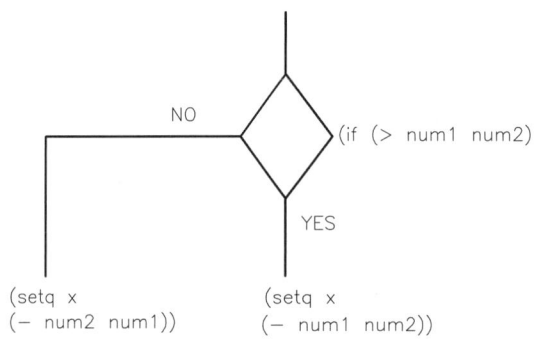

Figure 12-19 if function

Example 6

Write an AutoLISP program that will subtract a smaller number from a larger number. The program should also prompt you to enter two numbers.

Step 1: Understanding program algorithm and flowchart

Input
Number (num1)
Number (num2)

Output
x = num1 - num2
or
x = num2 - num1

Process
If num1 > num2 then x = num1 - num2
If num1 < num2 then x = num2 - num1

The flowchart in Figure 12-20 describes the process involved in writing the program using standard flowchart symbols.

Step 2: Writing the lisp program

Use any text editor to write the lisp file. The following file is a listing of the program for Example 6. **The line numbers are not a part of the file; they are for reference only.**

```
;This program subtracts smaller number                   1
;from larger number                                      2
;                                                        3
```

AutoLISP

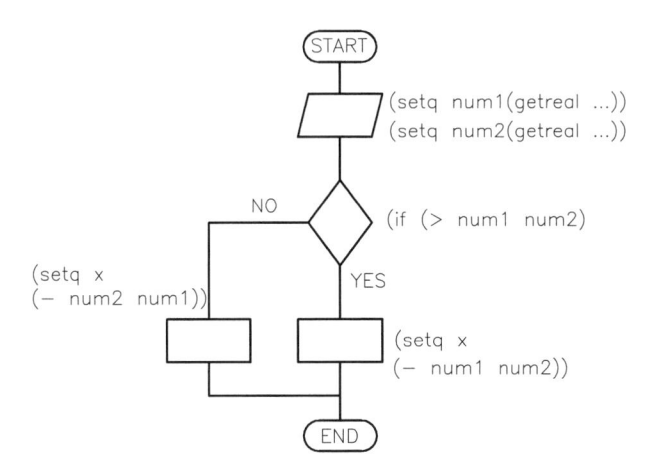

Figure 12-20 Flowchart for Example 6

```
(defun c:subnum( )                                    4
  (setvar "cmdecho" 0)                                5
  (setq num1 (getreal "\n Enter first number: "))     6
  (setq num2 (getreal "\n Enter second number: "))    7
  (if (> num1 num2)                                   8
     (setq x (- num1 num2))                           9
     (setq x (- num2 num1))                          10
  )                                                  11
  (setvar "cmdecho" 1)                               12
  (princ)                                            13
)                                                    14
```

Explanation

Line 8
(if (> num1 num2)
In this line, the **if** function evaluates the test expression **(> num1 num2)**. If the condition is true, it returns **T**. If the condition is not true, it returns nil.

Line 9
(setq x (- num1 num2))
This expression is evaluated if the test expression **(if (> num1 num2)** returns T. The value of variable num2 is subtracted from num1, and the resulting value is assigned to variable x.

Line 10
(setq x (- num2 num1))
This expression is evaluated if the test expression **(if (> num1 num2)** returns nil. The value of variable num1 is subtracted from num2, and the resulting value is assigned to variable x.

Line 11
)

The closing parenthesis completes the definition of the if function.

Step 3: Loading the lisp program
Save the file and then load the file using **APPLOAD** command as described in Example 1.

Example 7

Write an AutoLISP program that will enable you to multiply or divide two numbers (Figure 12-21). The program should prompt you to enter the choice of multiplication or division. The program should also display an appropriate message if you do not enter the right choice.

Step 1: Making the flowchart
A flowchart can be designed as shown in Figure 12-21.

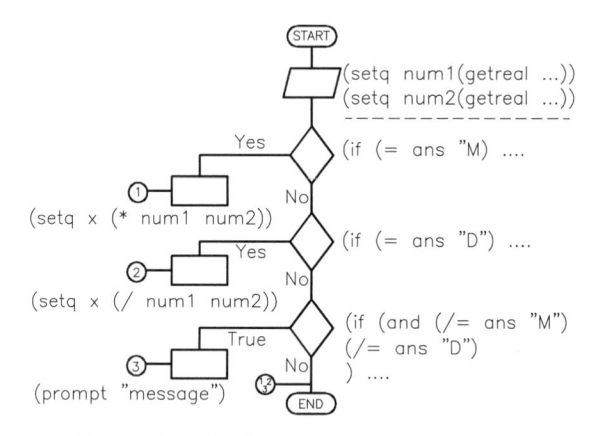

Figure 12-21 Flow diagram for Example 7

Step 2: Writing the lisp file
The following file is a listing of the AutoLISP program for Example 7:

```
;This program multiplies or divides two given numbers
(defun c:mdnum()
    (setvar "cmdecho" 0)
    (setq num1 (getreal "\n Enter first number: "))
    (setq num2 (getreal "\n Enter second number: "))
    (prompt "Do you want to multiply or divide. Enter M or D: ")
    (setq ans (strcase (getstring)))
    (if (= ans "M")
       (setq x (* num1 num2))
    )
    (if (= ans "D")
       (setq x (/ num1 num2))
    )
    (if (and (/= ans "D")(/= ans "M"))
```

```
        (prompt "Sorry! Wrong entry, Try again")
)
(setvar "cmdecho" 1)
(princ))
```

Step 3: Loading the lisp file
Save the file and then load the file using the **APPLOAD** command as described in Example 1.

progn

The **progn** function can be used with the **if** function to evaluate several expressions. The format of the **progn** function is:

(progn expression expression . . .)

The **if** function evaluates only one expression if the test condition returns "true." The **progn** function can be used in conjunction with the **if** function to evaluate several expressions. Here is an example to demonstrate the use of the **progn** function with the **if** function.

Example
```
(defun c:IFPRGN()
        (setq p1(getint "enter the integer"))
        (If (>= p1 5)
        (progn (command "line" "2,2" "3,3" "")
              (command "rec" "3,3" "6,6") )

        (progn (command "circle" "3,3" 1)
              (command "line" "3,3" "5,5" "") )
            )
    )      ; End of program
```

while

The **while** function (Figure 12-22) evaluates a test condition. If the condition is true (expression does not return nil), the operations that follow the while statement are repeated until the test expression returns nil. The format of the **while** function is:

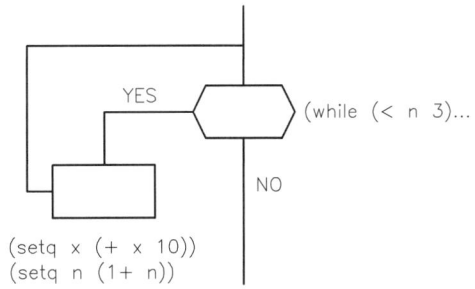

Figure 12-22 While function

(while testexpression operations)
        Where **Testexpression** ----------- Expression that tests a condition
                    Operations --------------- Operations to be performed until the test
                                     expression returns nil

Example
(while (= ans "yes")

```
    (setq x (+ x 1))
    (setq ans (getstring "Enter yes or no: "))
)
(while (< n 3)
    (setq x (+ x 10))
    (setq n (1+ n))
)
```

Example 8

Write an AutoLISP program that will find the nth power of a given number. The power is an integer. The program should prompt you to enter the number and the nth power (Figure 12-23).

Step 1: Understanding the program algorithm and flowchart

Input
Number x
nth power n

Output
product x^n

Process
1. Set the value of t = 1 and c = 1.
2. Multiply t * x and assign that value to the variable t.
3. Repeat the process until the counter c is less than or equal to n.

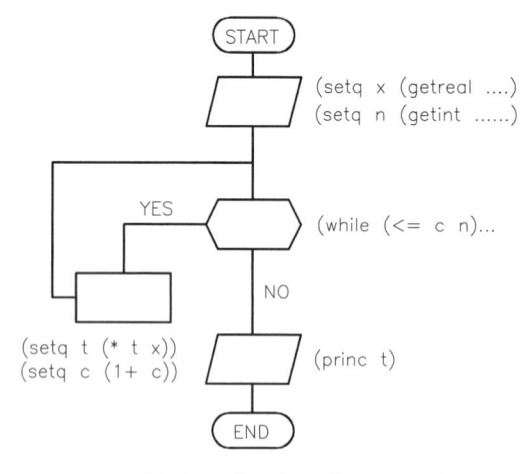

Figure 12-23 Flowchart for Example 8

Step 2: Writing the lisp program

The following file is a listing of the AutoLISP program for Example 8.

AutoLISP

```
;This program calculates the nth
;power of a given number
(defun c:npower()
    (setvar "cmdecho" 0)
    (setq x(getreal "\n Enter a number: "))
    (setq n(getint "\n Enter Nth power-integer number: "))
    (setq t 1) (setq c 1)
    (while (<= c n)
      (setq t (* t x))
      (setq c (1+ c))
    )
    (setvar "cmdecho" 1)
    (princ t)
)
```

Step 3: Loading the lisp file
Save the file and then load the file with the help of **APPLOAD** command.

Example 9

Write an AutoLISP program that will generate the holes of a bolt circle (Figure 12-24). The program should prompt you to enter the center point of the bolt circle, the bolt circle diameter, the bolt circle hole diameter, the number of holes, and the start angle of the bolt circles.

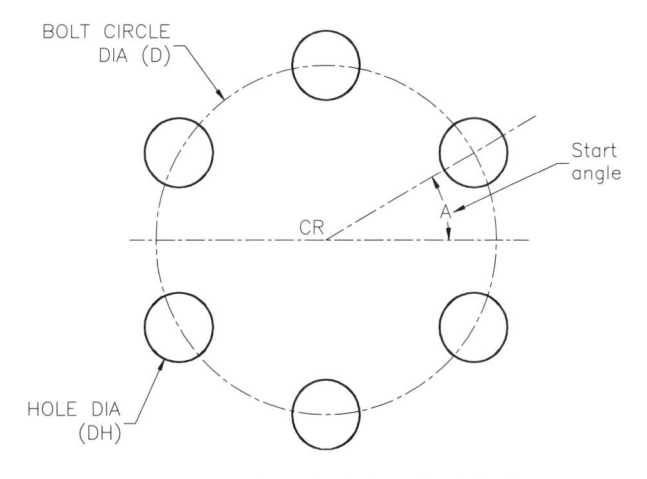

Figure 12-24 Bolt circle with six holes

Step 1: Writing the lisp program

```
;This program generates the bolt circles
;
(defun c:bc1( )
(graphscr)
(setvar "cmdecho" 0)
    (setq cr(getpoint "\n Enter center of Bolt-Circle: "))
```

```
(setq d(getdist "\n Dia of Bolt-Circle: "))
(setq n(getint "\n Number of holes in Bolt-Circle: "))
(setq a(getangle "\n Enter start angle: "))
(setq dh(getdist "\n Enter diameter of hole: "))
(setq inc(/ (* 2 pi) n))
(setq ang 0)
(setq r (/ dh 2))
(while (< ang (* 2 pi))
   (setq p1 (polar cr (+ a inc) (/ d 2)))
   (command "circle" p1 r)
   (setq a (+ a inc))
   (setq ang (+ ang inc))
   )
(setvar "cmdecho" 1)
(princ)
)
```

Step 2: Loading the lisp file
Save the file and then load the file using **APPLOAD** command as described in Example 1.

repeat
The **repeat** function evaluates the expressions **n** number of times as specified in the **repeat** function (Figure 12-25). The variable **n** must be an integer. The format of the **repeat** function is:

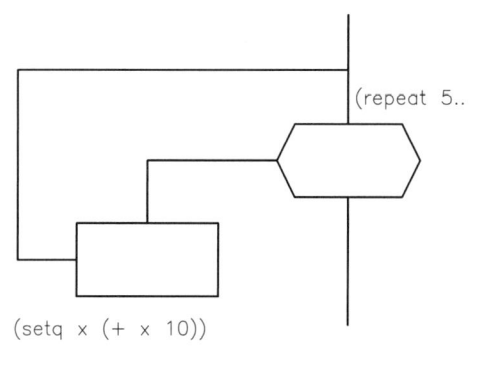

Figure 12-25 Repeat function

repeat n

Where ------------------------------ n is an integer that defines the number of times the expressions are to be evaluated

Example
(repeat 5
  (setq x (+ x 10))
)

Note
*AutoCAD allows you to load the specified AutoLISP programs automatically, each time you start AutoCAD. For example, if you are working on a project and you have loaded an AutoLISP program, the program will autoload if you start another drawing. You can enable this feature by adding the name of the file to the Startup Suite of the **Load/Unload Application** dialog box. For details, see the Load/Unload Application discussed earlier in this chapter.*

AutoLISP

Example 10

Write an AutoLISP program that will generate a given number of concentric circles. The program should prompt you to enter the center point of the circles, the start radius, and the radius increment (Figure 12-26).

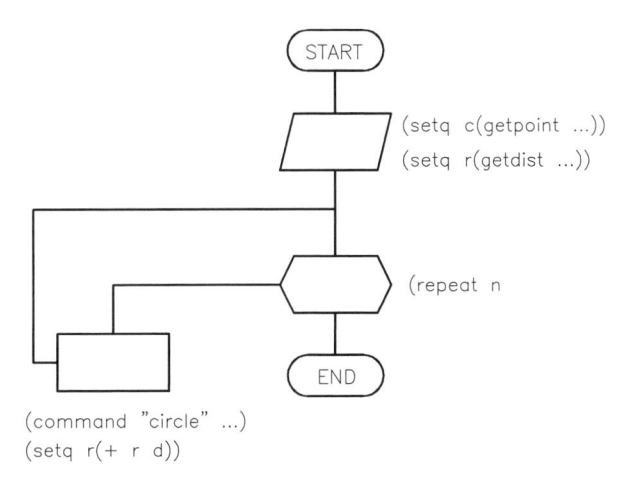

Figure 12-26 Flowchart for Example 10

The following file is a listing of the AutoLISP program for Example 10.

```
;This program uses the repeat function to draw
;a given number of concentric circles.
(defun c:concir( )
(graphscr)
(setvar "cmdecho" 0)
(setq c (getpoint "\n Enter center point of circles: "))
(setq n (getint "\n Enter number of circles: "))
(setq r (getdist "\n Enter radius of first circle: "))
(setq d (getdist "\n Enter radius increment: "))
(repeat n
  (command "circle" c r)
  (setq r (+ r d))
  )
  (setvar "cmdecho" 1)
  (princ)
  )
```

Self-Evaluation Test

Answer the following questions and then match your answers to the answers given at the end of the chapter.

1. Lisp is a leading programming language for _____.

2. The AutoLISP interpreter is embeded within the _____ software package.

3. The function _____ checks whether the two atoms are not equal.

4. AuoLISP contains some _____ functions.

5. If some built-in functions are used with setq then there is every possibility that reserved functions will be _____.

6. When using LOAD function, the AutoLISP file name and the optional path name must be enclosed in _____.

7. The command _____ is used to load a lisp file.

8. The strcase function converts the characters of a string into _____.

9. The **Progn** function can be used in conjunction with the _____.

10. The **setq** function is used to assign a value to _____.

Review Questions

1. Fill in the blanks:

Command: (+ 2 30 5 50)	returns _____
Command: (+ 2 30 4 55.0)	returns _____
(- 20 40)	returns _____
(- 30.0 40.0)	returns _____
(* 72 5 3 2.0)	returns _____
(* 7 -5.5)	returns _____
(/ 299 -5)	returns _____
(/ -200 -9.0)	returns _____
(1- 99)	returns _____
(1- -18.5)	returns _____
(abs -90)	returns _____
(abs -27.5)	returns _____

AutoLISP

(sin pi)	returns _____
(sin 1.5)	returns _____
(cos pi)	returns _____
(cos 1.2)	returns _____
(atan 1.1 0.0)	returns _____ radians
(atan -0.4 0.0)	returns _____ radians
(angtos 1.5708 0 5)	returns _____
(angtos -1.5708 0 3)	returns _____
(< "x" "y")	returns _____
(>= 80 90 79)	returns _____

2. The _____ function pauses to enable you to enter the X, Y coordinates or X, Y, Z coordinates of a point.

3. The _____ function is used to execute standard AutoCAD commands from within an AutoLISP program.

4. In an AutoLISP expression, the AutoCAD command name and the command options have to be enclosed in double quotation marks. (T/F).

5. The getdist function pauses for you to enter a _____ and it then returns the distance as a real number.

6. The _____ function assigns a value to an AutoCAD system variable. The name of the system variable must be enclosed in _____ .

7. The **cadr** function performs two operations, _____ and _____, to return the second element of the list.

8. The _____ function prints a new line on the screen just as \n.

9. The _____ function pauses for you to enter the angle, then it returns the value of that angle in radians.

10. The _____ function always measures the angle with a positive X-axis and in a counterclockwise direction.

11. The _____ function pauses for you to enter an integer. The function always returns an integer, even if the number that you enter is a real number.

12. The _____ function lets you retrieve the value of an AutoCAD system variable.

13. The _____ function defines a point at a given angle and distance from the given point.

14. The _____ function calculates the square root of a number and the value this function

returns is always a real number.

15. The _____ function changes a real number into a string and the function returns the real number as a string.

16. The **if** function evaluates the test expression **(> num1 num2)**. If the condition is true, it returns _____; if the condition is not true, it returns _____.

17. The _____ function can be used with the **if** function to evaluate several expressions.

18. The **while** function evaluates the test condition. If the condition is true (expression does not return nil) the operations that follow the while statement are _____ until the test expression returns _____.

19. The **repeat** function evaluates the expressions n number of times as specified in the **repeat** function. The variable n must be a real number. (T/F).

Exercises

Exercise 6 — *General*

Write an AutoLISP program that will draw three concentric circles with center C1 and diameters D1, D2, D3 (Figure 12-27). The program should prompt you to enter the coordinates of center point C1 and the circle diameters D1, D2, D3.

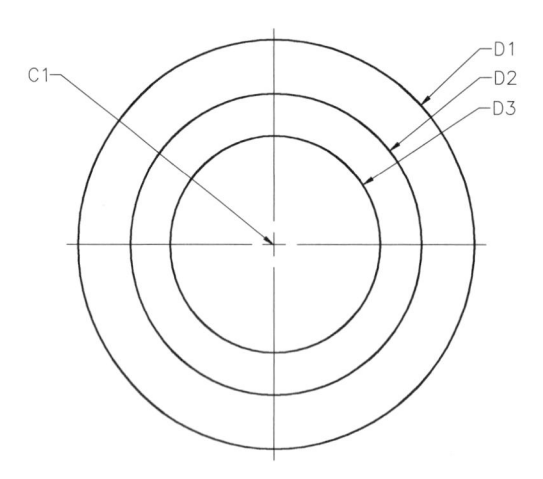

Figure 12-27 Three concentric circles with diameters D1, D2, D3

Exercise 7
General

Write an AutoLISP program that will draw a line from point P1 to point P2 (Figure 12-28). Line P1, P2 makes an angle A with the positive X-axis. Distance between the points P1 and P2 is L. The diameter of the circles is D1 (D1 = L/4).

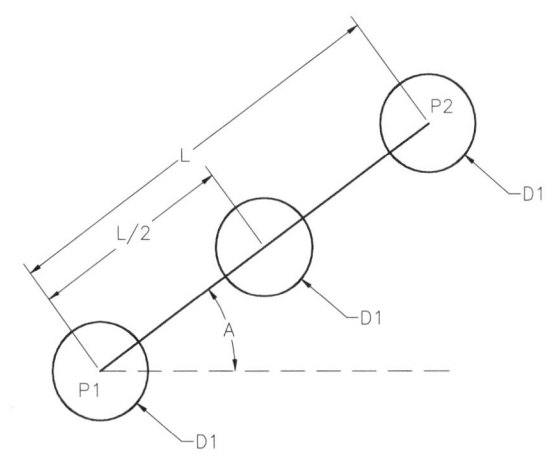

Figure 12-28 Circles and line making an angle A with X-axis

Exercise 8
General

Write an AutoLISP program that will draw an isosceles triangle P1, P2, P3 (Figure 12-29). The program should prompt you to enter the starting point P1, length L1, and the included angle A.

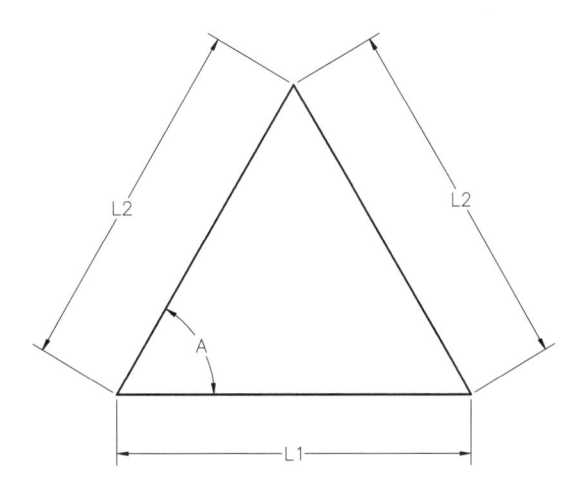

Figure 12-29 Isosceles triangle

Exercise 9
General

Write an AutoLISP program that will draw a parallelogram with sides S1, S2 and angle W as shown in Figure 12-30. The program should prompt you to enter the starting point PT1, lengths S1, S2, and the included angle W.

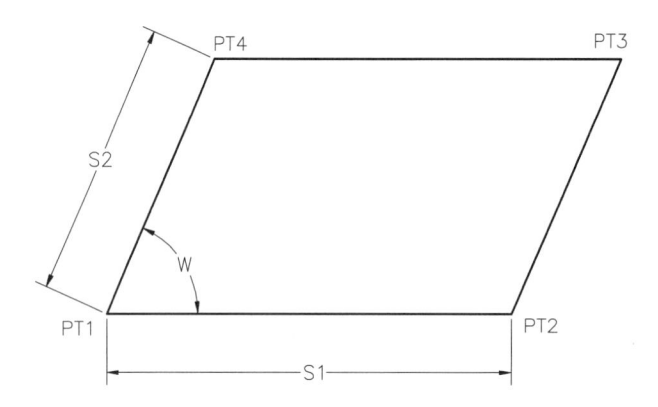

Figure 12-30 Parallelogram with sides S1, S2, and angle W

Exercise 10
General

Write an AutoLISP program that will draw a square of sides S and a circle tangent to the four sides of the square as shown in Figure 12-31. The base of the square makes an angle, ANG, with the positive X-axis. The program should prompt you to enter the starting point P1, length S, and angle ANG.

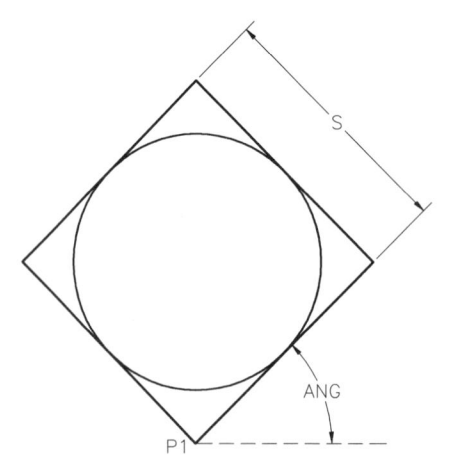

Figure 12-31 Square of side S at an angle ANG

AutoLISP

Exercise 11 — General

Write a program that will draw a slot with center lines. The program should prompt you to enter slot length, slot width, and the layer name for center lines (Figure 12-32).

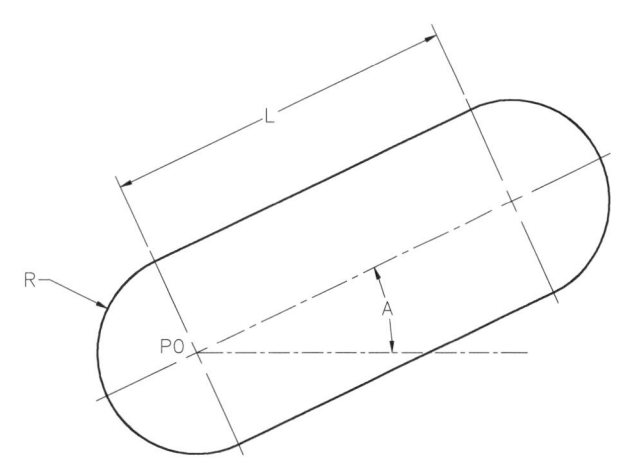

Figure 12-32 Slot of length L and radius R

Exercise 12 — General

Write an AutoLISP program that will draw a line and then generate a given number of lines (N), parallel to the first line (Figure 12-33).

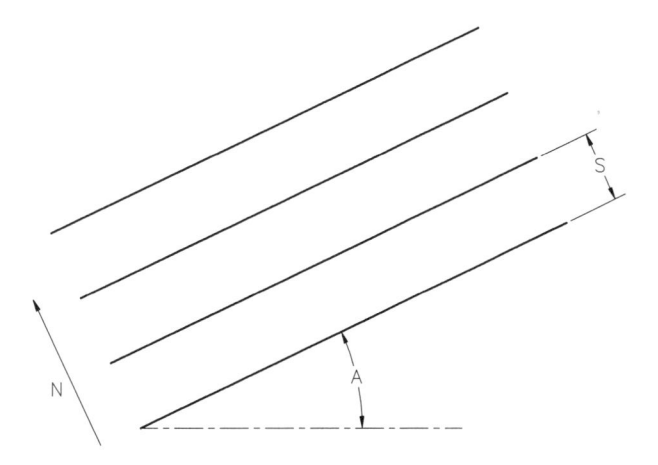

Figure 12-33 N number of lines offset at a distance S

Exercise 13 *General*

Write an AutoLISP program to draw a circle with center lines. The program should prompt for the diameter of circle, center of circle, and the angle of center lines as shown in Figure 12-34.

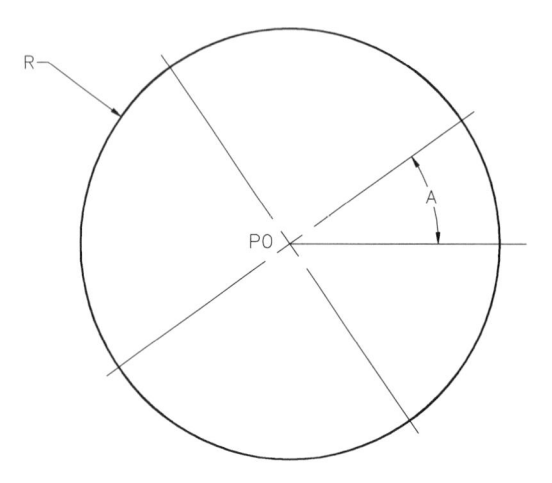

Figure 12-34 Circle with center lines at an angle A

Exercise 14 *General*

Write a program to draw a keyway slot. The program should prompt you to enter the width of slot, depth of slot, angle of slot, and starting point (Figure 12-35).

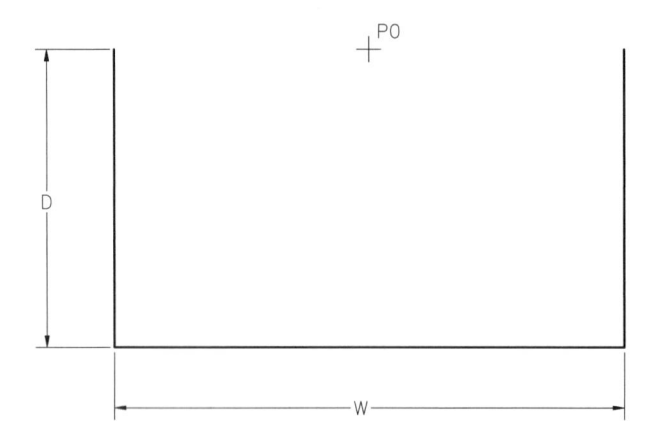

Figure 12-35 Keyway slot of width W and depth D

Exercise 15 — *Mechanical*

Write an AutoLISP program that will draw the two views of a bushing as shown in Figure 12-36. The program should prompt you to enter the starting point P0, lengths L1, L2, and the bushing diameters ID, OD, HD. The distance between the front view and the side view of bushing is DIS (DIS = 1.25 * HD). The program should also draw the hidden lines in the HID layer and center lines in the CEN layer. The center lines should extend 0.75 units beyond the object line.

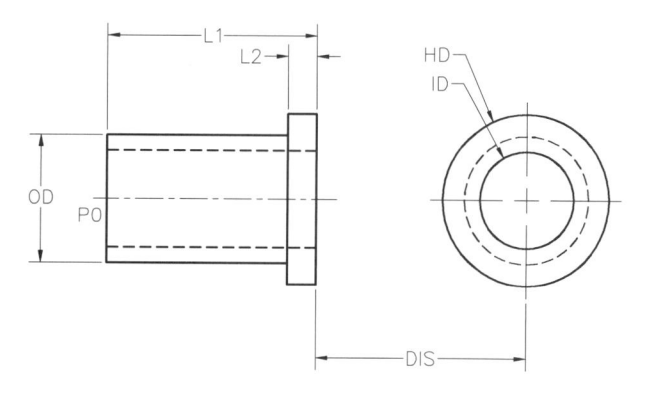

Figure 12-36 Two views of bushing

Exercise 16 — *General*

Write an AutoLISP program that will delete all objects contained within the upper (limmax) and lower (limmin) limits. Use AutoCAD's GETVAR and ERASE commands to delete the objects.

Answers to the Self-Evaluation test
1. Artificial Intelligence, 2. AutoCAD, 3. /=, 4. Built- in, 5. Redefined, 6. double quoates, 7. **Appload**, 8. Uppercase or Lowercase, 9. If, 10. Variable.

Chapter 13

Visual LISP

Learning Objectives
After completing this chapter, you will be able to:
- *Start Visual LISP in AutoCAD.*
- *Use the Visual LISP Text Editor.*
- *Load and run Visual LISP programs.*
- *Load existing AutoLISP files.*
- *Use the Visual LISP Console.*
- *Use the Visual LISP formatter and change formatting options.*
- *Debug a Visual LISP program and trace the variables.*

Visual LISP

Visual LISP is another step forward to extend the customizing power of AutoCAD. Since AutoCAD Release 2.0 in the mid 1980s, AutoCAD users have used AutoLISP to write programs for their applications; architecture, mechanical, electrical, air conditioning, civil, sheet metal, and hundreds of other applications. AutoLISP has some limitations. For example, when you use a text editor to write a program, it is difficult to locate and check parentheses, AutoLISP functions, and variables. Debugging is equally problematic, because it is hard to find out what the program is doing and what is causing an error in the program. Generally, programmers will add some statements in the program to check the values of variables at different stages of the program. Once the program is successful, the statements are removed or changed into comments. Balancing parentheses and formatting the code are some other problems with traditional AutoLISP programming.

Visual LISP has been designed to make programming easier and more efficient. It has a powerful text editor and a formatter. The text editor allows color coding of parentheses, function names, variables and other components of the program. The formatter formats the code in a easily readable format. It also has a watch facility that allows you to watch the values of variables and expressions. Visual LISP has an interactive and intelligent console that has made programming easier. It also provides context-sensitive help for AutoLISP functions and apropos features for symbol name search. The debugging tools have made it easier to debug the program and check the source code. These and other features have made Visual LISP the preferred tool for writing LISP programs. Visual LISP works with AutoCAD and it contains its own set of windows. AutoCAD must be running when you are working with Visual LISP.

Overview of Visual LISP

In this overview you will learn how to start Visual LISP, use the Visual LISP Text Editor to write a LISP Program, and then how to load and run the programs. As you go through the following steps, you will notice that Visual LISP has some new capabilities and features that are not available in AutoLISP.

Starting Visual LISP

Pull-down:	Tools > AutoLISP > Visual LISP Editor
Command:	VLIDE

1. Start AutoCAD.

2. In the **Tools** menu select **AutoLISP** and then select **Visual LISP Editor**. You can also start Visual LISP by entering the **VLIDE** command at the Command prompt. AutoCAD displays the Visual LISP window, Figure 13-1. If it is the first time in the current drawing session that you are loading Visual LISP, the Trace window might appear, displaying information about the current release of Visual LISP and errors that might be encountered when loading Visual LISP. The Visual LISP window has four areas: Menu, Toolbars, Console window, and Status bar, Figure 13-1. The following is a short description of these areas:

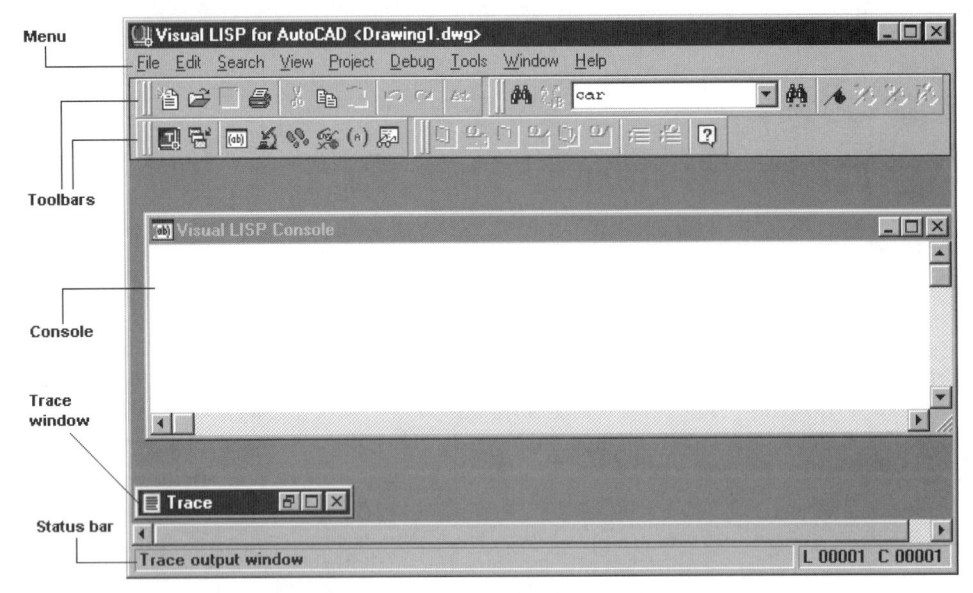

Figure 13-1 Visual LISP window

Menu. The menu bar is at the top of the Visual LISP window and lists different menu items. If you select a menu, AutoCAD displays the items in that menu. Also, the function of the menu item is displayed in the status bar that is located at the bottom of the Visual LISP window.

Toolbars. The toolbars provide a quick and easy way to invoke Visual LISP commands. Visual LISP has 5 toolbars; Debug, Edit, Find, Inspect, and Run. Depending on the window that is active, the display of the toolbars changes. If you move the mouse pointer over a tool and position it there for a few seconds, a tool tip is displayed that indicates the function of that tool. A more detailed description of the tool appears in the status bar.

Console window. The Visual LISP Console window is contained within the Visual LISP window. It can be scrolled and positioned anywhere in the Visual LISP window. The Console window can be used to enter AutoLISP or Visual LISP commands. For example, if you want to add two numbers, enter (+ 2 9.5) next to the $ sign and then press the **Enter** key. Visual LISP will return the result and display it in the same window.

Status bar. When you select a menu item or a tool in the toolbar, Visual LISP displays the related information in the status bar that is located near the bottom of the Visual LISP window.

Using the Visual LISP Text Editor

1. Select **New File** from the **File** menu. The Visual LISP Text Editor window is displayed, Figure 13-2. The default file name is Untitled-0 as displayed at the top of the editor window.

2. To activate the editor, click anywhere in the Visual LISP Text Editor.

3. Type the following program and notice the difference between the Visual LISP Text Editor and the text editors you have been using to write AutoLISP programs (like Notepad).

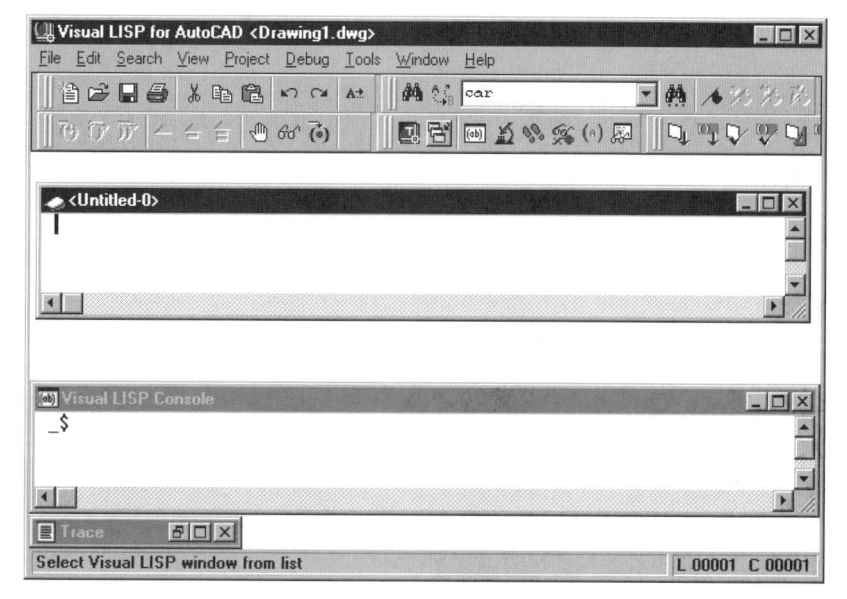

Figure 13-2 Visual LISP Text Editor

```
;;;This program will draw a triangle
(defun tr1 ()
(setq p1(list 2 2))
(setq p2(list 6.0 3.0))
(setq p3(list 4.0 7.0))
(command "line" p1 p2 p3 "c")
)
```

5. Choose the **File** menu and select the **Save** or **Save As** option. In the **Save As** dialog box enter the name of the file, **Triang1.lsp**. After you save the file, the file name is displayed at the top of the editor window, Figure 13-3.

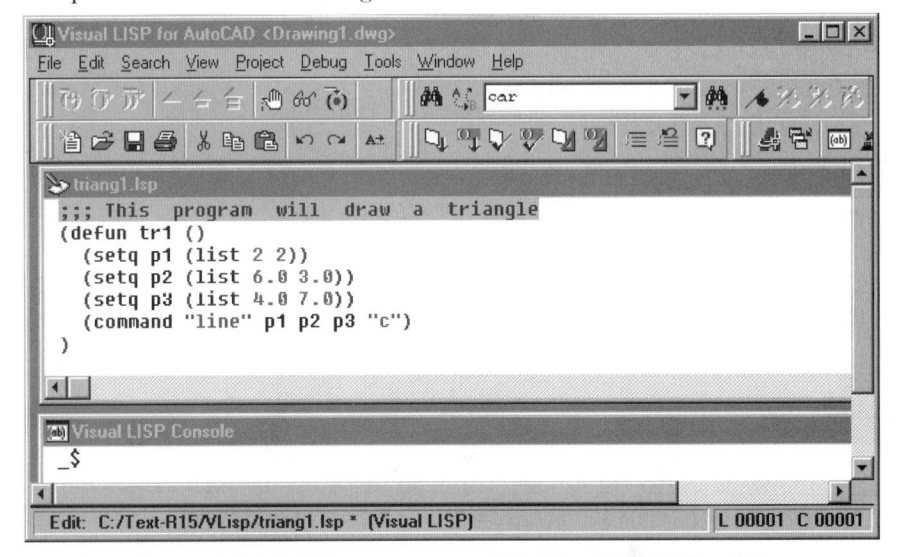

Figure 13-3 Writing program code in Visual LISP Text Editor

Loading and Running Programs

1. Make sure the Visual Lisp Text Editor window is active. If not, click a point anywhere in the window to make it active.

2. In the **Tools** menu select **Load Text in Editor**. You can also load the program by choosing **Load active edit window** in the **Tools** toolbar. Visual LISP displays a message in the Console window indicating that the program has been loaded. If there was a problem in loading the program, an error message will appear.

3. To run the program, enter the function name (**tr1**) at the console prompt (-$ sign). The function name must be enclosed in parentheses. The program will draw a triangle in AutoCAD. To view the program output, switch to AutoCAD by choosing **Activate AutoCAD** in the **View** toolbar. You can also run the program from AutoCAD. Switch to AutoCAD and enter the name of the function **TR1** at the Command prompt. AutoCAD will run the program and draw a triangle on the screen, Figure 13-4.

Visual LISP   13-5

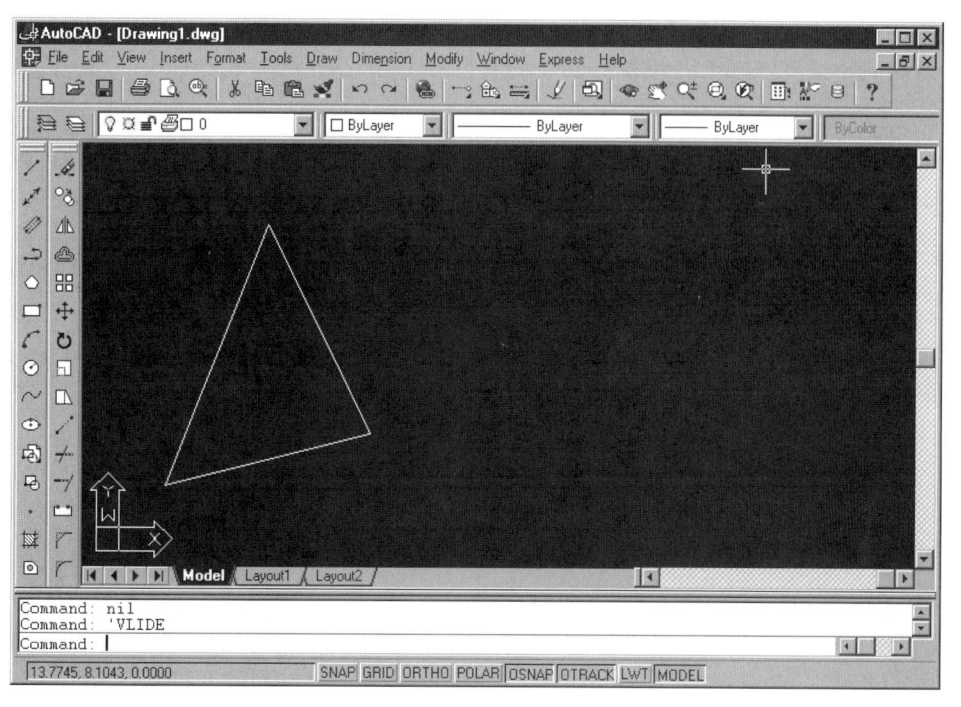

Figure 13-4 Program output in AutoCAD

Exercise 1   *General*

Write a LISP program that will draw an equilateral triangle. The program should prompt the user to select the starting point (bottom left corner) and the length of one of the sides of the triangle, Figure 13-5.

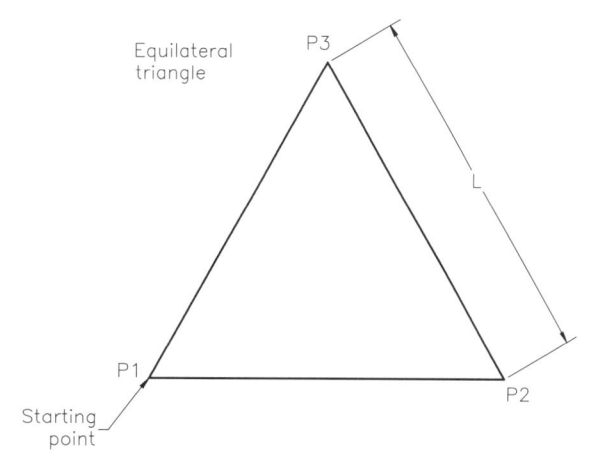

Figure 13-5 Drawing for Exercise 1

Loading Existing AutoLISP Files
You can also load the AutoLISP files in Visual LISP and then edit, debug, load, and run the programs from Visual LISP.

1. Start AutoCAD. In the **Tools** menu select **AutoLISP** and then select **Visual LISP Editor**. You can also start Visual LISP by entering the **VLIDE** command at the Command prompt.

2. In the Visual LISP Text Editor, choose **Open File** in the **File** menu. The **Open file to edit/view** dialog box is displayed on the screen.

3. Select the AutoLISP program that you want to load and then choose **Open**. The program is loaded in the Visual LISP Text Editor.

4. To format the program in the edit window, select the **Format edit window** tool from the **Tools** toolbar. The program will automatically get formatted.

5. To load the program, choose **Load active edit window** in the **Tools** toolbar or choose **Load Text in Editor** from the **Tools** menu.

6. To run the program, enter the function name at the console prompt ($) in the Visual LISP Console window and then press enter. If it gives nil, this means it has executed the drawing on the graphical screen.

Visual LISP Console
Most of the programming in Visual LISP is done in the Visual LISP Text Editor. However, you can also enter the code in the Visual LISP console and see the results immediately. For example, if you enter **(sqrt 37.2)** at the console prompt (after the -$ sign) and press the **Enter** key, Visual LISP will evaluate the statement and return the value 6.09918. Similarly, enter the code **(setq x 99.3)**, **(+ 38 23.44)**, **(- 23.786 35)**, **(abs -37.5)**, **(sin 0.333)** and see the results. Each statement must be entered after the $ sign as shown in Figure 13-6.

Features of the Visual LISP Console
1. In the Visual LISP console you can enter more than one statement on a line and then press the Enter key. Visual LISP will evaluate all statements and then return the value. Enter the following statement after the $ sign:
   -$(setq x 37.5) (setq y (/ x 2))
   Visual LISP will display the value of X and Y in the console, Figure 13-7

2. If you want to find the value of a variable in the code being developed, enter the variable name after the $ sign and Visual LISP will return the value and display it in the console. For example, if you want to find the value of X, enter X after the $ sign.

3. Visual LISP allows you to write an AutoLISP expression on more than one line by pressing **Ctrl+Enter** at the end of the line. For example, if you want to write the following two expressions on two lines, enter the first line and then press the **Ctrl+Enter** keys. You will notice that Visual LISP does not display the $ sign in the next line, indicating continuation

Visual LISP 13-7

of the statement on the previous line. Now enter the next line and press the **Enter** key, Figure 13-7.

$ (setq n 38)   Press the **Ctrl+Enter** keys
(setq counter (- n 1))

4. To retrieve the text that has been entered in the previous statement, press the **Tab** key

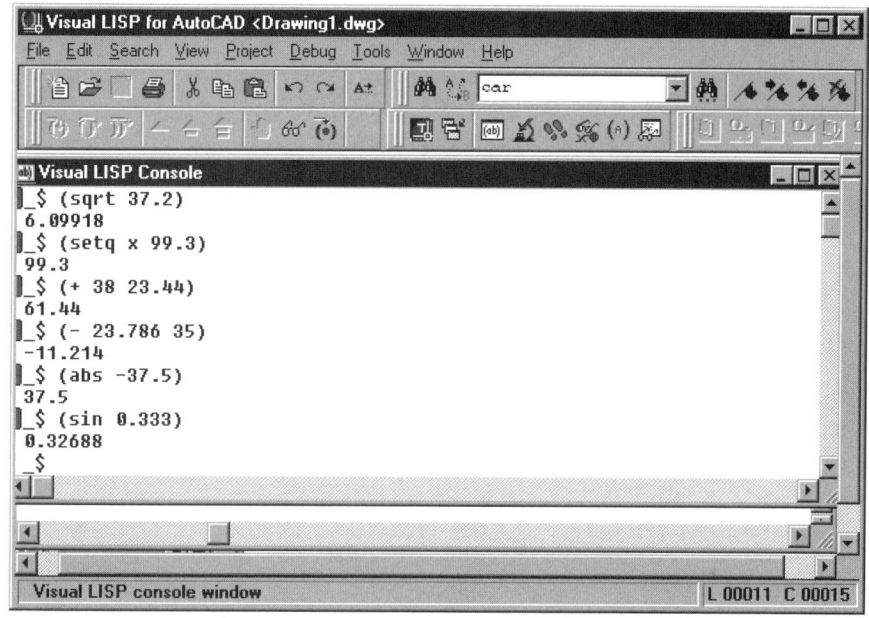

Figure 13-6 Entering LISP code in Visual LISP Console

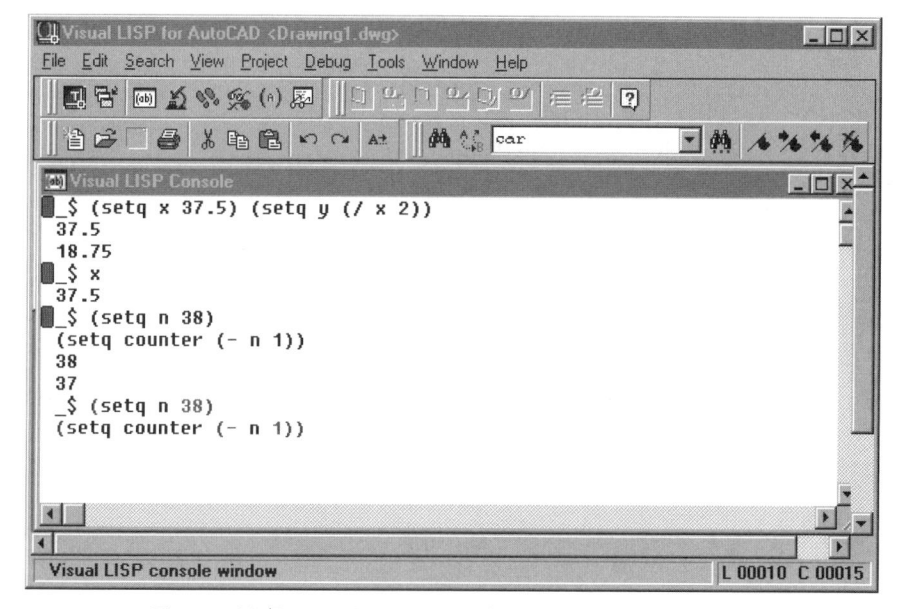

Figure 13-7 Entering LISP code in Visual LISP Console

while at the Console prompt (-$). Each time you press the **Tab** key, the previously entered text is displayed. You can continue pressing the **Tab** key to cycle through the text that has been entered in the Visual LISP Console. If the same statement has been entered several times, pressing the Tab key will display the text only once and then jump to the statement before that. Pressing the **Shift+Tab** key displays the text in the opposite direction.

5. Pressing the **Ese** key clears the text following the Console prompt (-$). However, if you press the **Shift+Esc** keys, the text following the Console prompt is not evaluated and stays as is. The cursor jumps to the next Console prompt. For example, if you enter the expression **(setq x 15)** and then press the **Esc** key, the text is cleared. However, if you press the **Shift+Esc** keys, the expression (setq x 15) is not evaluated and the cursor jumps to the next Console prompt without clearing the text.

6. The **Tab** key also lets you do an associative search for the expression that starts with the specified text string. For example, if you want to search the expression that started with (sin, enter it at the Console prompt (-$) and then press the **Enter** key. Visual LISP will search the text in the Console window and return the statement that starts with (sin. If it cannot find the text, nothing happens, except maybe a beep.

7. If you click the **right mouse** button (right-click) anywhere in the Visual LISP Console or press the **Shift+F10** keys, Visual LISP displays the context menu, Figure 13-8. You can use the context menu to perform several tasks like Cut, Copy Paste, Clear Console Window, Find, Inspect, Add Watch, Apropros Window, Symbol Service, Undo, Redo, AutoCAD Mode, and Toggle Console Log.

8. One of the important features of Visual LISP is the facility of context-sensitive help available for the Visual LISP functions. If you need help with any Visual LISP function, enter or select the function and then choose the Help button in the **Tools** toolbar. Visual LISP will display the help for the selected function, Figure 13-9.

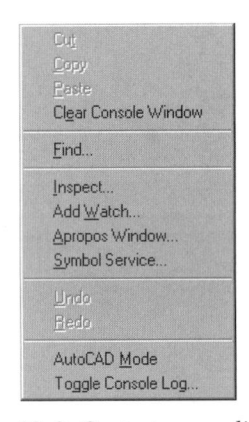

Figure 13-8 Context menu displayed when you click the right mouse button

9. Visual LISP also allows you to keep a record of the Visual LISP Console activities by saving them on the disk as a log files (.log). To create a log file, select **Toggle Console Log** in the **File** menu or select **Toggle Console Log** from the context menu. The Visual LISP Console window must be active to create a log file.

10. As you enter the text, Visual LISP automatically assigns a color to it. The colors are assigned based on the nature of the text string. For example, if the text is a built-in function or a protected symbol, the color assigned is blue. Similarly, text strings are magenta, integers are green, parentheses are red, real numbers are teal, and comments are magenta.

11. Although native AutoLISP and Visual LISP are separate programming environments, you

Visual LISP

can transfer any text entered in the Visual LISP Console to AutoCAD Command line. To accomplish this, enter the code at the Console prompt and then select **AutoCAD Mode** in the context menu or from the **Tools** menu. The Console prompt is replaced with the Command prompt. Press the **Tab** key to display the text that was entered at the Console prompt. Press the ENTER key to switch to the AutoCAD screen. The text is displayed at AutoCAD's Command prompt.

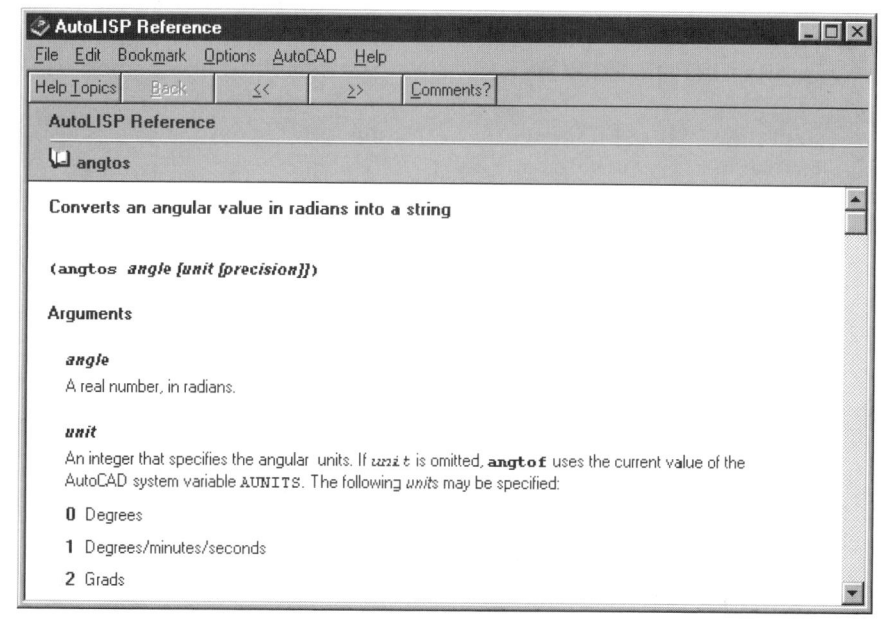

Figure 13-9 Context-sensitive help screen

Exercise 2 *General*

Enter the following statements in the Visual LISP Console at the console prompt (-$):
(+ 2 30 4 38.50)
(- 20 39 32)
(* 2.0 -7.6 31.25)
(/ -230 -7.63 2.15)
(sin 1.0472)
(atan -1.0)
(< 3 10)
(<= -2.0 0)
(setq variable_a 27.5)
(setq variable_b (getreal "Enter a value:"))

Using the Visual LISP Text Editor

You can use the **Visual LISP Text Editor** to enter the programming code. It has several features that are not available in other text editors. For example, when you enter text, it is automatically assigned a color based on the nature of the text string. If the text is in parenthesis, it is assigned red color. If the text is a Visual LISP function, it is assigned blue color. Visual

LISP Text Editor also allows you to execute an AutoLISP function without leaving the text editor, and also checks for balanced parentheses. These features make it an ideal tool for writing Visual LISP programs.

Color Coding

Visual LISP Text Editor provides color coding for files that are recognized by Visual LISP. Some of the files that it recognizes are LISP, DCL, SQL, and C language source files. When you enter text in the text editor, Visual LISP automatically determines the color and assigns it to the text, Figure 13-10. The following table shows the default color scheme:

Visual LISP Text Element	Color
Parentheses	Red
Built-in functions	Blue
Protected symbols	Blue
Comments	Magenta with gray background
Strings	Magenta
Real numbers	Teal
Integers	Green
Unrecognized items	Black (like variables)

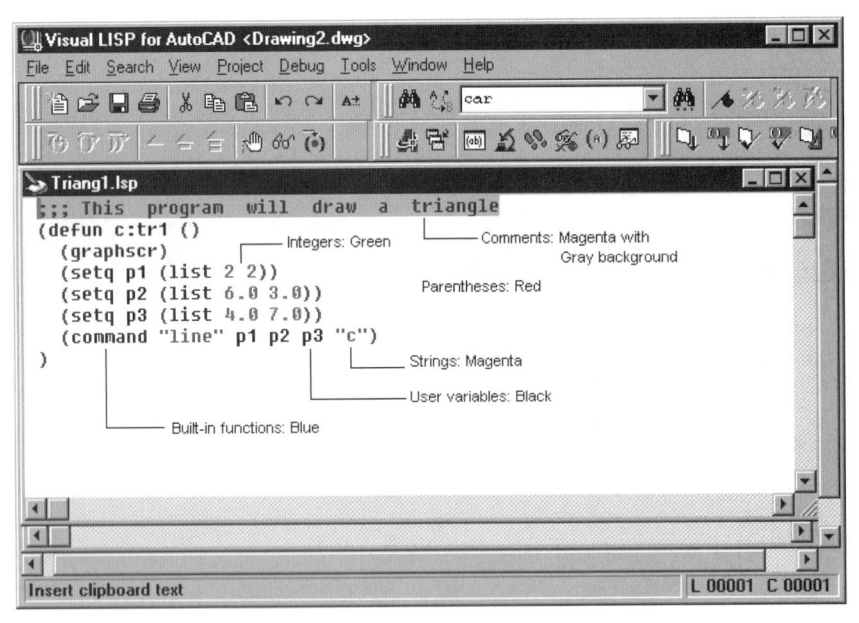

Figure 13-10 Color coding for LISP code

Words in the Visual LISP Text Editor

In most of the text editors the words are separated by space. In the Visual LISP Text Editor, the word has a different meaning. A word is a set of characters that are separated by one or more of the following special characters:

Visual LISP 13-11

Character Description	Visual LISP Special Characters
Space	
Tab	
Single quote	'
Left parenthesis	(
Right parenthesis	)
Double quotes	"
Semicolon	;
Newline	\n

For example, if you enter the text (Command"Line"P1 P2 P3"c), Visual LISP treats Command and Line as separate words because they are separated by double quotes. In a regular text editor (Command"Line" would have been treated as a single word. Similarly, in **(setq p1,** setq and p1 are separate words because they are separated by a space. If there was no space, Visual LISP will treat **(setqp1** as one word.

Context-Sensitive Help

One of the important features of Visual LISP is the facility of context-sensitive help available for the Visual LISP functions. If you need help with any Visual LISP function, enter or select the function in the text editor and then choose the **Help** button in the **Tools** toolbar. Visual LISP will display the help for the selected function.

Context Menu

If you click the **right mouse** button (right-click) anywhere in the Visual LISP Text Editor or press the **Shift+F10** keys, Visual LISP displays the context menu, Figure 13-11. You can use the context menu to perform several tasks like Cut, Copy, Paste, Clear Console Window, Find, Inspect, Add Watch, Apropos Window, Symbol Service, Undo, Redo, AutoCAD Mode, and Toggle Console Log. The following table contains a short description of these functions:

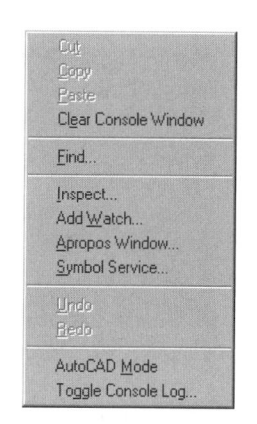

Figure 13-11 Context menu displayed when you click the right mouse button

Functions in the context menu	Description of the function
Cut	Moves the text to the clipboard
Copy	Copies the text to the clipboard
Paste	Pastes the contents of the clipboard at the cursor position
Find	Finds the specified text
Go to Last Edited	Moves the cursor to the last edited position
Toggle Breakpoint	Sets or removes a breakpoint at the cursor position
Inspect	Opens the **Inspect** dialog box
Add Watch	Opens the **Add Watch** dialog box
Apropos Window	Opens **Apropos options** dialog box

Symbol Service   Opens **Symbol Service** dialog box
Undo   Reverses the last operation
Redo   Reverses the previous Undo

Visual LISP Formatter

You can use the Visual LISP Formatter to format the text entered in the text editor. When you format the text, the formatter automatically arranges the AutoLISP expressions in a way that makes it easy to read and understand the code. It also makes it easier to debug the program when the program does not perform the intended function.

Running the Formatter

With the formatter you can format all the text in the Visual LISP text editor window or the selected text. To format the text, the text editor window must be active. If it is not, click a point anywhere in the window to make it active. To format all the text, select the **Format edit window** tool in the **Tools** toolbar or select **Format code in Editor** from the **Tools** menu. To format part of the text in the text editor, select the text and then select the **Format selection** tool in the **Tools** toolbar or select **Format code in Selection** from the **Tools** menu. Figure 13-12 shows the **Tools** menu and **Tools** toolbar.

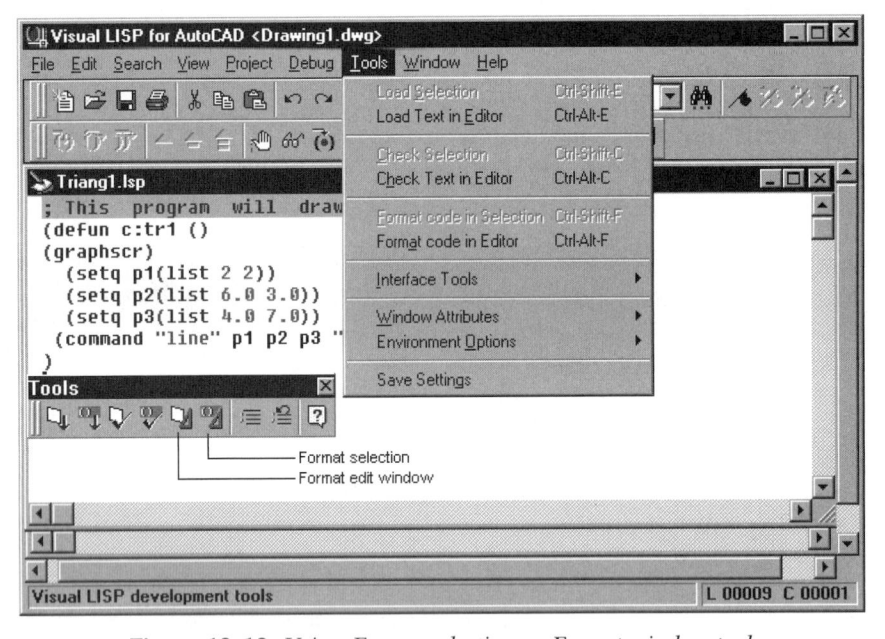

Figure 13-12 Using Format selection or Format window tools

Formatting Comments

In Visual LISP you can have five types of comments and depending on the comment in the program code, Visual LISP will format and position the comment text according to its type. Following is the list of various comment types available in Visual LISP:

Comment formatting   **Description**
; Single semicolon   The single semicolon is indented as specified in the

Visual LISP

	Single-semicolon comment indentation of AutoLISP Format option.
;; Current column comment	The comment is indented at the same position as the previous line of the program code.
;;; Heading comment	The comment appears without any indentation.
;| Inline comment |;	The Inline comments appear as any other expression.
;_ Function closing comment	The function closing comment appears just after the previous function, provided Insert form-closing comment is on in the AutoLISP Format options.

Two text editor windows are shown in Figure 13-13. The window at the top has the code without formatting. The text window at the bottom has been formatted. If you compare the two, you will see the difference formatting makes in the text display.

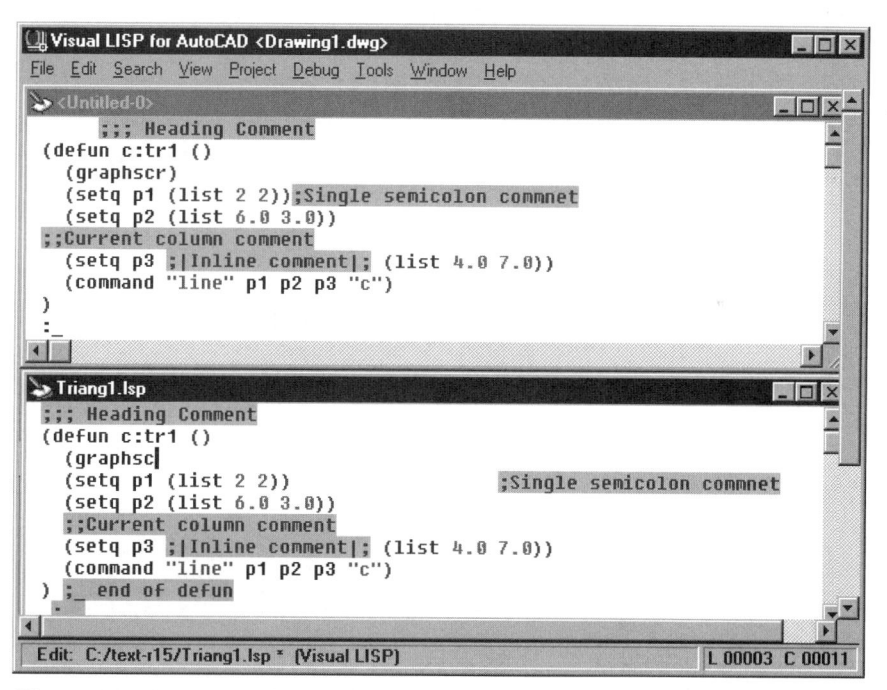

Figure 13-13 Top window shows code without formatting and the bottom window shows the same code after formatting the programming code

Changing Formatting Options

You can change the formatting options through the **Format options** dialog box (Figure 13-14) that can be invoked by choosing the **Tools** menu, **Environment Options**, and then **Visual LISP Format Options**. In this dialog box if you select **More** options button, Visual LISP will display another dialog box (Figure 13-15) listing more formatting options.

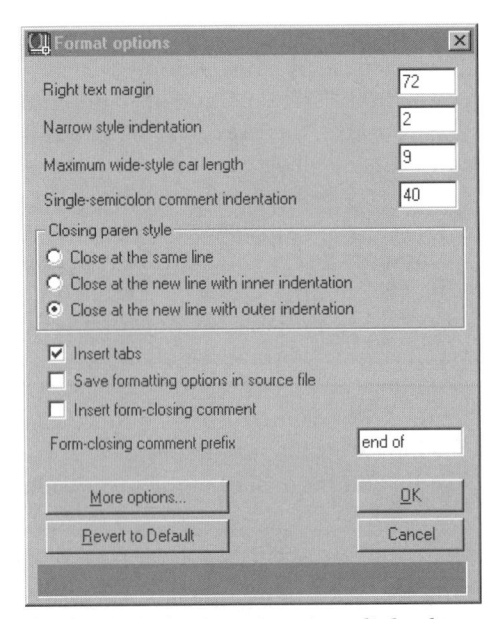

Figure 13-14 Format options dialog box

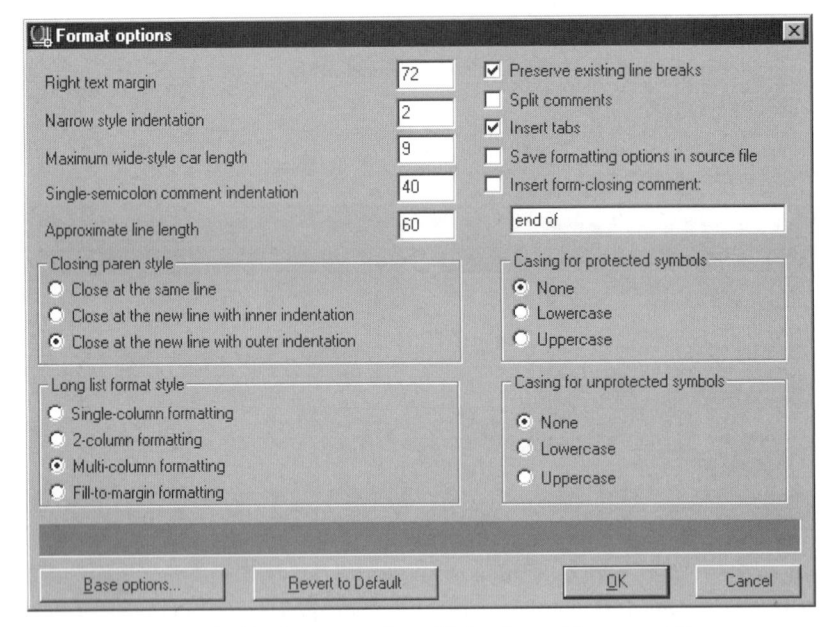

Figure 13-15 Format options dialog box with more options

DEBUGGING THE PROGRAM

When you write a program, it is rare that the program will run the first time. Even if runs, it may not perform the desired function. The error could be a syntax error, missing parentheses, a misspelled function name, improper use of a function, or simply a typographical error. It is time consuming and sometimes difficult to locate the source of the problem. Visual LISP

Visual LISP   13-15

comes with several debugging tools that make it easier to locate the problem and debug the code. Make sure the Debug toolbar is displayed on the screen. If it is not, in the Visual LISP window click on the **View** menu and then select **Toolbars** to display the **Toolbars** dialog box. Select the **Debug** check box, choose the **Apply** button, and then choose the **Close** button to exit the dialog box. The **Debug** toolbar will appear on the screen. Figure 13-16 shows the toolbar with tool tips.

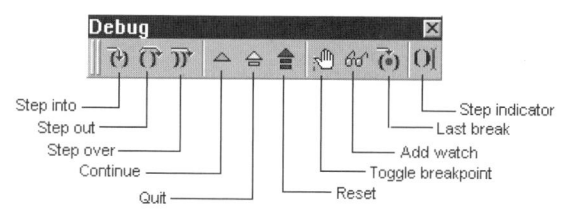

Figure 13-16 Tools in Debug toolbar

1. To try some of the debugging tools, enter the following program in the Visual LISP Text Editor and then save the file as **triang2.lsp**:

   ```
   ;;; This program will draw a triangle and an arc
   (defun tr2 ()
     (setq p1 (getpoint "\nEnter first point p1:"))
     (setq p2 (getpoint "\nEnter second point p2:"))
     (setq p3 (getpoint "\nEnter third point p3:"))
     (command "arc" p1 p2 p3)
     (command "line" p1 p2 p3 "c")
   )
   ```

2. To format the code, choose the **Format edit window** tool in the **Tools** toolbar. You can also format the code by selecting **Format code in Editor** in the Tools menu.

3. Load the program by choosing the **Load active edit window** tool in the **Tools** toolbar. You can also load the program from the **Tools** menu by selecting **Load Text in Editor**.

4. In the Visual LISP Text Editor, position the cursor in front of the following line and then choose the **Toggle breakpoint** tool in the **Debug** toolbar. You can also select it from the **Debug** menu. Visual LISP inserts a breakpoint at the cursor location.

   |(setq p2 (getpoint "\Enter second point p2:"))

5. In the Visual LISP Console, at the Console prompt, enter the name of the function and then press the **Enter** key to run the program.

   -$ **(tr2)**

The AutoCAD window will appear and the first prompt in the program (Enter first point p1:) will be displayed in the Command prompt area. Pick a point and the Visual LISP window will appear. Notice, the line that follows the breakpoint is highlighted as shown in Figure 13-17.

```
;;; This program will draw a triangle and an arc
(defun tr2 ()
   (setq p1 (getpoint "\nEnter first point p1:"))
   (setq p2 (getpoint "\nEnter second point p2:"))
   (setq p3 (getpoint "\nEnter third point p3:"))
   (COMMAND "arc" p1 p2 p3)
   (command "line" p1 p2 p3 "c")
)
```

Figure 13-17 Using Step Into command to start execution until it encounters the next expression

6. Choose the **Step Into** button in the **Debug** toolbar or select **Step Into** from the **Debug** menu. You can also press the function key **F8** to invoke the **Step Into** command. The program starts execution until it encounters the next expression. The next expression that is contained within the inside parentheses is highlighted, Figure 13-18. Notice the position of the vertical bar in the Step indicator tool in the Debug toolbar. It is on the left of the Step indicator tool. Also, the cursor is just before the expression ("getpoint "\nEnter second point:).

```
;;; This program will draw a triangle and an arc
(defun tr2 ()
   (setq p1 (getpoint "\nEnter first point p1:"))
   (setq p2 (getpoint "\nEnter second point p2:"))
   (setq p3 (getpoint "\nEnter third point p3:"))
   (COMMAND "arc" p1 p2 p3)
   (command "line" p1 p2 p3 "c")
)
```

Figure 13-18 Using Step Into command again, the cursor jumps to the end of the expression

7. Choose the **Step Into** button again. The AutoCAD window appears and the second prompt in the program (Enter second point p2:) is displayed in the Command prompt area. Pick a point, and the Visual LISP window returns, Figure 13-19. Notice the position of the vertical bar in the **Step indicator** tool in the **Debug** toolbar. It is just to the right of the Step Indicator. Also, the cursor is just after the expression ("getpoint "\nEnter second point:).

```
Triang2.LSP
;;; This program will draw a triangle and an arc
(defun tr2 ()
   (setq p1 (getpoint "\nEnter first point p1:"))
   (setq p2 (getpoint "\nEnter second point p2:"))
   (setq p3 (getpoint "\nEnter third point p3:"))
   (COMMAND "arc" p1 p2 p3)
   (command "line" p1 p2 p3 "c")
)
```

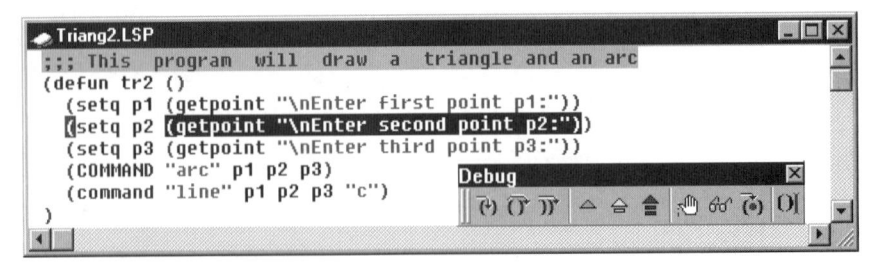

Figure 13-19 Using Step Into command again, the vertical bar is displayed to the right of Step Indicator symbol

8. Choose the **Step Into** button again to step through the program. The entire line is

Visual LISP

highlighted and the cursor moves to the end of the expression (getpoint "Enter second point:). Notice the position of the vertical bar in the **Step indicator** tool in the **Debug** toolbar.

9. Choose the **Step Into** button again. The next line of the program is highlighted and the cursor is at the start of the statement.

10. To step over the highlighted statement, choose the **Step over** tool in the **Debug** toolbar, or select **Step Over** in the Debug menu. You can also press the **Shift+F8** keys to invoke this command. The AutoCAD window is displayed and the prompt (Enter third point:) is displayed in the Command prompt area. Select a point, and the Visual LISP window appears.

11. Choose the **Step over** button again; Visual LISP highlights the next line. If you choose the Step over button again, the entire expression is executed.

12. Choose the **Step over** button until you go through all the statements and the entire program is highlighted.

General Recommendations for writing the LISP file

1. Built-in functions and arguments should be spaced.
2. Arguments should be spaced.
3. Every function must start with an open parenthesis.
4. Every function must end with a closed parenthesis.

Example 1

Write a LISP program to draw the I-section as shown in Figure 14-20. The starting point is p1. The point P1, length, width, and the thicknesses of flange and web are user-defined values that are to be entered at the command prompt.

Step 1: Understanding the program algorithm

Before writing the program it is very important to understand what is given, what is the desired output, and how we can get that output from the given information. Analyze the program as follows:

Input
- P1   Starting point of the I-section
- L   Length of the I-section
- W   Width of the I-section
- T1   Thickness of the flange
- T2   Thickness of the web

Output
I section as shown in Figure 13-20

Process
Define all points like p2, p3, p4, p5, p6, p7, p8, p9, p10, p11 and p12
Draw lines between points

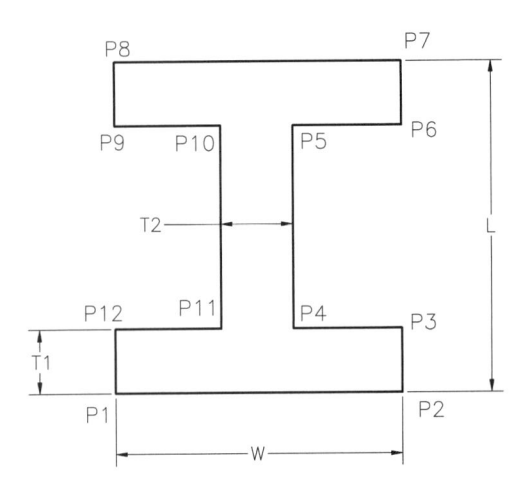

Figure 13-20 I-section for Example 1

Step 2: Writing the Visual LISP program
Select **Tools> AutoLISP> Visual LISP Editor** menu or enter **VLIDE** at the Command prompt to display the **Visual LISP** window. Select **New File** from the **File** menu to display the **Visual LISP Text Editor** window. Write the following program in the **Visual LISP Text Editor**.

```
(defun isec ()
 (setq   p1  (getpoint "\n Enter the starting point of the I-section: ")
         l   (getdist "\n Enter the length of the I-section: ")
         w   (getdist "\n Enter the width of the I-section: ")
         t1  (getdist "\n Enter the thickness of the flange: ")
         t2  (getdist "\n Enter the thickness of the web: ")
         p2  (list (+ (car p1) w) (cadr p1))
         p3  (list (car p2) (+ (cadr p2) t1))
         p4  (list (- (car p3) (/ (- w t2) 2)) (cadr p3))
         p5  (list (car p4) (+ (cadr p4) (- l (* 2 t1))))
         p6  (list (car p3) (cadr p5))
         p7  (list (car p6) (+ (cadr p6) t1))
         p8  (list (car p1) (+ (cadr p1) l))
         p9  (list (car p8) (- (cadr p8) t1))
         p10 (list (- (car p5) t2) (cadr p5))
         p11 (list (- (car p4) t2) (cadr p4))
         p12 (list (car p1) (cadr p11))
 )
 (command "PLINE" p1 p2 p3 p4 p5 p6 p7 p8 p9 p10 p11 p12 p1 "")
)
```

Step 3: Loading and running the Visual LISP program

Save the file as **isec.lsp** using the **Save** or **Save As** option in the **File** menu. To load the program, select **Load Text in Editor** in the **Tools** menu. You can also load the program by choosing **Load active edit window** in the **Tools** toolbar. **Visual LISP** displays a message in the **Console** window indicating that program has been loaded. If the Console window doesn't appear on the screen, select the **Tile Horizontally** option from the **Window** toolbar to display it on the screen. It also gives the error message, if there is any error in the program.

The step-by-step debugging can be done as described under the heading **"DEBUGGING THE PROGRAM."** The mismatch in the parentheses can be corrected by choosing the **Format Edit Window** in the **Tools** toolbar. The nature of the error and its meaning are given at the end of this chapter under the heading **"ERROR CODES AND MESSAGES."** To run the program, enter the function name **(isec)** at the console prompt adjacent to the $ sign. The function name must be enclosed in parentheses. When you enter the function name and press enter, the AutoCAD screen is displayed prompting you to enter the information about the I-section. After you enter the information, the I-section will be drawn on the screen.

Tip
*1. This program can also be entered using any text editor like Notepad. The program can be loaded by entering the **APPLOAD** command at the Command prompt as described in the AutoLISP chapter.*

*2. The above example can also be solved by defining only half of the I-section and then mirroring the I-section with the **MIRROR** command to get the complete I-section. You can also draw only one-fourth of the I-section and then use the **MIRROR** command to get the remaining section.*

Step 4: Writing the Visual LISP program using POLAR function

The points can also be calculated by using the POLAR function and then draw the lines to create the I-section. In this program d1=distance between P2 and P3, d2=distance between P3 and P4, and d3 = distance between P4 and P5. Notice the dtr function used to convert the degrees into radians. The following file is the listing of the program using POLAR function.

```
((defun dtr (a)
 (* a (/ pi 180.0))
)
(defun isec ()
  (setq  p1 (getpoint "\n Enter the starting point of the I-section: ")
         l  (getdist "\n Enter the length of the I-section: ")
         w  (getdist "\n Enter the width of the I-section: ")
         t1 (getdist "\n Enter the thickness of the flange: ")
         t2 (getdist "\n Enter the thickness of the web: ")
  )
  (setq  d1 t1
         d2 (- (/ w 2.0) (/ t2 2.0))
         d3 (- l (* 2.0 t1))
  )
  (setq  p2  (polar p1 (dtr 0) w)
```

```
          p3  (polar p2 (dtr 90) d1)
          p4  (polar p3 (dtr 180) d2)
          p5  (polar p4 (dtr 90) d3)
          p6  (polar p5 (dtr 0) d2)
          p7  (polar p6 (dtr 90) d1)
          p8  (polar p7 (dtr 180) W)
          p9  (polar p8 (dtr 270) d1)
          p10 (polar p9 (dtr 0) d2)
          p11 (polar p10 (dtr 270) d3)
          p12 (polar p11 (dtr 180) d2)
   )
   (command "PLINE" p1 p2 p3 p4 p5 p6 p7      p8 p9 p10 p11 p12 p1 "")
)
```

Example 2

Write a Visual LISP program that will extrude the I-section of Example 1 and create a 3D shape of the section, Figure 13-21. The additional input to the program is vpoint (v), extrusion height (h), and extrusion angle (a).

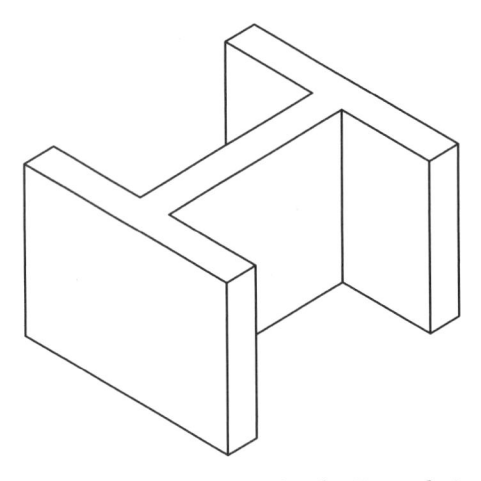

Figure 13-21 3D I-section for Example 2

Step 1: Understanding the program algorithm

In addition to the inputs of Example 1, add the Vpoint, Extrusion, and Angle of Extrusion as additional inputs in the program. Also add the commands corresponding to vpoint, extrusion, zoom, and hide. Vpoint, extrusion height, and angle of extrusion have been represented as **v**, **h**, and **a** respectively. Angle has been entered using **getreal** function to accept the angle in decimal degrees. Do not use **getangle** function because this function returns the angle in radians. The extrusion angle in the **EXTRUDE** command must be in degrees.

Step 2: Writing the Visual LISP program

Select **Tools> AutoLISP> Visual LISP Editor** menu or enter **VLIDE** at the Command prompt to display the **Visual LISP** window. Select **New File** from the **File** menu to display the **Visual LISP Text Editor** window. Now write the program as shown below in the **Visual LISP Text Editor** or open the LISP file of Example 1 and add the new lines to this program.

```
(defun dtr (a)
  (* a (/ pi 180.0)))
(defun isec ()
  (setq p1 (getpoint "\n Enter the starting point of the I-section: ")
        l  (getdist "\n Enter the length of the I-section: ")
        w  (getdist "\n Enter the width of the I-section: ")
        t1 (getdist "\n Enter the thickness of the flange: ")
        t2 (getdist "\n Enter the thickness of the web: ")
        h  (getdist "\n Enter Extrusion Height: ")
        v  (getpoint "\n Enter the Viewpoint: ")
        a  (getreal "\n Enter Angle of Extrusion: ")
  )
  (setq  d1 t1
         d2 (- (/ w 2.0) (/ t2 2.0))
         d3 (- l (* 2.0 t1))
  )
  (setq  p2  (polar p1 (dtr 0) w)
         p3  (polar p2 (dtr 90) d1)
         p4  (polar p3 (dtr 180) d2)
         p5  (polar p4 (dtr 90) d3)
         p6  (polar p5 (dtr 0) d2)
         p7  (polar p6 (dtr 90) d1)
         p8  (polar p7 (dtr 180) W)
         p9  (polar p8 (dtr 270) d1)
         p10 (polar p9 (dtr 0) d2)
         p11 (polar p10 (dtr 270) d3)
         p12 (polar p11 (dtr 180) d2)
  )
  (command "PLINE" p1 p2 p3 p4 p5 p6 p7   p8 p9 p10 p11 p12 p1 "")
  (command "VPOINT" v)
  (command "EXTRUDE" "All" "" h a)
  (command "ZOOM" "All")
  (command "HIDE")
)
```

Step 3: Loading and running the Visual LISP program

Save the file as **isec3d.lsp** using the **Save** or **Save As** option in the **File** menu. To load the program, select **Load Text in Editor** in the **Tools** menu. You can also load the program by choosing **Load active edit window** in the **Tools** toolbar. **Visual LISP** displays a message in the **Console** window indicating that program has been loaded. To run the program enter the function name (**isec3d**) at the console prompt adjacent to $ sign. The function name must be enclosed in parentheses. The program will prompt the user to enter the starting point followed by length, width, thicknesses, vpoint, extrusion height and angle at the Command prompt. After entering all the inputs, a 3D I-section will be displayed on the AutoCAD screen.

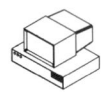

> **Tip**
> *Step by step debugging can be done by choosing the Toggle breakpoint tool in the Debug toolbar as described under the heading* **Debugging the Program (Page 13-14).**

Example 3

Write a Visual LISP program to draw a c-section as shown in the Figure 13-22. The program should prompt the user to enter the starting point, length, width, and thickness.

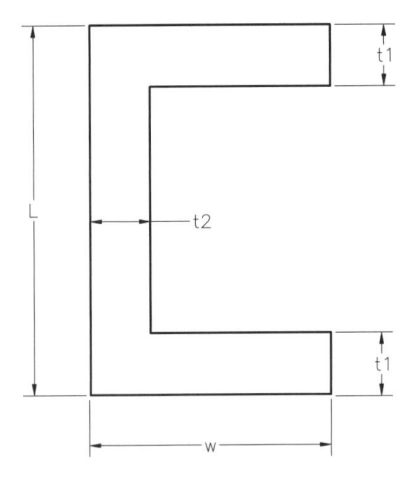

Figure 13-22 C-section

Step 1: Writing the Visual LISP program

Select **Tools> AutoLISP> Visual LISP Editor** menu or enter **VLIDE** at the Command prompt to display the **Visual LISP** window. Select **New File** from the **File** menu to display the **Visual LISP Text Editor** window. Now write the program in the **Visual LISP Text Editor**. You can also use any text editor like Notepad to write the LISP program. The following file is the listing of the program that will draw a c-section.

```
;;This program draws a C-Section
(defun csec ()
  (setq p1 (getpoint "\n enter the point:"))
  (setq L (getdist "\n enter the length:"))
  (setq W (getdist "\n enter the width:"))
  (setq t1 (getdist "\n enter the thickness of the flange:"))
  (setq t2 (getdist "\n enter the thickness of the web:"))
  (setq p2 (list (+ (car p1) w) (cadr p1)))
  (setq p3 (list (car p2) (+ (cadr p2) t1)))
```

Visual LISP 13-23

```
(setq p4 (list (- (car p3) (- w t2)) (cadr p3)))
(setq p5 (list (car p4) (+ (cadr p4) (- L (* 2 t1)))))
(setq p6 (list (+ (car p5) (- w t2)) (cadr p5)))
(setq p7 (list (car p6) (+ (cadr p6) t1)))
(setq p8 (list (car p1) (cadr p7)))
(command "pline" p1 p2 p3 p4 p5 p6 p7 p8 p1 "")
)
```

Step 2: Loading and Running the Visual LISP program

Save the file as **csec.lsp** using the **Save** or **Save As** option in the **File** menu. To load the program, select **Load Text in Editor** in the **Tools** menu. You can also load the program by choosing **Load active edit window** in the **Tools** toolbar. **Visual LISP** displays a message in the **Console** window indicating that program has been loaded. To run the program enter the function name **(isec3d)** at the console prompt adjacent to the $ sign. The function name must be enclosed in parentheses.

Exercise 3   *General*

Write a Visual LISP program to draw a T-section as shown in Figure 13-23. The program should prompt the user to specify the starting point, length (L), width (w), and thicknesses (t1 and t2) of the T-section.

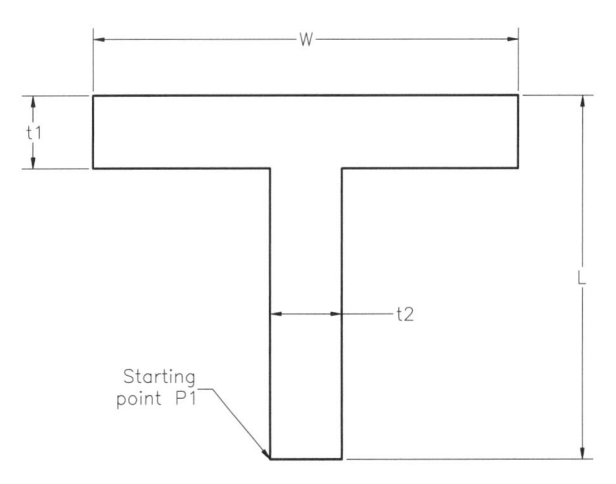

Figure 13-23 T-section for Exercise 3

Exercise 4   *General*

Write a Visual LISP program to draw a 3D T-section with the same specifications as Exercise 3. In addition to the inputs in Exercise 3, the program should also prompt the user to specify vpoint, extrusion height, and angle of extrusion.

Exercise 5   *General*

Write a Visual LISP program to draw an L-section as shown in Figure 13-24. The program should prompt the user to specify starting point, length, width, and thicknesses of the L-section.

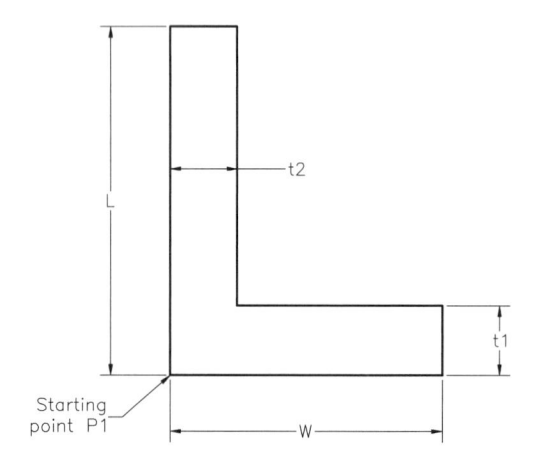

Figure 13-24 L-section for Exercise 5

Example 4

Write an Visual LISP program that will generate a flat layout drawing of a transition and then dimension the layout. The transition and the layout without dimensions are shown in Figure 13-25.

Step 1: Writing the Visual LISP program
Select **Tools> AutoLISP> Visual LISP Editor** menu or enter **VLIDE** at the Command prompt to display the **Visual LISP** window. Select **New File** from the **File** menu to display the **Visual LISP Text Editor** window. Now write the program in the **Visual LISP Text Editor** window. You can also use any text editor like Notepad to write the LISP program. The following file is the listing of the **LISP** program for Example 4. The LISP programs do not need to be in lowercase letters. They can be uppercase or a combination of uppercase and lowercase.

```
;This program generates flat layout of
;a rectangle to rectangle transition
;
(defun TRANA ()
(graphscr)
(setvar "cmdecho" 0)
(setq L (getdist "\n Enter length of bottom rectangle: "))
(setq W (getdist "\n Enter width of bottom rectangle: "))
(setq H (getdist "\n Enter height of transition: "))
(setq L1 (getdist "\n Enter length of top rectangle: "))
(setq W1 (getdist "\n Enter width of top rectangle: "))
```

Visual LISP

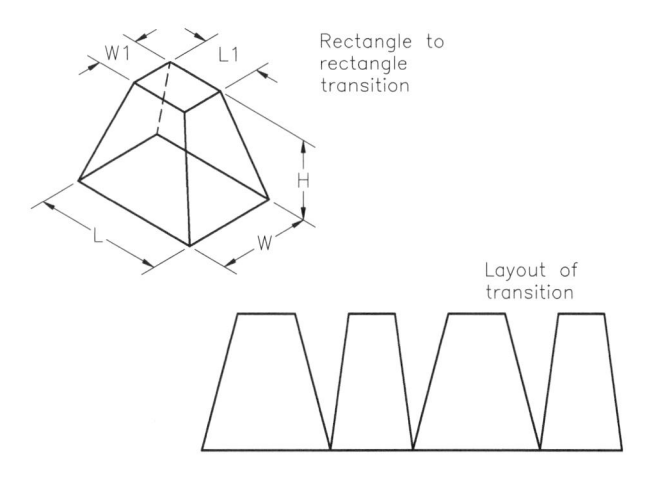

Figure 13-25 *Flat layout of a transition*

```
(setq x1 (/ (- w w1) 2))
(setq y1 (/ (- l l1) 2))
(setq d1 (sqrt (+ (* h h) (* x1 x1))))
(setq d2 (sqrt (+ (* d1 d1) (* y1 y1))))
(setq s1 (/ (- l l1) 2))
(setq p1 (sqrt (- (* d2 d2) (* s1 s1))))
(setq s2 (/ (- w w1) 2))
(setq p2 (sqrt (- (* d2 d2) (* s2 s2))))

(setq t1 (+ l1 s1))
(setq t2 (+ l w))
(setq t3 (+ l s2 w1))
(setq t4 (+ l s2))
(setq pt1 (list 0 0))
(setq pt2 (list s1 p1))
(setq pt3 (list t1 p1))
(setq pt4 (list l 0))
(setq pt5 (list t4 p2))
(setq pt6 (list t3 p2))
(setq pt7 (list t2 0))
(command "layer" "make" "ccto" "c" "1" "ccto" "")
(command "line" pt1 pt2 pt3 pt4 pt5 pt6 pt7 "c")

(setq sf (/ (+ l w) 12))
(setvar "dimscale" sf)
(setq c1 (list 0 (- 0 (* 0.75 sf))))
(setq c7 (list (- 0 (* 0.75 sf)) 0))
(setq c8 (list (- l (* 0.75 sf)) 0))
```

```
(command "layer" "make" "cctd" "c" "2" "cctd" "")
(command "dim" "hor" pt1 pt2 c1 "" "base" pt3 "" "base" pt4 "" "exit")
(command "dim" "hor" pt4 pt5 c1 "" "base" pt6 "" "base" pt7 "" "exit")
(command "dim" "vert" pt1 pt2 pt2 "" "exit")
(command "dim" "vert" pt4 pt5 pt5 "" "exit")
(command "dim" "aligned" pt1 pt2 c7 "" "exit")
(command "dim" "aligned" pt4 pt5 c8 "" "exit")
(setvar "cmdecho" 1)
(princ))
```

Step 2: Loading and Running the Visual LISP program

Save the file as **trana.lsp** using the **Save** or **Save As** option in the **File** menu. To load the program, select **Load Text in Editor** in the **Tools** menu. You can also load the program by choosing **Load active edit window** in the **Tools** toolbar. **Visual LISP** displays a message in the **Console** window indicating that the program has been loaded. To run the program enter the function name (**isec3d**) at the console prompt adjacent to the $ sign. The function name must be enclosed in parentheses.

Example 5

Write an Visual LISP program that can generate a flat layout of a cone as shown in Figure 13-26. The program should also dimension the layout.

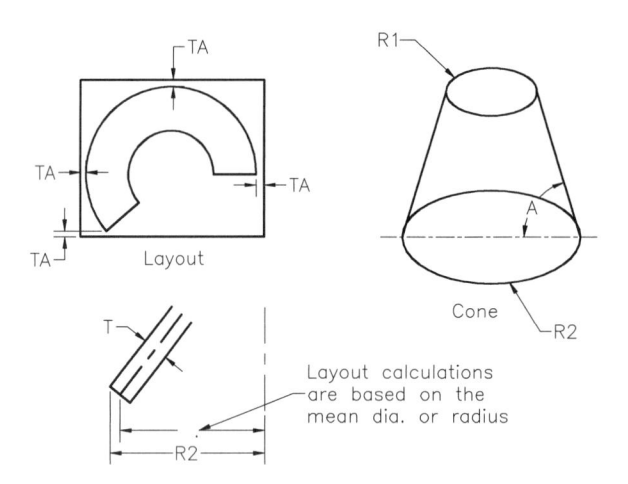

Figure 13-26 Flat layout of a cone

Step 1: Writing the Visual LISP program

```
;The following file is a listing of the Visual LISP program for Example 5:
;This program generates layout of a cone
;
;DTR function changes degrees to radians
(defun DTR (a)
 (* PI (/ A 180.0))
```

)
;RTD Function changes radians to degrees
(defun rtd (a)
(* a (/ 180.0 pi))
)

(defun tan (a)
(/ (sin a) (cos a))
)
(defun cone-1p ()
(graphscr)
(setvar "cmdecho" 0)
(setq r2 (getdist "\n enter outer radius at larger end: "))
(setq r1 (getdist "\n enter inner radius at smaller end: "))
(setq t (getdist "\n enter sheet thickness:-"))
(setq a (getangle "\n enter cone angle:-"))
;this part of the program calculates various parameters
;needed in calculating the strip layout
(setq x0 0)
(setq y0 0)
(setq sf (/ r2 3))
(setvar "dimscale" sf)
(setq ar a)
(setq tx (/ (* t (sin ar)) 2))
(setq rx2 (- r2 tx))
(setq rx1 (+ r1 tx))
(setq w (* (* 2 pi) (cos ar)))
(setq rl1 (/ rx1 (cos ar)))
(setq rl2 (/ rx2 (cos ar)))

;this part of the program calculates the x-coordinate
;of the points
  (setq x1 (+ x0 rl1)
     x3 (+ x0 rl2)
     x2 (- x0 (* rl1 (cos (- pi w))))
     x4 (- x0 (* rl2 (cos (- pi w))))
)

;this part of the program calculates the y-coordinate
;of the points
  (setq y1 y0
     y3 y0
     y2 (+ y0 (* rl1 (sin (- pi w))))
     y4 (+ y0 (* rl2 (sin (- pi w))))
)

  (setq p0 (list x0 y0)

```
            p1 (list x1 y1)
            p2 (list x2 y2)
            p3 (list x3 y3)
            p4 (list x4 y4)
            )
    (command "layer" "make" "ccto" "c" "1" "ccto" "")
    (command "arc" p1 "c" p0 p2)
    (command "arc" p3 "c" p0 p4)
    (command "line" p1 p3 "")
    (command "line" p2 p4 "")

    (setq f1 (/ r2 24))
    (setq f2 (/ r2 2))
    (setq d1 (list (+ x3 f2) y3))
    (setq d2 (list x0 (- y0 f2)))

    (command "layer" "make" "cctd" "c" "2" "cctd" "")
    (setvar "dimtih" 0)
    (command "dim" "hor" p0 p1 d2 "" "baseline" p3 "" "baseline" p2 "" "baseline" p4 ""
    "exit")
    (command "dim" "vert" p0 p2 d1 "" "baseline" p4 "" "exit")
    (setvar "dimscale" 1)
    (setvar "cmdecho" 1)
    (princ)
)
```

Step 2: Loading and running the Visual LISP program

Save the file as **cone-1p.lsp** using the **Save** or **Save As** option in the **File** menu. To load the program, select **Load Text in Editor** in the **Tools** menu. You can also load the program by choosing **Load active edit window** in the **Tools** toolbar. **Visual LISP** displays a message in the **Console** window indicating that program has been loaded. To run the program enter the function name **(isec3d)** at the console prompt adjacent to the $ sign. The function name must be enclosed in parentheses.

TRACING VARIABLES

Sometimes it becomes necessary to trace the values of the variables used in the program. For example, let us assume the program crashes or does not perform as desired. Now you want to locate where the problem occurred. One of the ways you can do this is by tracing the value of the variables used in the program. The following steps illustrate this procedure:

1. Open the file containing the program code in the Visual LISP Text Editor. You can open the file by selecting Open File from the File menu.

2. Load the program by selecting Load Text in Editor from the Tools menu or by selecting Load active edit window tool in the Tools toolbar.

Visual LISP 13-29

3. Run the program by entering the name of the function (tr2) at the Console prompt (-$) in the Visual LISP Console. The program will perform the operations as defined in the program. In this program, it will draw a triangle and an arc between the user-specified points. Now, we want to find out the coordinates of the points p1, p2, and p3. To accomplish this we can use the Watch window tool in the View toolbar.

4. Highlight the variable **p1** and then choose the **Watch window** tool in the **View** toolbar. You can also select **Watch Window** in the **View** menu. Visual LISP displays the **Watch** window that lists the X, Y, and Z coordinates of point p1.

5. Position the cursor near (in front, middle, or end) the variable p2 and then choose the **Add Watch** tool in the **Watch** window. Select the **OK** button in the **Add Watch** window. The value of the selected variable is listed in the **Watch** window, Figure 13-27. Similarly you can trace the value of other variables in the program.

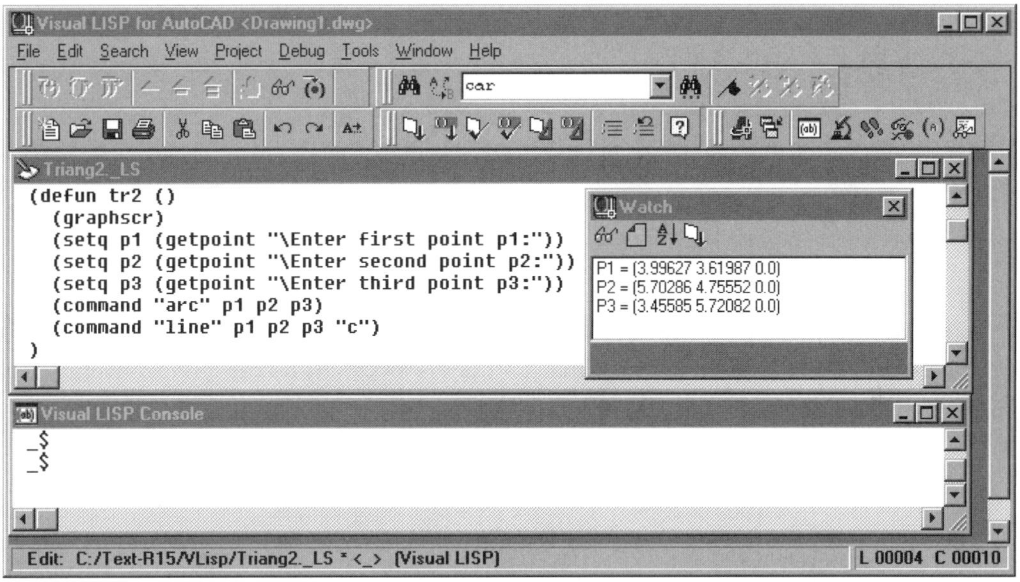

Figure 13-27 Using Add Watch to trace variables in the program

VISUAL LISP ERROR CODES AND MESSAGES

This section describes error messages that you may encounter at the console prompt while running the Visual LISP program and their corresponding meaning. The errors are arranged depending upon the frequency with which they occur during running the Visual LISP program. The common errors encountered have been listed first.

Malformed list
Premature ending of the list. Common cause is the mismatch between the number of opening and closing parentheses.

Malformed string
Premature ending of the string being read from a file.

Null function
A function was evaluated which has nil definition.

Too few arguments
The built-in function was given less arguments.

String too long
The system variable SETVAR has been given too long string.

Too many arguments
The built-in function has been given too many arguments.

Invalid argument
The type of argument is improper.

Invalid argument list
The function has been given an invalid argument list.

Incorrect number of arguments
It is expected to provide one argument within the quote function, but some other arguments have also been given.

Incorrect number of arguments to a function
Mismatch between the number of formal arguments specified and the number of arguments to the user-defined function.

Function cancelled
The user entered Ctrl+C or Esc in response to an input prompt.

Extra right paren
One or more extra right parentheses have been encountered.

Exceeded maximum string length
A string given to a function is greater than 132 characters.

Divide by zero
Division by zero is invalid.

Divide overflow
Division by a very small value has resulted in a invalid quotient.

Console break
The user entered Ctrl+C while a function is in progress.

Bad argument type
A function was passed an incorrect type of argument.

Bad association list
The list of the assoc function does not consist of key value lists.

Visual LISP stack overflow
The program has exceeded the Visual LISP stack storage space. This can be due to very large function arguments list.

Bad ENTMOD list
The argument passed to ENTMOD is not a proper entity data list.

Bad conversion code
An invalid space identifier was passed to the trans function.

Bad entmod list value
One of the sublists in the association list passed to entmod contains an improper value.

Bad function
The first element in the list is not a valid function name.

Bad list
An improper list was given to a function.

Bad ssget list
The argument passed to (ssget"x") is not a proper entity data.

Bad ssget list value
One of the sublists in the association list passed to (ssget"x") contains an improper value.

Bad ssget mode string
Caused when ssget was passed an invalid string in the mode argument.

Base point is required
The getcorner function was called without the required base point argument.

Can't evaluate expression
A decimal point was improperly placed, or some other expression was poorly formed.

Can't open file for input- LOAD failed
The file name in the load function could not be found, or the user does not have read access to the file.

Input aborted
An error or premature end of file condition has been detected, causing termination of the file input.

Invalid character
An expression contains an improper character.

Invalid dotted pair
You might get this error message if you begin a real number with a decimal point; you must use a leading zero in such a case.

Misplaced dot
A real number begins with a decimal point. You must use a leading zero in such a case.

Self-Evaluation Test

Answer the following questions and then compare your answers to the answer given at the end of the chapter.

1. Visual LISP has a powerful _____ and a _____ .

2. The text editor allows the _____ of parentheses, function names, variables and other components.

3. The formatter _____ the code in an easily readable format.

4. Values can be traced with the help of _____ tool in the view toolbar.

5. Step-by-step debugging can be done with the _____ tool in the Debug toolbar.

6. What are the different ways to start Visual LISP?
   1. _____ 2. _____

7. Name the four areas of the Visual LISP window.
   1. _____ 2. _____
   3. _____ 4. _____

8. Name the toolbars available in Visual LISP
   1. _____ 2. _____
   3. _____ 4. _____
   5. _____

Review Questions

Answer the following questions.

9. What are the different ways of loading a Visual LISP program?
   1. _____ 2. _____

Visual LISP 13-33

10. How can you run a Visual LISP program?

11. Can you open an AutoLISP file in the Visual LISP Text Editor?

12. Explain the function of the Visual LISP Console.

13. Can you enter several LISP statements in the Visual LISP Console? If yes, give an example.

14. In the Visual LISP Console, how can you enter a LISP statement in more than one line?

15. What is a context menu and how can you display it?

16. In the Visual LISP Text Editor, what are the default colors assigned to the following?

    Parantheses _____   LISP functions _____
    Comments _____   Integers _____

17. What is the function of the Visual LISP Formatter and how can you format the text?

18. How can you get the heading comment and function closing comment?

19. How can you change the formatting options?

20. How can you debug a Visual LISP program?

21. How can you trace the value of variables in the Visual LISP program?

Exercises

Exercise 6 *General*

Write a Visual LISP program that will draw the staircase shown in Figure 13-28. When you run the program, the user should be prompted to enter the rise and run of the staircase and the number of steps. The user should also be prompted to enter the length of the landing area.

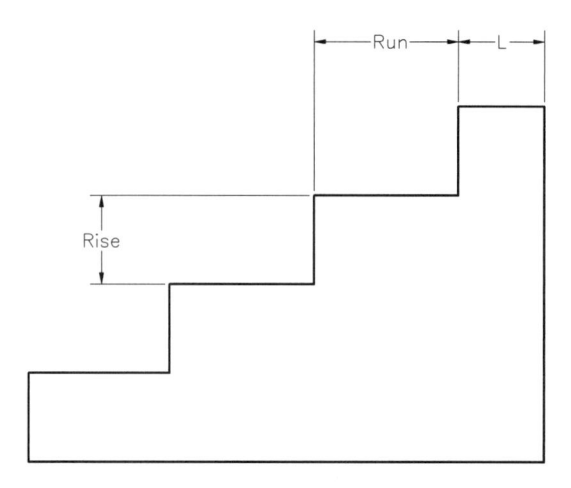

Figure 13-28 Drawing for Exercise 6

Exercise 7 *General*

Write a Visual LISP program that will draw the picture frame of Figure 13-29. The program should prompt the user to enter the length, width, and the thickness of the frame.

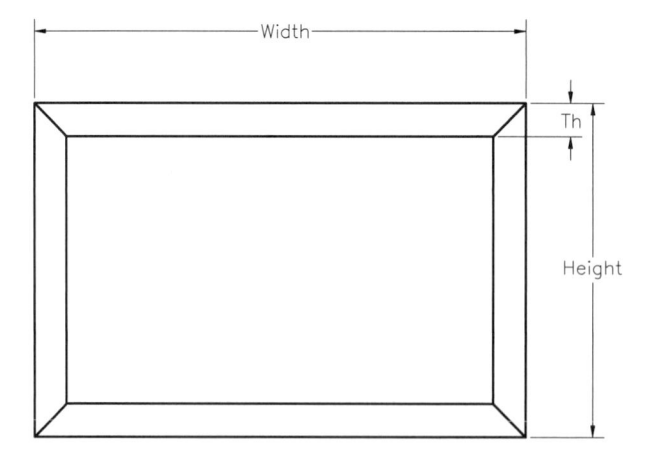

Figure 13-29 Drawing for Exercise 7

Visual LISP 13-35

Exercise 8 *General*

Write a Visual LISP program that will draw a three-dimensional table as shown in Figure 13-30. The program should prompt the user to enter the length, width, height, thickness of tabletop, and the size of the legs. The program should also use the VPOINT command to change the view direction to display the 3D view of the table.

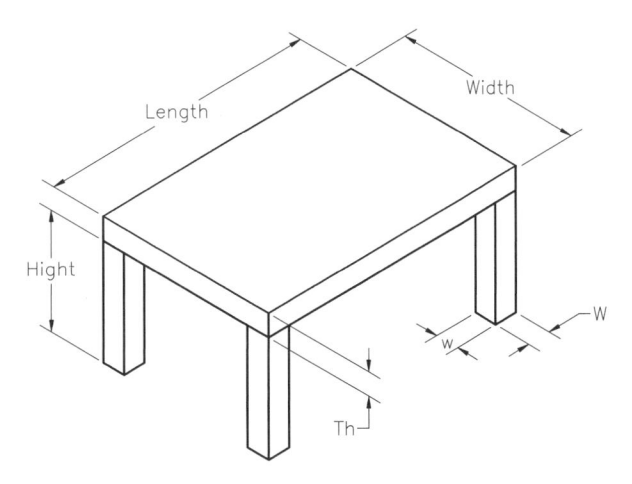

Figure 13-30 Drawing for Exercise 8

Exercise 9 *General*

Write a Visual LISP program that will draw the figure as shown in Figure 13-31 with center lines and dimensions. Assume, L5=D1, L3=1.5*D1, L6=10*D1, L1= L6-D1, L4=L3+D1.

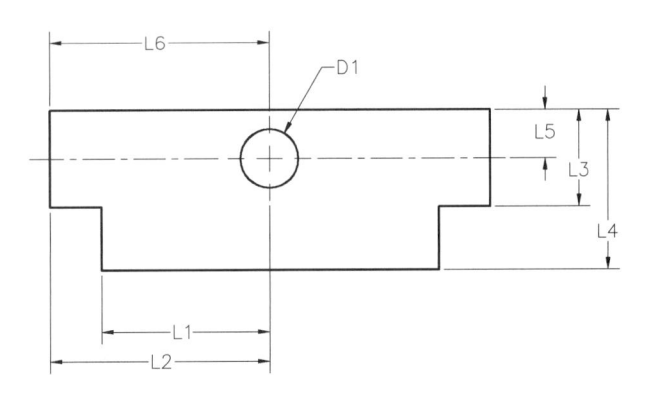

Figure 13-31 Drawing for Exercise 9

Exercise 10   *General*

Write a Visual LISP program that will draw a hub with the key slot as shown in Figure 13-32. The program should prompt the user to enter the values for P0 (Center of hub/shaft), D1 (Diameter of shaft), D2 (Outer diameter of hub), W (Width of key), and H (Height of key). The program should also draw the center lines in Center layer (Green color) and draw dimension T and W in Dim layer (Magenta color).

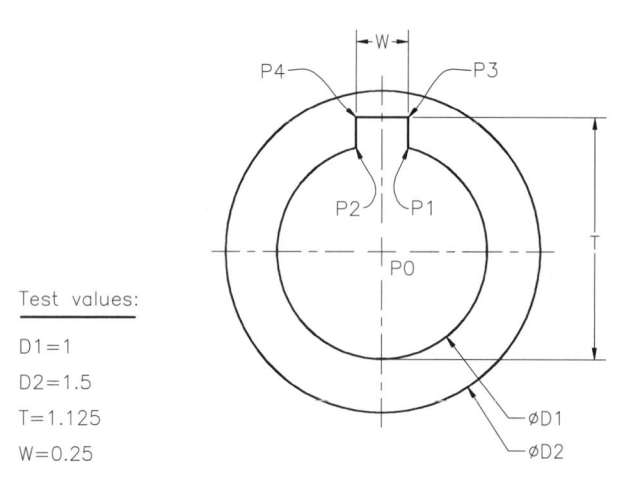

Test values:

D1=1
D2=1.5
T=1.125
W=0.25

Figure 13-32 Drawing for Exercise 10

Exercise 11   *General*

Write a Visual LISP program that will draw an equilateral triangle inside the circle (Figure 13-33). The program should prompt you to enter the radius and the center point of the circle.

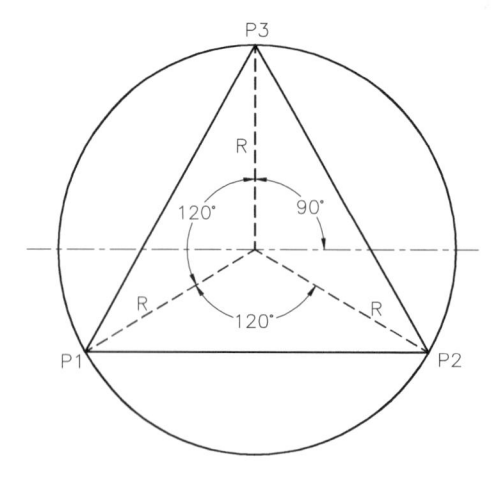

Figure 13-33 Equilateral triangle inside a circle

Exercise 12

General

Write a Visual LISP program that will draw two lines tangent to two circles as shown in Figure 13-34. The program should prompt you to enter the circle diameters and the center distance between the circles.

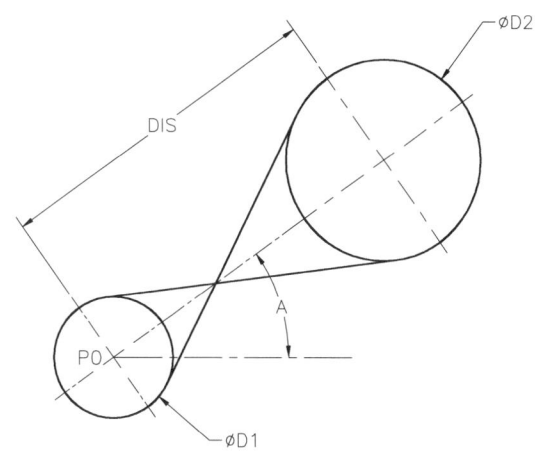

Figure 13-34 Circle with tangent lines at angle A

Answers to the Self-Evaluation test
1. **Text Editor**, **Formatter**, 2. Color Coding, 3. Formats, 4. **Watch Window**, 5. **Toggle Breakpoint**, 6. **Tools > AutoLISP > Visual LISP Editor, VLIDE**, 7. Menu, Toolbars, Console Window and Status bar, 8. Standard, Search, Tools, Debug, View

Chapter 14

Visual LISP: Editing the Drawing Database

Learning Objectives
After completing this chapter, you will be able to:
• *Edit the drawing database using Visual LISP.*
• *Use the* **ssget**, **sslength**, **ssname**, **ssadd**, **ssdel**, **entget**, **assoc**, **cons**, **subst**, **entnext**, **entlast** *and* **entmod** *functions.*
• *Retrieve information from the drawing database.*
• *Edit the database and substitute the values back into the drawing database.*

EDITING THE DRAWING DATABASE
In addition to writing programs to create new commands, you can use the Visual LISP programming language to edit the drawing database. This is a powerful tool to make changes in a drawing. For example, you can write a program that will delete all text objects in the drawing or change the layer and color of all circles just by entering one command. Once you understand how AutoCAD stores the information of the drawing objects and how it can be retrieved and edited, you can manipulate the database any way you want to, limited only by your imagination.

This chapter discusses some of the commands that are frequently used to edit the drawing database. For other commands, not discussed in this section, please refer to the online reference manual, Visual LISP Programmers Reference, published by Autodesk.

ssget
The **ssget** function enables you to select any number of objects in a drawing. The object

selection modes (window, crossing, previous, last, etc.) and the points that define the corners of the window can be included in the **ssget** assignment. The format of the **ssget** function is:

(ssget [<u>selection-mode</u>] [point1 point2])
        Where **selection-mode** ---------- Object selection mode (w, c, l, p, etc.)
                      point 1 ---------------------- First point of window (optional)
                      point 2 ---------------------- Second point of window (optional)

Examples

(ssget)	For general object selection
(ssget "L")	For selecting last object
(ssget "p")	For selecting previous selection set
(ssget "w" (list 0 0) (list 12.0 9.0))	Object selection using window object selection mode, where the window is defined by points 0,0 and 12.0,9.0.
(ssget "c" pt1 pt2)	Object selection using crossing object selection mode, where the window is defined by predefined points pt1 and pt2.

Example 1

Write a Visual LISP program that will erase all objects within the drawing limits, limmax, and limmin. Use the **ssget** function to select the objects. Assume that the objects have already been drawn within the limits.

Step 1: Writing the Visual LISP program

Select **Tools> AutoLISP> Visual LISP Editor** menu or enter **VLIDE** at the Command prompt to display the **Visual LISP** window. Select **New File** from the **File** menu to display the **Visual LISP Text Editor** window. Now write the program as shown below in the **Visual LISP Text Editor**.

The following file is a listing of the Visual LISP program for Example 1. The line numbers are not a part of the program; they are shown here for reference only.

```
;This program will delete all objects                    1
;that are within the drawing limits                      2
;                                                        3
(defun delall ()                                         4
   (setvar "cmdecho" 0)                                  5
   (setq pt1 (getvar "limmin"))                          6
   (setq pt2 (getvar "limmax"))                          7
   (setq ss1 (ssget "c" pt1 pt2))                        8
   (command "erase" ss1 "")                              9
   (command "redraw")                                    10
   (setvar "cmdecho" 1)                                  11
   (princ)                                               12
)                                                        13
```

Visual LISP: Editing the Drawing Database 14-3

Explanation
Lines 1-3
The first three lines are comment lines that describe the function of the program. Notice that all comment lines start with a semicolon (;).

Line 4
(defun c:delall()
In this line the **defun** function defines function delall.

Line 6
(setq pt1 (getvar "limmin"))
The **getvar** function secures the value of the lower left corner of the drawing limits (limmin;) and the **setq** function assigns that value to variable pt1.

Line 7
(setq pt2 (getvar "limmax"))
The **getvar** function secures the value of the upper right corner of the drawing limits (limmax) and the **setq** function assigns that value to variable pt2.

Line 8
(setq ss1 (ssget "c" pt1 pt2))
The **ssget** function uses the "crossing" objection selection mode to select the objects that are within or touching the window defined by points pt1 and pt2. The **setq** function then assigns this object selection set to variable ss1.

Line 9
(command "erase" ss1 "")
The **command** function uses AutoCAD's **ERASE** command to erase the predefined object selection set ss1.

Line 10
(command "redraw")
In this line the **command** function uses AutoCAD's **REDRAW** command to redraw the screen and get rid of the blip marks left after erasing the objects.

Step 2: Loading the Visual LISP program

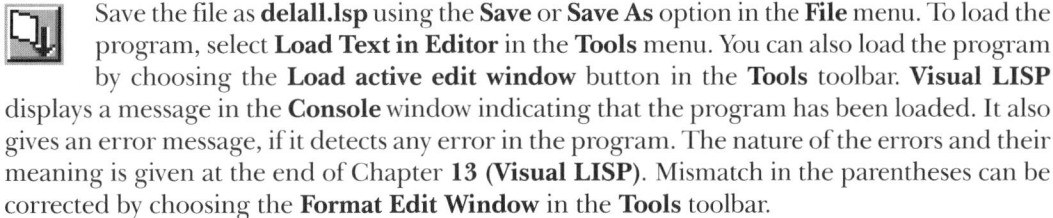

Save the file as **delall.lsp** using the **Save** or **Save As** option in the **File** menu. To load the program, select **Load Text in Editor** in the **Tools** menu. You can also load the program by choosing the **Load active edit window** button in the **Tools** toolbar. **Visual LISP** displays a message in the **Console** window indicating that the program has been loaded. It also gives an error message, if it detects any error in the program. The nature of the errors and their meaning is given at the end of Chapter 13 (**Visual LISP**). Mismatch in the parentheses can be corrected by choosing the **Format Edit Window** in the **Tools** toolbar.

To run the program, enter the function name (**delall**) at the console prompt $ sign. The function name must be enclosed in parentheses. After writing the function name, press Enter and then choose **Activate AutoCAD** from the **Window** menu to display the AutoCAD screen.

You can see the objects within the limits already erased.

ssget "X"

The **ssget "X"** function enables you to select specified types of objects in the entire drawing database, even if the layers are frozen or turned off. The format of the **ssget "X"** function is:

    (ssget "X" <u>specified-criteria</u>)
              Where   **X** ----------------------------- Filter mode of the ssget function
                        specified-criteria -------- List of the specified criteria for selecting the objects

Examples

(ssget "X" (list (cons 0 "TEXT")))   returns a selection set that consists of all TEXT objects in the drawing

(ssget "X" (list (cons 7 "ROMANC")))   returns a selection set that consists of all TEXT objects in the drawing with the text style name ROMANC

(ssget "X" (list (cons 0 "LINE")))   returns a selection set that consists of all LINE objects in the drawing

(ssget "X" (list (cons 8 "OBJECT")))   returns a selection set that consists of all objects in the OBJECT layer.

The **ssget "X"** function can contain more than one selection criterion. This option can be used to select a specific set of objects in a drawing. For example, if you want to select all the line objects in the OBJECT layer, there are two selection criteria. The first is that the object has to be a line, the second is that the line object has to be in the OBJECT layer.

    (ssget "X" (list (cons 0 "LINE")(cons 8 "OBJECT")))

Group Codes for ssget "X"

The following table is a list of AutoCAD group codes that can be used with the function ssget "X":

Group Code	Code Function
0	Object type
2	Block name for block reference
3	Dimension object DIMSTYLE name
6	Linetype name
7	Text style name
8	Layer name
38	Elevation
39	Thickness
62	Color number

Visual LISP: Editing the Drawing Database 14-5

        66   Attributes
       210   3D extrusion direction

Example 2

Write a Visual LISP program that will erase all text objects in a drawing on a specified layer. Use the filter option of the **ssget** function (ssget "X") to select text objects in the specified layer. Assume that some objects have already been drawn on any specified layer.

Step 1: Writing the Visual LISP program

Select **Tools> AutoLISP> Visual LISP Editor** menu or enter **VLIDE** at the Command prompt to display the **Visual LISP** window. Select **New File** from the **File** menu to display the **Visual LISP Text Editor** window. Now write the program as shown below.

The following file is a listing of the Visual LISP program for Example 2:

```
;This program will delete all text
;in the user-specified layer
;
(defun deltext ()
   (setvar "cmdecho" 0)
   (setq layer (getstring "\n Enter layer name: "))
   (setq ss1 (ssget "x" (list (cons 8 layer) (cons 0 "text"))))
   (command "erase" ss1 "")
   (command "redraw")
   (setvar "cmdecho" 1)
   (princ))
```

Step 2: Loading the Visual LISP program

Save the file as **deltext.lsp** using the **Save** or **Save As** option in the **File** menu. To load the program, select **Load Text in Editor** in the **Tools** menu. You can also load the program by choosing the **Load active edit window** button in the **Tools** toolbar. **Visual LISP** displays a message in the **Console** window indicating that the program has been loaded. It also gives an error message, if it detects any error in the program. To run the program enter the function name (**deltext**) at the console prompt $ sign. The function name must be enclosed in parentheses. After entering the function name, press Enter and then choose **Activate AutoCAD** from the **Window** menu to display AutoCAD screen. You can see that the text has been erased.

sslength

The **sslength** function determines the number of objects in a selection set and returns an integer corresponding to the number of objects found. The format of the **sslength** function is:

    (sslength selection-set)
                Where   **selection-set** --- Name of the selection set

Examples

(setq ss1 (ssget))
(setq num (sslength ss1))   returns the number of objects in the predefined selection set ss1

(setq ss2 (ssget "l"))
(setq num (sslength ss2))   returns the number of objects (1) in selection set ss2, where selection set ss2 has been defined as the last object in the drawing

ssname

The **ssname** function returns the name of the object, from a predefined selection set, as referenced by the index that designates the object number. The name of the object returned by this function is in the hexadecimal format (such as 60000014). The format of the **ssname** function is:

(ssname selection-set index)

    Where **selection-set** --- A predefined selection set
               index ------------- Index designates the object number in a selection set

Examples

(setq ss1 (ssget))
(setq index 0)
(setq entname (ssname ss1 index))   returns the name of the first object contained in the predefined selection set ss1

Note
If the index is 0, the ssname function returns the name of the first object in the selection set. Similarly, if the index is 1, it returns the name of the second object and so on.

ssadd

The **ssadd** function adds an object (entity) to a selection set, or creates a new selection set. The syntax for the **ssadd** function is:

(ssadd entity-name selection-set)

When **ssadd** is used without arguments, a new selection set is created with no member. If only entity name arguments are used, a selection set is created with one member.

Examples

(setq all(ssget))
    Select objects: All
    Selection set=30

Visual LISP: Editing the Drawing Database 14-7

A selection set has been created with the name **all** which contains specified objects, say two circles.

Now we want to add a new circle to the existing selection set.

    (setq b(entsel))
    Select objects: Use any object selection method to select the circle.
    (<entity name: 1ffo5a0>(10.977 4.99 0))
    (setq c(car b))
  entity name: 1ffo5a0
    (setq d(ssadd c all))
    (Command "erase" all)

A new circle is added to the existing selection set.

ssdel

The **ssdel** function enables the user to delete any object or entity from any selection set. It functions like **ssadd**. The syntax for the **ssdel** function is:

(ssdel entity name selection set)

Examples

Let us say you have created three circles and have converted them into a selection set. Now you want to delete one circle from this selection set.

    (setq all(ssget))
    Select objects: All
    (setq b(entsel))
    Select objects: Use any object selection method to select the circle to be deleted.
    (<entity name : 1ffo5a0>(2.455 5.99 0))
    (setq c(car b))
    Entity name : 1ffo5a0
    (setq d(ssdel c all))
    selection set=36
    (Command "erase" all)

Only two circles will be erased because one circle was removed from the selection set.

entget

The **entget** function retrieves the object list from the object name. The name of the object can be obtained by using the function ssname. The format of the **entget** function is:

    (entget <u>object-name</u>)
           Where **object-name** ---- Name of the object obtained by ssname

Examples
(setq ss1 (ssget))
(setq index 0)
(setq entname (ssname ss1 index))
(setq entlist (entget entname))   returns the list of the first object from the variable, entname, and assigns the list to the variable, entlist

assoc

The **assoc** function searches for a specified code in the object list and returns the element that contains that code. The format of the **assoc** function is:

(assoc code <u>object-list</u>)
 Where **code** -------------- AutoCAD's object code
        object-list ------- List of an object

Examples
(setq ss1 (ssget))
(setq index 0)
(setq entname (ssname ss1 index))
(setq entlist (entget entname))
(setq entasso (assoc 0 entlist))   returns the element associated with AutoCAD's object code 0, from the list defined by the variable entlist

cons

The **cons** function constructs a new list from the given elements or lists. The format of the **cons** function is:

(cons <u>first-element</u> <u>second-element</u>)
 Where **first-element** -------------- First element or list
        second-element ---------- Second element or list

Examples
(cons 'x 'y)   returns (X . Y)
(cons '(x y) 'z)   returns ((X Y) . Z)
(cons '(x y z) '(0.5 5.0)) returns ((X Y Z) 0.5 5.0)

subst

The **subst** function substitutes the new item in place of old items. The old items can be a single item or multiple items, provided they are in the same list. The format of the **subst** function is:

(subst <u>new-item</u> <u>old-item</u> <u>object-list</u>)
 Where **new-item** -------- New item that will replace old items
        old-item --------- Old items that are to be replaced
        object-list ------- Object list name

Visual LISP: Editing the Drawing Database 14-9

Examples
(setq entlist '(x y x))
(setq newlist (subst '(z) '(x) entlist)   returns (z y z); the subst function replaces x in the object list (entlist) by z

entmod

The **entmod** function updates the drawing by writing the modified list back to the drawing database. The format of the **entmod** function is:

(entmod <u>object-list</u>)
        Where   **object-list** ------- Name of the modified object list

Example 3

Write a Visual LISP program that will enable you to change the height of a text object. The program should prompt you to enter the new height of the text.

Step 1: Understanding the program algorithm

<u>Input</u>	<u>Output</u>
New text height	Text with new text height
Text object	

<u>Process</u>
1. Select the text object; obtain the name of the object using the function **ssname**.
2. Extract the list of the object using the function **entget**.
3. Separate the element associated with AutoCAD object code 0 from the list using the function **assoc**.
4. Construct a new element where the height of text is changed to a new height using the function **cons**.
5. Substitute the new element back into the original list using the **subst** function.
6. Update the drawing database using the **entmod** function.

Step 2: Writing the Visual LISP program

Select **Tools> AutoLISP> Visual LISP Editor** menu or enter **VLIDE** at the Command prompt to display the **Visual LISP** window. Select **New File** from the **File** menu to display the **Visual LISP Text Editor** window. Now write the program in the **Visual LISP Text Editor**. The following file is a listing of the Visual LISP program for Example 3.

```
;This program changes the height of the
;selected text, only one text at a time.
;
(defun chgtext1 ()
(setvar "cmdecho" 0)
(setq newht (getreal "\n Enter new text height: "))
(setq ss1 (ssget))
```

```
(setq name (ssname ss1 0))
(setq ent (entget name))
(setq oldlist (assoc 40 ent))
(setq conlist (cons (car oldlist) newht))
(setq newlist (subst conlist oldlist ent))
(entmod newlist)
(setvar "cmdecho" 1)
(princ)
)
```

Step 3: Loading the Visual LISP program

Save the file as **chgext1.lsp** using the **Save** or **Save As** option in the **File** menu. To load the program, select **Load Text in Editor** in the **Tools** menu. You can also load the program by choosing **Load active edit window** button in the **Tools** toolbar. **Visual LISP** displays a message in the **Console** window indicating that the program has been loaded. It also gives an error message, if it detects any error in the program. To run the program, enter the function name (**chgtext1**) at the console prompt $ sign. The function name must be enclosed in parentheses. Afetr writing the function name, press Enter. You will automatically get into the drawing editor and you will be prompted to enter the text height and select object. Select the text object and press Enter. You will notice that the text height has changed.

HOW THE DATABASE IS RETRIEVED AND EDITED

To change the objects in a drawing you need to understand the structure of the drawing database and how it can be manipulated. Once you understand this concept, it is easy and sometimes fun to edit the drawing database and the drawing. The following step-by-step explanation describes the process involved in changing the height of a selected text object in a drawing. Assume that the text that needs to be edited is "CHANGE TEXT" and that this text is already drawn on the screen. The height of the text is 0.3 units. Before going through the following steps, load the Visual LISP program from Example 3, and run it so that the variables are assigned a value.

Step 1

Select the text using the function **ssget** or **ssget "X"** and assign it to variable ss1. AutoCAD creates a selection set that could have one or more objects. In line 7 **(setq ss1 (ssget))** of the program for Example 3, the selection set is assigned to variable ss1. Use the following command to check the variable ss1:

Command: **!ss1**
<Selection set: 2>

Step 2

There could be several objects in a selection set and these objects need to be separated, one at a time, before any change is made to an object. This is made possible using the function **ssname** which extracts the name of an object. The index number used in the function **ssname** determines the object whose name is being extracted. For example, if the index is 0 the **ssname** function will extract the name of the first object, if the index is 1 the **ssname** function will extract the name of second object, and so on. In line 8 **(setq name (ssname ss1 0))** of the

Visual LISP: Editing the Drawing Database 14-11

program, the **ssname** extracts the name of the first object and assigns it to the variable name. Use the following command to check the variable name:

Command: **!name**
<Object name: 60000018>

Step 3

Extract the object list using the function **entget**. In line 9 **(setq ent (entget name))** of the program, the value of the list has been assigned to the variable ent. Use the following command to check the value of the variable ent.

Command: **!ent**
((-1.<Object name: 600000018> (0 . "TEXT") (8 . "0") (10 4.91227 5.36301 0.0) **(40 . 0.3)** (1 . "CHANGE TEXT") (50 .0.0) (41 . 1.0) (51 .0.0) (7 . "standard") (71 .0)) (72 . 1) (11 6.51227 5.36302 0.0) (210 0.0 0.0 1.0))

This list contains all the information about the selected text object (CHANGE TEXT), but you are only interested in changing the height of the text. Therefore, you need to identify the element that contains the information about the text height (40 . 0.3) and separate that from the list.

Step 4

Use the function **assoc** to separate the element that is associated with code 40 (text height). The statement in line 10 **(setq oldlist (assoc 40 ent))** of the program uses the assoc function to separate the value and assign it to the variable oldlist. Use the following command to check the value of this variable:

Command: **!oldlist**
(40 . 0.3)

Step 5

The **(40 . 0.3)** element consists of the code for text (40), and the text height (0.3). To change the height of the old text, the text height value (0.3) needs to be replaced by the new value. This is accomplished by constructing a new list as described in line 11 **(setq conlist (cons (car oldlist) newht))** of the program. This line also assigns the new element to variable conlist. For example, if the value assigned to variable newht is 0.5, the new element will be (40 . 0.5). Use the following command to check the value of conlist.

Command: **!conlist**
(40 . 0.5)

Step 6

After constructing the new element, use the subst function to substitute the new element back into the original list, ent. This is accomplished by line 12 **(setq newlist (subst conlist oldlist ent))** of the program. Use the following command to check the value of the variable newlist:

Command: **!newlist**
((-1.<Object name: 600000018> (0 . "TEXT") (8 . "0") (10 4.91227 5.36301 0.0)

(40 . 0.5) (1 . "CHANGE TEXT") (50 .0.0) (41 . 1.0) (51 .0.0) (7 . "standard") (71 .0))
(72 . 1) (11 6.51227 5.36302 0.0) (210 0.0 0.0 1.0))

Step 7
The last step is to update the drawing database. Do this by using the function **entmod** as shown in line 13 **(entmod newlist)** of the program.

SOME MORE FUNCTIONS TO RETRIEVE ENTITY DATA

entnext
The entnext function allows the programmer to extract main entity and subentity names from the database. Entnext returns the name of the next non-deleted entity or first drawn entity. (Entnext(Entnext)) returns the next to the next non-deleted entity or next drawn entity. The format of the Entnext function is:

    (entnext <entity name>)

entlast
This function returns the name of the last non-deleted entity from the database. The format of the Entlast function is:

    (entlast <entity name>)

entsel
The Entsel function permits the user to select one entity. The format of the Entsel function is:
    (entsel [promot])

Prompt is optional.

Example 4

Write a Visual LISP program that will enable you to change the height of all text objects in a drawing. The program should prompt you to enter the new text height.

Step 1: Making the flowchart
The following flowchart, Figure 14-1 gives the procedure to write the Visual LISP program (Figure 14-1).

Step 2: Writing the Visual LISP program
Select **Tools> AutoLISP> Visual LISP Editor** menu or enter **VLIDE** at the Command prompt to display the **Visual LISP** window. Select **New File** from the **File** menu to display the **Visual LISP Text Editor** window. Now write the program as shown below in the **Visual LISP Text Editor**. The line numbers are not a part of the program; they are shown here for reference only.

Visual LISP: Editing the Drawing Database 14-13

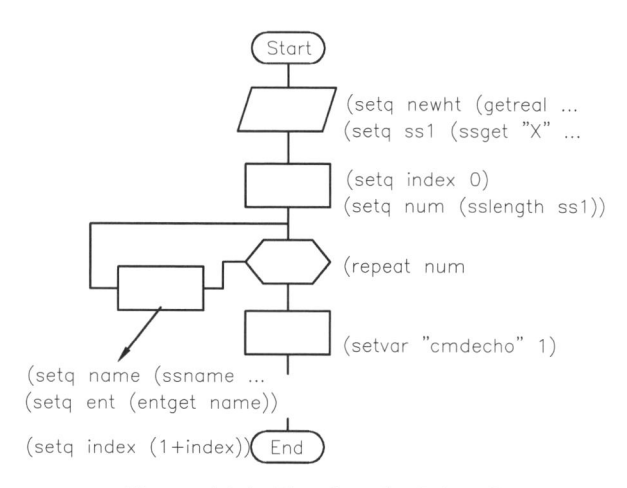

Figure 14-1 Flowchart for Example 4

```
;This program changes the height of                              1
;all text objects in a drawing.                                  2
;                                                                3
(defun chgtext2 ()                                               4
   (setvar "cmdecho" 0)                                          5
   (setq newht (getreal "\n Enter new text height: "))           6
   (setq ss1 (ssget "x" (list (cons 0 "text"))))                 7
   (setq index 0)                                                8
   (setq num (sslength ss1))                                     9
   (repeat num                                                  10
      (setq name (ssname ss1 index))                            11
      (setq ent (entget name))                                  12
      (setq oldlist (assoc 40 ent))                             13
      (setq conlist (cons (car oldlist) newht))                 14
      (setq newlist (subst conlist oldlist ent))                15
      (entmod newlist)                                          16
      (setq index (1+ index))                                   17
   )                                                            18
   (setvar "cmdecho" 1)                                         19
   (princ)                                                      20
)                                                               21
```

Explanation
Line 7
(setq ss1 (ssget "x" (list (cons 0 "text"))))
The **ssget "X"** function filters the text objects from the drawing database. The **setq** function assigns that selected set of text objects to variable **ss1**.

Line 8
(setq index 0)
The **setq** function sets the value of the **index** variable to 0. This variable is used later to select different objects.

Line 9
(setq num (sslength ss1))
The function **sslength** determines the number of objects in the selection set ss1 and the setq function assigns that number to the **num** variable.

Line 10
(repeat num
The **repeat** function will repeat the processes defined within the repeat function **num** number of times.

Step 3: Loading the Visual LISP program

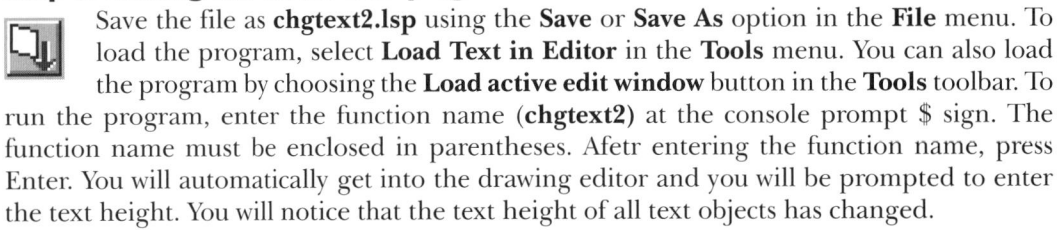

Save the file as **chgtext2.lsp** using the **Save** or **Save As** option in the **File** menu. To load the program, select **Load Text in Editor** in the **Tools** menu. You can also load the program by choosing the **Load active edit window** button in the **Tools** toolbar. To run the program, enter the function name (**chgtext2**) at the console prompt $ sign. The function name must be enclosed in parentheses. Afetr entering the function name, press Enter. You will automatically get into the drawing editor and you will be prompted to enter the text height. You will notice that the text height of all text objects has changed.

Example 5

Write a Visual LISP program that will enable you to change the text height of the selected text objects in the drawing.

Step 1: Making the flowchart
The flowchart can be designed as shown in the Figure 14-2.

Step 2: Writing the Visual LISP program
Select **Tools> AutoLISP> Visual LISP Editor** menu or enter **VLIDE** at the Command prompt to display the **Visual LISP** window. Select **New File** from the **File** menu to display the **Visual LISP Text Editor** window. Now write the program as shown below in the **Visual LISP Text Editor**.

```
(defun chgtext3 ()
(setvar "cmdecho" 0)
(setq newht (getreal "\n Enter new text height: "))
(setq ss1 (ssget))
(setq index 0)
(setq num (sslength ss1))
(repeat num
   (setq name (ssname ss1 index))
   (setq ent (entget name))
```

Visual LISP: Editing the Drawing Database 14-15

```
      (setq ass (assoc 0 ent))
      (setq index (1+ index))
      (If (= "TEXT" (cdr ass))
          (progn
          (setq oldlist (assoc 40 ent))
          (setq conlist (cons (car oldlist) newht))
          (setq newlist (subst conlist oldlist ent))
          (entmod newlist)
          )
        )
      )
(setvar "cmdecho" 1)
(princ)
)
```

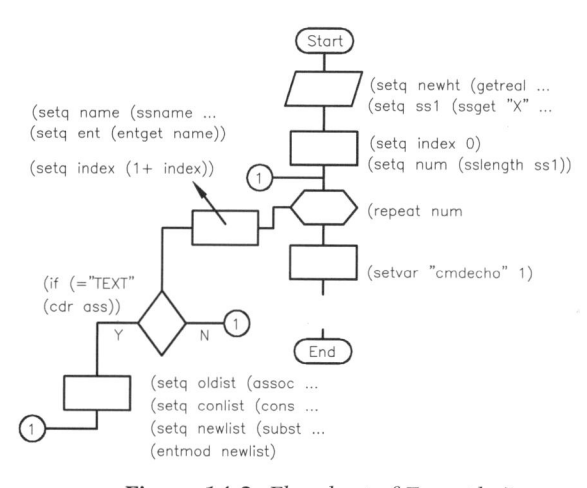

Figure 14-2 Flowchart of Example 5

Step 3: Loading the Visual LISP program

Save the file as **chgtext3.lsp** using the **Save** or **Save As** option in the **File** menu. To load the program, select **Load Text in Editor** in the **Tools** menu. You can also load the program by choosing the **Load active edit window** button in the **Tools** toolbar. To run the program, enter the function name (**chgtext3**) at the console prompt $ sign. After entering the function name, press ENTER. You will automatically get into the drawing editor where you will be prompted to enter the text height and the select objects. Give the text height and then select the object. The height of the selected text will change.

Example 6

Write a Visual LISP program to change the center point of the circle by manipulating the database of the circle. The prompt should ask the user to enter the new center point of the circle.

Step 2: Writing the Visual LISP program

Select **Tools > AutoLISP > Visual LISP Editor** menu or enter **VLIDE** at the Command prompt to display the **Visual LISP** window. Select **New File** from the **File** menu to display the **Visual LISP Text Editor** window. Now write the program as shown below in the **Visual LISP Text Editor**.

```
(Defun radcir ()
        (setq p1(entsel "select the circle")
        p2(car p1)
        p3(entget p2)
        p4(assoc 10 p3)
        p5(getpoint "\n enter the new center point of the circle")
        p6(cons 10 p5)
        p3(subst p6 p4 p3)
        );setq
    (entmod p3)
    );defun
```

Step 4: Loading the Visual LISP program

Save the file as **radcir.lsp** using the **Save** or **Save As** option in the **File** menu. To load the program, select **Load Text in Editor** in the **Tools** menu. You can also load the program by choosing the **Load active edit window** button in the **Tools** toolbar. To run the program, enter the function name (**radcir**) at the **Console** prompt $ sign. After writing the function name, press Enter. You will automatically get into the drawing editor where you will be prompted to select the circle and then enter the new center point. Select the circle and enter the new center point and its center point will change.

Self-Evaluation Test

Answer the following questions and then match your answers to the answers given at the end of this chapter.

1. In addition to writing programs to create new commands, you can use the Visual LISP programming language to edit the drawing database. (T/F)

2. The _____ function enables you to select any number of objects in a drawing.

3. The _____ function enables you to select specified types of objects in the entire drawing database, even if the layers are frozen or turned off.

4. The _____ function determines the number of objects in a selection set and returns an integer corresponding to the number of objects found.

5. The _____ function returns the name of the object, from a predefined selection set, as referenced by the index that designates the object number.

Visual LISP: Editing the Drawing Database

Review Questions

Answer the following questions.

1. The _____ function gives the name of the first drawn entity.

2. The _____ function gives the name of the last drawn entity.

3. The _____ function retrieves the object list from the object name.

4. The _____ function searches for a specified code in the object list and returns the element that contains that code.

5. The _____ function constructs a new list from the given elements or lists.

6. The _____ function substitutes the new item in place of old items.

Exercises

Exercise 1
General

Write an Visual LISP program that will enable you to change the layer of the selected objects in a drawing. The program should prompt you to enter the new layer name.

Exercise 2
General

Write an Visual LISP program that will change the text style name of the selected text objects in a drawing. The program should prompt you to enter the new text style.

Exercise 3
General

Write an Visual LISP program that will change the layer of the selected objects in a drawing to a new layer. You should be able to enter the new layer by selecting an object in that layer.

Answers to the Self-Evaluation Test
1. True, **2.** SSGET, **3.** SSGET "X", **4.** SSLENGTH, **5.** SSNAME

Chapter 15

Programmable Dialog Boxes Using Dialog Control Language

Learning Objectives
After completing this chapter, you will be able to:
- *Write programs using dialog control language.*
- *Use predefined attributes.*
- *Load a dialog control language (DCL) file.*
- *Display new dialog boxes.*
- *Use standard button subassemblies.*
- *Use AutoLISP functions to control dialog boxes.*
- *Manage dialog boxes with AutoLISP.*
- *Use tiles, buttons, and attributes in DCL programs.*

DIALOG CONTROL LANGUAGE

Dialog control language (**DCL**) files are ASCII files that contain the descriptions of dialog boxes. A DCL file can contain the description of single or multiple dialog boxes. There is no limit to the number of dialog box descriptions that can be defined in a DCL file. The suffix of a DCL file is **.DCL** (such as **DDOSNAP.DCL.**)

This chapter assumes that you are familiar with AutoCAD commands, AutoCAD system variables, and AutoLISP programming. You need not be a programming expert to learn to write programs for dialog boxes in DCL or to control the dialog boxes through AutoLISP programing. However, knowledge of any programming language should help you to understand

and learn DCL. This chapter introduces you to the basic concepts of developing a dialog box, frequently used attributes, and tiles. A thorough discussion of DCL functions and a step-by-step explanation of examples should make it easy for you to learn DCL. For those functions not discussed in this chapter, you can refer to the *AutoCAD Customization Guide from Autodesk*. To write programs in DCL, you need no special software or hardware. If AutoCAD is installed on your computer, you can write DCL files. To write DCL files, you can use any text editor.

DIALOG BOX

A dialog control language (DCL) file contains the description of how the dialog boxes will appear on the screen. These boxes can contain buttons, text, lists, edit boxes, rows, columns, sliders, and images. A sample dialog box is shown in Figure 15-1.

You do not need to specify the size and layout of a dialog box or its component parts. Sizing is done automatically when the dialog box is loaded on the screen. By itself, the dialog box cannot perform the functions it is designed for. The functions of a dialog box are controlled by a program written in the AutoLISP programming language or with the AutoCAD development system (ADS), or ARX. For

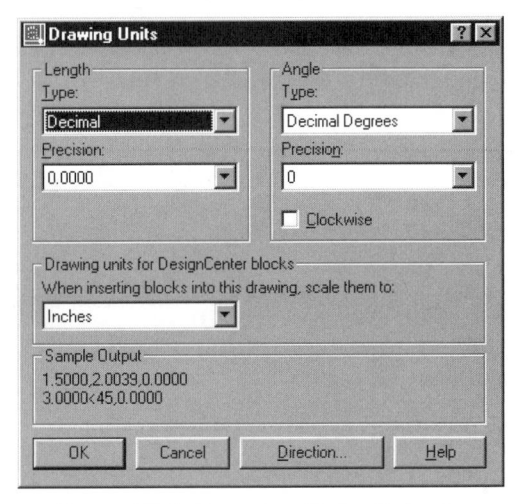

Figure 15-1 Drawing Units dialog box

example, if you load a dialog box and select the Cancel button, it will not perform the cancel operation. The instructions associated with a button or any part of the dialog box are handled through the functions provided in AutoLISP, ADS, or ARX. Therefore, AutoLISP or ADS are needed to control the dialog boxes, and you should have, in addition to an understanding of DCL, a good knowledge of AutoLISP or ADS to develop new dialog boxes or edit existing ones.

Dialog boxes are not dependent on the platform; therefore, they can run on any system that supports AutoCAD. However, depending on the graphical user interface (GUI) of the platform, the appearance of the dialog boxes might change from one system to another. The functions defined in the dialog box will still work without making any changes in the dialog box or the application program (AutoLISP or ADS) that uses these dialog boxes.

DIALOG BOX COMPONENTS

The two major components of a dialog box are the tiles and the box itself. The tiles can be arranged in rows and columns in any desired configuration. They can also be enclosed in boxes or borders to form subassemblies, giving them a tree structure, Figure 15-2(b). The basic tiles, such as buttons, lists, edit boxes, and images, are predefined by the programmable dialog box (PDB) facility of AutoCAD. These buttons are described in the file **base.dcl**. The layout and function of a tile is determined by the attribute assigned to it. For example, the height attribute controls the height of the tile. Similarly, the label attribute specifies the text that is associated with the tile. Some of the components of a dialog box are shown in Figure 15-2(a). Following this figure is a list of the predefined tiles and their format in DCL.

Programmable Dialog Boxes Using Dialog Control Language 15-3

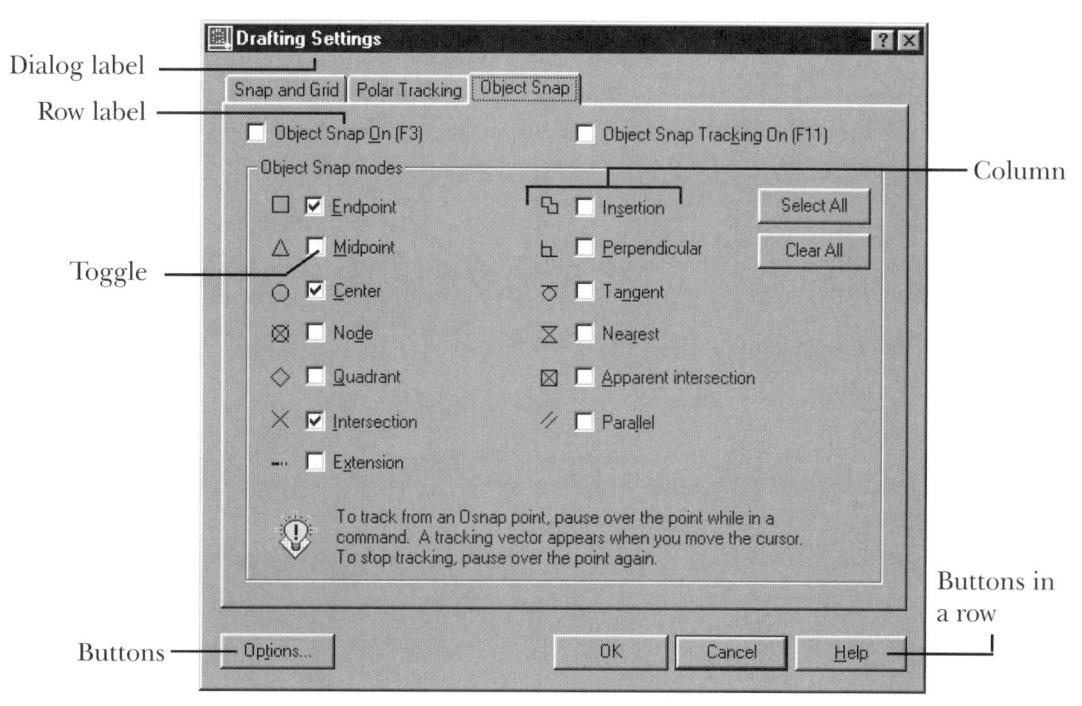

Figure 15-2(a) Components of a dialog box

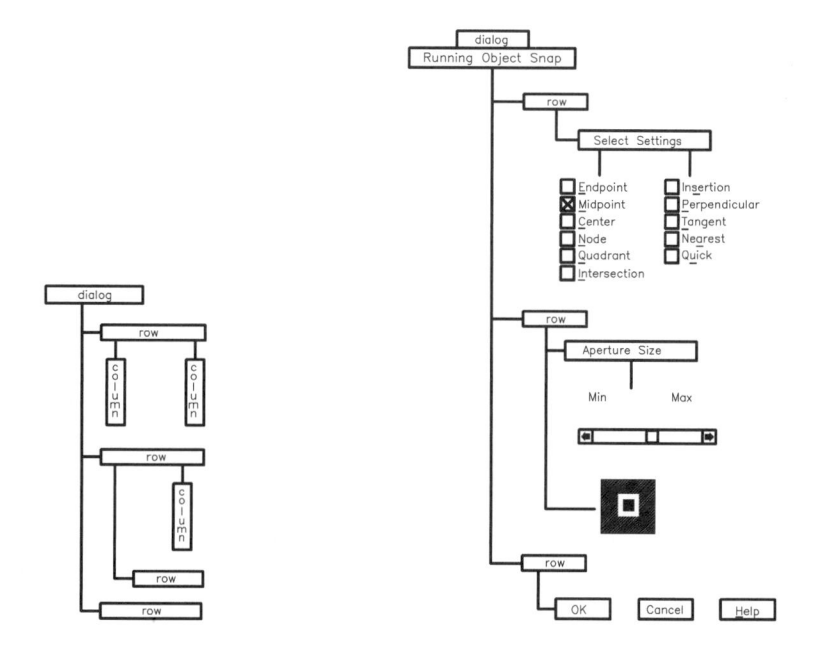

Figure 15-2(b) Tree structure of a dialog box

Predefined Tile	DCL Format
Button	button
Edit box	edit_box
Image button	image_button
List box	list_box
Pop-up list	popup_list
Radio button	radio_button
Slider	slider
Toggle	toggle
Column	column
Boxed column	boxed_column
Row	row
Boxed row	boxed_row
Radio column	radio_column
Boxed radio column	boxed_radio_column
Radio row	radio_row
Boxed radio row	boxed_radio_row
Image	image
Text	text
Spacer	spacer

BUTTON AND TEXT TILES

Button Tile

Format in DCL: **button**

The button tile consists of a rectangular box that resembles a push button. The button's label appears inside the button. For example, in the OK button of a dialog box, the label OK appears inside the button. If you select the OK button in a dialog box, it performs the functions defined in the dialog box and clears the dialog box from the screen. Similarly, if you select the cancel button, it cancels the dialog box without taking any action.

> **Note**
> *A dialog box should contain at least one OK button or a button that is equivalent to it. This allows you to exit the dialog box when you are done using it.*

Text Tile

Format in DCL: **text**

The text tile is used to display information or a title in a dialog box. It has limited application in the dialog boxes because most of the tiles have their own label attributes for tiling. However, if you need to display any text string in the dialog box, you can do so with the text tile. The text tile is used extensively in AutoCAD alert boxes to display warnings or error messages.

> **Note**
> *An alert box must contain an OK button or a Cancel button to end the dialog box.*

TILE ATTRIBUTES

The appearance, size, and function performed by a dialog box tile depend on the attributes that have been assigned to the tile. For example, if a button has been assigned the **fixed_width** attribute, the width of the box surrounding the button will not stretch through the entire length of the dialog box. Similarly, the **height** attribute determines the height of the tile, and the key attribute assigns a name to the tile that is then used by the application program. The tile attribute consists of two parts: the name of the attribute and the value assigned to the attribute. For example, in the expression **fixed_width = true**, **fixed_width** is the name of the attribute and **true** is the value assigned to the attribute. Attribute names are like variable names in programming, and the values assigned to these variables must be of a specific type.

Types of Attribute Values

Integer. Unlike integer values in programming (1, 15, 22), the numeric values assigned to attributes can be both integers and real numbers.

Examples
width = 15
height = 10

Real Number. The real values assigned to attributes should always be fractional real numbers with a leading digit.

Examples
aspect_ratio = 0.75

Quoted String. The string values assigned to an attribute consist of a text string that is enclosed in double quotes.

Examples
key = "accept"
label = "OK"

Reserved Word

Dialog control language uses some reserved words as identifiers. These identifiers are alphanumeric characters that start with a letter.

Examples
is_default = true
fixed_width = true

> **Note**
> *Reserved names are case-sensitive. For example, "is_default = true" is not the same as "is_default = True". The reserved word is "true," not "True" or "TRUE".*
>
> *Like reserved words, attribute names are also case-sensitive. For example, the attribute is "key", not "Key" or "KEY".*

PREDEFINED ATTRIBUTES

To facilitate writing programs in dialog control language, AutoCAD has provided some predefined attributes that are defined in the programmable dialog box (PDB) package that comes with the AutoCAD software. Some of these attributes can be used with any tile and some only with a particular type of tile. The values assigned to various attributes in a DCL file are used by the application program to handle the tiles or the dialog box. Therefore you must use the correct attributes and assign an appropriate value to these attributes. The following is a list of some of the frequently used predefined attributes defined in the PDB facility of AutoCAD:

action	key
alignment	label
allow_accept	layout
aspect_ratio	list
color	max_value
edit_limit	min_value
edit_width	mnemonic
fixed_height	multiple_select
fixed_width	small_increment
height	tabs
is_cancel	valuc
is_default	width

key, label, AND is_default ATTRIBUTES

key Attribute

Format in DCL
key

Examples
key = "accept"
key = "XLimit"

The **key** attribute assigns a name to a tile. The name must be enclosed in double quotes. This name can then be used by the application program to handle the tile. A dialog box can have any number of key values, but within a particular dialog box the values used for the key attributes must be unique. In the example key = "accept", a string value "accept" is assigned to the **key** attribute. If there is another key attribute in the dialog box, you must assign it a different value.

label Attribute

Format in DCL
label

Examples
label = "OK"
label = "Hello DCL Users"

Sometimes it is necessary to display a label in a dialog box. The label attribute can be used in a boxed column, boxed radio column, boxed radio row, boxed row, button tile, dialog box, or edit box. Some of the frequently used label attributes are described next.

Use of the label Attribute in a Dialog Box. When the label attribute is used in a dialog box, it is displayed in the top border or the title bar of the dialog box. The label must be a string enclosed in double quotes. Use of the label in a dialog box is optional; in case of default, no title is displayed in the dialog box.

Example
welcome : dialog {
    label = "Sample Dialog Box";

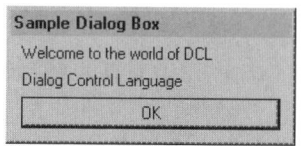

In this example, the label "Sample Dialog Box" is displayed at the top of the dialog box.

Use of the label Attribute in a Boxed Column. When the label attribute is used in a boxed column, the label is displayed within a box in the upper left corner of the column; the box consists of a single line at the top of the column. The label must be a quoted string, and the default is a set of a quoted string (" "). If the default is used, only the box is displayed, without any label.

Example
: text {
   label = "Welcome to the world of DCL";
}

In this example, the label "Welcome to the world of DCL" is displayed within a box in the upper left corner of the column.

Use of the label Attribute in a Button. When the label attribute is used in a button, the label is displayed inside the button. The label must be a quoted string and has no default.

Example
: button {
  key = "accept";
  label = "OK";
}

In this example, the label "OK" is displayed inside and in the center of the button tile.

is_default Attribute

Format in DCL
is_default

Example
is_default = true

The **is_default** attribute is used for the button of a dialog box. In the example, the value assigned to the **is_default** attribute is true. Therefore, this button will be automatically selected when you press ENTER. For example, if you load a dialog box on the screen, one way to exit and accept the values of the dialog box is to select the OK button. You can accomplish the same thing by pressing ENTER (accept key). This action is made possible by assigning the value, **true**, to the **is_default** attribute. In a dialog box the default button can be recognized by the thick border drawn around the text string.

> **Note**
> *In a dialog box, only one button can be assigned the true value for the is_default attribute.*

fixed_width AND alignment ATTRIBUTES

fixed_width Attribute

Format in DCL
fixed_width

Example
fixed_width = true

This attribute controls the width of the tile. If the value of this attribute is set to true, the width of the tile does not extend across the complete width of the dialog box. The width of the tile is automatically adjusted to the length of the text string that is displayed in the tile.

alignment Attribute

Format in DCL
alignment

Example
alignment = centered
alignment = right

The value assigned to the **alignment** attribute determines the horizontal or vertical position of the tile in a row or column. For a row, the values that can be assigned to this attribute are left, right, and centered. The default value of the alignment attribute is left; this forces the tile to be displayed left-justified. For a column, the possible values of the **alignment** attribute are top, bottom, and centered. The default value is centered.

Example 1

Using dialog control language (DCL), write a program for the following dialog box (Figure 15-3). The dialog box has two text labels and an OK button to end the dialog box.

The following file is a listing of the DCL file for the dialog box of Example 1. The name of this DCL file is **dclwe1.dcl. The line numbers are not a part of the file; they are for reference only.**

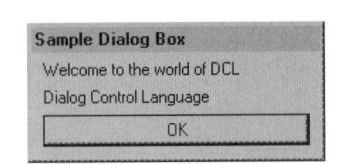

Figure 15-3 Dialog box for Example 1

```
welcome1 : dialog {                                          1
      label = "Sample Dialog Box";                           2
      : text {                                               3
         label = "Welcome to the world of DCL";              4
      }                                                      5
      : text {                                               6
         label = "Dialog Control Language";                  7
      }                                                      8
      : button {                                             9
         key = "accept";                                    10
```

Programmable Dialog Boxes Using Dialog Control Language 15-9

```
            label = "OK";                  11
            is_default = true;             12
        }                                  13
    }                                      14
```

Explanation

Line 1
welcome1 : dialog {
In this line, **welcome1** is the name of the dialog box and the definition of the dialog box is contained within the braces. The open brace in this line starts the definition of this dialog box.

Line 2
label = "Sample Dialog Box";
In this line, **label** is the label attribute, and "**Sample Dialog Box**" is the string value assigned to the label attribute. This string will be displayed in the title bar of the dialog box. The label description must be enclosed in quotes. If this line is missing, no title is displayed in the title bar of the dialog box.

Lines 3-5
: text {
label = "Welcome to the world of DCL";
}
These three lines define a text tile with the label description. In the first line, **text** refers to the text tile. The line that follows it, **label = "Welcome to the world of DCL";** defines a label for this tile that will be displayed left-justified in the dialog box. The closing brace completes the definition of this tile.

Lines 9-13
: button {
key = "accept";
label = "OK";
is_default = true;
}
These five lines define the attributes of the button tile. In the first line, **button** refers to the button tile. In the second line, **key = "accept";** specifies an ASCII name, "**accept**", that will be used by the application program to refer to this tile. The next line, **is_default = true;**, specifies that this button is the default button. It is automatically selected if you press ENTER at the keyboard.

The closing brace in line 14 completes the definition of the dialog box.

LOADING A DCL FILE

As with an AutoLISP file, you can load a DCL file from the AutoCAD drawing editor. A DCL file can contain the definition of one or several dialog boxes. There is no limit to the number of dialog boxes you can define in a DCL file. The format of the command for loading a dialog box is:

> **Note**
> *When you use the load_dialog and new_dialog functions to load and display a DCL program, the AutoCAD screen will freeze. To avoid this, it is recommended to use the AutoLISP program to load and display a DCL program. See* **"Using AutoLISP Function to Load a DCL File"** *at the end of this section.)*

(load_dialog filename)
        Where   **load_dialog** ---- Load command for loading a dialog file
                     filename -------- Name of the DCL file, with or without the file extension (.dcl)

Example
(load_dialog "dclwel1.dcl") or (load_dialog "dclwel1")

In this example, **dclwel1** is the name of the DCL file and **.dcl** is the file extension for DCL files.

The file name can be with or without the DCL file extension (**welcome1** or **welcome1.dcl**). The load function returns an integer value that is used as a handle in the **new_dialog** function and the **unload_dialog** function. This integer will be referred to as **dcl_id** in subsequent sections. You do not need to use the name dcl_id for this integer; it could be any name (dclid, or just id).

DISPLAYING A NEW DIALOG BOX

The load_dialog function loads the DCL file, but it does not display it on the screen. The format of the command used to display a new dialog box is:

(new_dialog dlgname dcl_id)
        Where   **new_dialog** ----- Command to display a dialog box
                     dlgname -------- Name of the dialog box
                     dcl_id ------------ Integer returned in the load_dialog function

Example
(new_dialog "welcome1" 1)

In this example, assume that dcl_id is 1, an integer returned by the load_dialog function. You can use the (load_dialog) and (new_dialog) commands to load the DCL file in Example 1. In the following command sequence, assume that the integer (dcl_id) returned by the (load_dialog) function is 3. Figure 15-4 shows the dialog box after entering the following two command lines:

    Command: (load_dialog "dclwel1.dcl")
    3
    Command: (new_dialog "welcome1" 3)

In this example, notice that the OK button stretches across the complete width of the dialog box. This button would look better if its width were limited to the width of the string "OK".

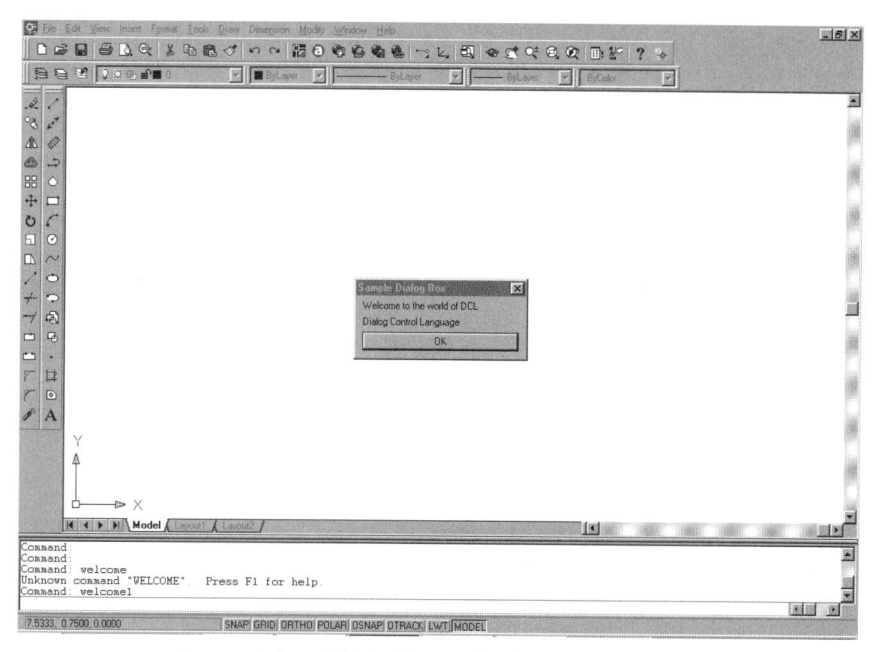

Programmable Dialog Boxes Using Dialog Control Language   15-11

Figure 15-4 Dialog box as displayed on the screen

This can be accomplished by using the **fixed_width = true;** attribute in the definition of this button. You can also use the **alignment = centered** attribute to display the OK button label center-justified. The following file is a listing of the DCL file where the OK label for the OK button is center-justified and the width of the box does not stretch across the width of the dialog box (Figure 15-5).

```
welcome2 : dialog {
    label = "Sample Dialog Box";
    : text {
        label = "Welcome to the world of DCL";
    }
    : text {
    label = "Hello - DCL";
    }
    : button {
    key = "accept";
    label = "OK";
        is_default = true;
    fixed_width = true;              // (Controls width)
    alignment = centered;            // (Controls justification)
    }
}
```

Using the AutoLISP Function to Load a DCL File

You can use the following AutoLISP program to load and display the dialog box without freezing the screen.

```
(defun c:load_dcl( / dcl_id )
   (setq dcl_id (load_dialog "dclwel1.dcl"))    (Loads the DCL file)
   (new_dialog "welcome2" dcl_id)               (Initializes the dialog box)
   (start_dialog)                               (Displays the dialog box)
   (princ)
)
```

Where -------------------- **load_dcl** is the name of the AutoLISP function.
       -------------------- **dclwel2** is the name of the DCL file that you want to load.
       -------------------- **welcome1** is the name of the dialog as defined in the DCL file.

Figure 15-5 Dialog box with fixed width for OK button

USE OF STANDARD BUTTON SUBASSEMBLIES

Some standard button subassemblies are predefined in the **base.dcl** file. You can use these standard buttons in your DCL file to maintain consistency among various dialog boxes. One such predefined button is **ok_cancel**, which displays the OK and Cancel buttons in the dialog box (Figure 15-6). The following file is the listing of the DCL file of Example 1, using the **ok_cancel** predefined standard button subassembly:

```
welcome3  : dialog {
    label = "Sample Dialog Box";
    : text {
       label = "Welcome to the world of DCL";
    }
    : text {
        label = "Dialog Control Language";
        alignment = right;
    }
    ok_cancel;
}
```

The following is a list of the standard button subassemblies that are predefined in the **base.dcl** file. These buttons are also referred to as **dialog exit buttons** because they are used to exit a dialog box.

OK Button 
   Format in DCL: **ok_only**
OK and Cancel

Programmable Dialog Boxes Using Dialog Control Language

Format in DCL: **ok_cancel**

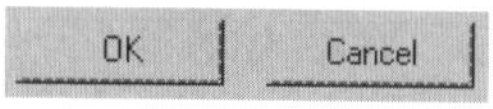

OK, Cancel, and Help Buttons
Format in DCL: **ok_cancel_help**

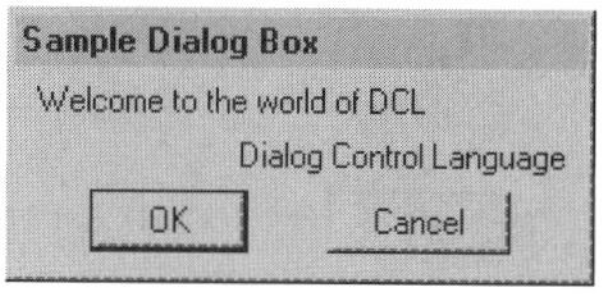

Figure 15-6 Dialog box with OK and Cancel

AUTOLISP FUNCTIONS

load_dialog

The AutoLISP function **load_dialog** is used to **load a DCL file** that is specified in the load_dialog function. In the following examples, the name of the file that AutoCAD loads is "**dclwel1**". The extension (**.dcl**) of the file is optional. When the DCL file is loaded successfully, AutoCAD returns an integer that identifies the dialog box.

Format in DCL
(load_dialog filename)

Examples
(load_dialog "dclwel1.dcl")
(load_dialog "dclwel1")

unload_dialog

The AutoLISP function **unload_dialog** is used to **unload a DCL file** that is specified in the unload_dialog function. The file is specified by the variable (dcl_id) that identifies a DCL file.

Format in DCL
(unload_dialog dcl_id)

Example
(unload_dialog dcl_id)

new_dialog

The AutoLISP function **new_dialog** is used to initialize a dialog box and then display it on the screen. In the following file, **"welcome1"** is the name of the **dialog box**. (Note: welcome1 is not the name of the DCL file.) The variable **dcl_id** contains an integer value that is returned when the DCL file is loaded.

Format in DCL
(new _dialog "dialogname" dcl_id)

Example
(new_dialog "welcome1" dcl_id)

start_dialog

The AutoLISP function **start_dialog** is used in an AutoLISP program to accept user input from the dialog box. For example, if you select OK from the dialog box, the start_dialog function retrieves the value of that tile and uses it to take action and to end the dialog box.

> Format in DCL
> **(start_dialog)**

done_dialog

The AutoLISP function **done_dialog** is used to terminate the display of the dialog box from the screen. This function must be defined within the action expression, as shown in the following example.

> Format in DCL
> **(done_dialog)**
>
> **Example**
> (action_tile "accept" "(done_dialog)")

action_tile

The AutoLISP function **action_tile** is used to associate an action expression with a tile in the dialog box. In the following example, the **action_tile** function associates the tile "accept" with the action expression (done_dialog) that terminates the dialog box. The "accept" is the name of the tile assigned to the OK button in the DCL file.

> Format in DCL
> **(action_tile tile-name action-expression)**
>
> **Example**
> (action_tile "accept" "(done_dialog)")

MANAGING DIALOG BOXES WITH AUTOLISP

When you load the DCL file of Example 1 and select the OK button, it does not perform the desired function (exit from the dialog box). This is because a dialog box cannot, by itself, execute the AutoCAD commands or the functions assigned to a tile. An application program is required to handle a dialog box. These application programs can be written in AutoLISP or ADS. The functions defined in AutoLISP and ADS can be used to load a DCL file, display the dialog box on the screen, prompt user input, set values in tiles, perform action associated with user input, and execute AutoCAD commands. Example 2 describes the use of an AutoLISP program to handle a dialog box.

Example 2

Write an AutoLISP program that will handle the dialog box and perform the functions shown in the dialog box of Example 1.

The following file is a listing of the DCL file of Example 1. This DCL file defines the dialog box **welcome1**, which contains only one action tile: **"OK"**. If you select the **OK** button, you must be able to exit the dialog box. As just mentioned, the dialog box will not perform by itself the function assigned to the **OK** button unless you write an application that will execute the functions defined in the dialog box.

```
welcome1    : dialog {
    label = "Sample Dialog Box";
    : text {
       label = "Welcome to the world of DCL";
    }
    : text {
    label = "Dialog Control Language";
    }
     : button {
        key = "accept";
        label = "OK";
        is_default = true;
     }
}
```

The following file is a listing of the AutoLISP program that loads the DCL file (**dclwel1**) of Example 1, displays the dialog box **welcome1** on, and defines the action for the **OK** button. **The line numbers are not a part of the program; they are for reference only.**

```
(defun C:welcome ( / dcl_id)                      1
(setq dcl_id (load_dialog "dclwel1.dcl"))         2
(new_dialog "welcome1" dcl_id)                    3
(action_tile                                      4
   "accept"                                       5
   "(done_dialog)")                               6
(start_dialog)                                    7
(unload_dialog dcl_id)                            8
(princ)                                           9
)                                                10
```

Explanation
Line 1
(defun C:welcome (/ dcl_id)
In this line, **defun** is an AutoLISP function that defines the function welcome. With the C: in front of the function name, the **welcome** function can be executed like an AutoCAD command. The **welcome** function has one local variable, **dcl_id.**

Line 2
(setq dcl_id (load_dialog "dclwel1.dcl"))
In this line, **(load_dialog "dclwel1.dcl")** loads the DCL file **dclwel1.dcl** and returns a positive integer. The **setq** function assigns this integer value to the local variable **dcl_id.**

Line 3
(new_dialog "welcome1" dcl_id)
In this line, the AutoLISP function **new_dialog** loads the dialog box **welcome1** that is defined in the DCL file (line 1 of DCL file). The variable **dcl_id** is an integer that identifies the DCL file.

> **Note**
> *The dialog name (welcome1) in the AutoLISP program must be the same as the dialog name in the DCL file (welcome1).*

Lines 4-6
(action_tile
  "accept"
  "(done_dialog)")
In the DCL file of Example 1, the **OK** button has been assigned an ASCII name, **accept** (key = "accept"). The first two lines associate the key (OK button) with the action expression. The action_tile initializes the association between the OK button and the action expression (done_dialog). If you select the **OK** button from the dialog box, the AutoLISP program reads that value and performs the function defined in the statement (done_dialog). The done_dialog function ends the dialog box.

Line 7
(start_dialog)
The **start_dialog** function enables the AutoLISP program to accept your input from the dialog box.

Line 8
(unload_dialog dcl_id)
This statement unloads the DCL file identified by the integer value of dcl_id.

Lines 9 and 10
(princ)
)
The **princ** function displays a blank on the screen. You use this to prevent display of the last expression in the command prompt area of the screen. If the **princ** function is not used, AutoCAD will display the value of the last expression. The closing parenthesis in the last line completes the definition of the welcome function.

ROW AND BOXED ROW TILES

Row Tile

Format in DCL: **row**

In a DCL file, several tiles can be grouped together to form a composite row or a composite column that is treated as a single tile. A row tile consists of several tiles grouped together in a horizontal row.

Boxed Row Tile

Format in DCL: **boxed_row**

In a boxed row, the tiles are grouped together in a row and a border is drawn around them, forming a box shape. If the boxed row has a label, it will be displayed left-justified at the top

Programmable Dialog Boxes Using Dialog Control Language 15-17

of the box, above the border line. If no label attribute is defined, only the box is displayed around the tile. On some systems, depending on the graphical user interface (GUI), the label may be displayed inside the border.

COLUMN, BOXED COLUMN, AND TOGGLE TILES

Column Tile

Format in DCL: **column**

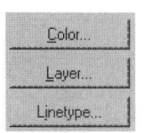

In a column tile, the tiles are grouped together in a vertical column to form a composite tile.

Boxed Column Tile

Format in DCL: **boxed_column**

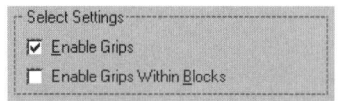

In a boxed column, the tiles are grouped together in a column and a border is drawn around the tiles. If the boxed column has a label, it will be displayed left-justified at the top of the box, above the border line. If there is no label attribute defined, only the box is displayed around the tile. On some systems, depending on the graphical user interface (GUI), the label may be displayed inside the border.

Toggle Tile

Format in DCL: **toggle**

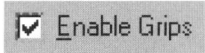

A toggle tile in a dialog box displays a small box on the screen with an optional label on the right of the box. Although the label is optional, you should label the toggle box so that users know what function is assigned to the toggle box. A toggle box has two states: on and off. When the function is turned on, a check mark is displayed in the box. Similarly, when a function is turned off, no check mark is displayed. The on or off state of the toggle box represents a Boolean value. When the toggle box is on, the Boolean value is 1; when the toggle box is off, the Boolean value is 0.

MNEMONIC ATTRIBUTE

Format in DCL
mnemonic

Example
mnemonic = "U"

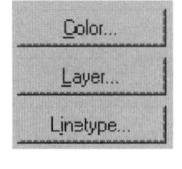

A dialog box can have several tiles with labels. One of the ways you can select a tile is by using the arrow to highlight the tile and then pressing the **accept** key. This is possible if you have a pointing device like a digitizer or a mouse. If you do not have a pointing device, it may not be possible to select a tile. However, you can select a tile by using the mnemonic key assigned to the tile. For example, the mnemonic character for the Color tile is C. If you press the C key at the keyboard, it will highlight the Color tile. The mnemonic character is designated by underlining one of the characters in the label. The underlining is done automatically once you define the mnemonic attribute for the tile. You can select only one character **in the label** as a mnemonic character and you must use different mnemonic characters for different tiles.

If a dialog box has two tiles with the same mnemonic character, only one tile will be selected when you press the mnemonic key. The mnemonic characters are not case-sensitive; therefore, they can be uppercase (C) or lowercase (c). However, the character in the label you select as a mnemonic character should be capitalized for easy identification.

> **Note**
> *Some graphical user interfaces (GUIs) do not support mnemonic attributes. On such systems, you cannot select a tile by pressing the mnemonic key. However, you can still use pointing devices to select the tile.*

Example 3 — *General*

Write a DCL program for the object snap dialog box shown in Figure 15-7(a). The object snap tiles are arranged in two columns in a boxed row.

Before writing a program, especially a DCL program for a dialog box, you should determine the organization of the dialog box. It is given in Example 3. But when you develop a dialog box yourself, you must be careful when organizing the tiles in the dialog box. The structure of the DCL program depends on the desired output. In Example 3, the desired output is shown in Figure 15-7(a). This dialog box has two rows and two columns. The first row has two columns and the second row has no columns, as shown in Figure 15-7(b).

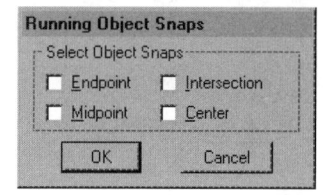

Figure 15-7(a) Dialog box for object snaps

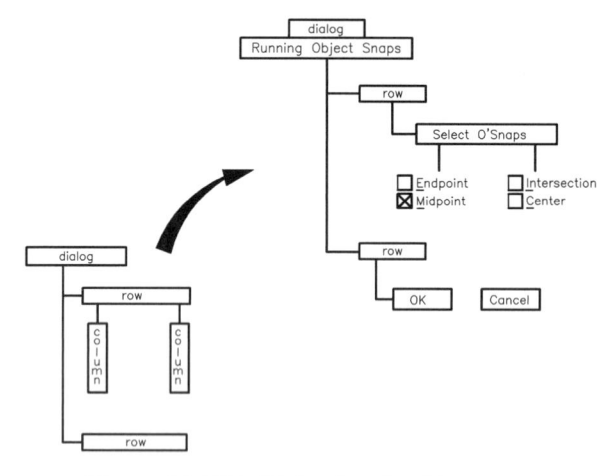

Figure 15-7(b) Dialog box for object snaps

The following file is a listing of the DCL file for Example 3. **The line numbers are not a part of the file; they are shown for reference only.**

```
osnapsh : dialog {                                              1
    label = "Running Object Snaps";                             2
    : boxed_row {                                               3
      label = "Select Osnaps";                                  4
      : column {                                                5
          : toggle {                                            6
            label = "Endpoint";                                 7
            key = "Endpoint";                                   8
            mnemonic = "E";                                     9
            fixed_width = true;                                10
          }                                                    11
          : toggle {                                           12
            label = "Midpoint";                                13
            key = "Midpoint";                                  14
            mnemonic = "M";                                    15
            fixed_width = true;                                16
          }                                                    17
      }                                                        18
      : column {                                               19
          : toggle {                                           20
            label = "Intersection";                            21
            key = "Intersection";                              22
            mnemonic = "I";                                    23
            fixed_width = true;                                24
          }                                                    25
          : toggle {                                           26
            label = "Center";                                  27
            key = "Center";                                    28
            mnemonic = "C";                                    29
            fixed_width = true;                                30
          }                                                    31
      }                                                        32
    }                                                          33
    ok_cancel;                                                 34
}                                                              35
```

Explanation

Lines 3 and 4
: boxed_row {
  label = "Select Osnaps";
The **boxed_row** is a predefined cluster tile that draws a border around the tiles. The second line, **label = "Select Osnaps";**, will display the label **(Select Osnaps)** left-justified at the top of the box. The **label** is a predefined DCL attribute.

Lines 5-7
: column {
  : toggle {

label = "Endpoint";

The **column** is a predefined tile that will arrange the tiles within it into a vertical column. The **toggle** is another predefined tile that displays a small box in the dialog box with an optional text on the right side of the box. The **label** attribute will display the label (**Endpoint**) to the right of the toggle box.

Lines 8-11
key = "Endpoint";
mnemonic = "E";
fixed_width = true;
}

The key attribute assigns a name (**Endpoint**) to the tile. This name is then used by the application program to handle the tile. The second line, **mnemonic = "E";**, defines the keyboard mnemonic for **Endpoint**. This causes the character **E** of **Endpoint** to be displayed underlined in the dialog box. The attribute **fixed_width** controls the width of the tile. If the value of this attribute is true, the tile does not stretch across the width of the dialog box. The closing brace (}) on the next line completes the definition of the toggle tile.

Lines 31-35
```
            }
        }
    }
   ok_cancel;
}
```

The closing brace on line 32 completes the definition of the toggle tile, and the closing brace on line 33 completes the definition of the column of line 20. The closing brace on line 34 completes the definition of the boxed_row of line 3. The predefined tile, **ok_cancel**, displays the **ok** and **cancel** tiles in the dialog box. The closing brace on the last line completes the definition of the dialog box.

After writing the program, use the following commands to load the DCL file and display the dialog box on the screen. The name of the file is assumed to be **osnapsh.dcl**. If AutoCAD is successful in loading the file, it will return an integer. In the following example, the integer AutoCAD returns is assumed to be 1. The screen display after loading the dialog box is shown in Figure 15-8. When you load the program, AutoCAD will freeze up. Use the Windows Task Manager to kill the AutoCAD program. See the note on page 15-10.

Command: (load_dialog "osnapsh.dcl")
1
Command: (new_dialog "osnapsh" 1)

AUTOLISP FUNCTIONS
logand and logior

These AutoLISP functions, **logand** and **logior**, are used to obtain the result of **logical bitwise AND and logical bitwise inclusive OR** of a list of numbers.

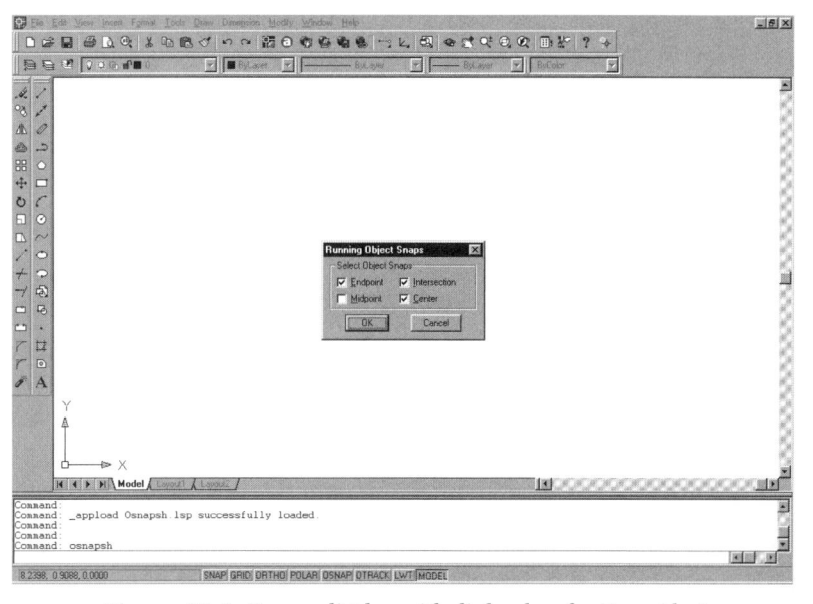

Programmable Dialog Boxes Using Dialog Control Language 15-21

Figure 15-8 Screen display with dialog box for Example 3

Examples of logand function

(logand 2 7)	will return	2
(logand 8 15)	will return	8
(logand 6 15 7)	will return	6
(logand 6 15 1)	will return	0
(logand 1 4)	will return	0
(logand 1 5)	will return	1

Examples of logior function

(logior 2 7)	will return	7
(logior 8 15)	will return	15
(logior 6 15 7)	will return	15
(logior 6 15 1)	will return	15
(logior 1 4)	will return	5
(logior 1 5)	will return	5

One application of bit-codes is found in the object snap modes. The following is a list of bit-codes assigned to different object snap modes:

Snap Mode	Bit Code	Snap Mode	Bit Code
None	0	Intersection	32
Endpoint	1	Insertion	64
Midpoint	2	Perpendicular	128
Center	4	Tangent	256
Node	8	Nearest	512
Quadrant	16	Quick	1024

The system variable **OSMODE** can be used to specify a snap mode. For example, if the value assigned to **OSMODE** is 4, the Object snap mode is center. The bit-codes can be combined to produce multiple Object snap modes. For example, you can get endpoint, midpoint, and center object snaps by adding the bit-codes of endpoint, midpoint, and center (1 + 2 + 4 = 7) and assigning that value (7) to the **OSMODE** system variable. In AutoLISP, the existing snap modes information can be extracted by using the logand function. For example, if **OSMODE** is set to 7, the logand function can be used to check different bit-codes that represent object snaps.

(logand 1 7)	will return	1	(Endpoint)
(logand 2 7)	will return	2	(Midpoint)
(logand 4 7)	will return	4	(Center)

atof and rtos Functions

The **atof** function converts a string into a real number and the rtos function converts a number into a string.

Examples
(atof "3.5") will return   3.5
(rtos 8 15) will return   "8.1500"

get_tile and set_tile Functions

The **get_tile** function retrieves the current value of a dialog box tile and the **set_tile** function sets the value of a dialog box tile.

Examples
(get_tile "midpoint")
(set_tile "xsnap" value)

Example 4

Write an AutoLISP program that will handle the dialog box and perform the functions as described in the dialog box of Example 3.

The following file is a listing of the AutoLISP program for Example 4. **The line numbers are not a part of the program; they are for reference only.**

```
;;Lisp program for setting Osnaps                                    1
;;Dialog file name is osnapsh.dcl                                    2
(defun c:osnapsh ( / dcl_id)                                         3
(setq dcl_id (load_dialog "osnapsh.dcl"))                            4
(new_dialog "osnapsh" dcl_id)                                        5
;;Get the existing value of object snaps and                         6
;;then write the values to the dialog box                            7
(setq osmode (getvar "osmode"))                                      8
(if (= 1 (logand 1 osmode))                                          9
   (set_tile "Endpoint" "1")                                        10
```

Programmable Dialog Boxes Using Dialog Control Language

```
                 )                                              11
              (if (= 2 (logand 2 osmode))                       12
                 (set_tile "Midpoint" "1")                      13
                 )                                              14
              (if (= 32 (logand 32 osmode))                     15
                 (set_tile "Intersection" "1")                  16
                 )                                              17
              (if (= 4 (logand 4 osmode))                       18
                 (set_tile "Center" "1")                        19
                 )                                              20
                                                                21
              ;;Read the values as set in the dialog box and    22
              ;;assign those values to AutoCAD variable osmode  23
              (defun setvars ()                                 24
              (setq osmode 0)                                   25
              (if (= "1" (get_tile "Endpoint"))                 26
                 (setq osmode (logior osmode 1))                27
                 )                                              28
              (if (= "1" (get_tile "Midpoint"))                 29
                 (setq osmode (logior osmode 2))                30
                 )                                              31
              (if (= "1" (get_tile "Intersection"))             32
                 (setq osmode (logior osmode 32))               33
                 )                                              34
              (if (= "1" (get_tile "Center"))                   35
                 (setq osmode (logior osmode 4))                36
                 )                                              37
                (setvar "osmode" osmode)                        38
                 )                                              39
                                                                40
              (action_tile "accept" "(setvars) (done_dialog)")  41
              (start_dialog)                                    42
              (princ)                                           43
              )                                                 44
```

Explanation

Lines 1 and 2
;;Lisp program for setting Osnaps
;;Dialog file name is osnapsh.dcl
These lines are comment lines, and all comment lines start with a semicolon. AutoCAD ignores the lines that start with a semicolon.

Lines 3-5
(defun c:osnapsh (/ dcl_id)
(setq dcl_id (load_dialog "osnapsh.dcl"))
(new_dialog "osnapsh" dcl_id)
In line 3, **defun** is an AutoLISP function that defines the **osnapsh** function. Because of the c:

in front of the function name, the **osnapsh** function can be executed like an AutoCAD command. The **osnapsh** function has one local variable, **dcl_id**. In line 4, the **(load_dialog "osnapsh.dcl")** loads the DCL file osnapsh.dcl and returns a positive integer. The **setq** function assigns this integer to the local variable, dcl_id. In line 5, the AutoLISP **new_dialog** function loads the dialog osnapsh that is defined in the DCL file (line 1 of the DCL file). The variable **dcl_id** has an integer value that identifies the DCL file.

Line 8
(setq osmode (getvar "osmode"))
In this line, **getvar "osmode"** has the value of the AutoCAD system variable **osmode**, and the **setq** function sets the **osmode** variable equal to that value. Note that the first **osmode** is just a variable, whereas the second osmode, in quotes (**"osmode"**), is the system variable.

Lines 9-11
(if (= 1 (logand 1 osmode))
   (set_tile "Endpoint" "1")
)

In line 9, the **(logand 1 osmode)** will return 1 if the bit-code of endpoint (1) is a part of the **OSMODE** value. For example, if the value assigned to **OSMODE** is 7 (1 + 2 + 4 = 7), (logand 1 7) will return 1. If **OSMODE** is 6 (2 + 4 = 6), then (logand 1 6) will return 0. The AutoLISP **if** function checks whether the value returned by **(logand 1 osmode)** is 1. If the function returns **T** (true), the instructions described in the second line are carried out. Line 10 sets the value of the **Endpoint** tile to 1; that displays a check mark in the toggle box. The closing parenthesis in line 11 completes the definition of the **if** function. If the expression **(if (= 1 (logand 1 osmode))** returns **nil**, the program skips to line 12 of the program.

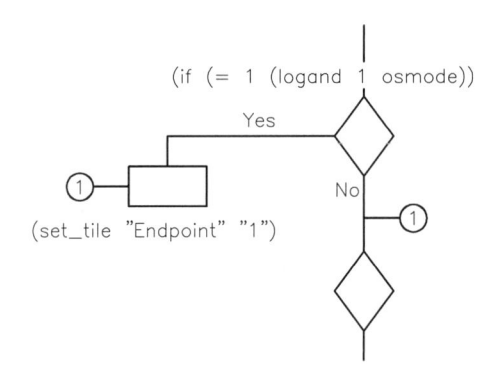

Lines 24 and 25
(defun setvars ( )
(setq osmode 0)
Line 24 defines a **setvars** function, and line 25 sets the value of the **osmode** variable to zero.

Lines 26-28
(if (= "1" (get_tile "Endpoint"))
   (setq osmode (logior osmode 1))
)
In line 26, the **(get_tile "Endpoint")** obtains the value of the toggle tile named **Endpoint**. If the Endpoint toggle tile is on, the value it returns is 1; if the toggle tile is off, the value it returns is 0. In line 27, the **setq** function sets the value of **osmode** to the value returned by **(logior osmode 1)**. For example, if the initial value of osmode is 0, then **(logior osmode 1)** will return 1. Similarly, if the initial value of **osmode** is 1, then **(logior osmode 2)** will return 3.

Lines 41 and 42
(action_tile "accept" "(setvars) (done_dialog)")
(start_dialog)
Line 42, **(start_dialog)**, starts the dialog box. In the dialog file the name assigned to the **OK** button is **"accept"**. When you select the **OK** button in the dialog box, the program executes the **setvars** function that updates the value of the **osmode** system variable and sets the selected object snaps.

PREDEFINED RADIO BUTTON, RADIO COLUMN, BOXED RADIO COLUMN, AND RADIO ROW TILES

Predefined Radio Button Tile

Format in DCL: **radio_button**

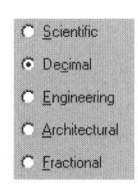

The **radio button** is a predefined active tile. Radio buttons can be arranged in a row or in a column. The unique characteristic of radio buttons is that only one button can be selected at a time. For example, if there are buttons for scientific, decimal, and engineering units, only one can be selected. If you select the decimal button, the other two buttons will be turned off automatically. Because of this special characteristic, radio buttons must be used only in a radio row or in a radio column. The label for the radio button is optional; in most systems the label appears to the right of the button.

Predefined Radio Column Tile

Format in DCL: **radio_column**

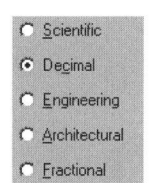

The radio column is an active predefined tile where the radio button tiles are arranged in a column and only one button can be selected at a time. When the radio buttons are arranged in a column, the buttons are next to each other vertically and are easy to select. Therefore, you should arrange radio buttons in a column to make it easy to select a radio button.

Predefined Boxed Radio Column Tile

Format in DCL: **boxed_radio_column**

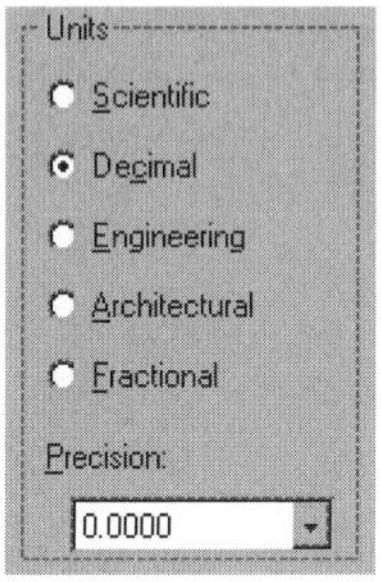

The **boxed radio column** is an active predefined tile where a border is drawn around the radio column.

Predefined Radio Row Tile

Format in DCL: **radio_row**

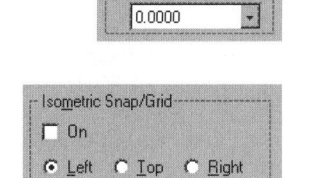

The **radio row** consists of radio button tiles arranged in a row. Only one button can be selected at a time. The radio row can become quite long if there are several radio button tiles and labels. In selecting a radio button, the cursor travel will increase because the buttons are not immediately next to each other. Therefore, you should avoid using radio button tiles in a row, especially if there are more than two.

Example 5

Write a DCL program for a dialog box that will enable you to select different units and unit precision as shown in Figure 15-9. Also, write an AutoLISP program that will handle the dialog box.

The following file is a listing of the DCL program for the dialog box shown in Figure 15-9. The name of the dialog box is **dwgunits. The line numbers are not a part of the file; they are shown here for reference only.**

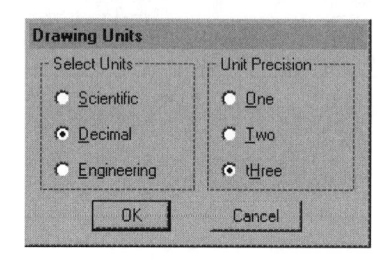

Figure 15-9 Dialog box for Example

```
dwgunits : dialog {                              1
  label = "Drawing Units";                       2
  : row {                                        3
    : boxed_column {                             4
      label = "Select Units";                    5
      : radio_column {                           6
        : radio_button {                         7
          key = "scientific";                    8
          label = "Scientific";                  9
          mnemonic = "S";                        10
        }                                        11
        : radio_button {                         12
          key = "decimal";                       13
          label = "Decimal";                     14
          mnemonic = "D";                        15
        }                                        16
        : radio_button {                         17
          key = "engineering";                   18
          label = "Engineering";                 19
          mnemonic = "E";                        20
        }                                        21
      }                                          22
    }                                            23
    : boxed_column {                             24
      label = "Unit Precision";                  25
      : radio_column {                           26
        : radio_button {                         27
          key = "one";                           28
          label = "One";                         29
          mnemonic = "O";                        30
        }                                        31
```

Programmable Dialog Boxes Using Dialog Control Language   15-27

```
            : radio_button {                                        32
                key = "two";                                        33
                label = "Two";                                      34
                mnemonic = "T";                                     35
            }                                                       36
            : radio_button {                                        37
                key = "three";                                      38
                label = "Three";                                    39
                mnemonic = "h";                                     40
            }                                                       41
        }                                                           42
    }                                                               43
  }                                                                 44
  ok_cancel;                                                        45
}                                                                   46
```

Explanation

Lines 4 and 5
: boxed_column {
  label = "Select Units";
The **boxed_column** is an active predefined tile that draws a border around the column. Line 5, **label = "Select Units";**, displays the label **(Select Units)** at the top of the column. The **label** is a predefined DCL attribute.

Lines 6-8
: radio_column {
  : radio_button {
    key = "scientific";
The **radio_column** is a predefined active tile that will arrange the tiles within it in a vertical column and draw a border around the column. The **radio_button** is another active predefined tile that displays a radio button in the dialog box, with an optional text to the right of the button. The key attribute assigns a name (scientific) to the tile. This name is then used by the application program to handle the tile.

Lines 9-11
   label = "Scientific";
   mnemonic = "S";
   }
The label attribute will display the label (Scientific) on the right of the toggle box. The second line, **mnemonic = "S";,** defines the keyboard mnemonic for **Scientific**. This causes the letter S of **Scientific** to be displayed underlined in the dialog box. The closing brace (}) on the next line completes the definition of the toggle tile.

Lines 41-46
```
            }
          }
        }
```

```
    }
  ok_cancel;
}
```

The closing brace on line 41 completes the definition of the radio button and the closing brace on the second line completes the definition of the radio column. The closing brace on the next line completes the definition of the boxed_column, and the closing brace on the next line completes the definition of the row on line 3 of the DCL file. The predefined tile, **ok_cancel**, displays the **OK** and **Cancel** tiles in the dialog box. The last closing brace completes the definition of the dialog box.

Use the following commands to load and display the dialog file on the screen. The name of the file is assumed to be **dwgunits.dcl**, and the name of the dialog is **dwgunits**. If AutoCAD is successful in loading the file, it will return an integer. In the following examples, the integer that AutoCAD returns is assumed to be 5. The screen display after loading the dialog box is shown in Figure 15-10.

Command: (load_dialog "osnapsh.dcl")
5
Command: (new_dialog "osnapsh" 5)

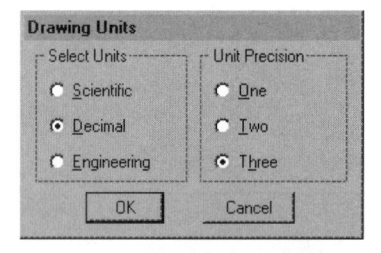

Figure 15-10 Dialog box for Example 5

The following file is a listing of the AutoLISP program that loads, displays, and handles the dialog box for Example 5. **The line numbers are not a part of the file; they are shown here for reference only.**

```
;;Lisp program dwgunits.lsp for setting units                      1
;;and precision. Dialog file name dwgunits.dcl                     2
;                                                                  3
(defun c:dwgunits ( / dcl_id)                                      4
(setq dcl_id (load_dialog "dwgunits.dcl"))                         5
(new_dialog "dwgunits" dcl_id)                                     6
;                                                                  7
;;Get the existing values of lunits and luprec                     8
;;and turn the corresponding radio_button on                       9
;                                                                 10
(setq lunits (getvar "lunits"))                                   11
(if (= 1 lunits)                                                  12
   (set_tile "scientific" "1")                                    13
```

```
          )                                                    14
     (if (= 2 lunits)                                          15
        (set_tile "decimal" "1")                               16
     )                                                         17
     (if (= 3 lunits)                                          18
        (set_tile "engineering" "1")                           19
     )                                                         20
     ;                                                         21
     (setq luprec (getvar "luprec"))                           22
     (if (= 1 luprec)                                          23
        (set_tile "one" "1")                                   24
     )                                                         25
     (if (= 2 luprec)                                          26
        (set_tile "two" "1")                                   27
     )                                                         28
     (if (= 3 luprec)                                          29
        (set_tile "three" "1")                                 30
     )                                                         31
                                                               32
     ;;Read the value of the radio_buttons and                 33
     ;;assign it to AutoCAD lunit and luprec variables         34
     ;                                                         35
     (action_tile "scientific" "(setq lunits 1)")              36
     (action_tile "decimal" "(setq lunits 2)")                 37
     (action_tile "engineering" "(setq lunits 3)")             38
     ;                                                         39
     (action_tile "one" "(setq luprec 1)")                     40
     (action_tile "two" "(setq luprec 2)")                     41
     (action_tile "three" "(setq luprec 3)")                   42
     (action_tile "accept" "(done_dialog)")                    43
     ;                                                         44
     (start_dialog)                                            45
     (setvar "lunits" lunits)                                  46
     (setvar "luprec" luprec)                                  47
     (princ)                                                   48
   )                                                           49
```

Explanation

Lines 1-3
;;Lisp program dwgunits.lsp for setting units
;;and precision. Dialog file name dwgunits.dcl
;
The first three lines of this program are comment lines, and all comment lines start with a semicolon. AutoCAD ignores them.

Lines 4-6
(defun c:dwgunits (/ dcl_id)

(setq dcl_id (load_dialog "dwgunits.dcl"))
(new_dialog "dwgunits" dcl_id)
In line 4, **defun** is an AutoLISP function that defines the **dwgunits** function, which has one local variable, **dcl_id**. The **c:** in front of the function name, **dwgunits**, makes the **dwgunits** function act like an AutoCAD command. In the next line, **(load_dialog "dwgunits.dcl")** loads the DCL file **dwgunits.dcl** and returns a positive integer. The **setq** function assigns this integer to the local variable, dcl_id. In the next line, the AutoLISP **new_dialog** function displays the **dwgunits** dialog box that is defined in the DCL file (line 1 of DCL file). The **dcl_id** variable is an integer that identifies the DCL file.

Line 11
(setq lunits (getvar "lunits"))
In this line, **getvar "lunits"** has the value of the AutoCAD system variable, **lunits**, and the **setq** function sets the **lunits** variable equal to that value. The first lunits is a variable, whereas the second **lunits**, in quotes (**"lunits"**), is a system variable.

Lines 12-14
(if (= 1 lunits)
   (set_tile "scientific" "1")
)
The **if** function (AutoLISP function) checks whether the value of the variable **lunits** is 1. If the function returns **T** (true), the instructions described in the next line are carried out. Line 13 sets the value of the tile named **"scientific"** equal to 1; that turns the corresponding radio button on. The closing parenthesis in line 14 completes the **if** function. If the **if** function, **(if (= 1 lunits)**, returns **nil**, the program skips to line number 14.

Line 22
(setq luprec (getvar "luprec"))
In this line, **getvar "luprec"** has the value of the AutoCAD system variable **luprec**, and the **setq** function sets the luprec variable equal to that value. The first **luprec** is a variable, whereas the second luprec, in quotes (**"luprec"**), is a system variable.

Lines 23-25
(if (= 1 luprec)
   (set_tile "one" "1")
)
In line 23, the **if** function checks whether the value of the **luprec** variable is 1. If the function returns **T** (true), the instructions described in the next line are carried out. This line sets the value of the tile named **"one"** to 1; that turns the corresponding

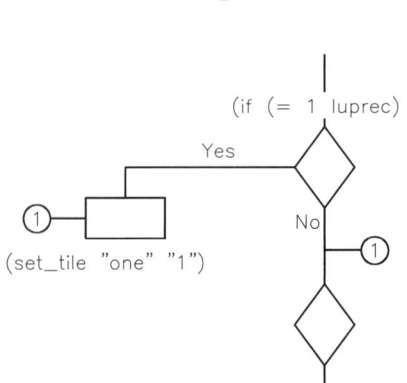

Programmable Dialog Boxes Using Dialog Control Language

radio button on. The closing parenthesis in the next line completes the definition of the **if** function. If the **if** function returns **nil**, the program skips to line 25.

Line 36
(action_tile "scientific" "(setq lunits 1)")
If the radio button tile named **"scientific"** is turned on, the **setq** function sets the value of the AutoCAD system variable **lunits** to 1. The **lunits** system variable controls the drawing units. The following is a list of the integer values that can be assigned to the **lunits** system variable:

 1 Scientific   4 Architectural
 2 Decimal   5 Fractional
 3 Engineering

Line 40
(action_tile "one" "(setq luprec 1)")
If the radio button tile named **"one"** is turned on, the **setq** function sets the value of the AutoCAD system variable **luprec** to 1. The **lunits** system variable controls the number of decimal places in a decimal number or the denominator of a fractional or architectural unit.

Lines 43-47
(action_tile "accept" "(done_dialog)")
;
(start_dialog)
(setvar "lunits" lunits)
(setvar "luprec" luprec)
In the dialog file, the name assigned to the **OK** button is **"accept"**. When you select the **OK** button in the dialog box, the program executes the function defined in the **dwgunits** dialog box. The **(start_dialog)** function starts the dialog box. The **(setvar "lunits" lunits)** and **(setvar "luprec" luprec)** set the values of the **lunits** and **luprec** system variables to lunits and luprec, respectively.

EDIT BOX TILE

Format in DCL: **edit_box**

The **edit box** is a predefined active tile that enables you to enter or edit a single line of text. If the text is longer than the length of the edit box, the text will automatically scroll to the right or left horizontally. The label for the edit box is optional, and it is displayed to the left of the edit box.

width AND edit_width ATTRIBUTES

width Attribute

Format in DCL: **width**   **Example**: width = 22

The **width** attribute is used to keep the width of the tile to a desired size. The value assigned to the **width** attribute can be a real number or an integer that represents the distance in character width. The value assigned to the **width** attribute defines the minimum width of the tile. For example, in this example the minimum width of the tile is 22. However, the tile will automatically stretch if more space is available. It will retain the size of 22 only if the **fixed_width** attribute is assigned to the tile. The **width** attribute can be used with any tile.

edit_width Attribute

Format in DCL: **edit_width**   **Example**: edit_width = 10

The **edit_width** attribute is used with the predefined edit box tiles, and it determines the size of the edit box in character width units. If the width of the edit box is 0, or if it is not specified and the fixed_width attribute is not assigned to the tile, the tile will automatically stretch to fill the available space. When the edit box is stretched, the PDB facility inserts spaces between the edit box and the label so that the box is right justified and the label is left justified.

Example 6   *General*

Write a DCL program for a dialog box (Figure 15-11) that will enable you to turn the snap and grid on and off. You should also be able to edit the X and Y values of snap and grid. Also, write an AutoLISP program that will load, display, and handle the dialog box.

The following file is a listing of the DCL file for Example 6.

```
dwgaids : dialog {
  label = "Drawing Aids";
  : row {
    : boxed_column {
      label = "SNAP";
      fixed_width = true;
      width = 22;
      : toggle {
        label = "On";
        mnemonic = "O";
        key = "snapon";
      }
      : edit_box {
        label = "X-Spacing";
        mnemonic = "X";
        key = "xsnap";
        edit_width = 10;
      }
      : edit_box {
        label = "Y-Spacing";
        mnemonic = "Y";
        key = "ysnap";
        edit_width = 10;
      }
```

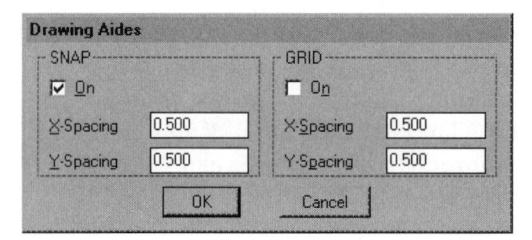

Figure 15-11 Dialog box for Example 6

Programmable Dialog Boxes Using Dialog Control Language 15-33

```
      }
    : boxed_column {
      label = "GRID";
      fixed_width = true;
      width = 22;
      : toggle {
        label = "On";
        mnemonic = "n";
        key = "gridon";
        }
      : edit_box {
        label = "X-Spacing";
        mnemonic = "S";
        key = "xgrid";
        edit_width = 10;
        }
      : edit_box {
        label = "Y-Spacing";
        mnemonic = "p";
        key = "ygrid";
        edit_width = 10;
        }
      }
    }
  ok_cancel;
}
```

The following file is a listing of the AutoLISP program for Example 6. When the program is loaded and run, it will load, display, and control the dialog box (Figure 15-12).

```
;;Lisp program for Drawing Aids dialog box
;;Dialog file name is dwgaids.dcl
;
(defun c:dwgaids( / dcl_id snapmode xsnap ysnap
 orgsnapunit gridmode gridsnap xgrid ygrid orggridunit)
(setq dcl_id (load_dialog "dwgaids.dcl"))
(new_dialog "dwgaids" dcl_id)

;;Get the existing value of snapmode and snapunit
;;and write those values to the dialog box
(setq snapmode (getvar "snapmode"))
(if (= 1 snapmode)
   (set_tile "snapon" "1")
   (set_tile "snapon" "0")
   )
(setq orgsnapunit (getvar "snapunit"))
(setq xsnap (car orgsnapunit))
```

```
      (setq ysnap (cadr orgsnapunit))
      (set_tile "xsnap" (rtos xsnap))
      (set_tile "ysnap" (rtos ysnap))
      ;
      ;;Get the existing value of gridmode and gridunit
      ;;and write those values to the dialog box
      (setq gridmode (getvar "gridmode"))
      (if (= 1 gridmode)
         (set_tile "gridon" "1")
         (set_tile "gridon" "0"))
      (setq orggridunit (getvar "gridunit"))
      (sctq xgrid (car orggridunit))
      (setq ygrid (cadr orggridunit))
      (set_tile "xgrid" (rtos xgrid))
      (set_tile "ygrid" (rtos ygrid))
      ;;Read the values set in the dialog box and
      ;;then change the associated AutoCAD variables
      (defun setvars ()
      (setq xsnap (atof (get_tile "xsnap")))
      (setq ysnap (atof (get_tile "ysnap")))
      (setvar "snapunit" (list xsnap ysnap))
      (if (= "1" (get_tile "snapon"))
         (setvar "snapmode" 1)
         (setvar "snapmode" 0))
      (setq xgrid (atof (get_tile "xgrid")))
      (setq ygrid (atof (get_tile "ygrid")))
      (setvar "gridunit" (list xgrid ygrid))
      (if (= "1" (get_tile "gridon"))
         (progn
            (setvar "gridmode" 0)
            (setvar "gridmode" 1))
         (setvar "gridmode" 0)
      )
      (action_tile "accept" "(setvars) (done_dialog)")
      (start_dialog)
      (princ)
      )
```

SLIDER AND IMAGE TILES

Slider Tile

Format in DCL
slider

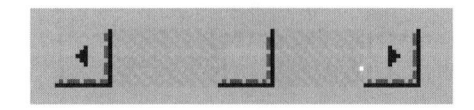

A slider tile is an active predefined tile that consists of a slider bar (rectangular strip), a small

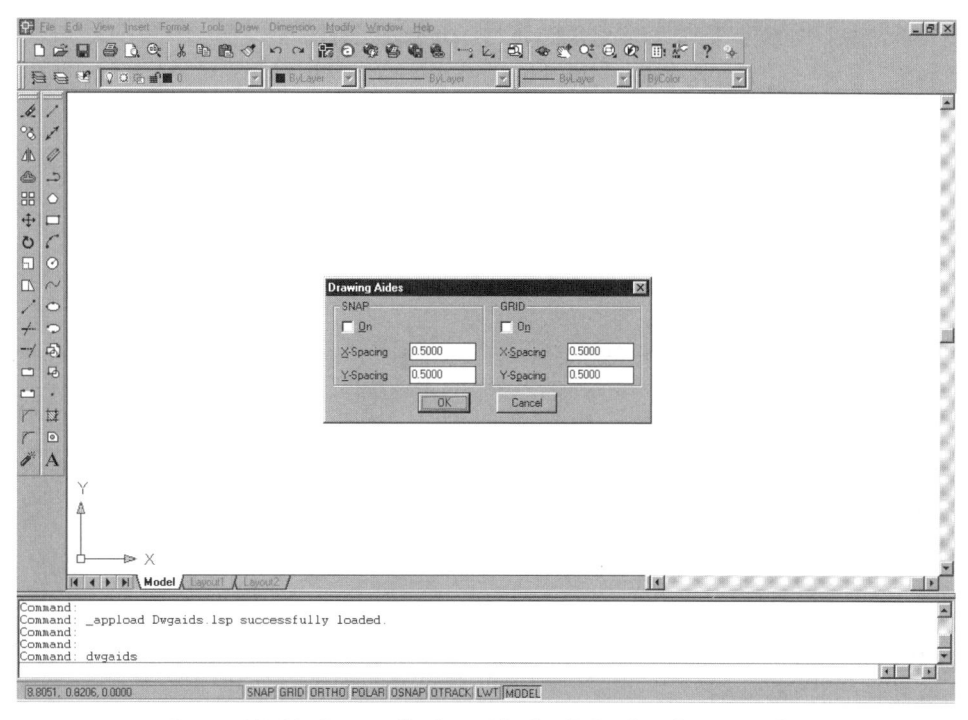

Programmable Dialog Boxes Using Dialog Control Language 15-35

Figure 15-12 Screen display with the dialog box for Example 6

indicator box, and the direction arrows at the ends of the slider bar. The slider tile can be used to obtain a string value. The value is determined by the position of the indicator box in the slider bar. The string value returned by the slider tile can then be used by the application program. For example, you can use the value returned by the aperture slider tile to set AutoCAD's **aperture** system variable. The indicator box can be dragged to the left or right by positioning the arrow on the indicator box and then moving the arrow while holding the pick button down. It can also be moved in increments by positioning the arrow to the left or right of the indicator box or in the arrow box of the slider, and clicking it. The slider tiles can be horizontal or vertical. If the slider tile is horizontal, the values increase from left to right; if the slider is vertical, the values increase from bottom to top. The values returned by the slider tile are always integers.

Image Tile

Format in DCL
image

The image tile is a predefined tile that is used to display graphical information. For example, it can be used to display the aperture box, linetypes, icons, and text fonts in dialog boxes. It consists of a rectangular box and the vector graphics that are displayed within the box.

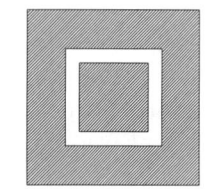

min_value, max_value, small_increment, and big_increment Attributes

min_value and max_value Attributes

Format in DCL	Examples
min_value	min_value = 2
max_value	max_value = 15

The **min_value** attribute and the **max_value** attribute are predefined attributes that specify the minimum and maximum value that the **slider** tile will return. In the previous example, the minimum value is 2 and the maximum value is 15. If you do not assign these attributes to a slider tile, the slider automatically assumes the default values. For the **min_value** attribute the default value is 0, for the **max_value** attribute the default value is 10,000.

small_increment and big_increment Attributes

Format in DCL	Examples
small_increment	small_increment = 1
big_increment	big_increment = 1

The value assigned to the **small_increment** attribute and the **big_increment** attribute determines the increment value of the slider incremental control. For example, if the increment is 1, the values returned by the slider will be in the increment of 1. If these attributes are not assigned to a slider tile, the slider automatically assumes the default values. The default value of **small_increment** is one one-hundredth (1/100) of the slider range and the default value of **big_increment** is one-tenth (1/10) of the slider range.

aspect_ratio and color Attributes

aspect_ratio Attribute

Format in DCL
aspect_ratio

Example
aspect_ratio = 1

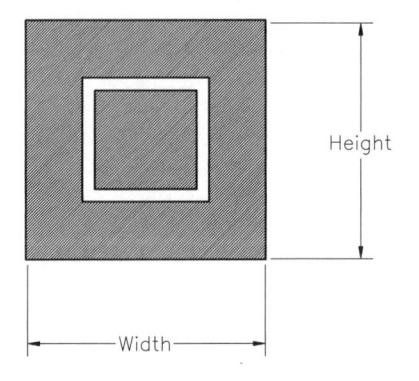

The aspect_ratio is a predefined attribute that can be used with an image. The aspect ratio is the ratio of the width of the image box to the height of the image box (width/height). You can control the size of the image box by assigning an integer value to the width attribute and the height attribute. You can also control the size of the image box by assigning an integer value to one of the attributes (width or height) and assigning a real or integer value to the aspect_ratio attribute. For example, if the value assigned to the height attribute is five and the value assigned to the aspect_ratio is 0.5, the width of the image box will be 2.5 (width = height x aspect_ratio). In this case, you do not need to specify the width attribute.

color Attribute

Format in DCL
color

Example
color = 2

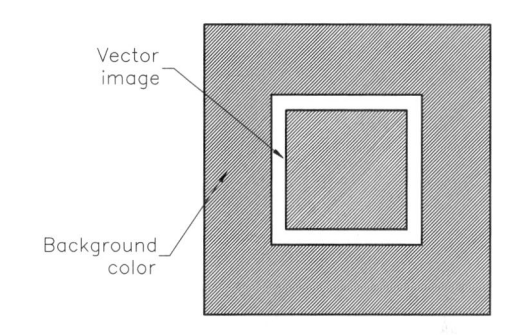

The color is a predefined attribute that can be used with the image box to control the background color. The integer number assigned to the color attribute specifies an AutoCAD color index. In the above example, the background color is yellow (color number 2). The color of the vector image that is drawn inside the image box is determined by the color specified in the vector_image attribute.

Example 7 *General*

Write a DCL program for the dialog box in Figure 15-13(a) that will enable you to change the size of the aperture box.

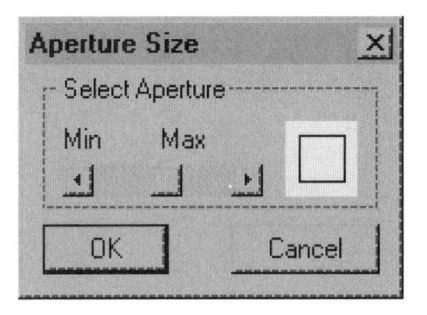

Figure 15-13(a) Dialog box for Example 7

The dialog box shown in Figure 15-13(a) has two rows and two columns. The first column has two items, label **(Min .... Max)** and the **slider bar**. The second column has only one item (the image box). The second row has the **OK** and **Cancel** buttons. As shown in Figure 15-13(b), the dialog box also has a dialog label (Aperture Size) and a row label (Select Aperture). The following file is a listing of the DCL file for Example 7. The line numbers are not a part of the file; they are shown here for reference only.

```
aprtsize : dialog {                                              1
  label = "Aperture Size";                                       2
```

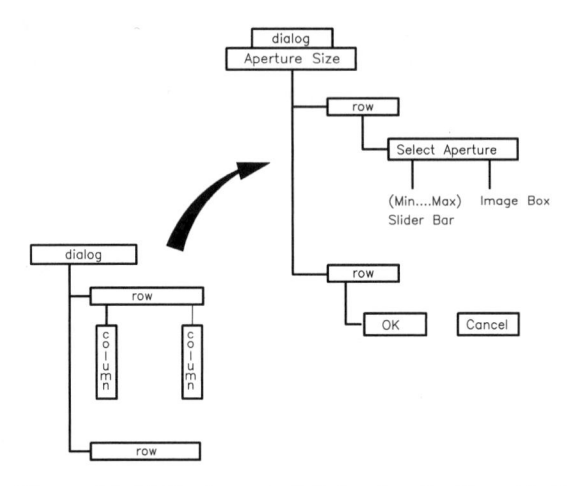

Figure 15-13(b) *Layout of dialog box for Example 7*

```
: boxed_row {                                    3
  label = "Select Aperture";                     4
  : column {                                     5
    fixed_width = true;                          6
    : text {                                     7
      label = "Min      Max";                    8
      alignment = centered;                      9
    }                                           10
    : slider {                                  11
      key = "aperture_slider";                  12
      min_value = 2;                            13
      max_value = 17;                           14
      width = 15;                               15
      height = 1;                               16
      big_increment = 1;                        17
      fixed_width = true;                       18
      fixed_height = true;                      19
    }                                           20
  }                                             21
  : image {                                     22
    key = "aperture_image";                     23
    aspect_ratio = 1;                           24
    width = 5;                                  25
    color = 2;                                  26
  }                                             27
}                                               28
ok_cancel;                                      29
}                                               30
```

Lines 11, 12
: slider {
key = "aperture_slider";
In the first line, **: slider** starts the definition of the slider tile. The **slider** is an active predefined tile that consists of a slider bar, a small indicator box, and the direction arrows at the ends of the slider bar. The second line, **key = "aperture_slider"**, assigns the name, **aperture_slider**, to this slider tile.

Lines 13, 14
min_value = 2;
max_value = 17;
The **min_value** is a predefined attribute that specifies the minimum value that the **slider** tile will return. Similarly, the **max_value** attribute specifies the maximum value that the slider tile will return. In the above two lines, the minimum value is 2 and the maximum value is 17.

Lines 15-17
width = 15;
height = 1;
big_increment = 1;
The first line specifies the width of the slider tile and the second line specifies the height of the slider tile. The third line, **big_increment = 1;**, defines the increment of the indicator box in the slider bar.

Lines 22, 23
: image {
  key = "aperture_image";
The first line, **: image {**, starts the definition of the image box and the second line assigns a name, **aperture_image**, to the image tile. The image is a predefined tile that is used to display graphical information.

Lines 24-26
aspect_ratio = 1;
width = 5;
color = 2;
The first line, **aspect_ratio = 1;**, defines the ratio of the width of the image box to the height of the image box. The second line, **width = 5;**, specifies the width of the image box. Since the aspect ratio is 1, the height of the image box is 5 (width/height = aspect_ratio). The third line, **color = 2;**, assigns the AutoCAD color number 2 (yellow) to the background of the image box.

AUTOLISP FUNCTIONS

dimx_tile and dimy_tile

The AutoLISP function, **dimx_tile**, obtains the width dimension (x_aperture) of the specified image box along the X-axis, and the **dimy_tile** function obtains the height dimension (y_aperture) along the Y-axis. In the following examples, "aperture_image" is the name of the image tile that is assigned by using the **key** attribute in the DCL file (key = "aperture_image").

Format
(dimx_tile tilename)
(dimy_tile tilename)

Examples
(dimx_tile "aperture_image")
(dimy_tile "aperture_image")

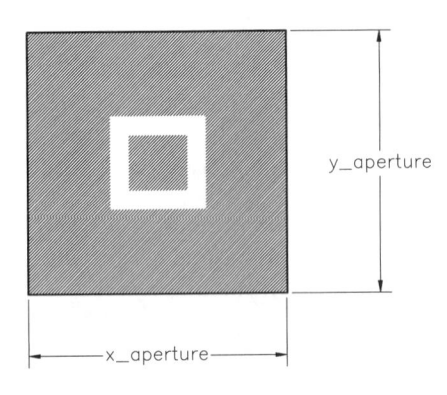

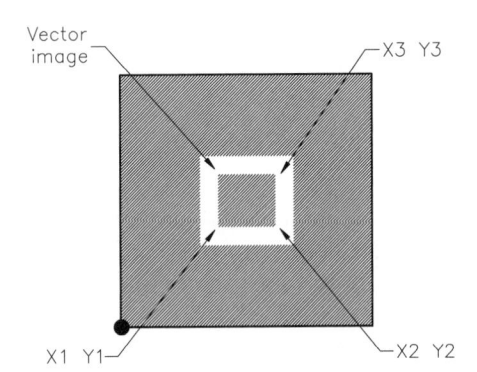

vector_image

The AutoLISP function **vector_image** draws a vector (line) in the active image box, between the points that are defined in the **vector_image** function. In the following example, AutoCAD will draw a line from point 1,1 to point 3,3 of the image box and the color of this vector will be AutoCAD's color number 1 (red).

Format
(vector_image x1 y1 x2 y2 color)

Example
(vector_image <u>1 1 3 3</u> 1)
Where   **1 1** ---------------- X and Y coordinates of first point
        3 ------------------- X and Y coordinates of second point
        3 ------------------- AutoCAD color index

fill_image

The AutoLISP function, **fill_image**, fills the rectangle in the active image box with the color specified in the **fill_image** function. For example, in the following example the active image box will be filled with yellow color (AutoCAD's color number 2). The filled rectangle is determined by the coordinates of the first point (x1 y1) and the second point (x2 y2), the two opposite corners of the rectangle.

Format
(fill_image x1 y1 x2 y2 color)

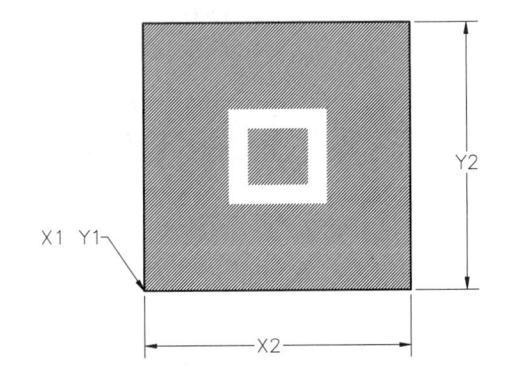

Programmable Dialog Boxes Using Dialog Control Language

Example
(fill_image 0 0 x_aperture y_aperture 2)

start_image

The AutoLISP function, **start_image**, starts the image whose name is specified in the **start_image** function. In the following example, the name of the image tile is **"aperture_image"** which is assigned by using the **key** attribute in the DCL file (key = "aperture_image").

Format   **Example**
(start_image)   (start_image "aperture_image")

end_image

The AutoLISP function **end_image** ends the active image whose name is specified in the **end_image** function. In the following example, the name of the active image tile is **"aperture_image"**.

Format   **Example**
(end_image)   (end_image "aperture_image")

$value

The **$value** function is an AutoLISP action expression that retrieves the string value from the tile of a dialog box. The tile could be an edit box or a toggle box. In Example 8, the **$value** function retrieves the current value of the active tile and the setq function assigns that value to the variable, aprt_size.

Format   **Example**
$value   (setq aprt_size $value)

Example 8

Write an AutoLISP program that will load, display, and control the dialog box of Example 7 (Figure 15-13(a)). The flowchart and screen display for this example are shown in Figures 15-14 and 15-15.

The following file is a listing of the AutoLISP file for Example 8. The line numbers are not a part of the file; they are shown here for reference only.

```
;;APRTSIZE.LSP, AutoLISP program for Aperture              1
;; Dialog Box. DCL file name — APRTSIZE.DCL                2
;                                                          3
(defun c:aprtsize ( )                                      4
  (setq dcl_id (load_dialog "aprtsize"))                   5
  (new_dialog "aprtsize" dcl_id)                           6
;                                                          7
;Obtain value of the system variable "aperture",           8
;calculate X and Y values of vector image, and draw        9
```

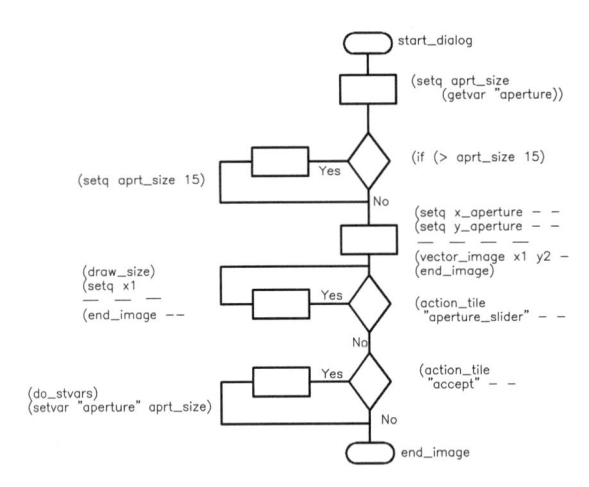

Figure 15-14 Flowchart for Example 8

```
;vector image in the image box.                                             10
  (setq aprt_size (getvar "aperture"))                                      11
  (if (> aprt_size 15)                                                      12
    (setq aprt_size 15)                                                     13
  )                                                                         14
  (setq x_aperture (dimx_tile "aperture_image"))                            15
  (setq y_aperture (dimy_tile "aperture_image"))                            16
  (set_tile "aperture_slider" (itoa aprt_size))                             17
  (setq x1 (- (/ x_aperture 2) aprt_size))                                  18
  (setq x2 (+ (/ x_aperture 2) aprt_size))                                  19
  (setq y1 (- (/ y_aperture 2) aprt_size))                                  20
  (setq y2 (+ (/ y_aperture 2) aprt_size))                                  21
  (start_image "aperture_image")                                            22
  (fill_image 0 0 x_aperture y_aperture 2)                                  23
  (vector_image x1 y1 x2 y1 1)                                              24
  (vector_image x2 y1 x2 y2 1)                                              25
  (vector_image x2 y2 x1 y2 1)                                              26
  (vector_image x1 y2 x1 y1 1)                                              27
  (end_image)                                                               28
  (action_tile "aperture_slider"                                            29
     "(draw_size (setq aprt_size (atoi $value)))")                          30
  (action_tile "accept" "(do_setvars)(done_dialog)")                        31
  (start_dialog)                                                            32
(princ)                                                                     33
)                                                                           34
;Set aperture variable "aperture" equal to aprt_size                        35
(defun do_setvars ( )                                                       36
  (setvar "aperture" aprt_size)                                             37
)                                                                           38
```

Programmable Dialog Boxes Using Dialog Control Language

```
  ;                                                          39
  ;Calculate the X and Y coordinates of vector image         40
  ;and draw the aperture image in the image box.             41
  (defun draw_size (aprt_size)                               42
    (setq x1 (- (/ x_aperture 2) aprt_size))                 43
    (setq x2 (+ (/ x_aperture 2) aprt_size))                 44
    (setq y1 (- (/ y_aperture 2) aprt_size))                 45
    (setq y2 (+ (/ y_aperture 2) aprt_size))                 46
    (start_image "aperture_image")                           47
    (fill_image 0 0 x_aperture y_aperture -2)                48
    (vector_image x1 y1 x2 y1 1)                             49
    (vector_image x2 y1 x2 y2 1)                             50
    (vector_image x2 y2 x1 y2 1)                             51
    (vector_image x1 y2 x1 y1 1)                             52
    (end_image)                                              53
  )                                                          54
```

Lines 11-13
(setq aprt_size (getvar "aperture"))
(if (> aprt_size 15)
  (setq aprt_size 15)

In the first line, the AutoLISP function **getvar** obtains the value of the **"aperture"** system variable and the **setq** function assigns that value to the **aprt_size** variable. The second line, **(if (> aprt_size 15)**, checks whether the value of the variable aprt_size is greater than 15. If it is, the third line, **(setq aprt_size 15)**, sets the value of the **aprt_size** variable equal to 15. The value of the "aperture" system variable must be equal to or less than 15 to display the vector image in the image box because the maximum size of the image box in this example is 15.

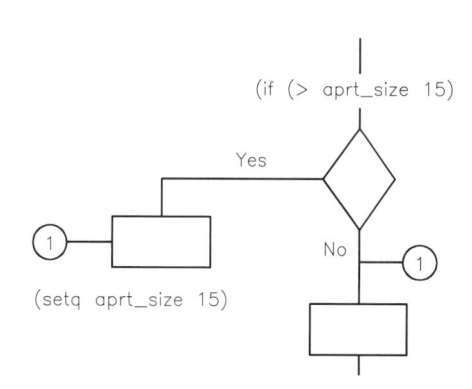

Lines 15, 16
(setq x_aperture (dimx_tile "aperture_image"))
(setq y_aperture (dimy_tile "aperture_image"))
In the first line, the AutoLISP function, **dimx_tile**, retrieves the X-dimension of the image tile and the setq function assigns that value to the **x_aperture** variable. Similarly, the **dimy_tile** function retrieves the Y-dimension of the image tile.

Line 17
(set_tile "aperture_slider" (itoa aprt_size))
The AutoLISP function **itoa** changes the integer value of the **aprt_size** variable into a string and the **set_tile** function assigns that value to **"aperture_slider"**. The **aperture_slider** is the name of the slider tile in the DCL file (DCL file of Example 7).

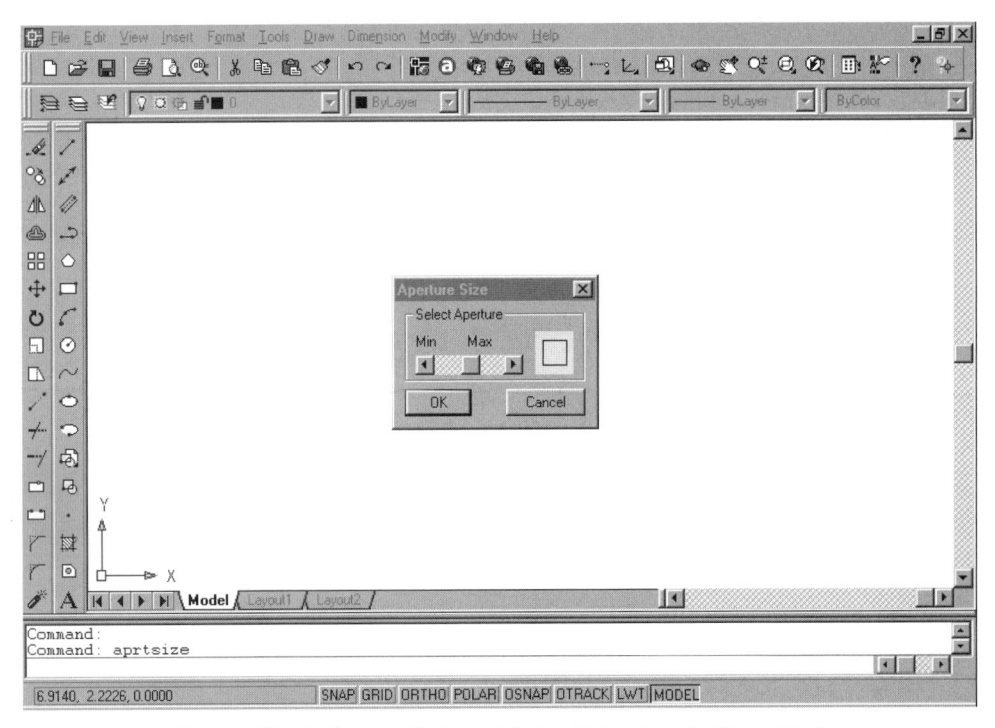

Figure 15-15 Screen display with the dialog box for Example 8

Lines 18-21
(setq x1 (- (/ x_aperture 2) aprt_size))
(setq x2 (+ (/ x_aperture 2) aprt_size))
(setq y1 (- (/ y_aperture 2) aprt_size))
(setq y2 (+ (/ y_aperture 2) aprt_size))
The first line calculates the X-coordinate of the lower left corner of the aperture box. The value is calculated by dividing the **x_aperture** distance by 2 and then subtracting the distance **aprt_size** from it. The aprt_size has the same value as that of the system variable, "aperture". Aperture is the distance of the sides of the aperture box from the crosshair lines, measured in pixels. Similarly, the second line calculates the X-coordinate of the lower-right corner. The next two lines calculate the Y-coordinates of these two points.

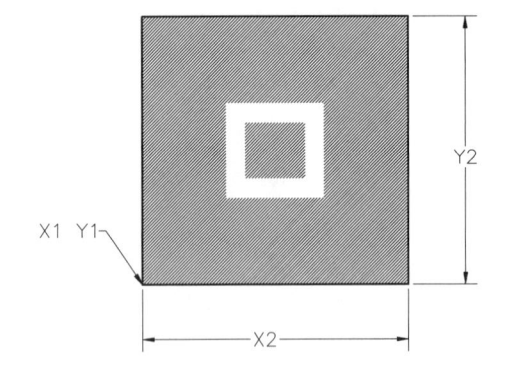

Lines 22, 23
(start_image "aperture_image")
(fill_image 0 0 x_aperture y_aperture 2)
The AutoLISP function **start_image** starts the image box and **aperture_image** is the name of

the image box as defined in the DCL file. The fill_image function fills the image within the specified coordinates with AutoCAD color index 2 (yellow).

Lines 27, 28
(vector_image x1 y2 x1 y1 1)
(end_image)
The AutoLISP function **vector_image** draws a vector (line) from the point with coordinates x1 y2 to the point with coordinates x1 y1. The **end_image** function ends the image.

Lines 29-31
(action_tile "aperture_slider"
  "(draw_size (setq aprt_size (atoi $value)))")
(action_tile "accept" "(do_setvars)(done_dialog)")
The AutoLISP action expression **$value** retrieves the string value of the active slider tile and the atoi function returns the integer value of the string. This value is then used by the **draw_size** function to draw the vector image of the aperture box. The **aperture_slider** is the name of the slider tile defined in the DCL file. The third line executes the **do_setvars** function and ends the dialog box when you select OK from the dialog box. The **draw_size** and **do_setvars** functions are defined in the program.

Lines 36, 37
(defun do_setvars ( )
  (setvar "aperture" aprt_size)
The defun function defines the **do_setvars** function. The **setvar** function sets the value of the AutoCAD system variable, **"aperture"**, equal to **aprt_size**.

Lines 42, 43
(defun draw_size (aprt_size)
  (setq x1 (- (/ x_aperture 2) aprt_size))
The first line defines the **draw_size** function with one argument, **aprt_size**, and the second line calculates the value of X-coordinate of the first vector.

Review Questions

Answer the following questions.

1. A dialog control language (DCL) file can contain descriptions of multiple files. (T/F)

2. In a DCL file, you do not need to specify the size of a dialog box. (T/F)

3. Dialog boxes are not dependent on the platform. (T/F)

4. Dialog boxes can contain several OK buttons. (T/F)

5. The numeric values assigned to the attributes in a DCL file can be both integers and real numbers. (T/F)

6. The reserved names in DCL are case-sensitive. (T/F)

7. The label attribute used in a button tile has no default value. (T/F)

8. In a dialog box, only one button can be assigned the true value for the is_default attribute. (T/F)

9. The load_dialog function displays the dialog box on the screen. (T/F)

10. You cannot select a tile by using the mnemonic key assigned to the tile. (T/F)

11. Mnemonic characters are case-sensitive. (T/F)

12. Bit-codes cannot be combined to produce multiple object snaps. (T/F)

13. The radio button is a predefined attribute. (T/F)

14. You should arrange the radio buttons in a column setting. (T/F)

15. The edit box active tile allows you to enter or edit multiple lines of text. (T/F)

16. The width attribute will make the tile stretch automatically to fill the entire width of the dialog box. (T/F)

17. The slider tile returns a string value. (T/F)

18. The aspect ratio is the ratio of the width of the image box to the height of the image box. (T/F)

19. The color attribute can be used with an image box to control the background color. (T/F)

20. The vector_image function can be used to draw vectors in the Drawing Editor. (T/F)

21. The AutoLISP function, $value, retrieves the real value from the tile of a dialog box. (T/F)

22. The basic tiles, such as buttons, edit boxes, and images, are predefined by the _____ facility of AutoCAD.

23. The label of a button appears _____ the button.

24. The _____ attribute assigns a name to the tile.

25. The label attribute in a boxed column must be a _____ string.

26. The _____ attribute controls the width of a tile.

27. The format of the command for loading a dialog box is _____.

Programmable Dialog Boxes Using Dialog Control Language 15-47

28. The AutoLISP function _____ is used to initialize a dialog box and then display it on the screen.

29. The AutoLISP function _____ is used to accept user input from the dialog box.

30. The AutoLISP function _____ is used to associate an action expression with a tile in the dialog box.

31. The AutoLISP function _____ can be used to obtain the result of logical bitwise AND.

32. The edit_width attribute determines the size of the edit box in _____ units.

33. The image tile is used to display _____ information.

34. The default value of the max_value attribute is _____.

35. The default value of the big_increment attribute is _____ of the slider range.

36. The AutoLISP function _____ can be used to fill the image box with a color.

Exercises

Exercise 1 *General*

Using dialog control language (DCL), write a program for the dialog box in Figure 15-16. Also, write an AutoLISP program that will load, display, and control the dialog box and perform the functions shown in the dialog box.

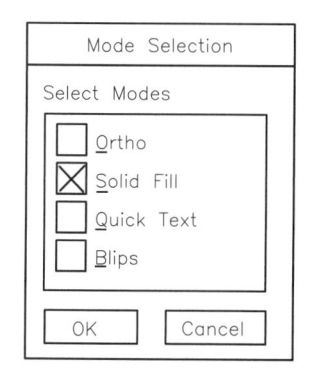

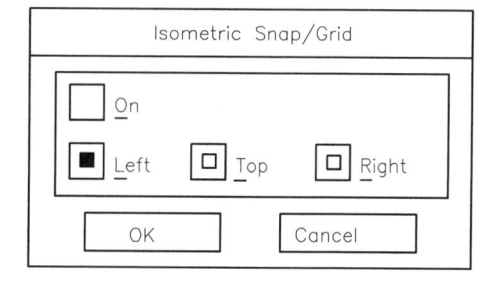

Figure 15-16 Dialog box for mode selection

Figure 15-17 Dialog box for isometric snap/grid

Exercise 2 — *General*

Write a DCL program for the isometric snap/grid dialog box shown in Figure 15-17. Also, write an AutoLISP program that will load, display, and control the dialog box and perform the functions shown in the dialog box.

Exercise 3 — *General*

Write a DCL program for the dialog box in Figure 15-18 that will enable you to insert a block. Also, write an AutoLISP program that will load, display, and control the dialog box. The values shown in the edit boxes for insertion point, scale, and rotation are the default values.

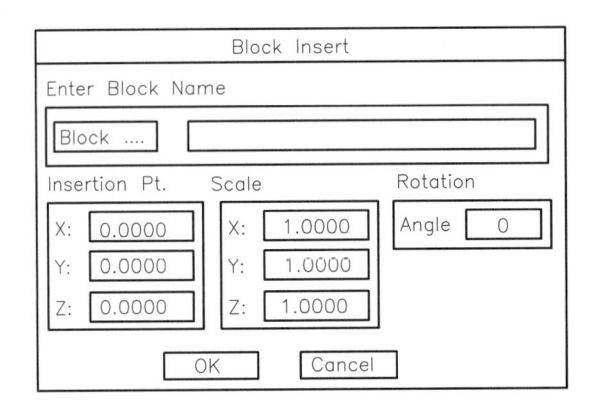

Figure 15-18 Dialog box for inserting blocks

Exercise 4 — *General*

Write a DCL program for the dialog box in Figure 15-19 that will enable you to select the angle and angle precision. Also, write an AutoLISP program that will load, display, and control the dialog box.

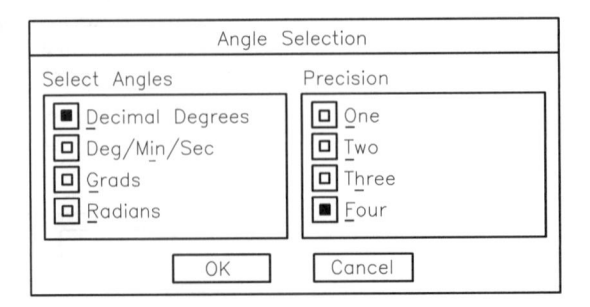

Figure 15-19 Dialog box for angle selection

Chapter 16

DIESEL:
A String Expression Language

Learning Objectives
After completing this chapter, you will be able to:
- *Use DIESEL to customize a status line.*
- *Use the **MODEMACRO** system variable.*
- *Write macro expressions using DIESEL.*
- *Use AutoLISP with **MODEMACRO**.*

DIESEL

DIESEL (Direct Interpretively Evaluated String Expression Language) is a string expression language. It can be used to display a user-defined text string (macro expression) in the status line by altering the value of the AutoCAD system variable **MODEMACRO**. The value assigned to **MODEMACRO** must be a string, and the output it generates will be a string. It is fairly easy to write a macro expression in DIESEL, and it is an important tool for customizing AutoCAD. However, DIESEL is slow, and it is not intended to function like AutoLISP or DCL. You can use AutoLISP to write and assign a value to the **MODEMACRO** variable, or you can write the definition of the **MODEMACRO** expression in the menu files. A detailed explanation of the DIESEL functions and the use of DIESEL in writing a macro expression is given later in this chapter.

STATUS LINE

When you are in AutoCAD, a status line is displayed at the bottom of the graphics screen (Figure 16-1). This line contains some useful information and tools that will make it easy to change the status of some AutoCAD functions. To change the status, you must click on the buttons. For example, if you want to display grid lines on the screen, click on the **GRID**

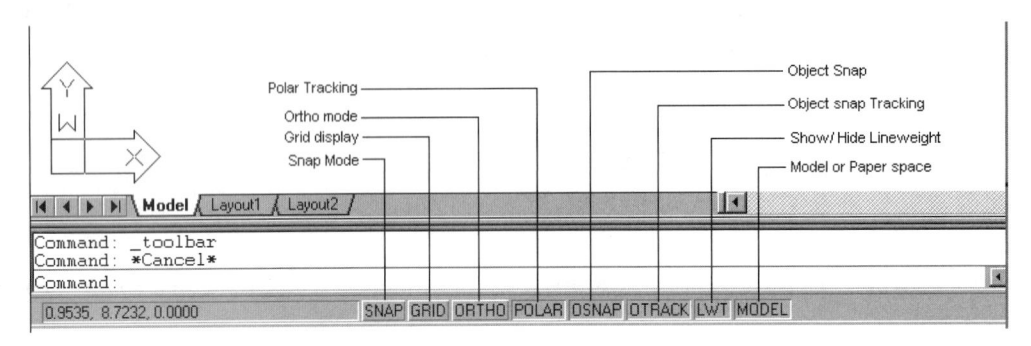

Figure 16-1 Default status line display

button. Similarly, if you want to switch to paper space, click on **MODEL**. The status line contains the following information:

Coordinate Display. The coordinate information displayed in the status line can be static or dynamic. If the coordinate display is static, the coordinate values displayed in the status line change only when you specify a point. However, if the coordinate display is dynamic (default setting), AutoCAD constantly displays the absolute coordinates of the graphics cursor with respect to the UCS origin. AutoCAD can also display the polar coordinates (length<angle) if you are in an AutoCAD command.

SNAP. If SNAP is on, the cursor snaps to the snap point.

GRID. If GRID is on, grid lines are displayed on the screen.

ORTHO. If ORTHO is on, a line can be drawn in a vertical or horizontal direction.

POLAR. When the POLAR snap is on, the cursor snaps to polar angles as set in the Polar Tracking tab of the **Drafting Settings** dialog box.

OSNAP. If OSNAP is on, you can use the running object snaps. If OSNAP is off, the running object snaps are temporarily disabled. The status of OSNAP (Off or On) does not prevent you from using regular object snaps.

OTRACK. When Object Tracking is on, you can track from a point. The object snap must be set before using Object Tracking.

LWT. When the Lineweight button is on, the objects are displayed with the assigned width.

MODEL/PAPER. AutoCAD displays MODEL in the status line when you are working in the model space. If you are working in the paper space, AutoCAD will display PAPER in place of MODEL.

MODEMACRO SYSTEM VARIABLE

The AutoCAD system variable **MODEMACRO** can be used to display a new text string in the status line. You can also display the value returned by a macro expression using the DIESEL language, which is discussed in a later section of this chapter. MODEMACRO is a system variable and you can assign a value to this variable by entering MODEMACRO at the Command prompt or by using the SETVAR command. For example, if you want to display **Customizing AutoCAD** in the status line, enter **SETVAR** at the Command: prompt and then press ENTER. AutoCAD will prompt you to enter the name of the system variable. Enter **MODEMACRO** and then press ENTER again. Now you can enter the text you want to display in the status line. After you enter **Customizing AutoCAD** and press ENTER, the status line will display the new text.

>   Command: **MODEMACRO**
>   or
>   Command: **SETVAR**
>   Variable name or ?: **MODEMACRO**
>   New value for MODEMACRO, or . for none<"">: **Customizing AutoCAD**

You can also enter MODEMACRO at the Command prompt and then enter the text that you want to display in the status line.

>   Command: **MODEMACRO**
>   New value for MODEMACRO, or . for none<"">: **Customizing AutoCAD**

Once the value of the **MODEMACRO** variable is changed, it retains that value until you enter a new value, start a new drawing, or open an existing drawing file. If you want to display the standard text in the status line, enter a period (.) at the prompt **New value for MODEMACRO, or . for none <"">:**. The value assigned to the **MODEMACRO** system variable is not saved with the drawing, in any configuration file, or anywhere in the system.

>   Command: **MODEMACRO**
>   New value for MODEMACRO, or . for none<"">:

CUSTOMIZING THE STATUS LINE

The information contained in the status line can be divided into two parts: toggle functions and coordinate display. The toggle functions part consists of the status of Snap, Grid, Ortho, Polar Tracking, Object Snap, Object Tracking, Lineweight, and Model Space (Figure 16-1). The coordinate display displays the X, Y, and Z coordinates of the cursor. The status line can be customized to your requirements by assigning a value to the AutoCAD system variable **MODEMACRO**. The value assigned to this variable is displayed left-justified in the status bar at the bottom of the AutoCAD window. The number of characters that can be displayed in the status line depends on the system display and the size of the AutoCAD window. The coordinate display field cannot be changed or edited.

The information displayed in the status line is a valuable resource. Therefore, you must be careful when selecting the information to be displayed in the status line. For example, when working on a project, you may like to display the name of the project in the status line. If you are using several dimensioning styles, you could display the name of the current dimensioning style (DIMSTYLE) in the status line. Similarly, if you have several text files with different fonts, the name of the current text file (TEXTSTYLE) and the text height (TEXTSIZE) can be displayed in the status line. Sometimes, in 3D drawings, if you need to monitor the viewing direction (VIEWDIR), the camera coordinate information can be displayed in the status line. Therefore, the information that should be displayed in the status line depends on you and the drawing requirements. AutoCAD lets you customize this line and have any information displayed in the status line that you think is appropriate for your application.

MACRO EXPRESSIONS USING DIESEL

You can also write a macro expression using DIESEL to assign a value to the **MODEMACRO** system variable. The macro expressions are similar to AutoLISP functions, with some differences. For example, the drawing name can be obtained by using the AutoLISP statement **(getvar dwgname)**. In DIESEL, the same information can be obtained by using the macro expression **$(getvar,dwgname)**. However, unlike the case with AutoLISP, the DIESEL macro expressions return only string values. The format of a macro expression is:

$(function-name,argument1,argument2, ...)

Example
$(getvar,dwgname)

Here, **getvar** is the name of the DIESEL string function and **dwgname** is the argument of the function. There must not be any spaces between different elements of a macro expression. For example, spaces between the $ sign and the open parentheses are not permitted. Similarly, there must not be any spaces between the comma and the argument, **dwgname**. All macro expressions must start with a **$** sign.

The following example illustrates the use of a macro expression using DIESEL to define and then assign a value to the **MODEMACRO** system variable.

Example 1

Using the AutoCAD MODEMACRO command, redefine the status line to display the following information in the status line:

Project name (Cust-Acad)
Name of the drawing (DEMO)
Name of the current layer (OBJ)

Note that in this example the project name is Cust-Acad, the drawing name is DEMO, and the current layer name is **OBJ**.

Before entering the **MODEMACRO** command, you need to determine how to retrieve the required information from the drawing database. For example, here the project name

DIESEL: A String Expression Language

(Cust-Acad) is a user-defined name that lets you know the name of the current project. This project name is not saved in the drawing database. The name of the drawing can be obtained using the DIESEL string function **GETVAR $(getvar,dwgname)**. Similarly, the **GETVAR** function can also be used to obtain the name of the current layer, **$(getvar,clayer)**. Once you determine how to retrieve the information from the system, you can use the **MODEMACRO** system variable to obtain the new status line. For Example 1, the following DIESEL expression will define the required status line.

    Command: **MODEMACRO**
    New value for MODEMACRO, or . for none<"">: **Cust-Acad**
    N:$(GETVAR,dwgname)
        L:$(GETVAR,clayer)

Explanation
Cust-Acad
Cust-Acad is assumed to be the project name you want to display in the status line.

N:$(GETVAR,dwgname)
Here, N: is used as an abbreviation for the drawing name. The GETVAR function retrieves the name of the drawing from the system variable **dwgname** and displays it in the status line, next to N:.

L:$(GETVAR,clayer)
Here L: is used as an abbreviation for the layer name. The GETVAR function retrieves the name of the current layer from the system variable **clayer** and displays it in the status line.

The new status line is shown in Figure 16-2(a).

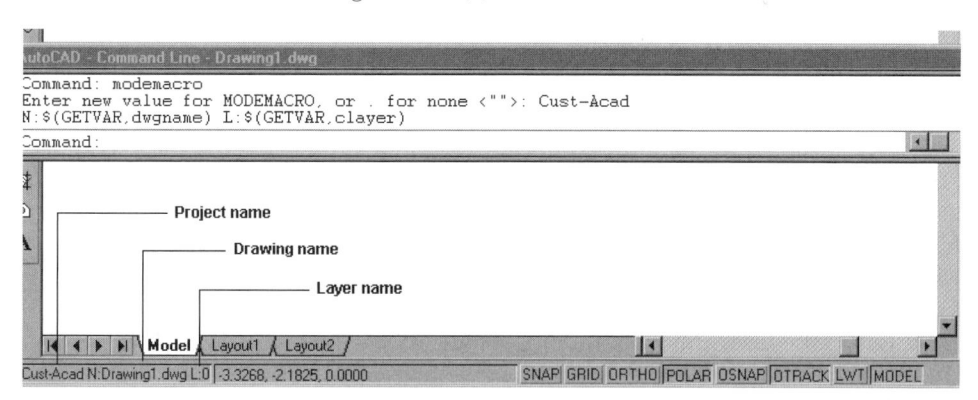

Figure 16-2(a) Status line for Example 1

Example 2

Using the AutoCAD **MODEMACRO** command, redefine the status line to display the following information in the status line, Figure 16-2(b):

    Name of the current textstyle
    Size of text

User-elapsed time in minutes

In this example the abbreviations for text style, text size, and user-elapsed time in minutes are TSTYLE:, TSIZE:, and ETM:, respectively.

Command: **MODEMACRO**
New value for MODEMACRO, or . for none<"">:
TSTYLE:$(GETVAR,TEXTSTYLE)TSIZE:$(GETVAR,TEXTSIZE)
ETM:$(FIX,$(*,60,$(*,24,$(GETVAR,TDUSRTIMER))))

Explanation
TSTYLE:$(GETVAR,TEXTSTYLE)
The **GETVAR** function obtains the name of the current textstyle from the system variable **TEXTSTYLE** and displays it next to TSTYLE: in the status line.

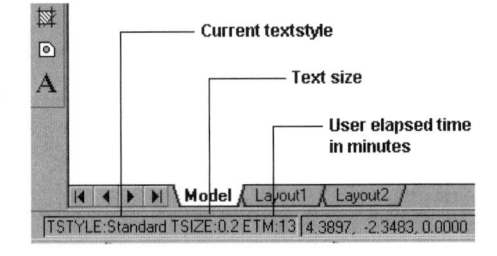

TSIZE:$(GETVAR,TEXTSIZE)
The **GETVAR** function obtains the current size of the text from the system variable **TEXTSIZE** and then displays it next to TSIZE: in the status line.

Figure 16-2(b) Status line for Example 2

ETM:$(FIX,$(*,60,$(*,24,$(GETVAR,TDUSRTIMER))))
The **GETVAR** function obtains the user-elapsed time from the system variable **TDUSRTIMER** in the following format:

<Number of days>.<Fraction>

Example
0.03206400 (time in days)

To change this time into minutes, multiply the value obtained from the system variable **TDUSRTIMER** by 24 to change it into hours, and then multiply the product by 60 to change the time into minutes. To express the minutes value without a decimal, determine the integer value using the DIESEL string function FIX.

Example
Assume that the value returned by the system variable TDUSRTIMER is 0.03206400. This time is in days. Use the following calculations to change the time into minutes, and then express the time as an integer:

0.03206400 days x 24 = 0.769536 hr
0.769536 hr x 60 = 46.17216 min
integer of 46.17216 min = 46 min

USING AUTOLISP WITH MODEMACRO

Sometimes the DIESEL expressions can be as long as those shown in Example 1 and Example 2. It takes time to type the DIESEL expression, and if you make a mistake in entering the

DIESEL: A String Expression Language 16-7

expression, you have to retype it. Also, if you need several different status line displays, you need to type them every time you want a new status line display. This can be time-consuming and sometimes confusing.

To make it convenient to change the status line display, you can use AutoLISP to write a DIESEL expression. It is easier to load an AutoLISP program, and it also eliminates any errors that might be caused by typing a DIESEL expression. The following example illustrates the use of AutoLISP to write a DIESEL expression to assign a new value to the **MODEMACRO** system variable.

Example 3

Using AutoLISP, redefine the value assigned to the **MODEMACRO** system variable to display the following information in the status line:

    Name of the current text style
    Size of text
    User-elapsed time in minutes

In this example the abbreviations for text style, text size, and user-elapsed time in minutes are TSTYLE:, TSIZE:, and ETM:, respectively.

The following file is a listing of the AutoLISP program for Example 3. The name of the file is **ETM.LSP**. The line numbers are not a part of the file; they are shown here for reference only.

```
(defun c:etm ( )                                    1
(setvar "MODEMACRO"                                 2
(strcat                                             3
   "TSTYLE:$(getvar,textstyle)"                     4
   " TSIZE:$(getvar,textsize)"                      5
   " ETM:$(fix,$(*,60,$(*,24,                       6
$(getvar,tdusrtimer))))"                            7
 )                                                  8
 )                                                  9
 )                                                 10
```

Explanation
Line 3
(strcat
The AutoLISP function **strcat** links the string value of lines 4 through 7 and returns a single string that becomes a DIESEL expression for the **MODEMACRO** command.

Line 4
"TSTYLE:$(getvar,textstyle)"
This line is a DIESEL expression in which getvar, a DIESEL string function, retrieves the value of the system variable **textstyle** and **$(getvar,textstyle)** is replaced by the name of the textstyle. For example, if the textstyle is STANDARD, the line will return "TSTYLE:STANDARD". This is

a string because it is enclosed in quotes.
Lines 6 and 7
" **ETM:$(fix,$(*,60,$(*,24,$(getvar,tdusrtimer))))"**
These two lines return **ETM:** and the time in minutes as a string. The **fix** is a DIESEL string function that changes a real number to an integer.

To load this AutoLISP file **(ETM.LSP)**, use the following commands. In this example, the file name and the function name are the same (ETM).

Command: **(load "ETM")**
ETM
Command: **ETM**

DIESEL EXPRESSIONS IN MENUS

You can also define a DIESEL expression in the screen, tablet, pull-down, or button menu. When you select the menu item, it will automatically assign the value to the MODEMACRO system variable and then display the new status line. The following example illustrates the use of the DIESEL expression in the screen menu.

Example 4

Write a DIESEL macro for the pull-down menu that displays the following information in the status line (Figure 16-3):

Macro-1	**Macro-2**	**Macro-3**
Project name	Pline width	Dimtad
Drawing name	Fillet radius	Dimtix
Current layer	Offset distance	Dimscale

The following file is a listing of the pull-down menu that contains the definition of three DIESEL macros for Example 4. This menu can be loaded using AutoCAD's MENU command and then entering the name of the file. If you select the first item, DIESEL1, it will display the new status line.

```
***MENUGROUP=MENU1
***POP1
[*DIESEL*]
[DIESEL1:]^C^CMODEMACRO;$M=Cust-Acad,N:$(GETVAR,DWGNAME)+
,L:$(GETVAR,CLAYER);
[DIESEL2:]^C^CMODEMACRO;$M=PLWID:$(GETVAR,PLINEWID),+
FRAD:$(GETVAR,FILLETRAD),OFFSET:$(GETVAR,OFFSETDIST),+
LTSCALE:$(GETVAR,LTSCALE);
[DIESEL3:]^C^CMODEMACRO;$M=DTAD:$(GETVAR,DIMTAD),+
DTIX:$(GETVAR,DIMTIX),DSCALE:$(GETVAR,DIMSCALE);
```

DIESEL: A String Expression Language

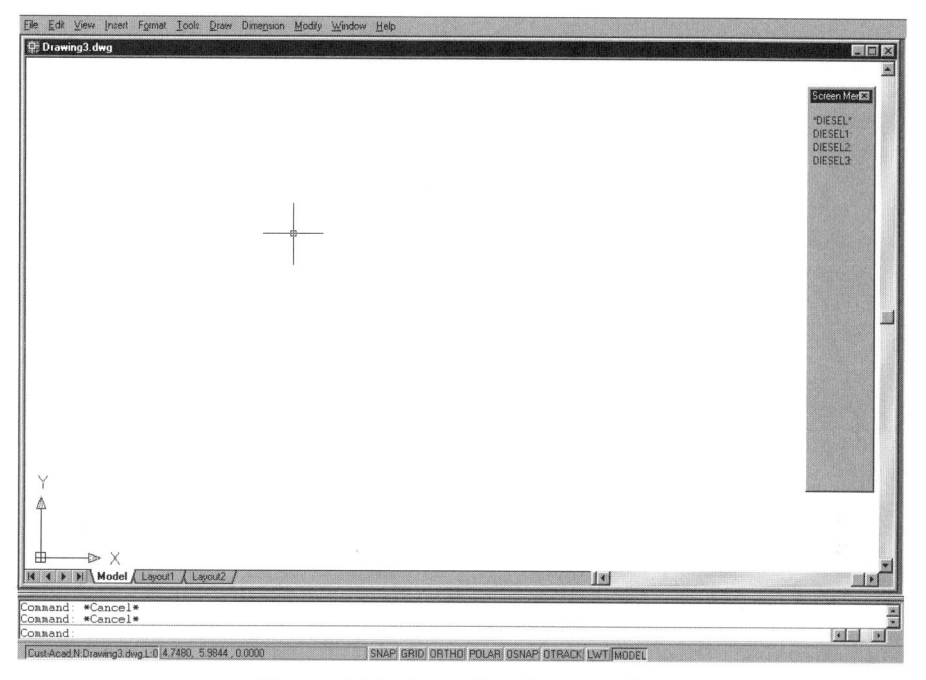

Figure 16-3 Status line for Example 4

MACROTRACE SYSTEM VARIABLE

The MACROTRACE variable is an AutoCAD system variable that can be used to debug a DIESEL expression. The default value of this variable is 0 (off). It is turned on by assigning a value of 1 (on). When on, the MACROTRACE system variable will evaluate all DIESEL expressions and display the results in the command prompt area. For example, if you have defined several DIESEL expressions in a drawing session, all of them will be evaluated at the same time and the messages, if any, will be displayed in the command prompt area.

Example
Command: **MODEMACRO**
New value for MODEMACRO, or . for none<"">: **$(getvar,dwgname),$(getvar clayer)**

Note that in this DIESEL expression a comma is missing between getvar and clayer. If the MACROTRACE system variable is on, the following message will be displayed in the Command prompt area:

Eval: $(GETVAR,DWGNAME)
=====>UNNAMED
Eval: $(GETVAR CLAYER)
Err: $(GETVAR CLAYER)??

This error message gives you an idea about the location of the error in a DIESEL expression.

In the previous example, the first part of the expression successfully returns the name of the drawing (unnamed) and the second part of the expression results in the error message. This confirms that there is an error in the second part of the DIESEL expression. You can further determine the cause of the error by comparing it with the error messages in the following table:

Error Message	Description
$?	Syntax error
$?(func,??)	Incorrect argument to function
$(func)??	Unknown function
$(++)	Output string too long

DIESEL STRING FUNCTIONS

Like AutoLISP, you can use DIESEL functions to do some mathematical operations, retrieve the values from the drawing database, and display the values in the status line in a predetermined order. For example, you can add, subtract, multiply, and divide the numbers. You can also obtain the values of some system variables and display them in the status line. The maximum number of parameters that the DIESEL expression can contain is 10. This number includes the name of the function. The following section discusses some frequently used DIESEL string functions.

Addition

Format **$(+,num1,num2,num3 - - -)**

This function (+) calculates the sum of the numbers that are to the right of the plus (+) sign. The numbers can be integers or real.

Examples
$(+,2,5)   returns 7
$(+,2,5,50.75)   returns 57.75

Note
You can test the calculations by assigning the DIESEL expression to the MODEMACRO system variable.

Command: **MODEMACRO**
New value for MODEMACRO, or . for none<"">: **$(+,2,5)**

AutoCAD will return a value of 7, which will be displayed in the status line. It is important to note that the values returned by the DIESEL string expressions are string values.

Subtraction

Format **$(-,num1,num2,num3,- - -)**

DIESEL: A String Expression Language

This function (-) subtracts the second number from the first number. If there are more than two numbers, the second and the subsequent numbers are added and their sum is subtracted from the first number.

Examples
$(-,28,14)   returns 14
$(-,25,7,11.5)   returns 6.5

Multiplication

Format **$(*,num1,num2,num3,- - -)**

This function (*) calculates the product of the numbers that are to the right of the asterisk.

Examples
$(*,2,5)   returns 10
$(*,2,5,3.0)   returns 30.0
$(*,2,5,3.25)   returns 32.5

Division

Format **$(/,num1,num2,num3 - - -)**

This function (/) divides the first number by the second number. If there are more than two numbers, the first number is divided by the product of the second and subsequent numbers.

Examples
(/ 3 2)   returns 1.5
(/ 3.0 2)   returns 1.5
(/ 200 5 4)   returns 10
(/ 200.0 5.5)   returns 36.363636
(/ 200 -5)   returns -40
(/ -200 -5.0)   returns 40.0

Relational Statements

Some DIESEL expressions involve features that test a particular condition. If the condition is true the expression performs a certain function; if the condition is not true, the expression performs another function. The following section discusses various relational statements used in DIESEL expressions.

Equal to

Format **$(=,num1,num2)**

This function (=) checks whether the two numbers are equal. If they are, the condition is true and the function will return **1**. Similarly, if the specified numbers are not equal, the condition is false and the function will return **0**.

Examples
$(=,5,5)	returns 1
$(=,5,4.9)	returns 0
$(=,5,-5)	returns 0

Not equal to

Format **$(!=,num1,num2)**

This function (!=) checks whether the two numbers are **not equal**. If they are not equal, the condition is true and the function will return **1**; otherwise the function will return **0**.

Examples
$(!=,50,4)	returns 1
$(!=,50,50)	returns 0
$(!=,50,-50)	returns 1

Less than

Format **$(<,num1,num2)**

This function (<) checks whether the first number (**num1**) is less than the second number (**num2**). If it is true, the function will return **1**; otherwise the function will return **0**.

Examples
$(<,3,5)	returns 1
$(<,5,3)	returns 0
$(<,3.0,5)	returns 1

Less than or equal to

Format **$(<=,num1,num2)**

This function (<=) checks whether the first number (**num1**) is less than or equal to the second number (**num2**). If it is true, the function will return **T**; otherwise the function will return **nil**.

Examples
$(<=,10,15)	returns 1
$(<=,19,10)	returns 0
$(<=,-2.0,0)	returns 1

DIESEL: A String Expression Language

Greater than

Format **$(>,num1,num2)**

This function (**>**) checks whether the first number (**num1**) is greater than the second number (**num2**). If it is true, the function will return **1**; otherwise the function will return **0**.

Examples
$(>,15,10)   returns 1
$(>,20,30)   returns 0

Greater than or equal to

Format **$(>=,num1,num2)**

This function (**>=**) checks whether the first number (**num1**) is greater than or equal to the second number (**num2**). If it is true, the function will return **1**; otherwise the function will return **0**.

Examples
$(>=,78,50)   returns 1
$(>=,78,88)   returns 0

eq function

Format **$(eq,value1,value2)**

This function (**eq**) checks whether the two string values are equal. If they are, the condition is true and the function will return **1**; otherwise the function will return **0**.

Examples
$(eq,5,5)   returns 1
$(eq,yes,yes)   returns 1
$(eq,yes,no)   returns 0

angtos

Format **$(angtos,angle[,mode,precision])**

The angtos function returns the angle expressed in radians in a string format. The format of the string is controlled by the mode and precision settings.

Examples
$(angtos,0.588003,0,4)   returns 33.6901
$(angtos,-1.5708,1,2)   returns 270d0'

Note
The following modes are available in AutoCAD.

ANGTOS MODE	EDITING FORMAT
0	Decimal degrees
1	Degrees/minutes/seconds
2	Grads
3	Radian
4	Surveyor's units

*Precision is an integer number that controls the number of decimal places. Precision corresponds to the AutoCAD system variable, **AUPREC**. The minimum value of **precision** is 0, the maximum is 4.*

eval function

Format **$(eval,string)**

The **eval** function passes the string to the DIESEL evaluator and the result obtained after evaluating the string is returned and displayed in the status line.

Examples
$(eval,welcome)	returns welcome
$(eval,$(getvar,dimscale))	returns value of dimscale

fix function

Format **$(fix,num)**

The **fix** function converts a real number into an integer by truncating the digits after the decimal.

Examples
$(fix,42.573)	returns 42
$(fix,-23.50)	returns -23

getvar function

Format **$(getvar,varname)**

The **getvar** function retrieves the value of an AutoCAD system variable.

Examples
$(getvar,dimtad)	returns value of dimtad
$(getvar,clayer)	returns current layer name

rtos function

Format   **$(rtos,number)**
or   **$(rtos,number,mode,precision)**

The **rtos** function changes a given number into a real number and the format of the real number is determined by the mode and precision values.

Examples
$(rtos,50)   returns 50
$(rtos,1.5,5,4)   returns 1 1/2

Note
The following linear unit modes are available in AutoCAD:

MODE VALUE	STRING FORMAT
1	Scientific
2	Decimal
3	Engineering
4	Architectural
5	Fractional

Precision *is an integer number that controls the number of decimal places. Precision corresponds to the AutoCAD system variable, LUPREC, and the mode corresponds to LUNITS. If the mode and precision values are omitted, AutoCAD uses the current values of LUNITS and LUPREC as set with the units command.*

if function

Format **$(if,condition,then,else)**

The **if** function evaluates the first expression (then), if the value returned by the specified condition is non zero. The second expression (else) is evaluated if the specified condition returns 0.

(if condition then [else])
    Where **condition** ------------------ Specified conditional statement
                then -------------------- Expression evaluated if the condition returns non zero
                else -------------------- Expression evaluated if the condition returns 0

Examples
$(if,$(=,7,7),true)   returns true
$(if,$(=,5,7),true,false)   returns false
$(if,1.5,true,false)   returns true

strlen function

Format **$(strlen,string)**

The **strlen** function returns an integer number that designates the number of characters contained in the specified string.

Example
$(strlen,Customizing AutoCAD)   returns 19

upper function

Format **$(upper,string)**

The **upper** function returns the specified string in uppercase.

Example
$(upper,Customizing)   returns CUSTOMIZING

edtime function

Format **$(edtime,date,display-format)**

The **edtime** function can be used to edit the date and display it in a format specified by display-format. For example, the Julian date obtained from AutoCAD's system variable, DATE, is obtained in the form 2449013.85156759. The **edtime** function can be used to display this date in an understandable form. The following table gives the format of the phrases that can be used with the **edtime** function:

FORMAT	OUTPUT	FORMAT	OUTPUT
D	5	H	2
DD	05	HH	02
DDD	Tue	MM	23
DDDD	Tuesday	SS	12
M	1	MSEC	325
MO	11	AM/PM	PM
MON	Nov	am/pm	am
MONTH	November	A/P	P
YY	92	a/p	p
YYYY	1992		

Example
$(edtime,$(getvar,date),DDDD","DD MONTH YY - HH:MMAM/PM)
                                  returns Monday,25 January 93 - 08:52PM

In the above example, the getvar function retrieves the date in Julian format. The edtime function displays the date as specified by the display-format. The following illustration shows the corresponding fields of the display-format and the value returned by the edtime function:

DDDD","DD MONTH YY - HH:MMAM/PM

Where **DDDD** -------------------- Monday
, -------------------- ,
DD -------------------- 25
MONTH -------------------- January
YY -------------------- 93
- -------------------- -
HH -------------------- 08
MMAM -------------------- 52
PM -------------------- PM

Review Questions

Answer the following questions.

1. DIESEL (direct interpretively evaluated string expression language) is a string expression language. (T/F)

2. The value assigned to the **MODEMACRO** variable is a string, and the output it generates is not a string. (T/F)

3. You cannot define a DIESEL expression in the screen menu. (T/F)

4. The coordinate information displayed in the status line can be dynamic only. (T/F)

5. Once the value of the **MODEMACRO** variable is changed, it retains that value until you enter a new value, start a new drawing, or open an existing drawing file. (T/F)

6. The number of characters that can be displayed in the layer mode field on any system is 38. (T/F)

7. The coordinate display field cannot be changed or edited. (T/F)

8. You can write a macro expression using DIESEL to assign a value to the **MODEMACRO** system variable. (T/F)

9. In DIESEL, the drawing name can be obtained by using the macro expression **$(getvar,dwgname)**. (T/F)

10. You cannot use AutoLISP to write a DIESEL expression. (T/F)

Fill in the blanks

11. All macro expressions must start with a _____ sign.

12. The display-format DDD will return Monday as _____.

13. The DIESEL expression, $(upper,AutoCAD), will return _____.

14. The DIESEL expression, $(strlen,AutoCAD), will return _____.

15. The DIESEL expression, $(if,$(=,3,2),yes,no), will return _____.

16. The DIESEL expression, $(fix,-17.75), will return _____.

17. The DIESEL expression, $(eq,Customizing,customizing), will return _____.

18. The DIESEL expression, $(/,81,9,9), will return _____.

19. The MACROTRACE system variable can be used to _____ a DIESEL expression.

Exercises

Exercise 1 — *General*

Using the AutoCAD **MODEMACRO** command, redefine the status line to display the following information in the status line:

    Your name
    Name of drawing

Exercise 2 — *General*

Using the AutoCAD **MODEMACRO** command, redefine the status line to display the following information in the status line:

    Name of the current dimension style
    Dimension scale factor (dimscale)
    User-elapsed time in hours

The abbreviations for dimstyle, dimscale, and user-elapsed time in hours are DIMS:, DIMFAC:, and ETH:, respectively.

Exercise 3   *General*

Using **AutoLISP**, redefine the status line to display the following information in the status line:

   Name of the current dimension style
   Dimension scale factor (dimscale)
   User-elapsed time in hours

Chapter 17

Visual Basic

Learning Objectives

After completing this chapter, you will be able to:
* *Load and run sample VBA projects.*
* *Utilize the Visual Basic Editor.*
* *Understand and use AutoCAD objects.*
* *Use object properties.*
* *Apply and use AutoCAD methods.*

ABOUT VISUAL BASIC

The original BASIC was developed in 1963. BASIC (Beginners All-purpose Symbolic Instruction Code) was intended to be an accessible, user-friendly programming language. BASIC was the language supplied with many of the early microcomputers starting in the late 1970s. It continued development until the advent of Windows. In 1991 the first version of **Visual Basic (VB)** appeared, followed by new versions spaced about a year apart. The current version 6, like its predecessors, lets even a beginning programmer create a user-friendly graphical user interface (GUI) with a small fraction of the effort required previously. It used to take a team of programmers working with a language like C, which came with a stack of reference materials nearly a foot thick, to do the same thing.

Another advantage inherent to Visual Basic is that engineers may have old BASIC programs used in sizing components. This code, minus old input/output, could be used as the core of new Visual Basic modules to parametrically design and draft the component

Autodesk first licensed **Visual Basic® for Applications (VBA)** from Microsoft for use in R14. Dozens of other large software companies have similarly adopted VBA. The VBA language tools are either identical or very similar to those in the stand-alone Visual Basic. Programmers who know how to write macros for Microsoft Office only have to learn AutoCAD-specific functionality to program VBA in AutoCAD.

A capability accompanying VBA is the ability to communicate with other applications such as Microsoft Excel, Microsoft Word, Microsoft Access, and Visual Basic using **ActiveX Automation**. Automation lets you access and manipulate the objects and functionality of AutoCAD from Excel or another application supporting ActiveX. Alternately the objects and functionality of Excel, for example, can be used inside AutoCAD VBA programming. This cross-application macro programming capability does not exist in AutoLISP. Of course, before you can use AutoCAD's objects in some other software supporting ActiveX, you must make the application aware that AutoCAD is installed along with its **Object Library** on the computer.

There are millions of programmers using Visual Basic. A great many more people have programmed in BASIC but not VB. This chapter assumes you have some familiarity with BASIC or programming concepts such as those covered in the AutoLISP chapter of this book and are comfortable looking up commands and syntax in the help system.

In this version of AutoCAD you can now reference a macro from another project and have more than one macro loaded. You can also create libraries of common functions and macros. The help system continues to improve.

OBJECTS

Visual Basic uses **objects** to facilitate what it can accomplish. Examples of objects include drawings or documents, geometric elements such as lines or circles, and user interface controls to handle input and output to programs or macros. The "visual" part of Visual Basic refers to familiar user interface controls such as check boxes, scroll bars, and command buttons as used to open or save files in Windows applications. These controls are screen objects that may be dragged from a "Toolbox" onto the background object called a form. An example of a simple form is the background window of a dialog box. In the course of dragging a control from the Toolbox onto a form, the coding that provides the functionality of the control is simultaneously added to the form object. Several aspects of true object-oriented programming (OOP) languages such as C++ are missing from Visual Basic. So VB is considered an object-based, rather than an object-oriented, programming language. The missing aspects are more than made up for by the development tools and environment available to the user.

Functions known as **methods** have been defined in the AutoCAD object library to perform an action on an object, for example, drawing a line in a drawing. The method AddLine adds a line object to a drawing. **Properties** are functions that set or return information about the state of an object. In the example of a line drawn in a drawing the Color, Layer, Linetype, and Start X would be some of the properties of the Line object. AutoCAD users are familiar with the drawing editor general toolbar icon called Properties that also can be accessed from the Modify pull down menu.

ADD METHOD

The way to draw in paper space, model space, or in a block is to use the **Add method** such as **AddCircle**, **AddLine**, **AddArc**, and **AddText**. Instead of thinking of drawing in the model space of the current drawing, think in terms of adding a geometric object. **Dot notation** is required to clarify on which object the method is acting. The dot notation reference starts with the most global object first narrowing to the right. Dot notation is used in a similar way with properties.

AddCircle

The **AddCircle** method requires a predefined center point and radius. Both of these arguments are required. Other input options used in the AutoCAD circle command such as three-points, two-points, or tangent-tangent-radius require additional programming in VBA to calculate the center point and radius for the AddCircle method. The format of the **AddCircle** method is:

ThisDrawing.ModelSpace.AddCircle centerpoint, radius
or
Set Circle1=ThisDrawing.ModelSpace.AddCircle(centerpoint, radius)
    where   **centerpoint:**   Center point, double precision vector
            radius: Radius, double precision number

Note
The arguments of the first form of the AddCircle method follow a required space. The second form of the add method, where the arguments are inside parentheses, is used where the circle object is assigned to a name for use later in the program. For the examples below, point1 and point2 are predefined points consisting of a vector of variant/double precision or double precision coordinate values. These may be defined with assignment statements as shown in Example 1 on page 17-8. The pound sign (#) signifies a double precision number in Basic.

Examples
ThisDrawing.ModelSpace.AddCircle point1, 3#
Set Circle2= ThisDrawing.ModelSpace.AddCircle(point2, 4#)

AddLine

The **AddLine method** requires two predefined endpoints. The **AddLine** method has a syntax similar to AddCircle:

ThisDrawing.ModelSpace.AddLine firstpoint, secondpoint
    where   **firstpoint**:   First point, double precision vector
            secondpoint: Second point, double precision vector

Examples
ThisDrawing.ModelSpace.AddLine point1, point2
Set Line1=ThisDrawing.ModelSpace.AddLine(point1, point2)

Note
The second form of the add method, where the arguments are inside parentheses, may be used for the AddLine method or any of the other Add methods below where the object is assigned to a name for use later in the program. A complication to using the Add method this way is that the object defined must be declared in a Dim statement in the General declarations.

AddArc

The **AddArc** method format is:

ThisDrawing.ModelSpace.AddArc ctrpt, radius, StartAng, EndAng
    where **ctrpt**:   Center point, double precision vector
               radius: Radius, double precision number
               StartAng: Arc starting angle in radians, double precision
               EndAng: Arc ending angle in radians, double precision

Example
ThisDrawing.ModelSpace.AddArc point1, 4#, 0#, 1.570796327

AddText

The **AddText** method requires a predefined string, insertion point and text height. The **AddText** method syntax is:

ThisDrawing.ModelSpace.AddText textString$, point1, textHeight
    where **textString$:** Actual text to be displayed
               point1:   Position point, double precision vector
               textHeight: Text Height, positive double precision number

Example
ThisDrawing.ModelSpace.AddText ".063 TYP, 4 PLACES", point1, 0.25#

FINDING HELP ON METHODS AND PROPERTIES

You will find excellent help in the VBA Integrated Development Environment with good examples showing the exact syntax necessary for the AddLine, AddCircle, or any of the particular methods or properties needed to write a parametric program. You can also find general information such as on Methods, help on Visual Basic key words, and many non-AutoCAD VBA programming examples taken from Excel, Word, or PowerPoint in AutoCAD 2002. The topics under "AutoCAD Help" from the Help pull-down menu include "ActiveX and VBA Developers Guide," shown in Figure 17-1.

Help is available through a shortcut from the AutoCAD drawing editor's help system or from the Visual Basic Editor Integrated Design Environment (IDE). Look under VBA and ActiveX Automation. The Index and Find tabs of this help menu are very useful. Another way to get help in the VBA IDE, is to use the **Object Browser** which can be accessed from the IDE **View** pull-down menu. Select **AcadProject** in the drop-down list box. Select object **ThisDrawing** in the left window and **ModelSpace** in the right one, as shown in Figure 17-2. The question mark help button on the toolbar or **F1** takes you to a screen where you can access the ModelSpace Collection of help screens on methods, properties, and examples. Alternately, the function key **F1** will get help on any word typed in the code window.

Visual Basic   17-5

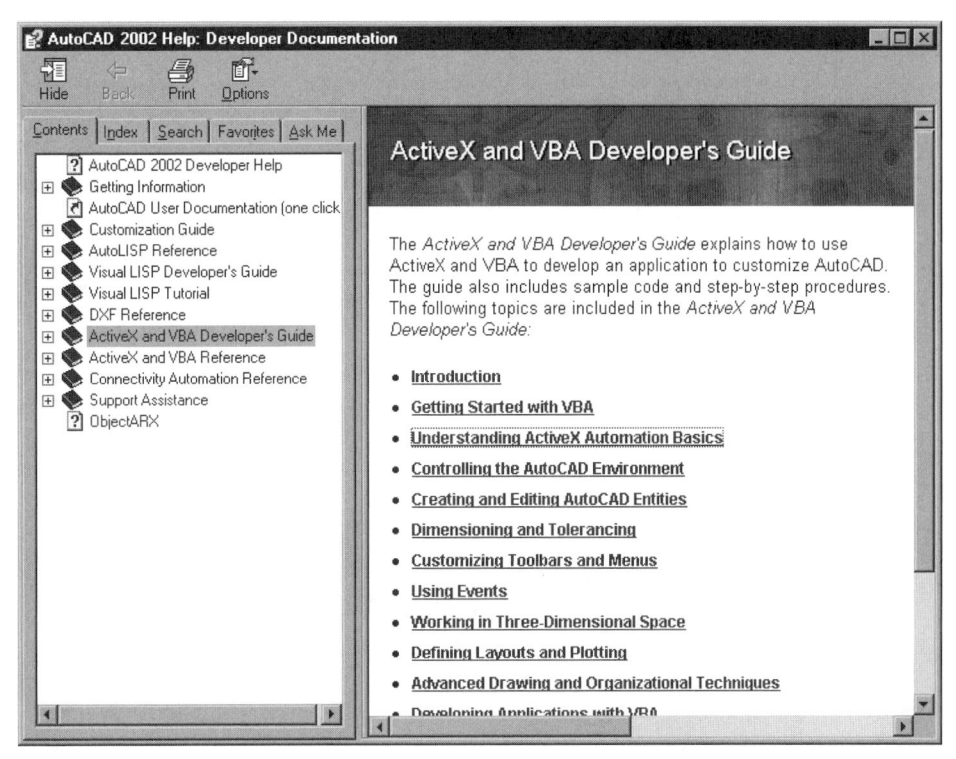

Figure 17-1 ActiveX and VBA Developer's Guide

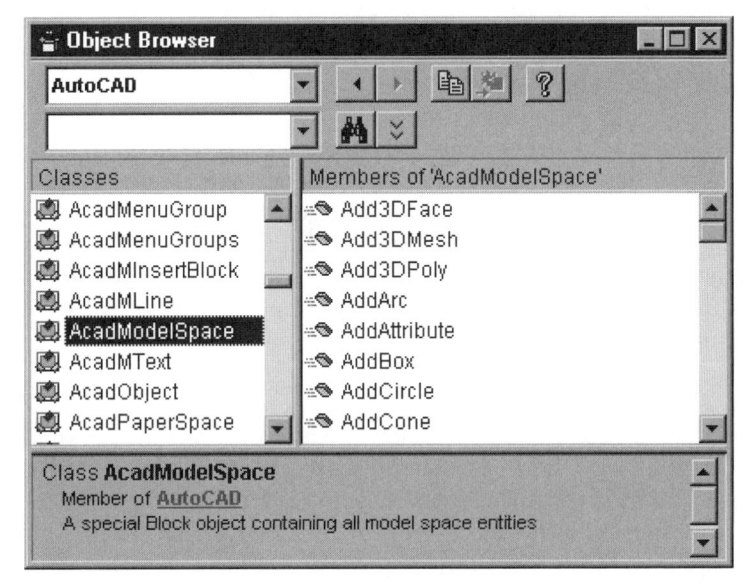

Figure 17-2 Object Browser

Loading and Saving VBA Projects

To start or load a VBA project in the AutoCAD Drawing Editor, choose **Macro > Visual Basic Editor** from the **Tools** menu, see Figure 17-3.

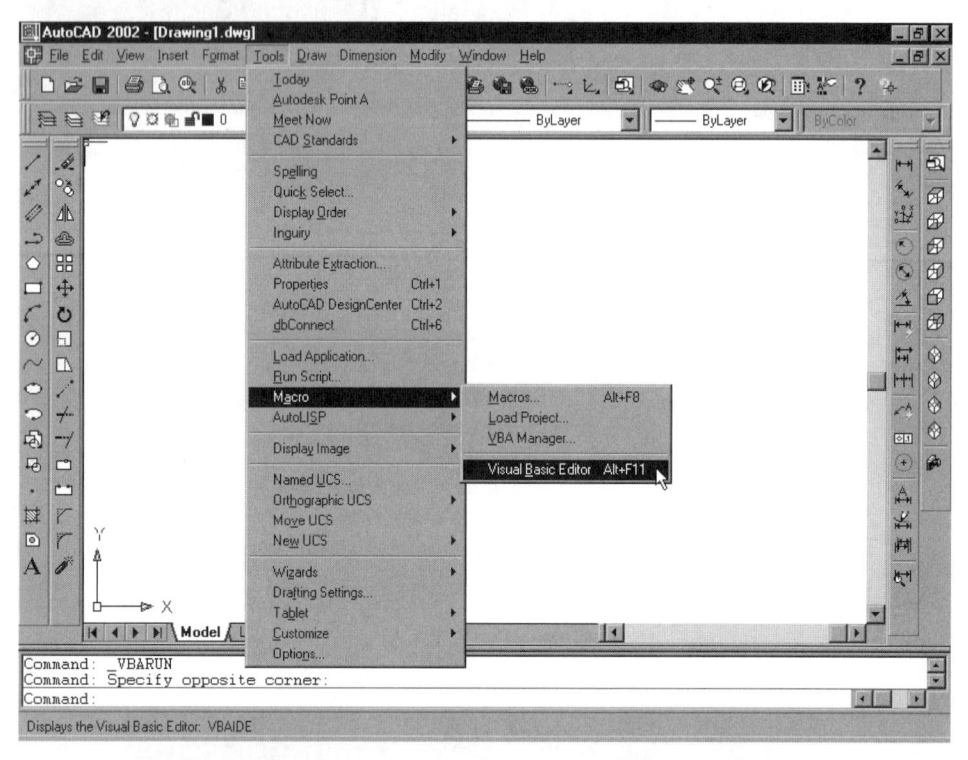

*Figure 17-3 Starting the Visual Basic Editor using the **Tools** menu*

You will enter the **Integrated Development Environment (IDE)**. The IDE consists of a number of useful windows that can be sized, shown, hidden, or otherwise customized to suit your needs. Figure 17-4 shows the IDE with the VBA project for Example 1 loaded. There are six windows shown. The top left window is the Project Explorer, which allows navigation among windows. Note the View Code, View Object, and Toggle Folders Icons at the top of the window. The bottom left menu is the Properties Window, which shows object properties and is a convenient way to change them.

The top center window is the UserForm Window where the "Visual" user interface is constructed using the Toolbox. The window below it is the UserForm code window where Basic language program code resides, which runs when triggered by events such as a mouse click on a control button. The next window down is the Module code window, where mathematical functions and procedures with the .bas extension are generally kept. The Immediate and the Watch Windows are used for debugging.

It is a good idea to save your VBA program before testing it. To save a project from the VBA Integrated Development Environment choose **Save** from the **File** menu. Menu items may be more quickly accessed with the combination of the Alt key and the underlined letter in the

Visual Basic 17-7

menu. AutoCAD VBA projects have the file extension DVB.

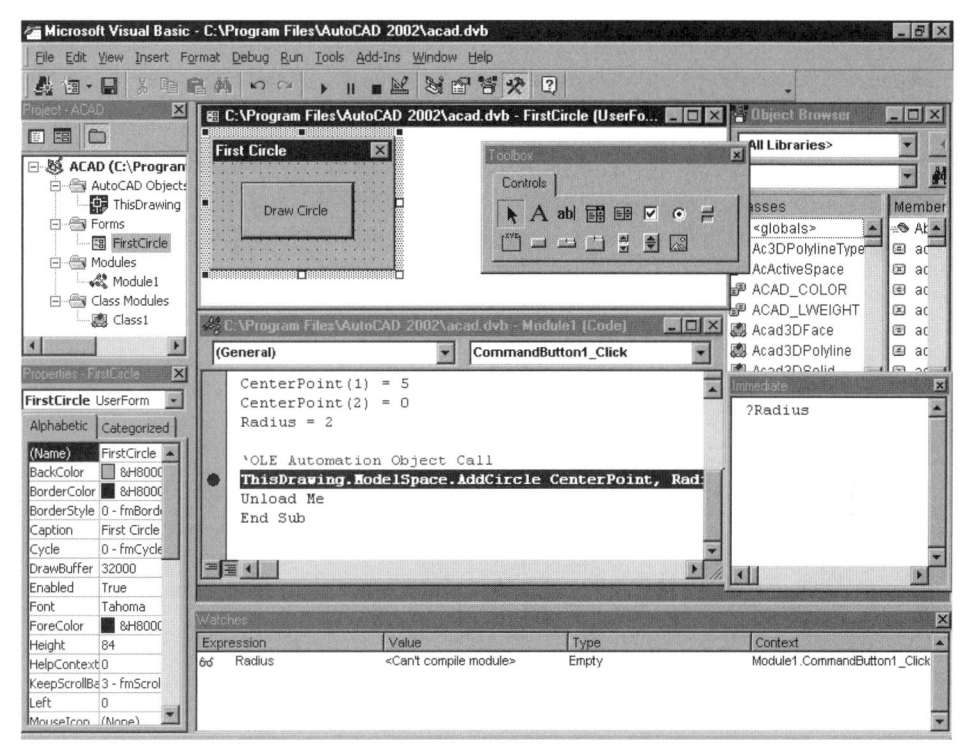

Figure 17-4 IDE with the VBA project for Example 1 loaded

Example 1

Write a program that will draw a circle centered at 5,5,0 with radius 2, as shown in Figure 17-5. The User Interface Form is shown in Figure 17-6.

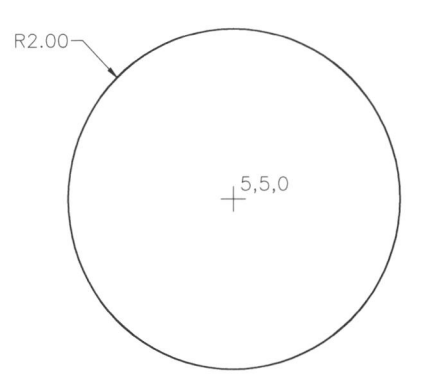

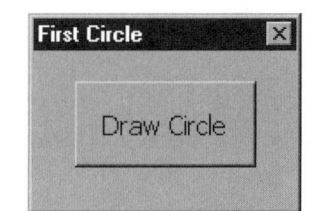

Figure 17-5 Circle for Example 1   *Figure 17-6 User Interface Form for Example 1*

Insert a Module and a UserForm by choosing **Module** and **UserForm** from the **Insert** menu, respectively. The default names will be Module1 and UserForm1. Click the form after insertion to make it the active window. To insert a control such as a command button onto the form, you choose the corresponding button on the toolbox. The symbol shown as the second button from the left in the second row of the toolbox in Figure 17-7 inserts the command button control.

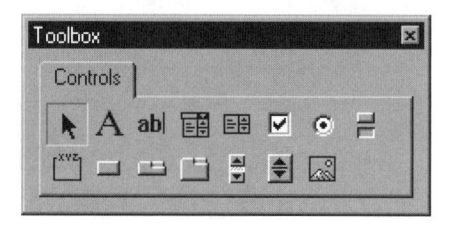

*Figure 17-7 The **Toolbox** for creating the Screen Interface Form*

Now, resize the UserForm and Command Button to the desired size and edit the caption on the button. This can also be done using the **Properties** window at the bottom left corner of screen, Figure 17-4. The name of the button should be **Draw Circle**.

A **project** is the name given to the forms, controls, modules, and programming making up a Visual Basic program. To finish a project it is necessary to write the code underlying the visual control(s). One way to get to the code window for the CommandButton control is to double-click on the button and enter the code for CommandButton1_Click(). Visual Basic is **event driven**. Lines 1 through 17 correspond to the code that runs when the command button is chosen with the mouse.

The following file is a listing of the Visual Basic program for Example 1. **The line numbers at the right are for reference only and are not a part of the programming.**

Command Button Code

```
'UserForm1 Code to Draw Circle, Radius 2                        1
'Centered at 5,5,0. Trigger is mouse click on                   2
'command button marked Draw Circle                              3
Private Sub CommandButton1_Click()                              4
Dim CenterPoint(0 To 2) As Double                               5
Dim Radius As Double                                            6

'Data                                                           7
CenterPoint(0) = 5                                              8
CenterPoint(1) = 5                                              9
CenterPoint(2) = 0                                             10
Radius = 2                                                     11
                                                               12
'OLE Automation Object Call                                    13
ThisDrawing.ModelSpace.AddCircle CenterPoint, Radius           14
Unload Me                                                      15
End Sub                                                        16
```

Explanation

Lines 1 to 3
The first lines are comments or remarks describing the function of the program. Comments make understanding and modifying a program easier and should be used liberally. Comments start with a Rem or use an apostrophe ('). These lines are ignored when the program is run.

Line 4
Private Sub CommandButton1_Click()
This line defines where Sub CommandButton1_Click() starts. The subroutine is executed when CommandButton1 is clicked with the mouse. It is the code that draws the circle when the command button is clicked. There is no need to type this line as VBA generates it automatically as soon as the command button control is added to the form.

Lines 5 and 6
Dim CenterPoint(0 To 2) As Double
Dim Radius As Double
These two lines are necessary to establish the double precision variable type for the arguments needed by the AddCircle method. The default type is Variant, which will not work with AddCircle.

Lines 8 through 11
CenterPoint(0) = 5
CenterPoint(1) = 5
CenterPoint(2) = 0
Radius = 2
Here the circle center and radius are assigned values.

Line 14
ThisDrawing.ModelSpace.AddCircle CenterPoint, Radius
This line applies the AddCircle method to the ModelSpace object, which is part of the ThisDrawing object. Note the required space between the key word AddCircle and the first argument, CenterPoint.

Line 15
Unload Me
This is a method that removes UserForm1 from memory and returns the focus back to AutoCAD.

Line 16
End Sub
Like line 4, this line is generated automatically.

Module1 Code

The following code is entered in the Module1 code window.

```
'Module 1 General Declarations                          1
Sub DrawCircle()                                        2
UserForm1.Show                                          3
End Sub                                                 4
```

Explanation

Line 1 through 4
Sub DrawCircle()
UserForm1.Show
End Sub
This subroutine's function is to create a Macro name, DrawCircle, which shows up in the lower list box of the **Macro** dialog box invoked by choosing **Macro > Macros** from the **Tools** menu. These four lines of code belong to the Module1 object. Another way to run a project is to choose **Run Sub/UserForm** from the **Run** menu in the VBA IDE.

GETPOINT, GETDISTANCE, AND GETANGLE METHODS
GetPoint Method

The **GetPoint** method allows you to enter the X, Y coordinates or X, Y, Z coordinates of a point. The coordinates of the point can be entered from the keyboard or by using the screen cursor. The format of the **GetPoint** method is:

    P = ThisDrawing.Utility.GetPoint([Point], [Prompt])
    Enter a point from the keyboard or select a point in the AutoCAD graphics editor
    where   **[Point]:**   Optional reference point, rubber-band origin
                [Prompt]: Optional prompt to be displayed on screen

    Example
    pnt1=ThisDrawing.Utility.GetPoint("Enter 1st Point")
    Pt2=ThisDrawing.Utility.GetPoint(Pnt1,"Enter 2nd Point")

GetDistance Method

The **GetDistance** method lets you enter a distance on the command line, a distance from a given point, or two points, and it then returns the distance as a double precision number. The format of the **GetDistance** method is:

    d = ThisDrawing.Utility.GetDistance([point], [prompt])
    where   **[point]**:   Optional reference point, rubber band origin
                [prompt]: Optional prompt to be displayed on screen

GetAngle Method

The **GetAngle** method allows you to enter an angle, either from the keyboard in degrees or by selecting two points. In the case of selecting points, the positive horizontal direction is taken as one leg of the angle, the first point selected as the vertex and the second point defines the second leg. If the point argument is specified, AutoCAD uses this point as the first point or angle vertex. The GetAngle method returns the value of the angle in radians as a double

Visual Basic 17-11

precision value. The format of the **GetAngle** method is:

ang = ThisDrawing.Utility.GetAngle([point], [prompt])
    where **ang**:   Angle in radians
                [point]:   Optional vertex point
                [prompt]:   Optional screen prompt to clarify angle selection

Examples
a1 = ThisDrawing.Utility.GetAngle(, "Enter taper angle in degrees")
ang = ThisDrawing.Utility.GetAngle(pt1) 'pt1 is a predefined point
ThisDrawing.Utility.GetAngle pt1, "Enter second point of angle"

Note
*The angle you enter is affected by the angle setting. The angle settings can be changed by changing the value of the AutoCAD system variables **ANGBASE** and **ANGDIR**. The default settings for measuring an angle are as follows:*

*The angle is measured with respect to the positive X-Axis (3 o'clock position). The value for 3 o'clock corresponds to the current value of **ANGBASE**, the AutoCAD system variable being 0. **ANGBASE** could be set in any of four 90-degree quadrant directions*

*The angle is positive if it is measured in the counterclockwise direction and negative if it is measured in the clockwise direction. The value of this setting is saved in the AutoCAD system variable **ANGDIR**. The **GetOrientation** method has the same syntax as the GetAngle method but ignores **ANGBASE** and **ANGDIR** system variables. The 0 angle is always at 3 o'clock, and angles are always positive counterclockwise.*

Example 2

Write a program that will draw a triangle with user supplied vertices P1, P2, and P3 as in Figure 17-8. This program is to use the GetPoint and AddLine methods. The User Interface Form for this example is shown in Figure 17-9

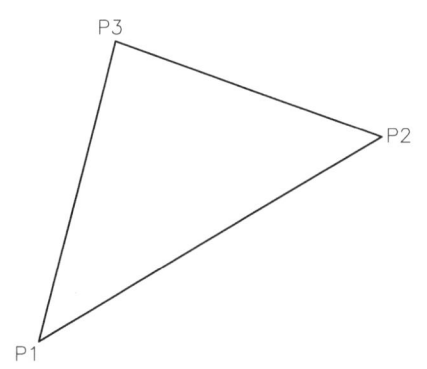

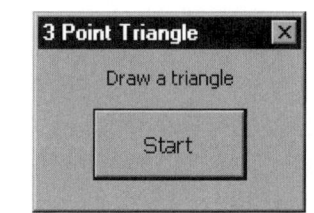

Figure 17-8 *Triangle with user-defined points*   ***Figure 17-9*** *User Interface Form for Example 2*

Choose **Module** and **UserForm** from the **Insert** menu to insert a Module and a UserForm respectively. The **Command** button is inserted or dragged over from the toolbar to the UserForm with the mouse to produce an interface form similar to Figure 17-9. The label above the command button is similarly inserted after clicking on the Toolbox Label icon thet looks like a capital A and the caption typed in. The following file is a listing of the project for Example 2. **The line numbers at the right are for reference only and are not a part of the program**.

Command Button Code

```
'The function of this routine is to draw                              1
'a triangle from 3 user specified points                              2
'The trigger is a mouse click on the command                          3
'button labeled Start.                                                4
'pnt1, pnt2, pnt3 are variant by default                              5
'Returns a point in WCS                                               6
Private Sub CommandButton1_Click()                                    7
UserForm2.Hide                                                        8
pnt1 = ThisDrawing.Utility.GetPoint(, "Provide the First Point: ")    9
'Returns a variant vector since GetPoint returns a point in WCS      10
'And draws a rubber-band line from the optional first point          11
pnt2 = ThisDrawing.Utility.GetPoint(pnt1, "Second Point? ")          12
'Draw first side of triangle                                         13
ThisDrawing.ModelSpace.AddLine pnt1, pnt2                            14
pnt3 = ThisDrawing.Utility.GetPoint(pnt2, "3rd Point? ")             15
ThisDrawing.ModelSpace.AddLine pnt2, pnt3                            16
ThisDrawing.ModelSpace.AddLine pnt3, pnt1                            17
Unload Me                                                            18
End Sub                                                              19
```

Explanation

Lines 1-6, 10, 11, and 13
These lines are comments or remarks describing the function of the following line or lines. Comments make understanding and modifying a program easier and should be used liberally. Comments start with a Rem or use an apostrophe ('). These lines are ignored when the program is run.

Line 7
Private Sub CommandButton1_Click()
This line defines where Sub CommandButton1_Click() starts. The subroutine code is executed when CommandButton1 is chosen with the mouse.

Line 8
UserForm1.Hide
This statement hides the user interface form and returns focus to AutoCAD. If UserForm from the previous example was not loaded, the UserForm for this example would be UserForm1 by default. In this event you should change the code in line 8 to **Userform1.Hide**.

Line 9, 12 and 15
pnt1 = ThisDrawing.Utility.GetPoint(, "Provide the First Point:")
pnt2 = ThisDrawing.Utility.GetPoint(pnt1, "Second Point?")
pnt3 = ThisDrawing.Utility.GetPoint(pnt2, "3rd Point?")
The **GetPoint** method is used without a reference point in line 15. The point may be specified either with the keyboard or mouse. When it is used with a reference point we see a rubberband line attached from the reference point to the mouse cursor.

Lines 14, 16, and 17
ThisDrawing.ModelSpace.AddLine pnt1, pnt2
ThisDrawing.ModelSpace.AddLine pnt2, pnt3
ThisDrawing.ModelSpace.AddLine pnt3, pnt1
The **AddLine** method requires the two point arguments to be double precision vectors with 3 components. Note the required space before the first point.

Line 18 and 19
Unload Me
End Sub
These lines remove UserForm1 from memory, return the focus back to AutoCAD, and end the subroutine.

Module2 Code

The following code is entered in the Module2 code window.

Option Explicit	1
Sub Triangle()	2
UserForm2.Show	3
End Sub	4

Explanation

Line 1
Option Explicit
The inclusion of this line forces explicit declaration of all variable types. This minimizes common inconsistent variable usage errors and typographic errors as they are quickly caught at run time. Otherwise, undeclared variable types would be Variant by default.

Lines 2-4
Sub Triangle()
UserForm2.Show
End Sub
This subroutine's function is to create a Macro name, Triangle, that can be run from AutoCAD by choosing **Macro > Macros** from the **Tools** menu. Lines 1-4 are a part of Module1. The remaining lines of code below are a part of the UserForm2 object. Again, if UserForm1 from the previous example are not still loaded, the UserForm for this example would be named UserForm1 and Module1 similarly would be named Module1, by default. In this event, you should change the names in lines 2-4 accordingly.

POLARPOINT AND ANGLEFROMXAXIS METHODS
PolarPoint Method
The **PolarPoint** method defines a point at a given angle and distance from a given point. It has the syntax:

> **P = ThisDrawing.Utility.PolarPoint(Point, Angle, Distance)**
> where **Point**: Reference point, rubber-band origin
> **Angle**: Angle in radians, double precision
> **Distance**: Distance from point, double precision

AngleFromXAxis Method
The **AngleFromXAxis** method calculates the angle of a line defined by two points from the horizontal axis in radians. The format of the **AngleFromXAxis method** is:

> **ang = ThisDrawing.Utility.AngleFromXAxis(point1, point2)**
> where **point1**: Start point of the line
> **point2**: End point of the line

Exercise 1
Write a Visual Basic program that will draw a line between two points, as in Figure 17-10. The user may either enter coordinates or choose points with the mouse. The graphical screen should draw a rubber-band on the screen to the current position of the mouse cursor if the second point is entered with the mouse. Use a graphical user interface form similar to the one shown in Figure 17-11.

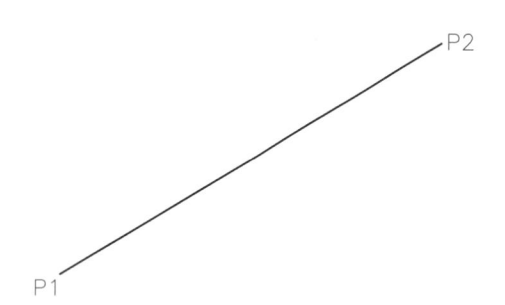

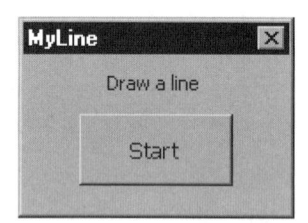

Figure 17-10 Line with user-defined endpoints   *Figure 17-11* User Interface Form for Exercise 1

Example 3
Write a Visual Basic program that will draw a triangle based on a given line produced from two points P1 and P2, on an included angle and on a length of the second side shown in Figure 17-12. To create the **Text Boxes** on the user interface form, Figure 17-13, use the **TextBox**

Visual Basic   **17-15**

button. The labels and commnad button on the form are created by clicking or dragging the respective icons from the toolbox. Insert a Module3 also.

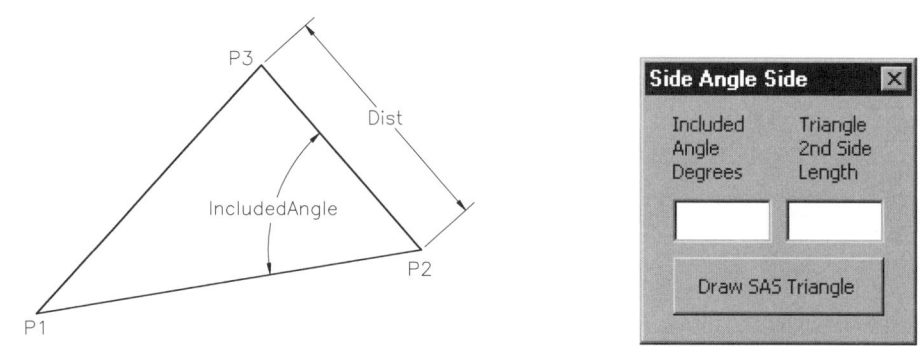

Figure 17-12 Side Angle Side Triangle   *Figure 17-13* User Interface Form for Example 3

The following file is a listing of the Visual Basic program for Example 3. **The line numbers at the right are for reference only and are not a part of the program.**

Command Button Code

```
'(Declarations) (General)                                              1
Const PI = 3.141592654                                                 2
Public IncludedAngle As Double 'pi-converted input angle               3
Public Angle As Double   'included angle                               4
Public Dist As Double    'length of 2nd side                           5
                                                                       6
'This procedure draws a triangle from 2 sides and an included angle SAS  7
'Program trigger is the command button labeled "Draw SAS Triangle"     8
'Additional feature is use of text boxes for included angle,
2nd side length input                                                  9
'Included angle is angle between base and 2nd side                    10
                                                                      11
Private Sub CommandButton1_Click()                                    12
UserForm1.Hide                                                        13
'p1 is 1st point of base, variant type by default                     14
'p2 is 2nd point of base, variant type by default                     15
p1 = ThisDrawing.Utility.GetPoint(, "Enter or select 1st base point:")  16
p2 = ThisDrawing.Utility.GetPoint(p1, "Enter  select 2nd base point:")  17
ThisDrawing.ModelSpace.AddLine p1, p2 'draw base line                 18
If TextBox1.Text = "" Then                                            19
Angle = ThisDrawing.Utility.GetAngle(p2,"Key ang. from +horiz.
or select incl. angle:")                                              20
Else                                                                  21
Angle = ThisDrawing.Utility.AngleFromXAxis(p1, p2) + IncludedAngle    22
End If                                                                23
If TextBox2.Text = "" Then                                            24
```

Dist = ThisDrawing.Utility.GetDistance(p2, "Enter select dist.	25
from base point:")	
End If	26
p3 = ThisDrawing.Utility.PolarPoint(p2, Angle, Dist)	27
ThisDrawing.ModelSpace.AddLine p2, p3	28
ThisDrawing.ModelSpace.AddLine p1, p3	29
Unload Me	30
End Sub	31
	32
Private Sub textBox1_Change()	33
IncludedAngle = PI - Val(TextBox1.Text) * PI / 180	34
End Sub	35
	36
Private Sub textBox2_Change()	37
Dist = Val(TextBox2.Text)	38
End Sub	39

Explanation

Line 2
Const PI = 3.141592654
The constant pi used in this routine is declared to demonstrate this type of statement.

Lines 3-5
Public IncludedAngle As Double
Public Angle As Double
Public Dist As Double
These variables are declared to be public, which means any procedure can access them from any module in the project (file) without passing the argument in a parameter list. Also three variables are declared to be double precision as required by the AutoCAD methods in which they will be employed.

Lines 7-12
'This module draws a triangle from 2 sides and an included angle SAS
'Program trigger is the command button labeled "Draw SAS Triangle"
'Additional feature is use of text box for included angle, 2nd side length input
'Included angle is angle between base and 2nd side
Private Sub CommandButton1_Click()
These lines give the purpose and trigger of the procedure CommandButton1_Click, which starts with line 8. Purpose and trigger remarks are useful for all event-driven subroutines.

Line 13
UserForm3.Hide
This statement hides the screen interface form and returns focus to AutoCAD. Otherwise the user would have to close the form with the button at the top right corner of the window to proceed with entering input in AutoCAD.

Visual Basic 17-17

Lines 14-17
'p1 is 1st point of base, variant type by default
'p2 is 2nd point of base, variant type by default
p1 = ThisDrawing.Utility.GetPoint(, "Enter or select 1st base point:")
p2 = ThisDrawing.Utility.GetPoint(p1, "Enter or select 2nd base point:")
These lines get input from the AutoCAD graphic screen or command line for the two points defining the base side of the triangle.

Lines 19-23 and 33-35
If textBox1.Text = "" Then
Angle=ThisDrawing.Utility.GetAngle(p2,"Enter angle from Horiz. or select included angle:")
Else
Angle=ThisDrawing.Utility.AngleFromXAxis(FirstPoint,SecondPoint)+IncludedAngle
End If
Private Sub textBox1_Change()
IncludedAngle = PI - Val(textBox1.Text) * PI / 180 'in radians
End Sub
The If ... Then ... Else statement checks TextBox1 for an entry. If no angle has been entered on the UserForm, the **GetAngle** method is used. If a numeric angle was entered on the form it is converted by the subroutine textBox1_Change() to a radian-included angle. Adding the radian angle returned by the **AngleFromXAxis** method gives the AutoCAD polar angle for the vertex (third) point.

Lines 24-26
If textBox2.Text = "" Then
Dist = ThisDrawing.Utility.GetDistance(p2, "Enter or select dist. from base point:")
End If
If no numeric entry for the second side length was entered on the user form, these lines use the **GetDistance** method.

Lines 37-39
Private Sub textBox2_Change()
Dist = Val(textBox2.Text)
End Sub
If a numeric entry for the second side length was entered on the user form, these lines convert from text to numeric form.

Lines 18, 28 and 29
ThisDrawing.ModelSpace.AddLine p1, p2
ThisDrawing.ModelSpace.AddLine p2, p3
ThisDrawing.ModelSpace.AddLine p1, p3
These methods draw the base, second, and third sides of the triangle.

Module3 Code
The following code is entered in the Module3 code window.

```
Option Explicit                                              1
Sub SAS()                                                    2
    UserForm1.Show                                           3
End Sub                                                      4
```

Exercise 2

Write a program in the AutoCAD VBA IDE that will draw a triangle based on a given line produced from two points P1 and P2, on an adjacent angle at one end, and another angle at the other end as shown in Figure 17-14. Employ a UserForm similar to the one shown in Figure 17-15

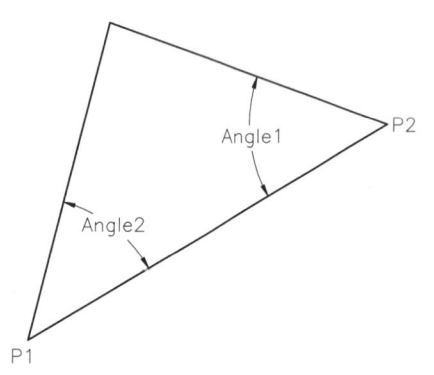

Figure 17-14 *Angle Side Angle Triangle with user-defined points defining the base*

Figure 17-15 *User Interface Form for Exercise 3*

Exercise 3

Write a program in the AutoCAD VBA IDE that will draw a triangle based on a given line produced from two points P1 and P2, and two distances, which are the lengths of the sides adjoining each end as in Figure 17-16. Employ a UserForm similar to the one shown in Figure 17-17.

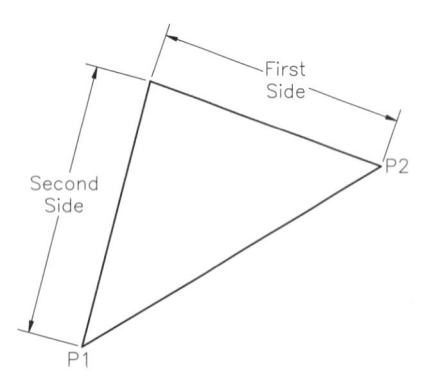

Figure 17-16 *Side Side Side Triangle with user-defined points defining the base*

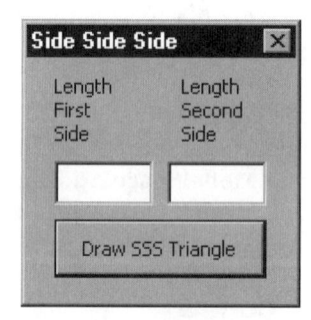

Figure 17-17 *User Interface Form for Exercise 3*

Visual Basic 17-19

ADDITIONAL VBA EXAMPLES

More VBA examples are included with AutoCAD 2002 beyond the large number in the help files. These are considered sample files and are located in the **Sample/VBA** subdirectory of the ACAD2002 directory and the AutoCAD 2002 CD ROM. Another resource for learning VBA is the Web.

Review Questions

Answer the following questions.

1. _____ lets you access and manipulate the objects and functionality of AutoCAD from Excel or another application supporting ActiveX.

2. Before you can use AutoCAD's objects in some other software supporting ActiveX, you must make the application aware that the AutoCAD _____ Library is available on the computer.

3. There are _____ programmers using Visual Basic.

4. Visual Basic is considered an object-_____, rather than an object-oriented, programming language.

5. Functions known as _____ have been defined in the AutoCAD object library to perform an action on an object, for example, drawing a line in a drawing.

6. _____ are functions that set or return information about the state of an object.

7. The term **IDE**, which is the Visual Basic Editor, stands for _____.

8. A _____ (extension .dvb) is the name given to the forms, controls, modules, and programming making up a saved AutoCAD Visual Basic file.

9. _____ forces explicit declaration of all variable types, which minimizes common inconsistent variable usage errors.

10. _____ is the variable type returned by the GetPoint method.

11. The _____ method always measures the angle with a positive X-axis (3 o'clock position) and in a counterclockwise direction.

12. The _____ method allows you to enter the X, Y coordinates or X, Y, Z coordinates of a point.

13. The _____ method lets you retrieve the value of an AutoCAD system variable.

14. The _____ method defines a point at a given angle and distance from the given point.

15. The _____ method lets you enter a distance on the command line, a distance from a given point, or two points.

16. The _____ method allows you to enter an angle, either from the keyboard in degrees or by selecting two points.

17. The _____ method calculates the angle of a line defined by two points from the horizontal axis in radians

Exercises

Exercise 4

Write a Visual Basic program that will draw an equilateral triangle inside the circle (Figure 17-18).

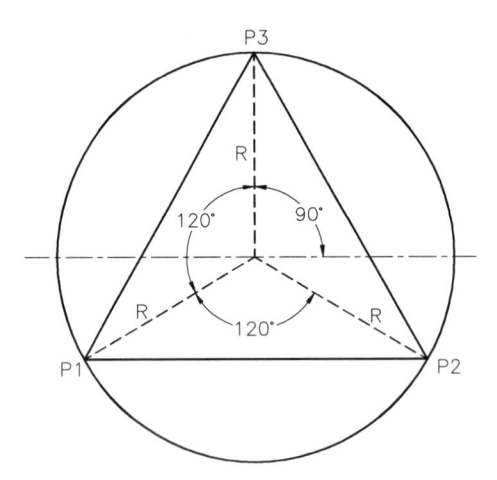

Figure 17-18 Equilateral triangle inside a circle

Visual Basic 17-21

Exercise 5

Write a Visual Basic program that will draw a square of sides S and a circle tangent to the four sides of the square as shown in Figure 17-19. The base of the square makes an angle, ANG, with the positive X-axis. The program should allow you to enter the starting point P1, length S, and angle ANG on a user interface form or in the AutoCAD graphical interface.

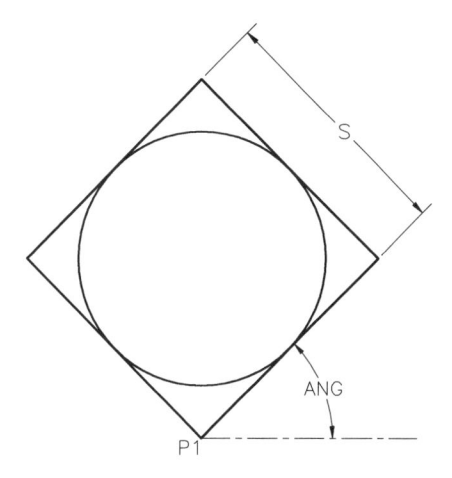

Figure 17-19 Square of side S and at an angle ANG

Exercise 6

Write a Visual Basic program that will draw a line at an angle A and then generate the given number of lines (N), parallel to the first line with an offset distance S as entered on a user form or from the keyboard.

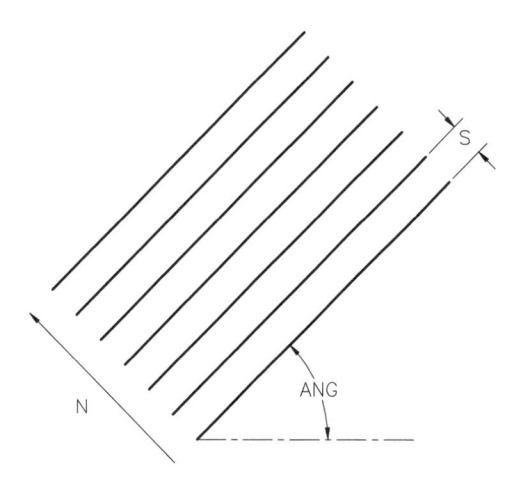

Figure 17-20 N number of lines offset at a distance of S

Exercise 7

Write a program that will draw a slot (Figure 17-21) with center lines. The program should allow you to enter slot length, slot width, and the layer name for center lines on a user interface form or in the AutoCAD graphical interface.

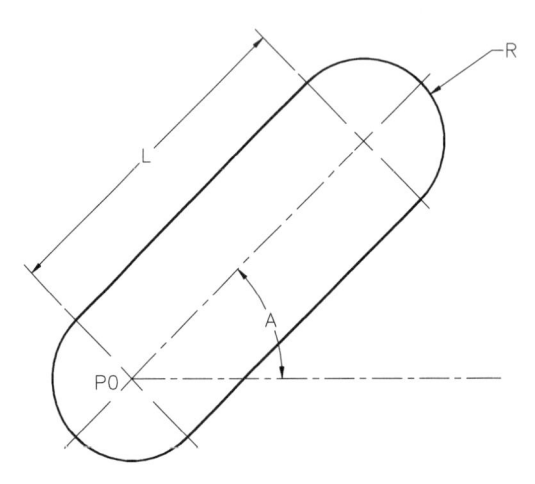

Figure 17-21 Slot of length L and radius R

Exercise 8

Write a Visual Basic program that will draw two lines tangent to two circles, as shown in Figure 17-22. The program should allow you to enter the circle diameters and the center distance between the circles on a user interface form or in the AutoCAD graphical interface.

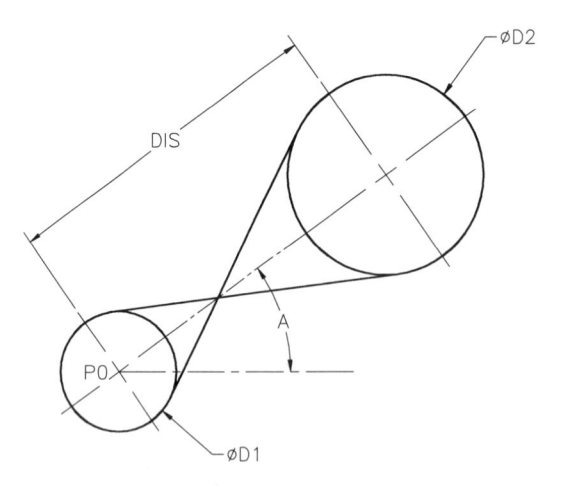

Figure 17-22 Circles with tangent lines

Visual Basic

Exercise 9

Write a VBA program that will draw a hub with the key slot, as shown in Figure 17-23. The user should enter the four parameters in the note below on a user form or in the AutoCAD graphical interface. Use the program inside a circle of Diameter D1 to produce a keyed bushing with the test dimensions.

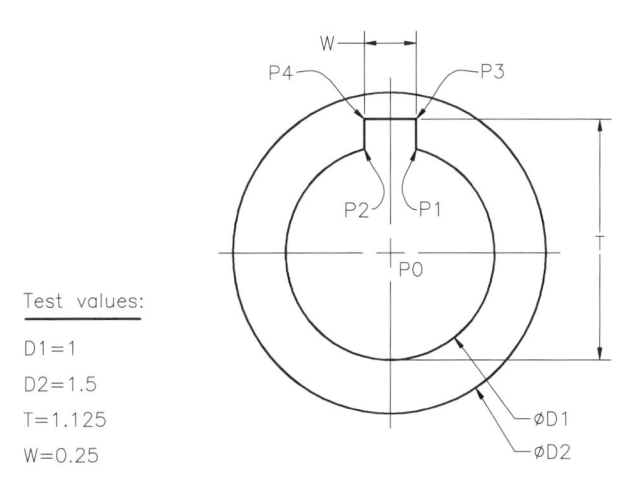

Test values:

D1=1
D2=1.5
T=1.125
W=0.25

Figure 17-23 Hub with Keyway

Exercise 10

Write a VBA program to draw the tangent arc cam shown in Figure 17-24.

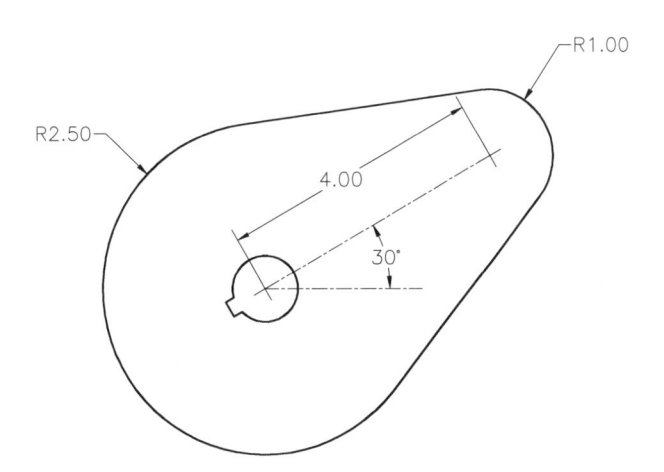

Figure 17-24 Tangent Arc Flat Plate Cam

Exercise 11

Write a VBA program to draw the circular arc cam shown in Figure 17-25.

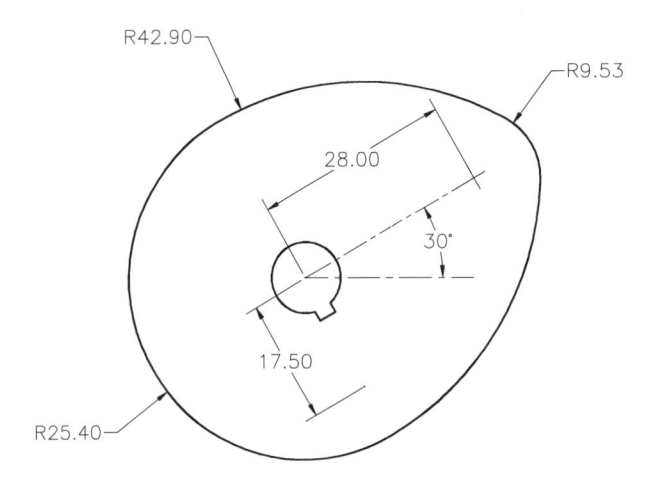

Figure 17-25 Circular Arc Cam

Project Exercise 1

Write a Visual Basic program that will draw the two views of a bushing as shown in Figure 17-26. The program should allow you to enter the starting point P0, lengths L1, L2, and the bushing diameters ID, OD, HD on a user interface form or in the AutoCAD graphical interface. The distance between the front view and the side view of bushing is DIS (DIS = 1.25 * HD). The program should also draw the hidden lines in the HID layer and center lines in the CEN layer. The center lines should extend 0.75 units beyond the object line.

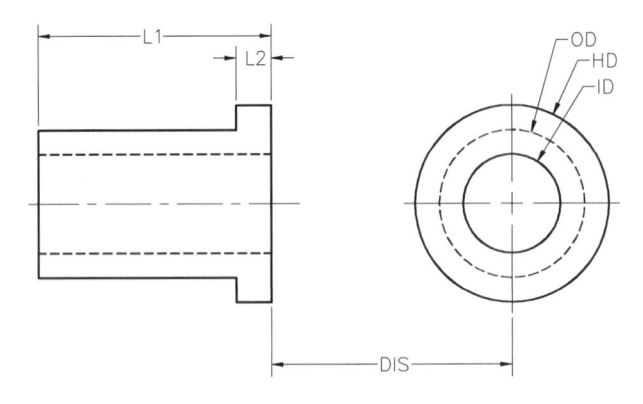

Figure 17-26 Two views of bushing

Project Exercise 2

Draw the following parametric drawing in a rectangle 320 mm wide by 200 mm high. The block depends on one number, Q, called the data set number. The dimensions as shown in Figure 17-27 are given by: X = Q + 160, Z = 240 - X, Y = 120 - Z, and R = Z/4. There are 40 mm between views and 20 mm between a view and the surrounding rectangle at the closest points. The circle is centered in the block.

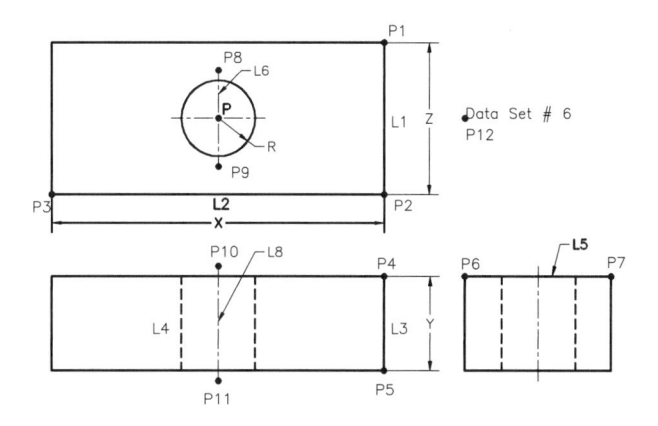

Figure 17-27 Parametric Block with Hole

Chapter 18

Accessing External Databases

Learning Objectives

After completing this chapter, you will be able to:
* *Understand databases, and the database management system (DBMS).*
* *Understand the **AutoCAD database connectivity** feature.*
* *Configure external databases.*
* *Access and edit a database using **dbConnect Manager**.*
* *Create **Links** with graphical objects.*
* *Create and display **Labels** in a drawing.*
* *Understand **AutoCAD SQL Environment** (ASE) and create queries using **Query Editor**.*
* *Form selection sets using **Link Select**.*
* *Convert ASE links into AutoCAD 2002 format.*

UNDERSTANDING DATABASES

Database

A **database** is a collection of data that is arranged in a logical order. For example, there are six computers in an office, and we want to keep a record of these computers on a sheet of paper. In a database, the columns are known as **fields**, the individual rows are called **records**, and the entries in the database tables are known as **cells** that store data for a particular variable. One of the ways of recording the computer information is to make a table with rows and columns as shown in Figure 18-1. Each column will have a heading that specifies a certain feature of the computer, such as COMP_CFG, CPU, HDRIVE, or RAM. Once the columns are labeled, the computer data can be placed in the columns for each computer. By doing this, we have created a database on a sheet of paper that contains information on our computers. The same information can be stored in a computer, generally known as a computerized database.

Figure 18-1 A table containing computer information

Most of the database systems are extremely flexible and any modifications to or additions of fields or records can be done easily. Database systems also allow you to define relationships between multiple tables, so that if the data of one table is altered, automatically the corresponding values of another table with predefined relationships will change accordingly.

Database Management System

The database management system (DBMS) is a program or a collection of programs (software) used to manage the data in the database. For example, PARADOX, dBASE, INFORMIX, and ORACLE are database management systems.

Components of a Table

A database **table** is a two-dimensional data structure that consists of rows and columns as shown in Figures 27-2 and 27-3.

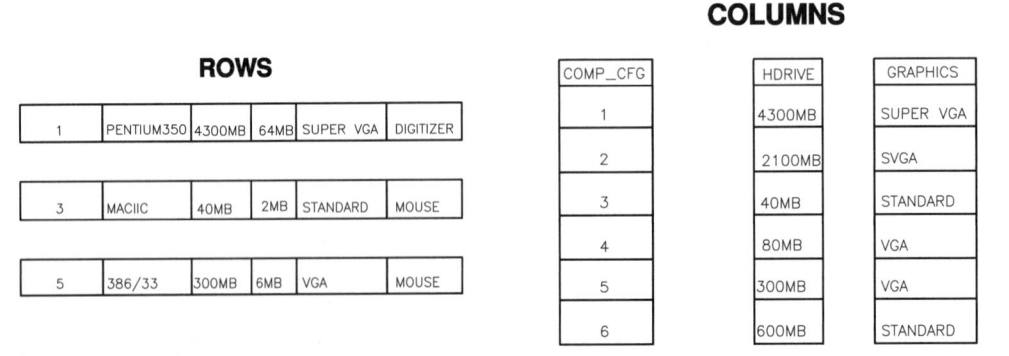

Figure 18-2 Rows in a table (horizontal group)

Figure 18-3 Columns in a table (vertical group)

Row

The horizontal group of data is called a **row**. For example, Figure 18-2 shows three rows of the table. Each value in a row defines an attribute of the item. For example, in Figure 18-2, the attributes assigned to COMP_CFG (1) include PENTIUM350, 4300MB, 64MB, and so on. These attributes are arranged in the first row of the table.

Column

A vertical group of data (attribute) is called a **column**. (See Figure 18-3.) HDRIVE is the column heading that represents a feature of the computer, and the HDRIVE attributes of each computer are placed vertically in this column.

AUTOCAD DATABASE CONNECTIVITY

AutoCAD can be effectively used in associating data contained in an external database table with the AutoCAD graphical objects by linking. The **Links** are the pointers to the database tables from which the data can be referred. AutoCAD can also be used to attach **Labels** that will display data from the selected tables as text objects. AutoCAD database connectivity offers the following facilities:

1. A **dbConnect Manager** that can be used to associate links, labels, and queries with AutoCAD drawings.
2. An **external configuration utility** that enables AutoCAD to access the data from a database system.
3. A **Data View window** that displays the records of a database table within the AutoCAD session.
4. A **Query Editor** that can be used to construct, store, and execute SQL queries. **SQL** is an acronym for **structured query language**.
5. A **migration tool** that converts links and other displayable attributes of files created by earlier releases to AutoCAD 2002.
6. A **Link Select operation** that creates iterative selection sets based on queries and graphical objects.

DATABASE CONFIGURATION

An External database can be accessed within AutoCAD only after configuring AutoCAD using Microsoft **ODBC** (Open Database Connectivity) and **OLE DB** (Object Linking and Embedding Database) programs. AutoCAD is capable of utilizing data from other applications, regardless of the format and the platform on which the file is stored. Configuration of a database involves creating a new **data source** that points to a collection of data and information about the required drivers to access it. A **data source** is an individual table or a collection of tables created and stored in an environment, catalog, or schema. Environments, catalogs, and schemas are the hierarchical database elements in most of the database management systems and they are analogous to Window-based directory structure in many ways. Schemas contain a collection of tables, while Catalogs contain subdirectories of schemas and Environment holds subdirectories of catalogs. The external applications supported by AutoCAD 2002 are **dBASE® V** and **III**, **Oracle®8.0** and **7.3**, **Microsoft® Access®, PARADOX 7.0, Microsoft Visual FoxPro®6.0**, **SQL Server 7.0** and **6.5**. The configuration process varies slightly from one database system to other.

DBCONNECT MANAGER

Menu:	Tools > dbConnect
Command:	DBCONNECT

When you invoke this command, the **dbConnect Manager** will be displayed as shown in Figure 18-4. Also, the **dbConnect** menu is added to the menu bar when you invoke this command. The **dbConnect Manager** enables you to access information from external databases more effectively. The **dbConnect Manager** is dockable as well as resizable and contains a set of buttons and a tree view showing all the configured and available databases. You can use **dbConnect Manager** for associating various database objects with an AutoCAD drawing. You can invoke **dbConnect Manager** (Figure 18-4) by entering **DBCONNECT** at the AutoCAD Command prompt or choosing **dbConnect** from the **Tools** menu. The **dbConnect Manager** contains two nodes in the tree view. The **Drawing nodes** displays all the open drawings and each node will display all the associated database objects with the drawing. **Data Sources node** displays all the configured data on your system.

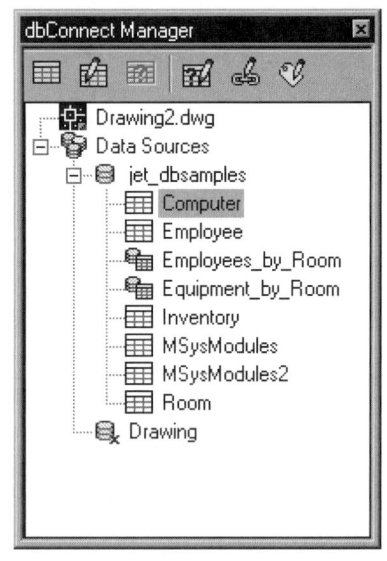

Figure 18-4 dbConnect Man-

AutoCAD contains several Microsoft Access sample database tables and a direct driver (*jet_dbsamples.udl*). The following example shows the procedure to configure a database with your drawing using the **dbConnect Manager**.

Example 1   *General*

Configure a data source of Microsoft Access database with your diagram. Update and use the **jet_samples.udl** configuration file with new information.

1. Invoke the **DBCONNECT** command to display the **dbConnect Manager**. The **dbConnect** menu will be inserted between the **Modify** and the **Window** menus.

2. The **dbConnect Manager** will display **jet_dbsamples** under **Data Sources**. Right-click on **jet_dbsamples** to display the shortcut menu. In the shortcut menu choose **Configure** as shown in Figure 18-5.

3. The **Data Link Properties** dialog box will be displayed with the **Connection** tab as the current tab. (See Figure 18-6.)

4. In the **Data Link Properties** (**Connection** tab) dialog box, choose the [...] button adjacent to the **Select or**

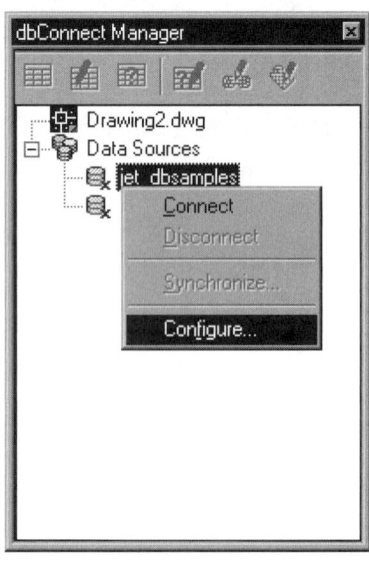

Figure 18-5 dbConnect Man-

Accessing External Databases 18-5

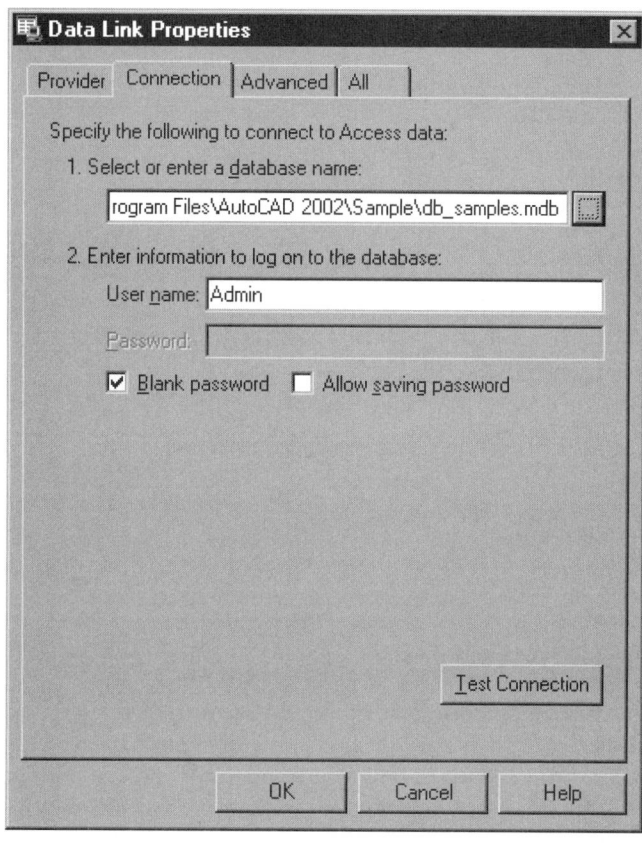

Figure 18-6 Data Link Properties dialog box

enter a database name text box. AutoCAD will display the **Select Access Database** dialog box (Figure 18-7). Select **db_samples** (**.mdb file**) and choose the **Open** button.

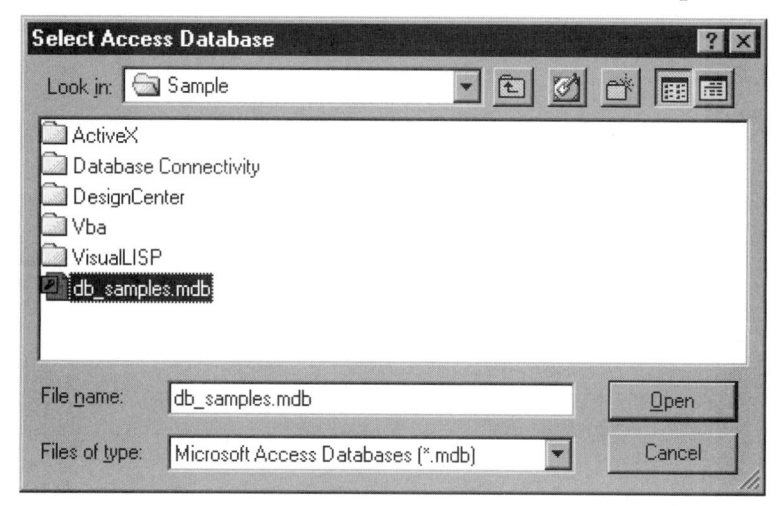

Figure 18-7 Select Access Database dialog box

Note

*If you are configuring a driver other than **Microsoft Jet** for database linking, you should consult the appropriate documentation for the configuring process.*

*The other tabs in the **Data Links Properties** dialog box are required for configuring different Database Providers supported by AutoCAD.*

5. Once the database name is selected, choose the **Test Connection** button to ensure that the database source has been configured correctly. If the source is configured correctly, AutoCAD will display the **Microsoft Data Link** message box (Figure 18-8). Choose **the OK** button to end the message and again choose the **OK** button to complete the configuration process.

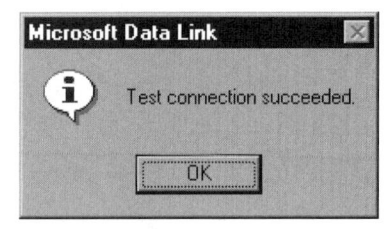

Figure 18-8 Microsoft Data Link message box

After configuring a data source when you double-click on **jet_dbsamples** in **dbConnect Manager**, all the sample tables will be displayed in the tree view and you can connect any table and link its records to your diagram.

VIEWING AND EDITING TABLE DATA FROM AUTOCAD

After you have configured a data source using the **dbConnect Manager**, you can view as well as edit its tables within the AutoCAD session. The **Data View window** can be used for viewing and editing a database table. You can open tables in **Read-only** mode to view its content. But you cannot edit its records in **Read-only** mode. Some database systems may require a valid username and password before connecting to AutoCAD drawing files. The records of a database table can be edited by opening in **Edit** mode. The procedures to open a table in various modes follow.

Read-only mode

The tables can be opened in the **Read-only** mode by choosing the **View Table** button from the **dbConnect Manager**. You can also choose **View Table** from the shortcut menu displayed upon right-clicking on the selected table. This can also be done by choosing **View Data > View External Table** from the **dbConnect** menu.

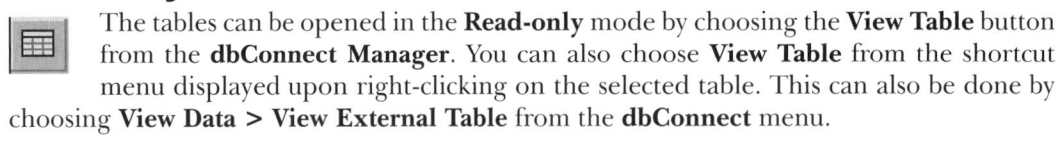

Edit mode

The tables can be opened in the **Edit** mode by choosing the **Edit Table** button from the **dbConnect Manager**. You can also choose **Edit Table** from the shortcut menu displayed upon right-clicking on the selected table. This can also be done by choosing **View Data > Edit External Table** from the **dbConnect** menu.

Note

*The **db_samples.mdb** file is available in the directory \AutoCAD 2002\Samples.*

Accessing External Databases

18-7

Example 2
General

In this example, you will select **Computer** table from the **jet_dbsamples** data source and edit each row in the table. Add a new computer and replace one by editing the table and save your changes.

1. Select **Computer** table from the **jet_dbsamples** in the **dbConnect Manager** and right-click to invoke the shortcut menu. Choose **Edit Table** from the shortcut menu (Figure 18-9). You can also do the same by double clicking on the Computer table.

2. AutoCAD will display the **Data View** window with the **Computer** table in it. In the **Data View** window, you can resize, sort, hide, or freeze the columns according to your requirements.

3. To add a new item in the table, double-click on the first empty row. In **Tag_Number** column, type **24675**. Then type the following data into the columns:

Manufacturer:	**IBM**
Equipment_Description:	**PIII/450, 4500GL, NETX**
Item_Type:	**CPU**
Room:	**6035**

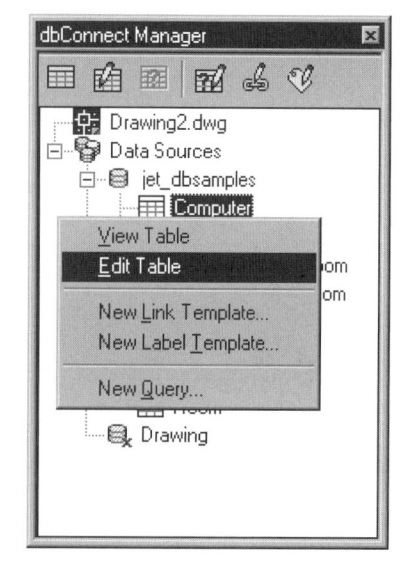

*Figure 18-9 Choosing **Edit Table** from the shortcut menu*

4. To edit the record for **Tag_Number 60298**, and display the following in **Data View** window (Figure 18-10), double-click on each cell and type:

MANUFACTURER:	**CREATIVE**
Equipment_Description:	**INFRA 6000, 40XR**
Item_Type:	**CD DRIVE**
Room:	**6996**

5. After you have made all the changes in the database table, you have to save the changes for further use. To save the changes in the table, right-click on the **Data View grid header**, a triangle mark available on the left of the **Tag_Number** column and choose **Commit** from the shortcut menu. The changes in the current table will be saved. The new record that you have entered does not necessarily get added at the end of the table, so if you close the table and then re-open it you might have to search for the new record in the table. Note that **if you quit Data View window without Committing, AutoCAD will automatically commit all the changes you have made during the editing session.**

   Note
   *After making changes, if you do not want to save the changes, choose **Restore** from the above shortcut menu and AutoCAD will restore the original values.*

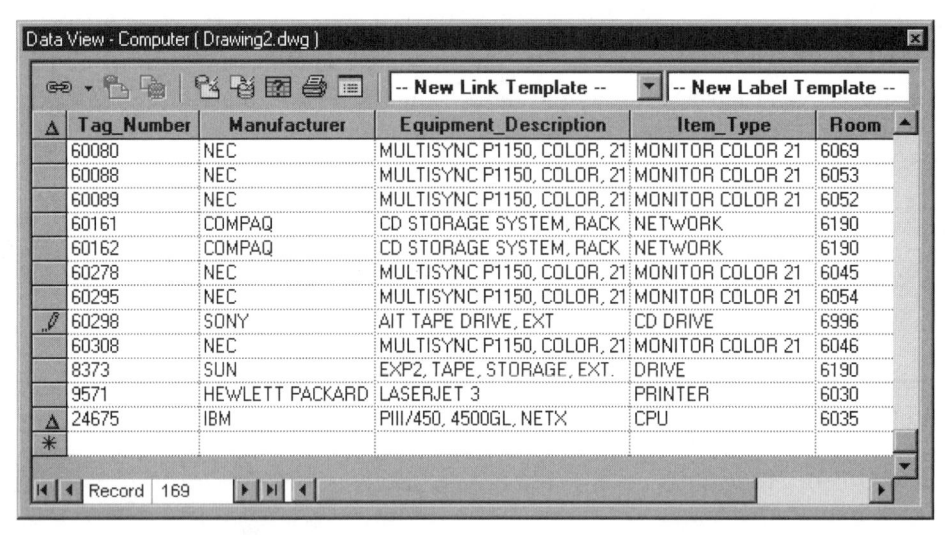

Figure 18-10 Data View window after editing

CREATING LINKS WITH GRAPHICAL OBJECTS

The main function of the database connectivity feature of AutoCAD is to associate data from external sources with its graphical objects. You can establish the association of the database table with the drawing objects by developing a **link**, which will make reference to one or more records from the table. But you **cannot** link nongraphical objects such as layers or linetypes with the external database. Links are very closely related with graphical objects and change with the change in graphical objects.

To develop links between database tables and graphical objects, you must create a **Link Template**, which will identify the fields of the tables with which links are associated to share the template. For example, you can create a link template that uses the **Tag_number** from the **COMPUTER database table**. Link template also acts as a shortcut that points to the associated database tables. You can associate multiple links to a single graphical object using different link templates. This is useful in associating multiple database tables with a single drawing object. The following example will describe the procedure of linking using link template creation.

Example 3   *General*

Create a link template between **Computer** database table from **jet_dbsamples** and your drawing, and use **Tag_Number** as the **key field** for linking.

1. Open your drawing that has to be linked with the **Computer** database table.

2. From the **Tools** menu, choose **dbConnect** to invoke **dbConnect Manager**. Select **Computer** from the **jet_dbsamples** data source and right click on it to invoke the shortcut menu.

3. In the shortcut menu, choose **New Link Template** to invoke the **New Link Template** dialog box (Figure 18-11). You can also invoke this dialog box by selecting the table in the

Accessing External Databases 18-9

tree view and choosing the **New Link Template** button from the toolbar in the **dbConnect Manager**.

4. Choose the **Continue** button to accept the default link template named **ComputerLink1**. AutoCAD will display the **Link Template** dialog box. Select the check box adjacent to **Tag_Number** to accept it as the **Key field** for associating the template with the block reference in the diagram. (See Figure 18-12.)

5. Choose the **OK** button to complete the new link template. The name of the link template will be displayed under the Drawing name node of the tree view in **dbConnect Manager**.

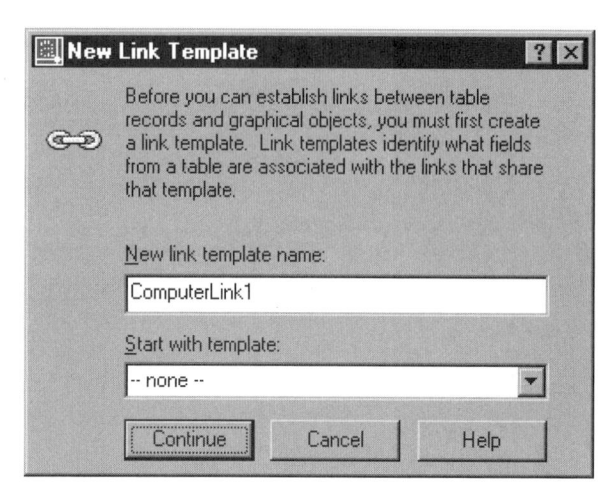

Figure 18-11 New Link Template dialog box

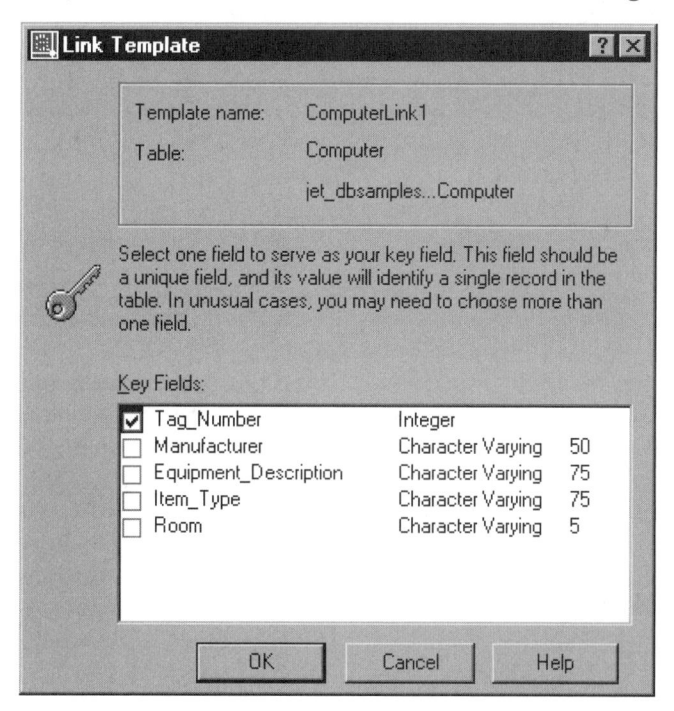

Figure 18-12 Link Template dialog box

6. To **link a record with the required object**, double-click on the **Computer** table to invoke the **Data View window** (Edit mode). In the table, go to Tag_Number **24675** and highlight the record (row).

7. Right-click on the row header and choose **Link!** from the shortcut menu (Figure 18-13).

You can also link by choosing the **Link!** button from the toolbar in **Data View** window.

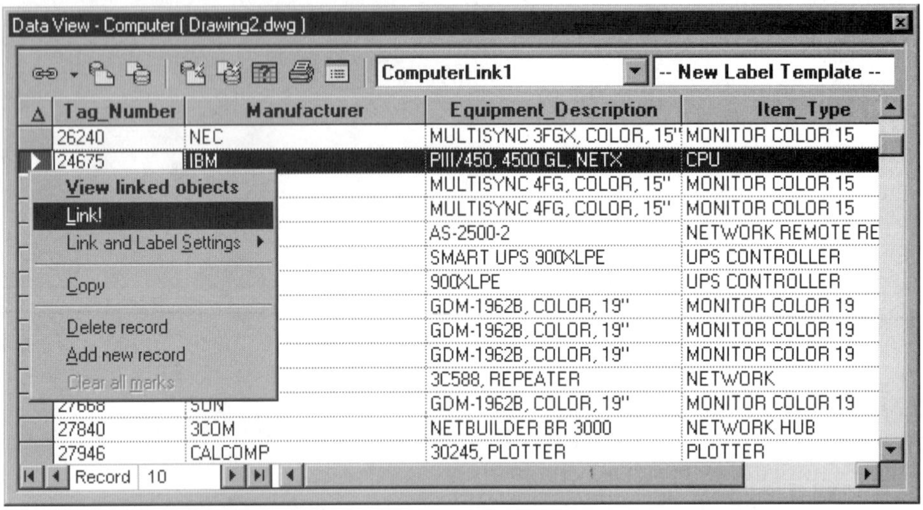

*Figure 18-13 Selecting **Link** from the shortcut menu*

8. AutoCAD prompts you to select the objects. Select the required objects you want to link with the record. Repeat the process to link all the records to the corresponding block references in the diagram.

9. After linking, you can **view the linked objects**, by choosing the **View Linked Objects in Drawing** button in the **Data View** dialog box, and the linked objects will get highlighted in the drawing area.

Additional Link Viewing Settings

You can set a number of viewing options for linked graphical objects and linked records by using the **Data View and Query Options** dialog box (Figure 18-14). This dialog box can be invoked by choosing the **Options** button from the **Query Editor** dialog box.

You can set the **Automatically Pan Drawing** option so that AutoCAD will pan the drawing automatically to display objects linked with the current set of selected records in the **Data View** window. You can also set options for **Automatically zoom drawing** and the **Zoom factor**. You can also change the Record Indication settings and the indication marking color.

> **Note**
> *Query Editor dialog box is discussed later in this chapter.*

Editing Link Data

After linking data with the drawing objects, you may need to edit the data or update their **Key field** values. For example, you may need to reallocate the Tag-Number for the computer equipment or Room for each of the linked items. **Link Manager** can be used for changing the Key values. The next example describes the procedure of editing linked data.

Accessing External Databases 18-11

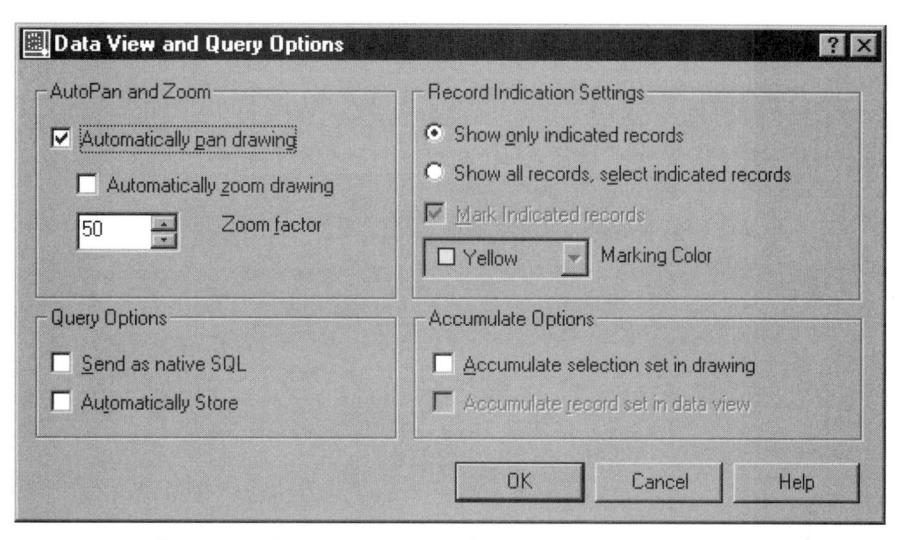

Figure 18-14 Data View and Query Options dialog box

Example 4   *General*

Use **Link Manager** to edit linked data from the Computer table and change the Key Value from **24352** to **24675**.

1. Open the diagram that was linked with the Computer table. Choose **Links > Link Manager** from the **dbConnect** menu. AutoCAD will prompt you to select the linked object.

2. Select the linked object to invoke the **Link Manager** dialog box for the **Computer** table (Figure 18-15).

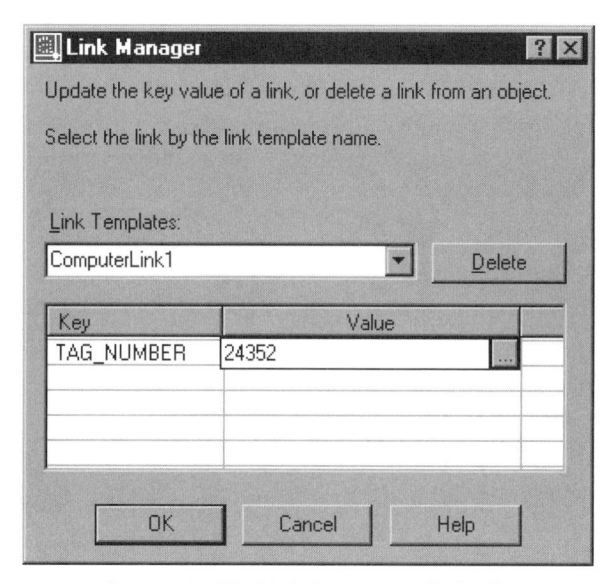

Figure 18-15 Link Manager dialog box

3. Choose **ComputerLink1** from the **Link Template** drop-down list.

4. Select **24352** field in the **Value** column and choose the [...] button to invoke the **Column Values** dialog box (Figure 18-16). Select **24675** from the list and choose the **OK** button.

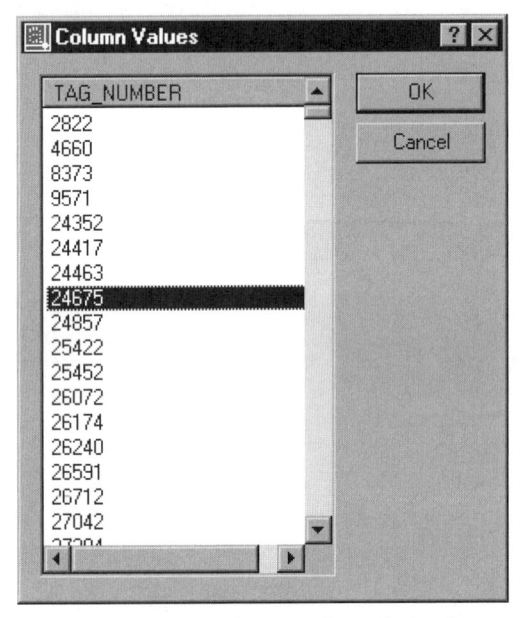

Figure 18-16 Column Values dialog box

5. Again choose the **OK** button in **Link Manager** to accept the changes.

CREATING LABELS

You can use linking as a powerful mechanism to associate drawing objects with external database tables. You can directly access associated records in the database table by selecting linked objects in the drawing. But linking has some limitations. Suppose you want to include the associated external data with the drawing objects. Since during printing, the links are only the pointers to the external database table, they will not appear in the printed drawing. In such situations, AutoCAD provides a feature called **Labels** that can be used for visible representation of external data in drawing.

Labels are the multiline text objects that display data from the selected fields in the AutoCAD drawing. The labels are of the following two types:

Freestanding labels

Freestanding labels exist in the AutoCAD drawing independent of the graphical objects. Their properties do not change with any change of graphical objects in the drawing.

Attached labels

The Attached labels are connected with the graphical objects they are associated with. If the

Accessing External Databases 18-13

graphical objects are moved, then the labels also move and if the objects are deleted, then the labels attached with them also get deleted.

Labels associated with the graphical objects in AutoCAD drawing are displayed with a leader. Labels are created and displayed by **Label Templates** and all the properties of label display can be controlled by using the **Label Template** dialog box. The next example will demonstrate the complete procedure of creating and displaying labels in AutoCAD drawing using the label template.

Example 5   *General*

Create a new label template in the **Computer** database table and use the following specifications for the display of labels in the drawing:

1. The label includes **Tag_Number**, **Manufacturer** and **Item_Type** fields.
2. The fields in the label are **0.25** in height, **Times New Roman** font, **black** (**Color 18**) in color and **Middle-left** justified.
3. The label offset starts with **Middle Center** justified and leader offset is **X=1.5** and **Y=1.5**.

1. Open the drawing where you want to attach a label with the drawing objects.

2. Choose **Tools menu > dbConnect** to invoke **dbConnect Manager**. Select **Computer** table from **jet_dbsamples** data source.

3. Right-click on **Computer** and choose **New Label Template** from the shortcut menu to invoke the **New Label Template** dialog box (Figure 18-17). You can also invoke the dialog box by choosing the **New Label Template** button from the **dbConnect Manager** toolbar.

4. In the **New Label Template** dialog box, choose the **Continue** button to accept the default label template name **Computer Label1**. AutoCAD will display the **Label Template** dialog box.

5. In the **Label Template** dialog box, choose the **Label Fields** tab. From the **Field** drop-down list, one by one add **Tag_Number**, **Manufacturer**, **Item_Type** fields by using the **Add** button (Figure 18-18).

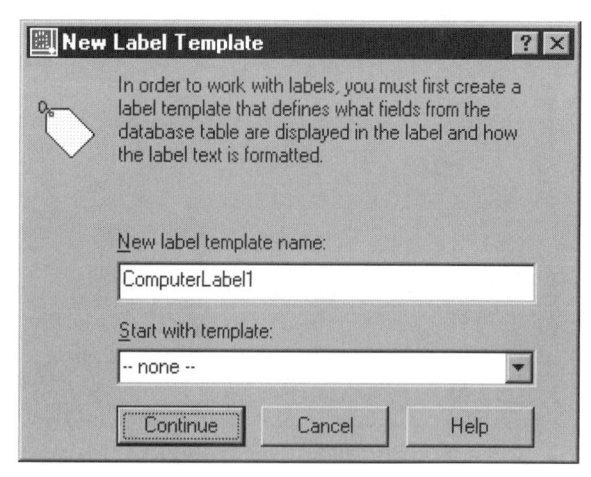

Figure 18-17 New Label Template dialog box

6. Highlight the field names by selecting them in the edit box and choose the **Character** tab. In the **Character** tab, select the **Times New Roman** font and font height to **0.25**. Set the color to **Color 18**. Also select **Middle-left** justification in the **Properties** tab (Figure 18-19).

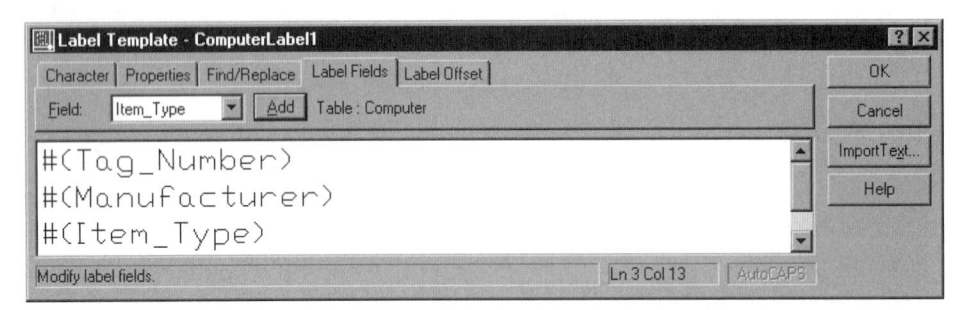

*Figure 18-18 Label Template dialog box (**Label Fields** tab)*

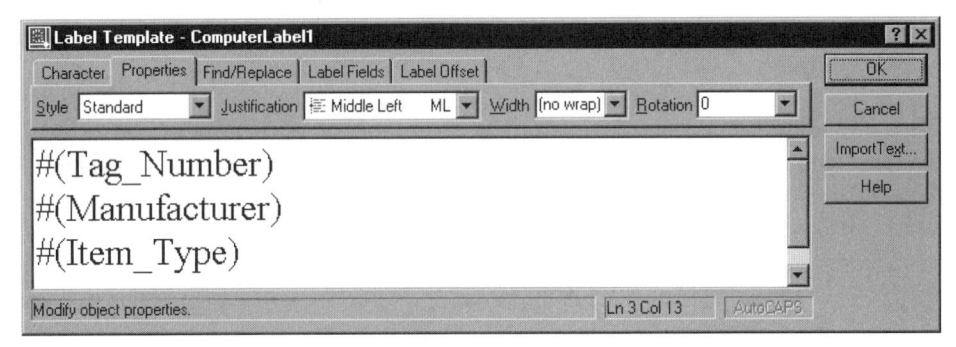

*Figure 18-19 Label Template dialog box (**Properties** tab)*

7. Choose the **Label Offset** tab and select **Middle-center** in the **Start**: drop-down list. Set the **Leader offset:** value to **X: 1.5** and **Y: 1.5** (Figure 18-20). Choose **OK**.

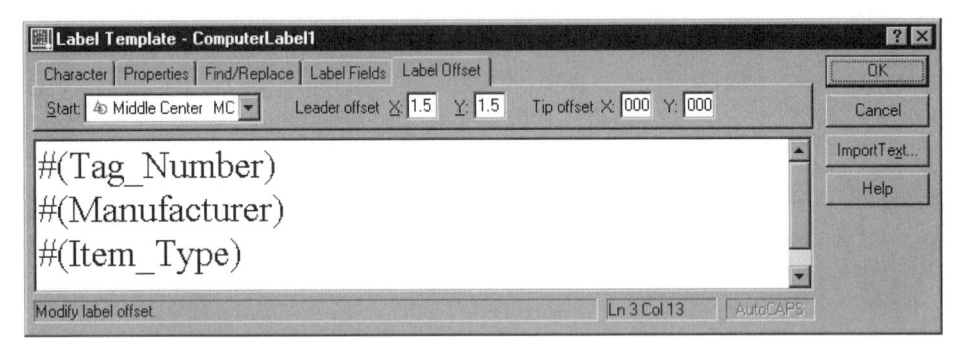

*Figure 18-20 Label Template dialog box (**Label Offset** tab)*

8. To display the label in the drawing, select the Computer table in the tree view of **dbConnect Manager**. Right-click on the table and choose **Edit Table** from the shortcut menu.

9. Select **Computer link1** (created in an earlier example) from the **link name** drop-down list and **Computer label1** from the **label name** drop-down list in the **Data View** window.

10. Select the record (row) you want to use as a label. Then choose the down arrow button provided on the right side of the **Links** button to display a shortcut menu. Choose **Create**

Accessing External Databases 18-15

Attached Labels from the menu (Figure 18-21). Then choose the **Create Attached Label** button that replaces the **Links** button after you choose the **Create Attached Labels** from the shortcut menu. Select the drawing object you want to label. AutoCAD will display the Label with given specifications in the drawing.

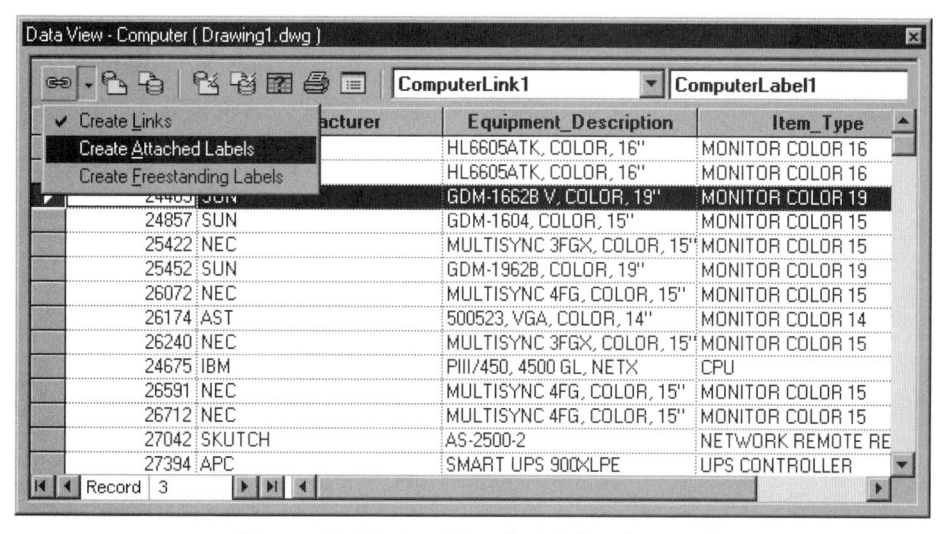

Figure 18-21 Attaching the label to the record

Note
*You can create **Freestanding Labels** in the same way by selecting **Create Freestanding Label** from the **Link and Label setting** menu in the **Data View** window.*

Updating Labels with New Database Values

There is a strong possibility that you may have to change the data values in the database table after adding a label in the AutoCAD drawing. Therefore, you should update the labels in the drawing after making any alteration in the database table the drawing is linked with. The following is the procedure for updating all label values in the AutoCAD drawing:

1. After editing the database table, open the drawing that has to be updated.

2. From the **dbConnect** menu, choose **Labels > Reload Labels**.

3. You will be prompted to specify the name of the label template to be reloaded. Enter the name of the label template.

Importing and Exporting Link and Label Templates

You may want to use the link and label templates that have been developed by some other AutoCAD users. This is very useful when developing a set of common tools to be shared by all of the team members in a project. AutoCAD is capable of importing as well as exporting all the link and label templates that are associated with a drawing. The following is the procedure to export a set of templates from the current drawing:

1. From the **dbConnect** menu, choose **Templates > Export Template Set** to **Export Template Set** dialog box.

2. In the dialog box in the **Save In list**, select the directory where you want to save the template set.

3. Under **File Name**, specify a name for the template set, and then choose the **Save** button to save the template in the specified directory.

The following is the procedure to import a set of templates into the current drawing:

1. From the **dbConnect** menu, choose **Templates > Import Template Set** to invoke the **Import Template Set** dialog box, see Figure 18-22.

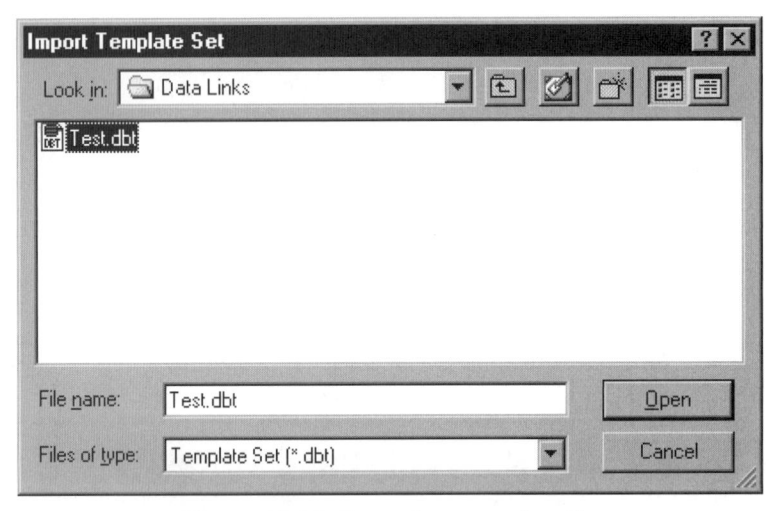

Figure 18-22 Importing a template file

2. In the dialog box, select the template set to be imported.

3. Choose **Open** to import the template set into the current drawing.

AutoCAD will display an **alert box** that can be used to provide a unique name for the template if there is a link or label template with the same name already associated with the current drawing.

AUTOCAD SQL ENVIRONMENT (ASE)

SQL is an acronym for **structured query language**. It is often referred to as **sequel**. SQL is a format in computer programming that lets the user ask questions about a database according to specific rules. The **AutoCAD SQL environment** (ASE) lets you access and manipulate the data that is stored in the external database table and link data from the database to objects in a drawing. Once you access the table, you can manipulate the data. The connection is made through a database management system (DBMS). The DBMS programs have their own methodology for working with databases. However, the ASE commands work the same way regardless of the database being used. This is made possible by the ASE drivers that come with AutoCAD software. In AutoCAD 2002, SQL has been incorporated. For example, you want to prepare a report that lists all the computer equipment that costs more then $25. **AutoCAD Query Editor** can be used to easily construct a query that returns a subset of records or linked graphical objects that follow the previously mentioned criterion.

AUTOCAD QUERY EDITOR

The **AutoCAD Query Editor** consists of four tabs that can be used to create new queries. The tabs are arranged in order of increasing complexities. For example, if you are not familiar with **SQL** (Structured Query Language), you can start with **Quick Query** and **Range Query** initially to get familiar with query syntax.

You can start developing a query in one tab and subsequently add and refine the query conditions in the next tabs. For example, if you have created a query in the **Quick Query** tab and then decide to add an additional query using the **Query Builder** tab, when you choose the **Query Builder** tab, all the values initially selected in the previous tabs are displayed in this tab and additional conditions can be added to the query. But it is not possible to go backwards through the tabs once you have created queries with one of the advanced tabs. The reason for this is that the additional functions are not available in the simpler tabs. AutoCAD will prompt a warning indicating that the query will be reset to its default values if you attempt to move backward through the query tabs.

The AutoCAD **Query Editor** has the following tabs for building queries:

Quick Query

This tab provides an environment where your simple queries can be developed based on a single database field, single operator, and a single value. For example, you can find all records from the current table where the value of the **'Item_type'** field equals **'CPU'**.

Range Query

This tab provides an environment where a query can be developed to return all records that fall within a given range of values. For example, you can find all records from the current table where the value of the **'Room'** field is greater than or equal to 6050 and less than or equal to 6150.

Query Builder

This tab provides an environment where more complicated queries can be developed based on multiple search criteria. For example, you can find all records from the current table where the **'Item_type'** equals **'CPU'** and **'Room'** number is greater than **6050**.

SQL Query

This tab provides an environment where sophisticated queries can be developed that confirm with the SQL 92 protocol. For example, you can select * from **Item type.Room. Tag_Number** where:

    Item_type = 'CPU' ; Room >= 6050 and <= 6150

and
Tag_number > 26072

The following example describes the complete procedure of creating a new query using all the tabs of the **Query Editor**.

Example 6   *General*

Create a new Query for the **Computer** database table and use all the tabs of the **Query Editor** to prepare a SQL query.

1. Right-click on the **Computer** in the **dbConnect Manager** dialog box to display the shortcut menu. Choose **New Query** from the shortcut menu to invoke the **New Query** dialog box (Figure 18-23). You can also invoke the dialog box by selecting **Computer** and choosing the **New Query** button from the toolbar in the **dbConnect Manager** dialog box.

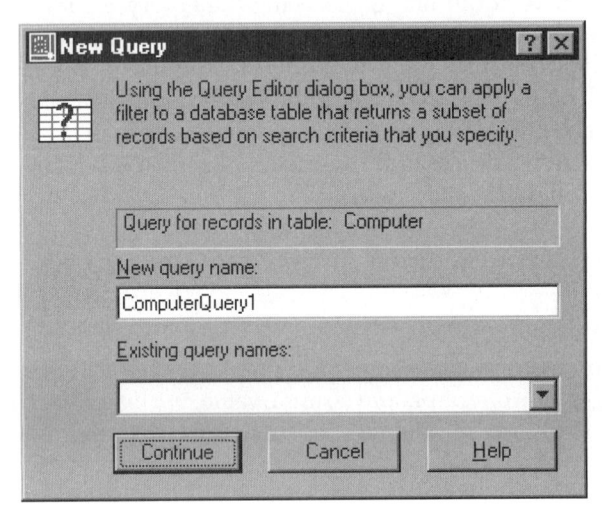

Figure 18-23 New Query dialog box

2. In the **New Query** dialog box, choose the **Continue** button to accept the default query name **ComputerQuery1**. AutoCAD will display the **Query Editor**.

3. In the **Quick Query** tab of **Query Editor** (Figure 18-24), select '**Item_Type**' from the **Field** list box, '**=Equal**' from the **Operator** drop-down list, and type '**CPU**' in the **Value** text box. You can also select '**CPU**' from the **Column Values** dialog box by choosing the **Look up values** button. Choose the **Store** button to save the query.

> **Note**
> *If you want to view the query choose the **Execute** button from the **Query Editor**, however you will not be able to continue with the example using same **Query** dialog box.*

4. Choose the **Range Query** tab (Figure 18-25), and select **Room** from the **Field** list box. Enter **6050** in the **From** edit box and **6150** in the **Through** edit box. You can also select

Accessing External Databases

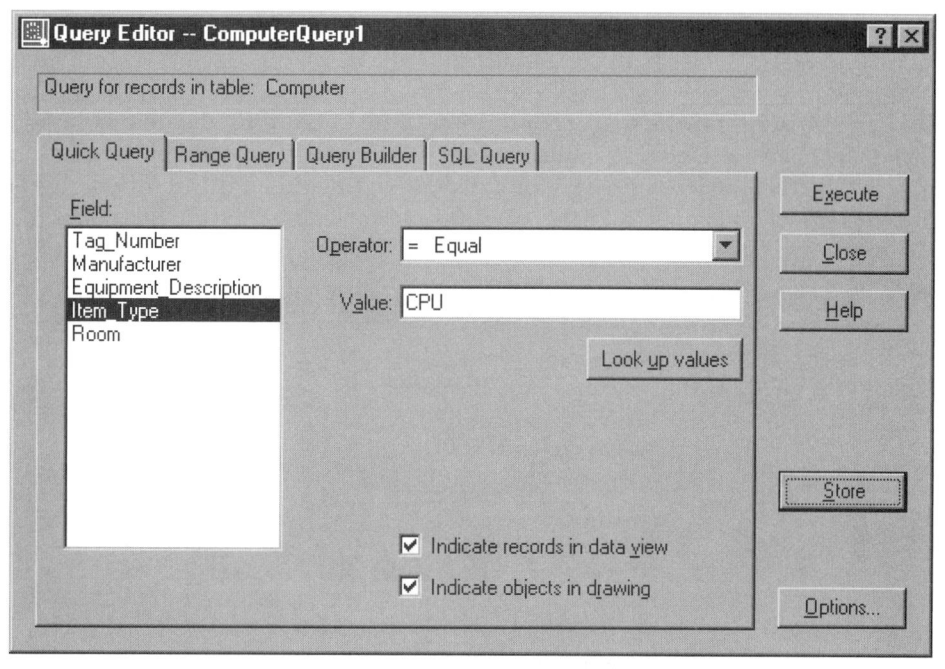

*Figure 18-24 Query Editor (**Quick Query** tab)*

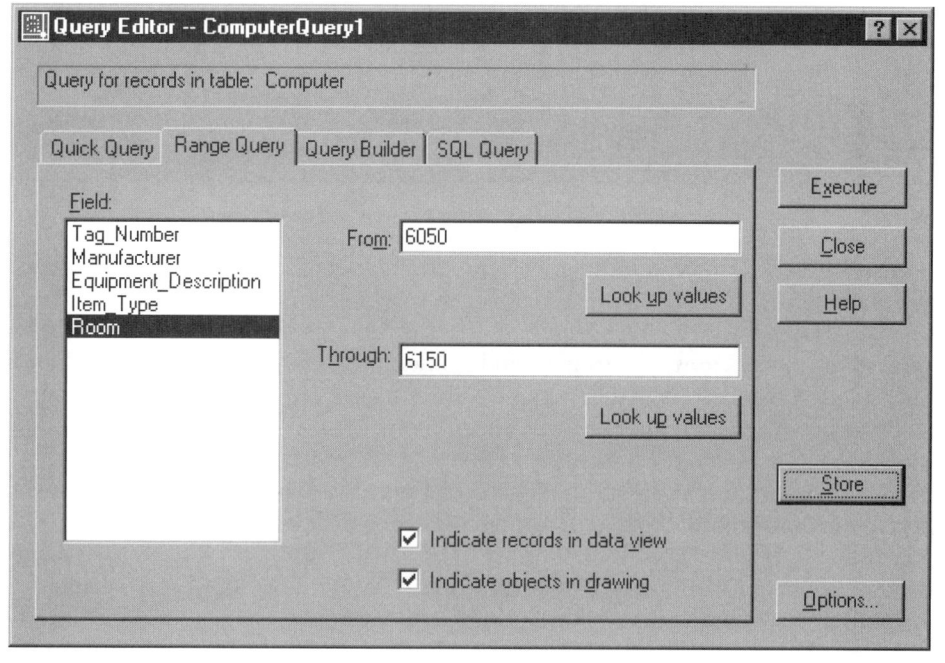

*Figure 18-25 Query Editor (**Range Query** tab)*

the values from the **Column Values** dialog box by choosing the **Look up values** button. Choose the **Store** button to save the query.

5. Choose the **Query Builder** tab, and add **Item_Type** in **Show fields:** list box by Selecting **Item_Type** from the **Fields in table:** list box and choosing the **Add** button. In the table area, change the entries of fields, Operator, Value, Logical, and Parenthetical grouping criteria as shown in Figure 18-26, by selecting each cell and selecting the values from the respective drop-down list or options lists. Choose the **Store** button to save the query.

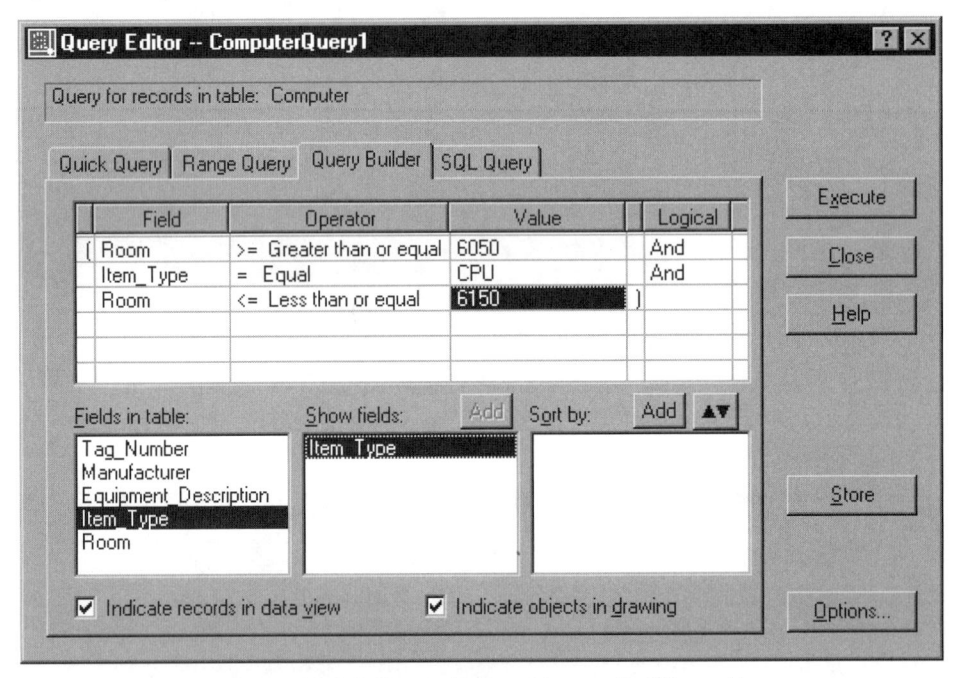

Figure 18-26 Query Editor (Query Builder tab)

6. Choose the **SQL Query** tab; the query conditions specified earlier will be carried to the tab automatically. Here you can create a query using multiple tables. But select **'Computer'** from the **Table** list box, **'Tag_Number'** from the **Fields** list box, **'>= Greater than or equal to'** from the **Operator** drop-down list and enter **26072** in the **Values** edit box. You can also select from the available values by choosing the [...] button (Figure 18-27). Choose the **Store** button to save the query.

7. To check whether the SQL syntax is correct, choose the **Check** button and AutoCAD will display the **Information box** to determine whether the syntax is correct.

8. Choose the **Execute** button to display the **Data View** window showing a subset of records matching the specified query criteria.

Note
You can view the subset of records conforming to your criterion with any of the four tabs during the creation of queries.

Accessing External Databases   18-21

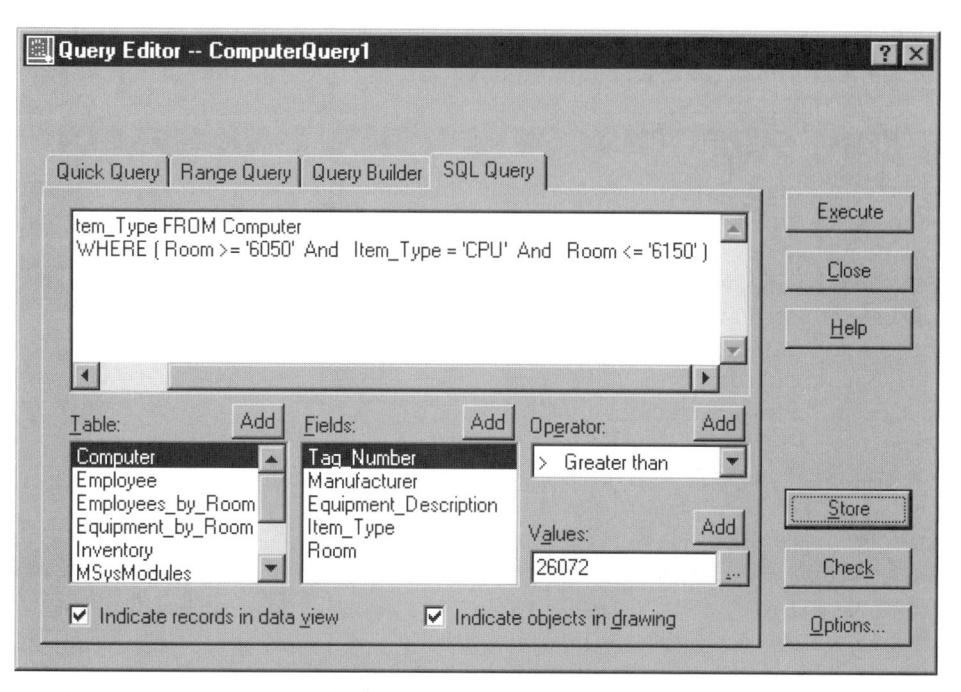

Figure 18-27 Query Editor (SQL Query tab)

Importing and Exporting SQL Queries

You may occasionally be required to use the queries made by some other user in your drawing or vice-versa. AutoCAD allows you to import or export stored queries. Sharing queries is very useful when developing common tools used by all team members on a project. Following is the procedure of **exporting a set of queries** from the current drawing:

1. From the **dbConnect** menu, choose **Queries > Export Query Set** to invoke the **Export Query Set** dialog box.

2. In the dialog box, select the directory you want to save the query set to from the **Save In** list.

3. Under the **File Name** box, specify a name for the query set, and then choose the **Save** button.

Following is the procedure of **importing a set of queries** into the current drawing:

1. From the **dbConnect** menu, choose **Queries > Import Query Set** to invoke the **Import Query Set** dialog box.

2. In the dialog box, select the query set to be imported.

3. Choose the **Open** button to import the query set into the current drawing.

FORMING SELECTION SETS USING LINK SELECT

AutoCAD displays an alert box that you can use to provide a unique name for the query, if there is a query with the same name that is already associated with the current drawing.

It is possible to locate objects on the drawing on the basis of the linked nongraphic information. For example, you can locate the object that is linked to the first row of the **Computer** table or to the first and second rows of the **Computer** table. You can highlight specified objects or form a selection set of the selected objects. The **Link Select** is an advanced feature of the **Query Editor** that can be used to construct iterative selection sets of AutoCAD graphical objects and the database records. You can start constructing a query or selecting AutoCAD graphical objects for iterative selection process. The initial selection set is referred to as **set A**. Now you can select an additional set of queries or graphical objects to further refine your selection set. The second selection set is referred to as **set B**. To refine your final selection set you must establish a relation between set A and set B. The following is the list of available relationships or set operations (Figure 18-28):

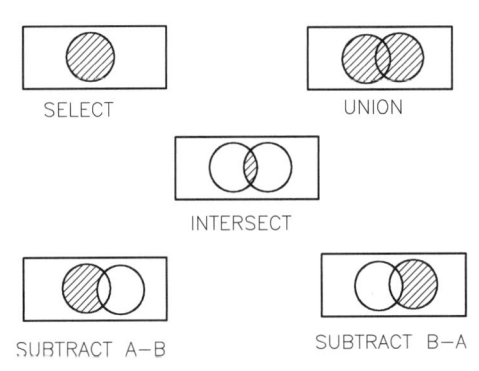

Figure 18-28 Available relationships or Set operations

Select

This creates an initial query or graphical objects selection set. This selection set can be further refined or modified using subsequent Link Select operations.

Union

This operation adds the outcome of a new selection set or query to the existing selection set. Union returns all the records that are members of set A **or** set B.

Intersect

This operation returns the intersection of the new selection set and the existing or running selection set. Intersection returns only the records that are common to both set A **and** set B.

Subtract A - B

This operation subtracts the result of the new selection set or query from the existing selection set.

Subtract B - A

This operation subtracts the result of the existing selection set or query from the new selection set.

After any of the Link Selection operation is executed, the result of the operation becomes the new running selection set and is assigned as set A. You can extend refining your selection set

Accessing External Databases 18-23

by creating a new set B and then continuing with the iterative process.

Following is the procedure to use **Link Select** for refining a selection set:

1. Choose **Links > Link Select** from the **dbConnect** menu to invoke the **Link Select** dialog box (Figure 18-29).

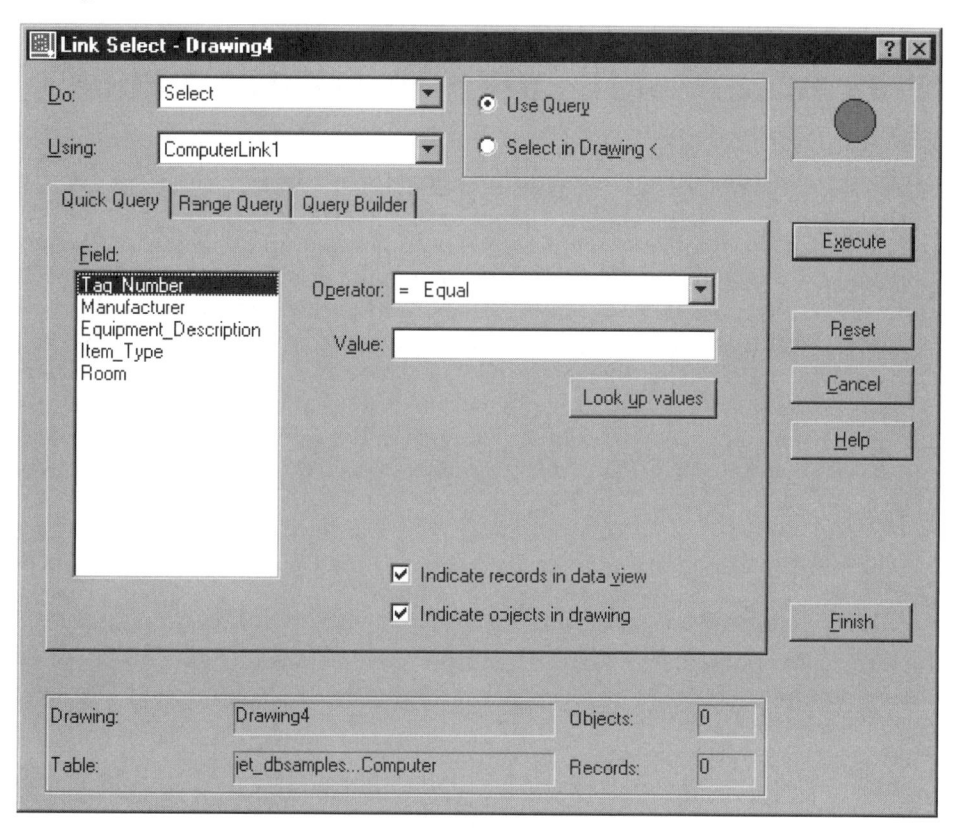

Figure 18-29 Link Select dialog box

2. Select the **Select** option from the **Do** drop-down list for creating a new selection set. Also select a link template from the **Using** drop-down list.

3. Choose either the **Use Query** or **Select in Drawing <** option for creating a new query or drawing object selection set.

4. Choose the **Execute** button to execute the refining operation or **Select** to add your query or graphical object selection set.

5. Again choose any **Link Selection** operation from the **Do:** drop-down list.

6. Repeat steps 2 through 4 for creating a set B to the **Link Select** operation and then choose the **Finish** button to complete the operation.

Note

*You can choose the **Use Query** option to construct a new query or **Select in Drawing** < to select a graphical object from the drawing as a selection set.*

*If you select **Indicate Records in Data View**, then the Link Select operation result will be displayed in the **Data View** window and if you select **Indicate objects in drawing**, then AutoCAD displays a set of linked graphics objects in the drawing.*

CONVERSION OF ASE LINKS TO AUTOCAD 2002 FORMAT

AutoCAD 2002 stores links in a different format than the previous releases. So if you want to work with your links in AutoCAD 2002, you need to convert them to the AutoCAD 2002 format. You will also have to create an AutoCAD 2002 configuration file that will point to the data source referred by the legacy links. When you open a drawing in AutoCAD 2002 where links were created by earlier releases, AutoCAD attempts to perform an automatic conversion of legacy information. But in some cases, where automatic conversion is not possible or a legacy data source might not match exactly the data source in AutoCAD 2002 format, you can use the **Link Conversion** dialog box. During the conversion process, AutoCAD writes the data source mapping information in **asi.ini** file and applies the information for further usage of the same data source. You can use the following steps to Convert AutoCAD R13 and R14 links to the AutoCAD 2002 format:

1. To invoke the **Link Conversion** dialog box, choose **Link Conversion** from the **dbConnect** menu.

2. The **Link Conversion** dialog box will be displayed as shown in Figure 18-30. In the **Old Link Format** area, enter the following specifications:

    Select R13/R14 link format to be converted.
    Name of the Environment, Catalog, Schema, and Table.
    Link path name of the link that has to be formatted.

3. In the **New Link Format**, enter the following specifications:

    Name of the AutoCAD 2002 data source (for example, **jet_dbsamples**).
    Name of the new AutoCAD 2002 Environment, Catalog, Schema, and Table.
    Name of the AutoCAD 2002 link template.

4. Choose **Apply** and then the **OK** button to complete the conversion procedure.

5. Open the drawing you want to convert and save it in AutoCAD 2002 format drawing.

Saving AutoCAD 2002 Links to R13/R14 Formats

You can convert Links created in AutoCAD 2002 format as R13 or R14 formats, but you cannot save AutoCAD 2002 links in R12 format. When you select SAVE AS for saving the links

Accessing External Databases

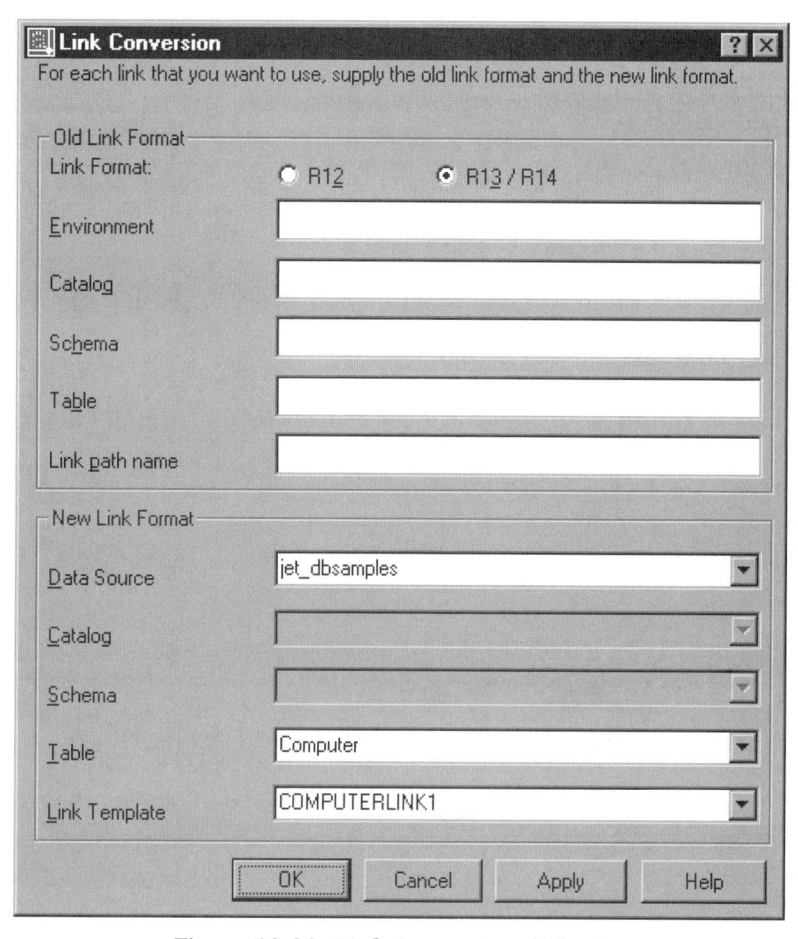

Figure 18-30 Link Conversion dialog box

with drawings, select links to R13 or R14 formats and AutoCAD will automatically convert into the format of earlier release.

Self-Evaluation Test

Answer the following questions and then compare your answers with the answers given at the end of this chapter.

1. The horizontal group of data is called a _____.

2. A vertical group of data (attribute) is called a _____.

3. You can connect to an external database table by using _____ Manager.

4. You can edit an external database table within AutoCAD session. (T/F)

5. You can resize, dock, as well as hide the **Data View** window. (T/F)

6. Once you define a link, AutoCAD does not store that information with the drawing. (T/F)

7. To display the associated records with the drawing objects, AutoCAD provides_____.

8. Labels are of two types, _____ and _____.

9. It is possible to import as well as export Link and label templates. (T/F)

10. ASE stands for _____.

Review Questions

Answer the following questions.

1. The SQL statements let you search through the database and retrieve the information as specified in the SQL statements. (T/F)

2. What are the various components of a table?

3. What is a database?
4. What is a database management system (DBMS)?

5. A row is also referred to as a _____.

6. A column is sometimes called a _____.

7. The _____ acts like an identification tag for locating and linking a row.

8. Only one row can be manipulated at a time. This row is called the _____.

9. **AutoCAD Query Editor** has four tabs, namely _____, _____, _____, and _____.

10. It is possible to go back to the previous query tab without resetting the values. (T/F)

11. You can execute query after any tab in **Query Editor.** (T/F)

Accessing External Databases 18-27

12. Link select is an _____ implementation of AutoCAD **Query Editor**, which constructs selection sets of graphical objects or database records.

13. AutoCAD writes data source mapping information in the _____ file during the conversion process of links.

14. AutoCAD 2002 links _____ be converted into AutoCAD R12 format.

Exercises

Exercise 1   *General*

In this exercise, you will select **Employee** table from **Jet_dbsamples** data source in **dbConnect Manager**, then edit the sixth row of the table (EMP_ID=1006, Keyser). You will also add a new row to the table, set the new row current, and then view it.

The row to be added has the following values:

EMP_ID	1064
LAST_NAME	Joel
FIRST_NAME	Billy
Gender	M
TITLE	Marketing Executive
Department	Marketing
ROOM	6071

Exercise 2   *General*

From the **Inventory** table in **Jet_dbsamples** data source, build an SQL query step-by-step using all the tabs of the AutoCAD Query Editor. The conditions to be implemented are as follows:

1. Type of Item = **Furniture**   and
2. Range of Cost = **200 to 650**   and
3. Manufacturer = **Office master**

Answers to the Self-Evaluation Test

1 - Rows, **2** - Columns, **3** - **dbConnect**, **4** - T, **5** - T, **6** - F, **7** - Leaders, **8** - Freestanding, Attached, **9** - T, **10** - **AutoCAD SQL Environment**

Chapter 19

Geometry Calculator

Learning Objectives
After completing this chapter, you will be able to:
• *Understand how the geometry calculator functions.*
• *Use real, integer, vector, and numeric expressions.*
• *Use snap modes in the geometry calculator.*
• *Obtain the radius of an object and locate a point on a line.*
• *Understand applications of the geometry calculator.*
• *Use AutoLISP variables and filter X, Y, and Z coordinates.*

GEOMETRY CALCULATOR
The **Geometry Calculator** is an ADS application that can be used as an online calculator. The calculator can be used to evaluate vector, real, and integer expressions. It can also access the existing geometry by using the first three characters of the standard AutoCAD object snap functions (MID, CEN, END). You can use the calculator to evaluate arithmetic and vector expressions. For example, you can use the calculator to evaluate an expression like 3.5 ^ 12.5*[234*log(12.5) - 3.5*cos(30)]. The results of a calculation can be returned as input to the current AutoCAD prompt.

Another application of the calculator is in assigning a value to an AutoLISP variable. For example, you can use an AutoLISP variable in the arithmetic expression, and then assign the value of the expression to an AutoLISP variable. You can invoke the **CAL** command by entering **CAL** or **'CAL** (for transparent use) at the AutoCAD Command prompt.

REAL, INTEGER, AND VECTOR EXPRESSIONS
Real and Integer Expressions
A **real expression** consists of real numbers and/or functions that are combined with numeric

operators. Similarly, an **integer expression** consists of integer numbers and/or functions combined with numeric operators. The following is the list of numeric operators:

Operator	Operation	Example
+	Adds numbers	2 + 3
-	Subtracts numbers	15.5 - 3.754
*	Multiplies numbers	12.34 * 4
/	Divides numbers	345.5/2.125
^	Exponentiation of numbers	25.5 ^ 2.5
( )	Used to group expressions	4.5 + (4.35 ^ 2)

Example
Command: **CAL**
Initializing...>> Expression: **(4.5 + (4.35 ^ 2))**
23.4225

Vector Expression

A vector expression consists of points, vectors, numbers, and functions that are combined with the following operators:

Operator	Operation / Example
+	Adds vectors [a,b,c] + [x,y,z] = [a + x, b + y, c + z] [2,4,3] + [5,4,7] = [2 + 5, 4 + 4, 3 + 7] = [7.0 8.0 10.0]
-	Subtracts vectors [a,b,c] - [x,y,z] = [a - x, b - y, c - z] [2,4,3] - [5,4,7.5] = [2 - 5, 4 - 4, 3 - 7.5] = [-3.0 0.0 -4.5]
*	Multiplies a vector by a real number a * [x,y,z] = [a * x, a * y, a * z] 3 * [2,8,3.5] = [3 * 2, 3 * 8, 3 * 3.5] = [6.0 24.0 10.5]
/	Divides a vector by a real number [x,y,z] / a = [x/a, y/a, z/a] [4,8,4.5] / 2 = [4/2, 8/2, 4.5/2] = [2.0 4.0 2.25]
&	Multiplies vectors [a,b,c] & [x,y,z] = [(b * z)-(c * y), (c * x)-(a * z), (a * y)-(b * x)] [2,4,6] & [3,5,8] = [(4 * 8)-(6 * 5), (6 * 3)-(2 * 8), (2 * 5)-(4 * 3)] = [2.0 2.0 -2.0]
( )	Used to group expressions a + (b ^ c)

Example
Command: **CAL**
Initializing...>> Expression: **[2,4,3] - [5,4,7.5]**
-3.0 0.0 4.5

NUMERIC FUNCTIONS

The **geometry calculator** (**CAL**) supports the following numeric functions:

Function	Description
sin(angle)	Calculates the **sine** of an angle
cos(angle)	Calculates the **cosine** of an angle
tang(angle)	Calculates the **tangent** of an angle
asin(real)	Calculates the **arcsine** of a number (The number must be between -1 and 1)
acos(real)	Calculates the **arccosine** of a number (The number must be between -1 and 1)
atan(real)	Calculates the **arctangent** of a number
ln(real)	Calculates the **natural log** of a number
log(real)	Calculates the **log, to the base 10**, of a number
exp(real)	Calculates the **natural exponent** of a number
exp 10(real)	Calculates the **exponent, to the base 10**, of a number
sqr(real)	Calculates the **square** of a number
sqrt(real)	Calculates the **square root** of a number
abs(real)	Calculates the **absolute value** of a number
round(real)	Rounds the number to the **nearest integer**
trunc(real)	Returns the **integer portion** of a number
r2d(angle)	Converts the **angle in radians** to degrees
d2r(angle)	Converts the **angle in degrees** to radians
pi	pi has a **constant value** (**3.14159**)

Example
Command: **CAL**
Initializing...>> Expression: **Sin (60)**
0.866025

USING SNAP MODES

You can use AutoCAD's snap modes with **CAL** functions to evaluate an expression. When you use snaps in an expression, AutoCAD will prompt you to select an object, and the returned value will be used in the expression. For example, if the **CAL** function is **(cen+end)/2**, the calculator will first prompt you to select an object for CENter snap mode and then select another object for ENDpoint snap mode. The corresponding coordinates of the two point values will be added and divided by **2**. The returned value is a point located midway between the center of the circle and the endpoint of the selected object. Following is the list of **CAL snap modes** and the corresponding AutoCAD snap modes:

CAL Snap Modes	AutoCAD Snap Modes
END	ENDpoint
EXT	EXTension
INS	INSert
INT	INTersection
MID	MIDpoint
CEN	CENter
NEA	NEArest
NOD	NODe
QUA	QUAdrant
PAR	PARallel
PER	PERpendicular
TAN	TANgent

Example 1

In this example, you will use the **CAL snap modes** to retrieve the point values (coordinates), and then use these values to draw a line (P3,P4). It is assumed that the circle and line (P1,P2) are already drawn (Figure 19-1).

Command: Choose the **Line** button.
Specify first point: **'CAL**
>> Expression: **(cen+end)/2**
>> Select entity for CEN snap: *Select the circle.*
>> Select entity for END snap: *Select one end of line (P1,P2).*
Resuming **LINE** command.
Specify next point or [Undo]: **'CAL**
>> Expression: **(cen+end)/2**
>> Select entity for CEN snap: *Select the circle.*
>> Select entity for END snap: *Select the other end of line (P1,P2).*

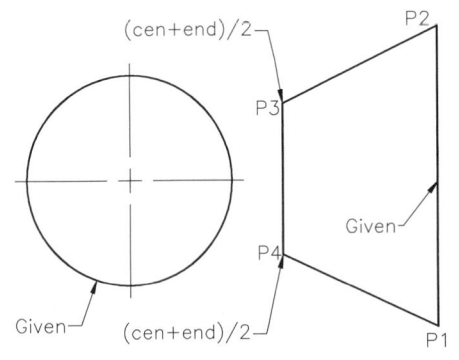

Figure 19-1 Using CAL snap modes

Now, join point (P4) with point (P1) and point (P3) with point (P2) to complete the drawing, as shown in Figure 19-1. The '**CAL** function initializes the geometry calculator transparently. The single quote in front of the cal function (') makes the cal function transparent. The expression **(cen+end)/2** will prompt you to select the objects for **CEN snap** and **END snap**. After you select these objects, the calculator will add the corresponding coordinate of these point values and then divide the sum of each coordinate by 2. The point value that this function returns is the midpoint between the center of the circle and the first endpoint of the line.

OBTAINING THE RADIUS OF AN OBJECT

You can use the **rad** function to obtain the radius of an object. The object can be a circle, an arc, or a 2D polyline arc.

Example 2

In this example, you are given a circle of certain radius (R). You will draw a second circle whose radius is 0.75 times the radius of the given circle (0.75 * R) (Figure 19-2).

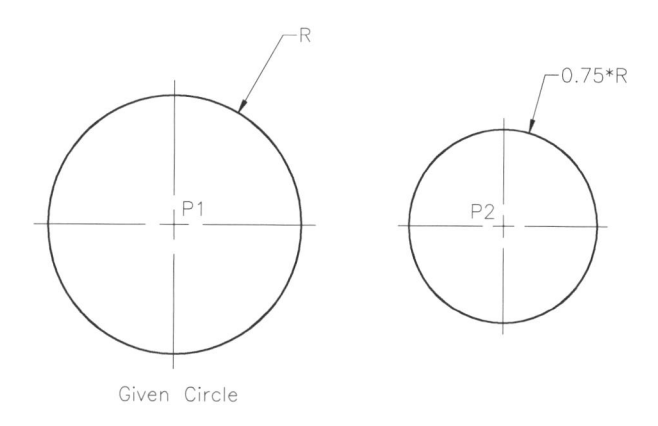

Figure 19-2 Obtaining the radius of an object

Command: **CIRCLE**
Specify center point for circle or [3P/2P/Ttr (tan tan radius)]: *Select a point (P2).*
Specify radius of circle or [Diameter] <current>: **'CAL**
>> Expression: **0.75*rad**
>> Select circle, arc or polyline segment for RAD function: *Select the given circle.*

Now you can enter the function name in response to the calculator prompt (**>>Expression**). In this example, the expression is **0.75*rad**. The **rad** function prompts the user to select an object, and it retrieves its radius. This radius is then multiplied by 0.75. The product of rad and 0.75 determines the radius of the new circle.

LOCATING A POINT ON A LINE

You can use the functions **pld** and **plt** to locate a point at a specified distance along a line between two points. The format of the **pld** function is **pld(p1,p2,dist)**. This function will locate a point on line (P1,P2) that is at a distance of **dist** from point (P1). For example, if the function is pld(p1,p2,0.7) and the length of the line is 1.5, the calculator will locate a point at a distance of 0.7 from point (P1) along line (P1,P2).

The format of the **plt** function is **plt(p1,p2,t)**. This function will locate a point on line (P1,P2) at a proportional distance as determined by the parameter, **t**. If **t = 0**, the point that this function will locate is at P1. Similarly, if **t = 1**, the point is located at (P2). However, if the value of **t** is greater than 0 and less than 1 (0> t <1), then the location of the point is determined by the value of **t**. For example, if the function is plt(p1,p2,0.3) and the length of the line is 1.5, the calculator will locate the point at a distance of 0.3 * 1.5 = 0.45 from point (P1).

Example 3

In this example, you will use the **pld** and **plt** functions to locate the centers of the circles. Lines (P1,P2) and (P3,P4) are given (Figure 19-3).

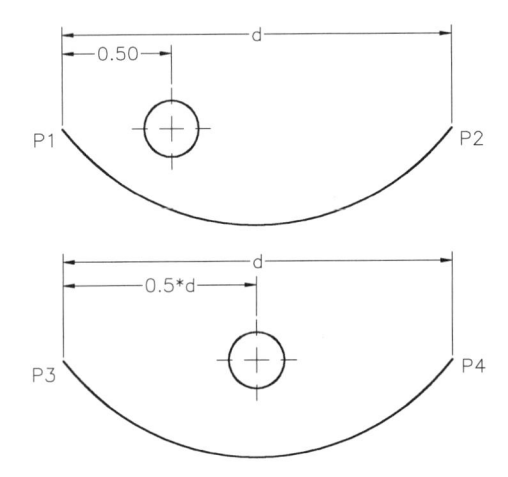

Figure 19-3 Locating a point using calculator

Figure 19-3 illustrates the use of the **plt** and **pld** functions. The **pld** function, **pld(end,end,0.5)**, will prompt the user to select the two endpoints of the arc (P1,P2), and the function will return a point at a distance of 0.5 from point P1.

> Command: Choose the **Circle** command.
> Specify center point for circle or [3P/2P/Ttr (tan tan radius)]: **'CAL**
> \>> Expression: **pld(end,end,0.5)**
> \>> Select entity for END snap: *Select the point P1.*
> \>> Select entity for END snap: *Select the point P2.*
> Specify radius of circle or [Diameter] <current>: *Enter radius.*

Similarly, the **plt** function, **plt(end,end,0.5)**, will prompt the user to select the two endpoints of line (P3,P4); it will return a point that is located at a distance of 0.5*d units from point (P3).

> Command: Choose the **Circle** command.
> Specify center point for circle or [3P/2P/Ttr (tan tan radius)]: **'CAL**
> \>>Expression: **plt(end,end,0.5)**
> \>>Select entity for END snap: *Select the point P3.*
> \>>Select entity for END snap: *Select the point P4.*
> Specify radius of circle or [Diameter] <current>: *Enter radius.*

OBTAINING AN ANGLE

You can use the **ang** function to obtain the angle between two lines. The function can also be used to obtain the angle that a line makes with the positive X axis. The function has the

Geometry Calculator

following formats (Figure 19-4):

   ang(v)
   ang(p1,p2)
   ang(apex,p1,p2)
   ang(apex,p1,p2,p)

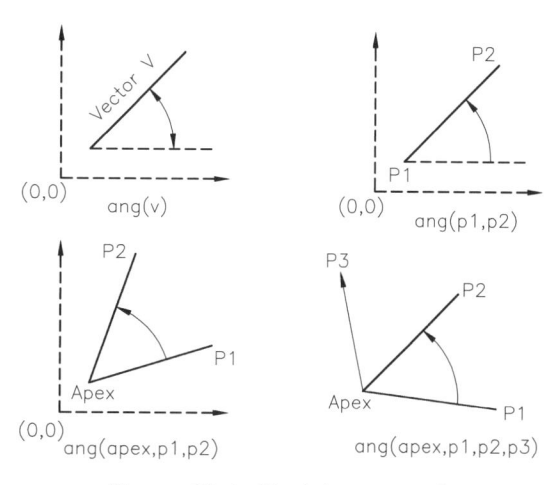

Figure 19-4 Obtaining an angle

ang(v)

The **ang(v)** function can be used to obtain the angle that a vector makes with the positive X axis. Assume a vector [2,2,0], and obtain its angle. You can use the ang(v) function to obtain the angle.

   Command: **CAL**
   \>\>Expression: **v=[2,2,0]**   *(Defines a vector v.)*
   Command: **CAL**
   \>\>Expression: **ang(v)**   *(v is a predefined vector.)*
   45.0

The vector v makes a 45-degree angle with the positive X axis.

ang(p1,p2)

The **ang(p1,p2)** function can be used to obtain the angle that a line (P1,P2) makes with the positive X axis. For example, if you want to obtain the angle of a line with start point and endpoint coordinates of (1,1,0) and (4,4,0), respectively:

   Command: **CAL**
   \>\>Expression: **p1=[1,1,0]**   *(Defines a vector p1.)*
   Command: **CAL**
   \>\>Expression: **p2=[4,4,0]**   *(Defines a vector p2.)*

Command: **CAL**
>>Expression: **ang(p1,p2)**
45.00

If the line exists, you can obtain the angle by using the following function:

Command: **CAL**
>>Expression: **ang(end,end)**
>>Select entity for END snap: *Select first endpoint of line (P1,P2).*
>>Select entity for END snap: *Select second endpoint of line (P1,P2).*
31.7134   *(This is the angle that the function returns.)*

ang(apex,p1,p2)

The **ang(apex,p1,p2)** function can be used to obtain the angle that a line (apex,P1) makes with (apex,P2). For example, if you want to obtain the angle between the two given lines as shown in the third drawing of Figure 19-4, use the following commands:

Command: **CAL**
>>Expression: **ang(end,end,end)**
>>Select entity for END snap: *Select first endpoint (apex).*
>>Select entity for END snap: *Select second endpoint (P1).*
>>Select entity for END snap: *Select third endpoint (P2).*
51.41459   *(This is the angle that the function returns.)*

ang(apex,p1,p2,p)

The **ang(apex,p1,p2,p)** function can be used to obtain the angle that a line (apex,P1) makes with (apex,P2). The last point, **p**, is used to determine the orientation of the angle.

LOCATING THE INTERSECTION POINT

You can obtain the intersection point of two lines (P1,P2) and (P3,P4) by using the following function:

   ill(p1,p2,p3,p4)

(P1,P2) are two points on the first line, and (P3,P4) are two points on the second line, as shown in Figure 19-5.

Example 4

In this example, you will draw a circle whose center point is located at the intersection point of two lines (P1,P2) and (P3,P4). It is assumed that the two lines are given as shown in Figure 19-5.

Use the following commands to obtain the intersection point and to draw a circle with the intersection point as the center of the circle:

Geometry Calculator

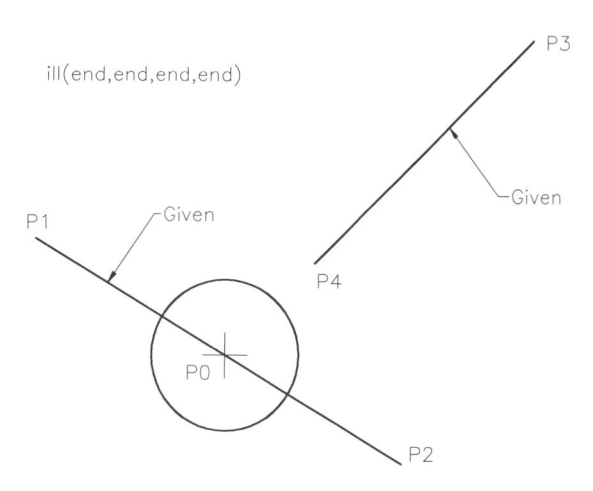

19-9

Figure 19-5 Obtaining intersection point

Command: Choose the **Circle** button.
Specify center point for circle or [3P/2P/Ttr (tan tan radius)]: **'CAL**
>>Expression: **ill(end,end,end,end)**
>>Select entity for END snap: *Select point P1.*
>>Select entity for END snap: *Select point P2.*
>>Select entity for END snap: *Select point P3.*
>>Select entity for END snap: *Select point P4.*
Specify radius of circle or [Diameter] <current>: *Enter radius.*

> **Tip**
> *The expression ill (end,end,end,end) can be replaced by the shortcut function **ille**. When used, it will automatically prompt for four endpoints to locate the point of intersection.*

APPLICATIONS OF THE GEOMETRY CALCULATOR

The following examples illustrate some additional applications of the geometry calculator.

Example 5

In this example, you will draw a circle whose center (P0) is located midway between endpoints (P4) and (P2), see Figure 19-6.

The center of the circle can be located by using the calculator snap modes. For example, to locate the center you can use the expression **(end+end)/2**. The other way of locating the center is by using the shortcut function **mee**, as shown here (Figure 19-6):

Command: Choose the **Circle** command.
Specify center point for circle or [3P/2P/Ttr (tan tan radius)]: **'CAL**
>>Expression: **mee**

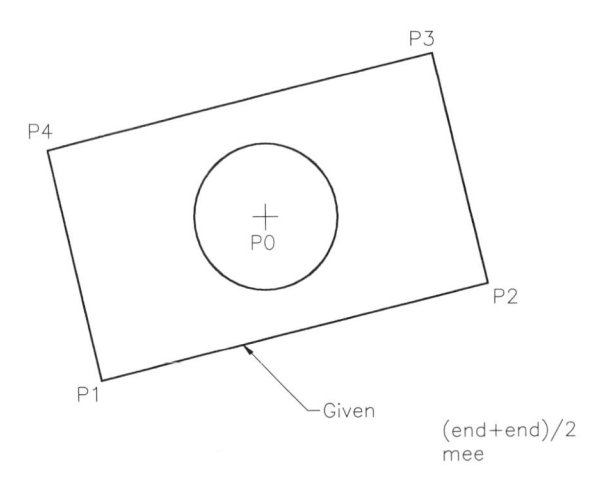

*Figure 19-6 Using the shortcut function **mee***

>>Select entity for END snap: *Select the first endpoint (P2).*
>>Select entity for END snap: *Select the second endpoint (P4).*
Specify radius of circle or [Diameter] <current>: *Enter the radius.*

Example 6

In this example, you will draw a circle that is tangent to a given line. The radius of the circle is 0.5 units, and the circle must pass through the selected point shown in Figure 19-7.

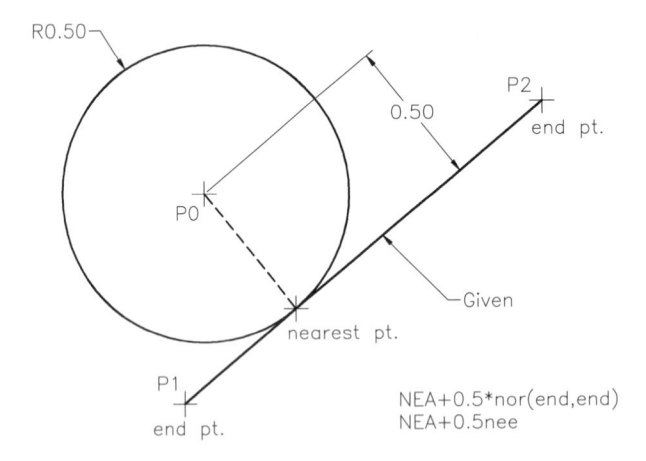

*Figure 19-7 Using the shortcut function **nee***

To draw a circle that is tangent to a line, you must first locate the center of the circle that is at a distance of 0.5 units from the selected point. This can be accomplished by using the function **nor(p1,p2)**, which returns a unit vector normal to the line (P1,P2). You can also use the shortcut function **nee**. The function will automatically prompt you to select the two endpoints of the

Geometry Calculator

19-11

given line. The unit vector must be multiplied by the radius (0.5) to locate the center of the circle.

> Command: Choose the **Circle** command.
> Specify center point for circle or [3P/2P/Ttr (tan tan radius)]: **'CAL**
> \>>Expression: **NEA+0.5*nee**
> \>>Select entity for NEA snap: *Select a point on the given line.*
> \>>Select one endpoint for NEE: *Select the first endpoint on the given line.*
> \>>Select another endpoint for NEE: *Select the second endpoint on the given line.*
> Specify radius of circle or [Diameter] <current>: **0.5**

Example 7

In this example, you will draw a circle with its center on a line. The radius of the circle is 0.25 times the length of the line, and it is assumed that the line is given (Figure 19-8).

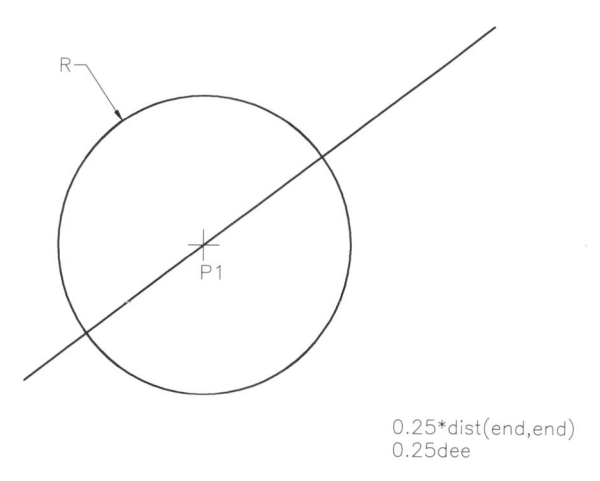

*Figure 19-8 Using the shortcut function **dee***

The radius of the circle can be determined by multiplying the length of the line by 0.25. The length of the line can be obtained by using the function dist(p1,p2) or by using the shortcut function **dee**. When you use the function **dee**, the calculator will automatically prompt you to select the two endpoints of the given line. It is equivalent to using the function dist(end,end).

> Command: Choose the **Circle** command.
> Specify center point for circle or [3P/2P/Ttr (tan tan radius)]: *Select a point on the line.*
> Specify radius of circle or [Diameter] <current>: **'CAL**
> \>>Expression: **0.25*dee**
> \>>Select entity for END snap: *Select the first endpoint on the given line.*
> \>>Select entity for END snap: *Select the second endpoint on the given line.*

USING AUTOLISP VARIABLES

The geometry calculator allows you to use an AutoLISP variable in the arithmetic expression. You can also use the calculator to assign a value to an AutoLISP variable. The variables can be integer, real, or a 2D or 3D point. The next example illustrates the use of AutoLISP variables.

Example 8

In this example, you will draw two circles that are in the middle and are offset 0.5 units from the center. It is assumed that the other two circles are given.

To locate the center of the top circle, you must first determine the point that is midway between the centers of the two given circles. This can be accomplished by defining a variable **midpoint** where **midpoint=(cen+cen)/2**. Similarly, you can define another variable for the offset distance: **offset=[0,0.5]**. The center of the circle can be obtained by adding these two variables **(midpoint+offset)** (Figure 19-9).

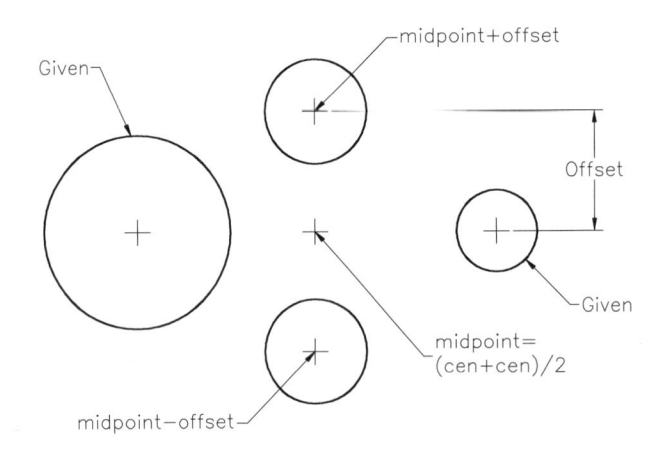

Figure 19-9 Adding two predefined vectors

Command: **CAL**
\>\>Expression: **midpoint=(cen+cen)/2**
\>\>Select entity for CEN snap: *Select the first circle.*
\>\>Select entity for CEN snap: *Select the second circle.*

Command: **CAL**
\>\>Expression: **offset=[0,0.5]**

Command: Choose the **Circle** button.
Specify center point for circle or [3P/2P/Ttr (tan tan radius)]: **'CAL**
\>\>Expression: **(midpoint+offset)**
Specify radius of circle or [Diameter] <current>: *Enter radius.*

Geometry Calculator

To locate the center point of the bottom circle, you must subtract offset from midpoint:

Command: Choose the **Circle** button.
Specify center point for circle or [3P/2P/Ttr (tan tan radius)]: **'CAL**
>>Expression: **(midpoint-offset)**
Specify radius of circle or [Diameter] <current>: *Enter radius.*

The same results can be obtained by using AutoLISP expressions, as follows:

Command: **(Setq NEWPOINT "(CEN+CEN)/2+[0,0.5]")**
Command: Choose the **Circle** button.
Specify center point for circle or [3P/2P/Ttr (tan tan radius)]: **(cal NEWPOINT)**
(Recalls the expression.)
>>Select entity for CEN snap: *Select the first circle.*
>>Select entity for CEN snap: *Select the second circle.*

FILTERING X, Y, AND Z COORDINATES

The following functions are used to retrieve the coordinates of a point.

Function	Description
xyof(p)	Retrieves the **X** and **Y** coordinates of a point (p) and returns a point; the Z coordinate is automatically set to 0.0
xzof(p)	Retrieves the **X** and **Z** coordinates of a point (p) and returns a point; the Y coordinate is automatically set to 0.0
yzof(p)	Retrieves the **Y** and **Z** coordinates of a point (p) and returns a point; the X coordinate is automatically set to 0.0
xof(p)	Retrieves the **X coordinate** of a point (p) and returns a point; the Y and Z coordinates are automatically set to 0.0
yof(p)	Retrieves the **Y coordinate** of a point (p) and returns a point; the X and Z coordinates are automatically set to 0.0
zof(p)	Retrieves the **Z coordinate** of a point (p) and returns a point; the X and Y coordinates are automatically set to 0.0
rxof(p)	Retrieves the **X coordinate** of a point (p)
ryof(p)	Retrieves the **Y coordinate** of a point (p)
rzof(p)	Retrieves the **Z coordinate** of a point (p)

Example 9

In this example, you will draw a line by using filters to extract coordinates and points. It is assumed that the two lines are as shown in Figure 19-10.

To draw a line, you need to determine the coordinates of the two endpoints of the line. The X coordinate of the first point can be obtained from point (P1), and the Y coordinate from point (P2). To obtain these coordinate points, you can use the filter function **rxof(end)** to extract the X coordinate and **ryof(end)** to extract the Y coordinate. To determine the coordinates of the

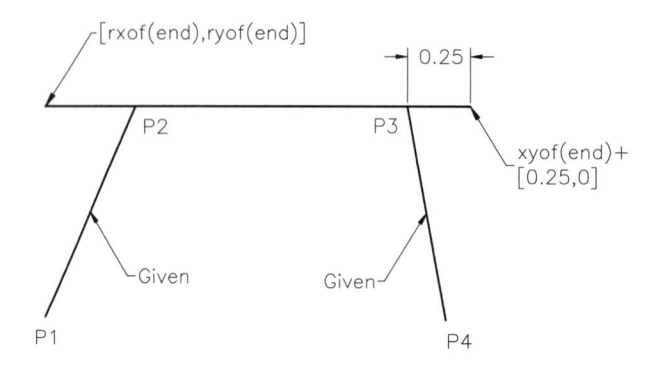

Figure 19-10 Using filters to extract points and coordinates

endpoint of the line, you can filter the XY coordinates of point (P3) and then add the offset of 0.25 units by defining a vector [0.25,0,0].

> Command: Choose the **Line** button.
> Specify first point: 'cal
> \>\>Expression: **[rxof(end),ryof(end)]**
> \>\>Select entity for END snap: *Select point P1.*
> \>\>Select entity for END snap: *Select point P2.*
> Specify next point or [Undo]: **'CAL**
> \>\>Expression: **xyof(end)+[0.25,0,0]**
> \>\> Select entity for END snap: *Select point P3.*

CONVERTING UNITS

You can use the calculator function **cvunit** to change a given value from one system of units to another. You can also use this function to change the unit format. For example, you can change the units from feet to inches, meters to centimeters, and vice versa. The value can be a number or a point. The units available are defined in the ACAD.UNT file which is an ASCII file that you can examine. The format of the cvunit expression is:

> **cvunit(value, units from, units to)**

> **Examples**
> Command: **CAL**
> \>\>Expression: **cvunit(100,cm,inch)**
> \>\>Expression: **cvunit(100,feet,meter)**
> \>\>Expression: **cvunit(1,feet,inch)**

ADDITIONAL FUNCTIONS

The following is the list of additional calculator functions. The description given next to the

Geometry Calculator

function summarizes the application of the function.

Function	Description
abs(real)	Calculates the **absolute value** of a number
abs(v)	Calculates the **length of vector v**
ang(v)	Calculates the **angle** between the X axis and vector v
ang(p1,p2)	Calculates the **angle** between the X axis and line (P1,P2)
cur	Retrieves **coordinates of a point** from the location of the graphics cursor
cvunit(val,from,to)	**Converts the given value** (val) from one unit measurement system to another
dee	Measures the **distance between two endpoints**; equivalent to the function dist(end,end)
dist(p1,p2)	**Measures distance** between two specified points (P1,P2)
getvar(var_name)	Retrieves the value of the AutoCAD **system variable**
ill(p1,p2,p3,p4)	Returns the **intersection point** of lines (P1,P2) and (P3,P4)
ille	Returns the **intersection point** of lines defined by four endpoints; equivalent to the function ill(end,end,end,end)
mee	Returns the **midpoint** between two endpoints; equivalent to the function (end+end)/2
nee	Returns a **unit vector** normal to two endpoints; equivalent to the function nor(end,end)
nor	Returns a **unit vector** that is normal to a circle or an arc
nor(v)	Returns a **unit vector** in the **xy** plane that is normal to vector v
nor(p1,p2)	Returns a **unit vector** in the **xy** plane that is normal to line (P1,P2)
nor(p1,p2,p3)	Returns a **unit vector** that is normal to the specified plane defined by points (P1,P2,P3)
pld(p1,p2,dist)	**Locates a point** on the line (P1,P2) that is **dist** units from point (P1)
plt(p1,p2,t)	Locates a point on the line (P1,P2) that is **t*dist** units from point (P1) (Note: when t = 0, the point is (P1). Also, when t = 1, the point is (P2))
rad	**Retrieves the radius** of the selected object
rot(p,org,ang)	Returns a point that is rotated through angle **ang** about point org
u2w(p)	Locates a point with respect to WCS from the current UCS
vec(p1,p2)	**Calculates a vector** from point (P1) to point (P2)
vec(p1,p2)	**Calculates a unit vector** from point (P1) to point (P2)
vee	**Calculates a vector** from two endpoints; equivalent to the function vec(end,end)
vee1	Calculates a unit vector from two endpoints; equivalent to the function vec1(end,end)
w2u(p)	Locates a point with respect to the current UCS from WCS

Self-Evaluation Test

Answer the following questions, and then compare your answers to the correct answers given at the end of this chapter.

1. A real expression consists of real numbers and/or functions that are combined with numeric operators. (T/F)

2. The AutoCAD snap modes can be used with calculator functions to evaluate an expression. (T/F)

3. The **ill** function can be used to locate the intersection of two lines. (T/F)

4. The **dee** function is used to locate the intersection of two lines. (T/F)

5. The length of a line can be obtained by using the function **dist(p1,p2)** or by using the shortcut function _____ .

6. The function **xzof(p)** retrieves the _____ coordinates of a point (P) and returns a point. The Y coordinate is automatically set to (0,0).

7. The function **nor(v)** returns a _____ in the XY plane that is normal to vector v.

8. The **ang(v)** function can be used to obtain the angle that a vector makes with the _____ axis.

9. The _____ function can be used to obtain the angle that a line (P1,P2) makes with the positive X axis.

10. The shortcut function **nee** is equivalent to _____ function.

Review Questions

Answer the following questions.

1. The calculator can also access the existing geometry by using standard AutoCAD object snap functions. (T/F)

2. The calculator cannot be used to assign a value to an AutoLISP variable. (T/F)

3. The **rad** function retrieves the radius of the selected object. (T/F)

4. The **ang(v)** function calculates the angle between the X axis and the line (P1,P2). (T/F)

Geometry Calculator 19-17

5. Which function is used to calculate the absolute value of a number?

  (a) **abs(real)**   (b) **abs(v)**
  (c) **abs(P1,P2)**   (d) **None**

6. Which function is used to retrieve the X and the Y coordinates of a point P and return a point?

  (a) **xyof(p)**   (b) **zxof(p)**
  (c) **xy(p)**   (d) **None**

7. Which function is used to locate a point on line (P1,P2) that is at a distance of d from P1?

  (a) **plt(p1,p2,d)**   (b) **pld(p1,p2,d)**
  (c) **plt(p2,p1,d)**   (d) **None**

8. Which function is used to retrieve the Y coordinates of a point P and return a point?

  (a) **xyof(p)**   (b) **xof(p)**
  (c) **yof(p)**   (d) **None**

9. What is the short form of the function vec(end,end)?

  (a) **vee**   (b) **eve**
  (c) **vem**   (d) **None**

10. The _____ function returns a unit vector normal to the two endpoints.

11. The _____ can be used to obtain the radius of an object.

12. The _____ function returns a point that is rotated through angle ang about a point org.

13. The format of the **pld** function is _____.

14. The format of the **plt** function is _____.

15. The angle between two given lines can be obtained by _____.

Exercises

Exercise 1 — General

Make the drawing shown in Figure 19-11. Use the real operators of the geometry calculator to calculate the value of L, H, and TL.

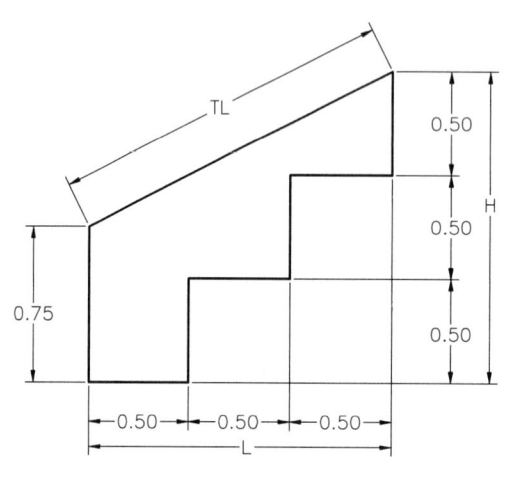

Figure 19-11 Drawing for Exercise 1

Exercise 2 — General

Make the drawing shown in Figure 19-12; assume the dimensions. Draw a circle whose center is at point (P3). Use the calculator function to locate the center of the circle (P3) that is midway between (P1) and (P2). The points (P1) and (P2) are the midpoints on the top and bottom lines, respectively.

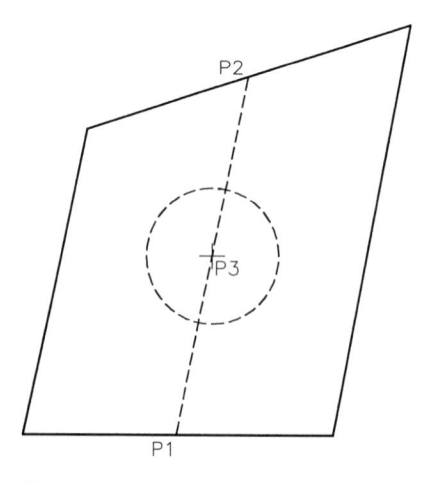

Figure 19-12 Drawing for Exercise 2

Geometry Calculator

Exercise 3 — *General*

Make the drawing shown in Figure 19-13; assume the dimensions.

1. Use the calculator function to locate the point (P3). Point (P3) is midway between (P1) and (P2). The points (P1) and (P2) are the midpoints on the top and the bottom lines, respectively.

2. Use the calculator function to locate the point (P4) that is normal to the line (P1,P2) at a distance of 0.25 units.

3. Draw a circle whose center is located at (P4).

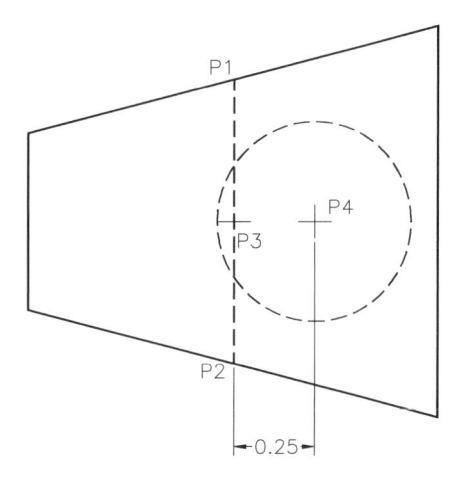

Figure 19-13 Drawing for Exercise 3

Problem Solving Exercise 1 — *General*

Make the drawing shown in Figure 19-14; assume the dimensions.

1. Draw a circle whose center is at point (P3). Use the calculator function to locate the center of the circle (P3) that is midway between (P1) and (P2). Points (P1) and (P2) are the midpoints on the top and bottom lines, respectively.

2. Draw a circle whose radius is 0.75 times the radius of the circle in part 1. The center of the circle (P5) is located midway between the line (P3,P4).

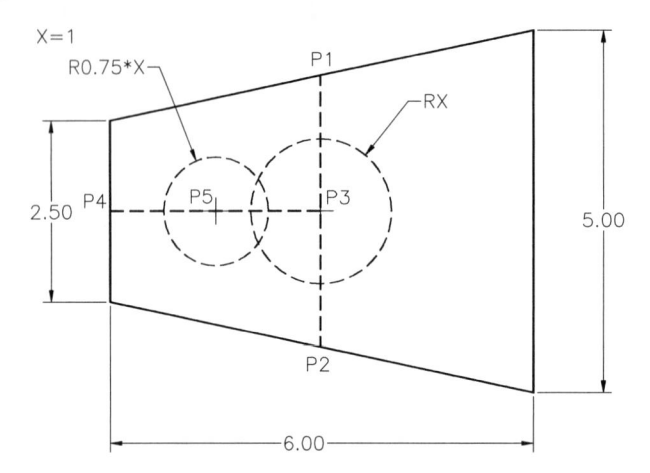

Figure 19-14 Drawing for Problem Solving Exercise 1

Answers to Self-Evaluation Test
1 - T, **2** - T, **3** - T, **4** - F, **5** - dee, **6** - X and Z, **7** - unit vector, **8** - positive X, **9** - ang(p1,p2), **10** - nor(p1,p2)

Index

Symbols

$value 15-41
***AUX1 10-38, 10-39
***BUTTONS 10-37
***BUTTONS1 7-3, 10-36
***HELPSTRING 5-39
***IMAGE 6-3, 6-6, 6-13, 10-51
***POP1 6-6, 6-13
***TABLET1 10-23, 10-26
**aliasname 5-35
3D drawing 1-12

A

ABOUT Visual Basic 17-1
Absolute Number 12-4
ACAD.LIN 3-1, 3-13
ACAD.MNU 5-1, 5-2, 5-10, 6-9, 8-7, 9-6, 10-1, 10-21, 10-34
ACAD.MNU file 5-3
ACAD.PAT 3-25, 3-28
ACAD.PGP 2-1
accelerator keys 5-32
action_tile 15-14
ActiveX Automation 17-2
Add method 17-2
Add to History 12-15
AddArc 17-4
AddCircle 17-3
Addition 12-2, 16-10
additional VBA examples 17-19
AddLine method 17-3, 17-13
AddText 17-4
algorithm 12-38
alias 5-35
alignment Attribute 15-8
alignment definition 3-8
alignment field specification 3-2
ALIGNMENT SPECIFICATION 3-9

ALTERNATE LINETYPES 3-12
ang function 19-6
ang(apex,p1,p2) function 19-8
ang(apex,p1,p2,p) function 19-8
ang(p1,p2) function 19-7
ang(v) function 19-7
ANGBASE 12-26, 12-28, 17-11
ANGDIR 12-26, 12-28, 17-11
AngleFromXAxis method 17-14
angtos 12-6, 16-13
APPLICATIONS OF THE GEOMETRY CALCULATOR 19-9
ASCII code 11-20
ASCII codes 11-20
ASCII control character 9-37
ASCII control characters 9-35
ASCII files 11-1
aspect_ratio and color Attributes 15-36
aspect_ratio Attribute 15-36
ASSIGNING COMMANDS TO A TABLET 8-12
Assigning Keyboard Shortcuts to Commands 5-44
assoc 8, 9, 11, 14-8, 14-9, 14-11
atan 12-5
atof and rtos Functions 15-22
AutoCAD 2002 Today 1-5
AutoCAD 2002 Today dialog box 1-7
AutoCAD Database Connectivity 18-3
AutoCAD menu file 8-2
AutoCAD Object Library 17-2
AutoCAD Query Editor 18-17
AutoCAD SQL Environment (ASE) 18-16
AutoCAD tablet template 8-2
AutoLISP 9-39, 9-40, 10-45, 1, 16, 14-1, 14-16
AUTOLISP FUNCTIONS 15-13, 15-20, 15-39
AUTOMATIC MENU SWAPPING 8-13, 9-33

B

BASE 1-2
BASIC 17-1
big_increment 15-36
BLIPMODE 1-2
Boxed Column Tile 15-17
Boxed Radio Column Tile 15-25
Boxed Row Tile 15-16
btnname 5-35
bulge factor 11-14
BUTTON AND TEXT TILES 15-4
button menu 7-5, 7-6
button tile 15-4
BUTTONS 10-35
buttons and auxiliary 10-34
Bylayer 1-2

C

CADR 9-39
cadr 12-21, 12-25
CAL snap modes 19-3, 19-4
CAR 9-39
car 12-20, 12-25
car, cdr, AND cadr FUNCTIONS 12-20
cdr 12-21
CELTSCALE 3-16
CHAMFERA 1-2, 12-17
chamfera 12-19
CHAMFERB 1-2, 12-17
chamferb 12-19
CMDECHO 9-39
cmdecho 12-24
Code 000: End of Shape Definition 11-6
Code 001: Activate Draw Mode 11-7
Code 002: Deactivate Draw Mode 11-7
Code 003: Divide Vector Lengths by Next Byte 11-9
Code 004: Multiply Vector Lengths by Next Byte 11-9
Code 007: Subshape 11-11
Code 008: X-Y Displacement 11-11
Code 009: Multiple X-Y Displacements 11-11
Code 00A or 10: Octant Arc 11-11
Code 00C or 12: Arc Definition by Displacement and 11-14
Code 00D or 13: Multiple Bulge-Specified Arcs 11-14
Code 00E or 14: Flag Vertical Text 11-15
Codes 005 and 006: Location Save/Restore 11-9
COLOR 1-2
color Attribute 15-37
COLUMN, BOXED COLUMN, AND TOGGLE TILES 15-17
Column Tile 15-17
Combining Slides and Rendered Iages 2-30
Command 12-10, 12-38
command 3, 14-3
Command Aliases 4-8
command definition 5-8, 6-4
COMMAND DEFINITION WITHOUT ENTER OR SPACE 9-37
Comments 4-7
COMPILE 3-23
components of a dialog box 15-2
CONDITIONAL FUNCTIONS 12-38
cons 8, 9, 14-8, 14-9
Context menus 5-18
control characters 9-36
Conversion Of ASE Links To AutoCAD 2002 Format 18-24
CONVERTING UNITS 19-14
Coordinate Display 16-2
Copying a Tool Icon 5-43
cos 12-5
Create New Dimension Style 1-10
Creating a New Image and Tooltip for an Icon 5-41
Creating a Shape Complex Linetype 3-22
Creating a String Complex Linetype 3-17
Creating Custom Toolbars with Flyout Icons 5-43
Creating Labels 18-12
CREATING LINETYPES 3-3
Creating Links with Graphical Objects 18-8
CREATING TEMPLATE DRAWINGS 1-1
CURRENT LINETYPE SCALING (CELTSCALE) 3-16
CUSTOM HATCH PATTERN FILE 3-38
customize the toolbars 5-39
CUSTOMIZING A TABLET MENU 8-3
customizing AutoCAD 6-2
CUSTOMIZING BUTTONS AND AUXILIARY MENUS 10-34
CUSTOMIZING DRAWINGS ACCORDING TOPLOT SIZE AND DRA 1-12
CUSTOMIZING DRAWINGS WITH LAYERSAND DIMENSIONING S 1-7

Index

CUSTOMIZING DRAWINGS WITH VIEWPORTS 1-22
CUSTOMIZING IMAGE TILE MENUS 10-45
CUSTOMIZING PULL-DOWN AND CURSOR MENUS 10-40
CUSTOMIZING TABLET AREA-1 10-21
CUSTOMIZING TABLET AREA-2 10-30
CUSTOMIZING TABLET AREA-3 10-30
CUSTOMIZING TABLET AREA-4 10-33
CUSTOMIZING THE SCREEN MENU 10-49
CUSTOMIZING THE STATUS LINE 16-3
CUSTOMIZING THE TOOLBARS 5-39
Customizing theACAD.PGP File 4-1

D

Data byte 11-17
data byte 11-8
data bytes 11-2
Database Configuration 18-3
Database Management System 18-2
DBCONNECT Manager 18-4
Decremented Number 12-4
default setup values 1-1
DEFBYTES 11-2
definition of the line pattern 3-2
DEFUN 9-39
defun 12-8, 12-24, 3, 14-3
DELAY COMMAND 2-9
Deleting a Toolbar 5-43
Deleting the Icons from a Toolbar 5-43
Descriptive Text 3-8
design of tablet template 8-3
design of the menu 9-9
design of the screen menu 9-9
DIALOG BOX 15-2
dialog box 6-2, 6-3, 10-45
DIALOG BOX COMPONENTS 15-2
DIALOG CONTROL LANGUAGE 15-1
Dialog control language 15-1
DIESEL 16-1, 16-6
DIESEL EXPRESSION IN MENUS 9-40
DIESEL EXPRESSIONS IN MENUS 16-8
DIESEL STRING FUNCTIONS 16-10
digitizing tablet 8-2
DIMALT 1-2
DIMALTD 1-2
DIMALTF 1-2

DIMASO 1-2
DIMASZ 1-2
DIMPOST 1-2
DIMSCALE 1-19, 2-3
dimx_tile and dimy_tile 15-39
direction code 11-3
Direction Control dialog box 1-6
direction vector 11-8, 11-17
direction vectors 11-5
DISPLAYING A NEW DIALOG BOX 15-10
Displaying a Submenu 5-23, 6-3
Division 12-3, 16-11
done_dialog 15-14
Dot notation 17-3
draw mode 11-7, 11-8
Drawing Limits 1-5
Drawing Units dialog box 1-6

E

edit a slide 2-26
EDIT BOX TILE 15-31
edit_width Attribute 15-32
Editing Link Data 18-10
EDITING THE DRAWING DATABASE 1, 14-1
edtime function 16-16
EFFECT OF ANGLE AND SCALE FACTOR ON HATCH 3-29
element 5-36
ELEMENTS OF LINETYPE SPECIFICATION 3-3
end of a shape definition 11-6
END OFFSET 11-12
end_image 15-41
entget 7, 9, 11, 14-7, 14-9, 14-11
entmod 9, 12, 14-9, 14-12
eq function 16-13
Equal to 12-6, 16-11
eval function 16-14
event driven 17-8
Explanation 11-8
EXPLODE command 3-38
External Command 4-7

F

FILEDIA 1-7
fill_image 15-40
FILLETRAD 1-2
FILLMODE 1-2

FILTERING X, Y, AND Z COORDINATES 19-13
fix function 16-14
fixed_width AND alignment ATTRIBUTES 15-8
fixed_width Attribute 15-8
floating viewports 2-21
flowchart 12-38
Flowchart symbols 12-38
flyname 5-35
Forming Selection Sets Using Link Select 18-22
FORTRAN 12-1

G

GEOMETRY CALCULATOR 19-1
get_tile and set_tile Functions 15-22
getangle 12-19, 12-26
GetAngle method 17-10
getcorner 12-16
getcorner, getdist, AND setvar FUNCTIONS 12-16
GETDIST 9-39
getdist 12-17, 12-19
GetDistance method 17-10
getint 12-28
getorient 12-27
GetOrientation 17-11
GETPOINT 9-39
getpoint 12-10, 12-24
GetPoint method 17-10
getreal 12-29
getstring 12-29
getvar 12-29, 3, 14-3
getvar function 16-14
graphical user interface 15-2
graphscr 12-22, 12-24
graphscr, textscr, princ, AND terpri FUNCTIONS 12-22
Greater than 12-7, 16-13
Greater than or equal to 12-8, 16-13
GRID 1-2, 1-6, 2-2, 2-3, 9-31, 16-2
GRIDMODE 1-2
Group Codes for ssget "X" 4, 14-4

H

hatch an area 3-27
hatch angle 3-26
hatch boundaries 3-27
HATCH command 3-29
hatch description 3-25

hatch descriptors 3-26
hatch name 3-25
hatch pattern 3-26
HATCH PATTERN DEFINITION 3-25
hatch pattern definition 3-28, 3-29
hatch pattern library file 3-25
hatch pattern specification 3-28
HATCH PATTERN WITH DASHES AND DOTS 3-30
hatch spacing 3-29
HATCH WITH MULTIPLE DESCRIPTORS 3-33
header 11-1
header line 3-2, 3-25, 11-3
hexadecimal notation 11-18
hexadecimal number 11-2
HIGHRADIUS 11-12
History List tab 12-15
HOW HATCH WORKS 3-27
HOW THE DATABASE IS RETRIEVED AND EDITED 10, 14-10

I

icon 5-35
id_big 5-35
id_small 5-35
if 12-39, 12-41, 12-42, 12-43
if function 16-15
image 10-45
Image Tile 15-35
image tile 6-2
image tile menu section 6-2, 10-46
image tile menus 10-45
Importing and Exporting Link and Label Templates 18-15
Importing and Exporting SQL Queries 18-21
Incremented Number 12-4
initial drawing setup 2-1
integer expression 19-2
integer values 15-5
Integrated Development Environment (IDE) 17-6
INVOKING A SCRIPT FILE WHEN LOADING AUTOCAD 2-11
is_default 15-7
ISOPLANE 1-2
itoa, rtos, strcase, AND prompt FUNCTIONS 12-33

Index

K
key attribute 15-6
key, label, AND is_default ATTRIBUTES 15-6

L
label attribute 15-6
Layer Properties Manager 1-11
LAYOUT 1-13
Layout Settings tab 1-13
length specification 11-3
Less than 12-7, 16-12
Less than or equal to 12-7, 16-12
LIMITS 1-5, 1-9, 1-21, 2-2, 2-3, 9-31
Limits 1-19
LIMMAX 1-2
limmax 3, 14-3
LIMMIN 1-2
Line Feed 11-20
line pattern 3-8
line style 3-8
LINETYPE 3-9
LINETYPE DEFINITION 3-2
linetype description 3-2
linetype name 3-2
linetype specification 3-2
Lineweight 16-2
LISP 12-1
LIST 9-39
list FUNCTION 12-20
Load 12-14
Load Application tab 12-14
Load/Unload Application dialog box 12-14
load_dialog 15-13
LOADING A DCL FILE 15-9
Loading a Pull-down Menu 7-10
LOADING A TEMPLATE DRAWING 1-7
LOADING AN AUTOLISP PROGRAM 12-13
Loading an Image Menu 7-10
Loading an Image Tile Menu 5-24
Loading Image Tile Menus 10-20
LOADING MENUS 5-23, 6-9, 7-9, 8-7, 9-6
Loading Pull-down Menus 10-20
Loading Screen Menus 5-24, 7-9, 10-19
Loading the hatch pattern 3-29
logand and logior 15-20
LONG MENU DEFINITIONS 9-30
LOWRADIUS 11-12

LTSCALE 1-2, 1-9, 1-19, 2-2, 2-3, 3-10, 3-17
Ltscale and Dimscale 1-20
LTSCALE COMMANDs 3-9
LTSCALE factor 3-11
LTSCALE FACTOR FOR PLOTTING 3-12
LWT 16-2

M
macro 5-35, 5-36, 10-45
MACRO EXPRESSIONS USING DIESEL 16-4
macros 7-2
MACROTRACE 16-9
MACROTRACE SYSTEM VARIABLE 16-9
MANAGING DIALOG BOXES WITH AUTOLISP 15-14
MATHEMATICAL OPERATIONS 12-2
max_value 15-36
maximum length of the vector 11-3
maximum number of saves and restores 11-10
Memory Reserve 4-7
menu bar area 5-3
menu bar titles 5-8
MENU command 5-10, 6-10, 8-3, 9-6, 10-25
MENU COMMAND REPETITION 9-32
menu item label 9-1, 9-2
menu item repetition 6-8
MENU ITEMS WITH SINGLE OBJECT SELECTION MODE 9-38
Menu-Specific Help 5-38
MENUCMD 5-30
MENUECHO 9-34
MENULOAD 5-29, 5-31, 5-36
MENUUNLOAD 5-31
methods 17-2
min_value 15-36
min_value and max_value Attributes 15-36
min_value, max_value, small_increment, and big_incr 15-36
MIRRTEXT 1-2
MNEMONIC ATTRIBUTE 15-17
mnemonic key 5-29
MODEL and PAPER Space 16-2
model space 1-12, 2-21
MODEMACRO 16-1, 16-3, 16-4
MODEMACRO SYSTEM VARIABLE 16-3
Modify Dimension Style 1-10
MODIFYING LINETYPES 3-13

More VBA examples 17-19
MSLIDE 2-21
MSLIDE COMMAND 2-20
MSPACE 1-14
multibutton pointing device 7-1
MULTIPLE SUBMENUS 9-15
Multiplication 12-3, 16-11
MVIEW 1-13

N

negative displacement 11-11
Nested Submenus 9-9, 10-50
new hatch pattern library file 3-25
new_dialog 15-13
nonstandard fractional arc 11-12
nonstandard vectors 11-11
Not equal to 12-7, 16-12
NUMERIC FUNCTIONS 19-3

O

Object Browser 17-4
Object Tracking 16-2
objects 17-2
OBTAINING THE RADIUS OF AN OBJECT 19-4
octant 11-11
octant arc 11-12, 11-18
octant boundary 11-11
online help 5-38
Options 3-32
orient 5-34
ORTHO 2-2, 2-3, 16-2
OS Command Name. 4-7
OSNAP 16-2
OTRACK 16-2

P

Paper space 1-12
partial menu 5-28
PARTIAL MENUS 5-27
Pattern Line 3-2
pattern line 3-2
pick button 7-2, 10-35
pld function 19-5
PLINE 1-9, 1-15, 1-22
PLINEWID 1-9
plt function 19-5
pointing device 7-2, 8-2, 10-35

POLAR 9-39, 16-2
polar 12-30, 12-37
PolarPoint method 17-14
POLYLINE 9-31
POP1 5-7
Position stack overflow in shape 11-10
Position stack underflow in shape 11-10
positive displacement 11-11
PREDEFINED ATTRIBUTES 15-6
predefined attributes 15-6
PREDEFINED RADIO BUTTON, RADIO COLUMN, BOXED RADIO 15-25
PRELOADING SLIDES 2-23
PRINC 9-39
princ 12-22
progn 12-43
project 17-8
prompt 12-34
Properties 17-2
prototype drawing 1-15
prototype drawings 1-12
PSPACE 1-14
pull-down and cursor menus 10-40
pull-down menu 5-22, 10-44
pull-down menu or cursor menu 5-12

R

rad function 19-4, 19-5
Radio Button Tile 15-25
Radio Column Tile 15-25
Radio Row Tile 15-25
real expression 19-1, 19-16
real values 15-5
REINIT COMMAND 4-10
RELATIONAL STATEMENTS 12-6
Relational Statements 16-11
Remove 12-15
rendered images 2-28
repeat 12-46, 14, 14-14
REPLAY 2-30
reserved words 15-5
RESTRICTIONS 6-10
restrictions 11-11
RESUME COMMAND 2-10
RESUME command in transparent mode 2-10
root menu 9-14
ROW AND BOXED ROW TILES 15-16

Index

Row Tile 15-16
rows 5-34
RSCRIPT COMMAND 2-9
rtos 12-33
rtos function 16-15
rxof(end) 19-13

S

Saving AutoCAD 2002 Links to R13/R14 Formats 18-24
SAVING HATCH PATTERNS IN A SEPARATE FILE 3-37
SCREEN 9-3
SCREEN MENU 9-1
script file name 2-4
section label 6-3, 7-3
SECTIONS OF THE ACAD.PGP FILE 4-7
SETQ 9-39
setq 12-9
SETVAR 1-6, 1-9, 1-21, 2-2, 2-3, 9-39
setvar 12-17, 12-24
Shape Definition 11-20
SHAPE DESCRIPTION 11-1
SHAPE FILES 11-1
shape files 11-19
SHAPE NAME 11-2
shape name 11-19
shape number 11-2, 11-8, 11-17, 11-19, 11-22
shape specification 11-2, 11-3, 11-8, 11-17
Shortcut menu 5-22
Shortcut menus 5-18
SHORTCUTMENU 5-19
SIMPLE HATCH PATTERN 3-28
sin 12-4
Single 9-38
SLIDE LIBRARIES 2-26
slide library file 2-26
slide presentation 2-20
slide show 2-1
SLIDE SHOWS WITH RENDERED IMAGES 2-28
SLIDELIB 2-25, 2-27
SLIDER AND IMAGE TILES 15-34
Slider Tile 15-34
SLIDES FOR IMAGE TILE MENUS 6-4
small_increment 15-36
small_increment and big_increment Attributes 15-36
SMLayout; 12-2

SNAP 1-6, 2-2, 2-3, 9-31, 16-2
SPECIAL CODES 11-5
SPECIAL HANDLING FOR BUTTON MENUS 7-5
Specifying a New Path for Hatch Pattern Files 3-31
ssget 1, 2, 3, 10, 14-1, 14-2, 14-3, 14-10
ssget "X" 4, 13, 14-4, 14-13
sslength 5, 14, 14-5, 14-14
ssname 6, 9, 10, 14-6, 14-9, 14-10
standard AutoCAD menu 10-49
Standard Codes 11-6
STANDARD PULL-DOWN MENUS 5-2
STANDARD TABLET MENU 8-2
standard template drawings 1-2
START OFFSET 11-12
start_dialog 15-14
start_image 15-41
Startup Suit 12-15
STATUS LINE 16-1
strcat 16-7
string values 15-5
strlen function 16-16
submenu 7-10
Submenu Definition 5-22, 7-8, 9-7, 10-19
Submenu Reference 5-23, 6-3, 7-9, 9-8, 10-19
SUBMENUS 5-22, 6-2, 7-8, 9-7, 10-18, 10-26, 10-44, 10-46, 10-49
submenus 5-26
subshape 11-11
subshape code 11-11
subst 8, 9, 11, 14-8, 14-9, 14-11
Subtraction 12-3, 16-10
Support File Search Path 3-32
Swapping Pull-down Menus 10-44

T

TABLET command 8-6
TABLET CONFIGURATION 8-6
tablet menu 8-2
TABLET MENUS WITH DIFFERENT BLOCK SIZES 8-9
tablet template 8-2
TABLET1 8-2
TABLET2 8-2
TABLET3 8-2
TABLET4 8-2, 10-33
tbarname 5-34

template designs 8-4
template drawing 1-7
template drawings 1-1
template overlay 10-30
terpri 12-22
text font description 11-19
TEXT FONT FILES 11-19
text fonts 11-19
Text height 1-20
text tile 15-4
textscr 12-22
TEXTSIZE 2-3
THE STANDARD AUTOCAD MENU 10-1
THE STANDARD TEMPLATE DRAWINGS 1-2
TILE ATTRIBUTES 15-5
TILEMODE 1-2, 1-13, 1-15
title of the image tile menu 6-2, 10-45
toggle functions 9-36
Toggle Tile 15-17
Toolbar 5-34
toolbar 5-34
Toolbar Definition 5-34
TRACEWID 1-2
TRIGONOMETRIC FUNCTIONS 12-4

U

Understanding Databases 18-1
UNITS 12-26
Unload 12-15
unload_dialog 15-13
Updating Labels with New Database Values 18-15
upper function 16-16
USE OF AUTOLISP IN MENUS 9-39
USE OF CONTROL CHARACTERS IN MENU ITEMS 9-35
USE OF STANDARD BUTTON SUBASSEMBLIES 15-12
Use of the label Attribute in a Boxed Column 15-7
Use of the label Attribute in a Button 15-7
Use of the label Attribute in a Dialog Box 15-7
Using AutoLISP Function to Load a DCL File 15-11
USING AUTOLISP VARIABLES 19-12
USING AUTOLISP WITH MODEMACRO 16-6
USING SNAP MODES 19-3

V

vector 11-11, 11-17
Vector Expression 19-2
vector length 11-3
VECTOR LENGTH AND DIRECTION ENCODING 11-2
vector_image 15-40
vectors 11-3
Viewing and Editing Table Data From AutoCAD 18-6
Viewports toolbar 1-16
virtual keys 5-33
visible 5-34
Visual Basic (VB) 17-1
Visual Basic for Applications (VBA) 17-1
VPOINT 1-14
VPORTS 1-16
VSLIDE COMMAND 2-21

W

WHAT ARE SCRIPT FILES? 2-1
WHAT ARE SLIDES? 2-20
WHAT IS A SLIDE SHOW? 2-20
WHAT IS THE ACAD.PGP FILE? 4-1
while 12-43, 12-45, 12-46
width AND edit_width ATTRIBUTES 15-31
width Attribute 15-31
WRITING A PULL-DOWN MENU 5-3
WRITING A TABLET MENU 8-4
WRITING AN IMAGE TILE MENU 6-3
WRITING BUTTON AND AUXILIARY MENUS 7-2

X

xval 5-34

Y

yval 5-34

Z

ZOOM 2-2, 2-3, 9-31